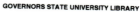

Chaos and Nonlinear Dynamics

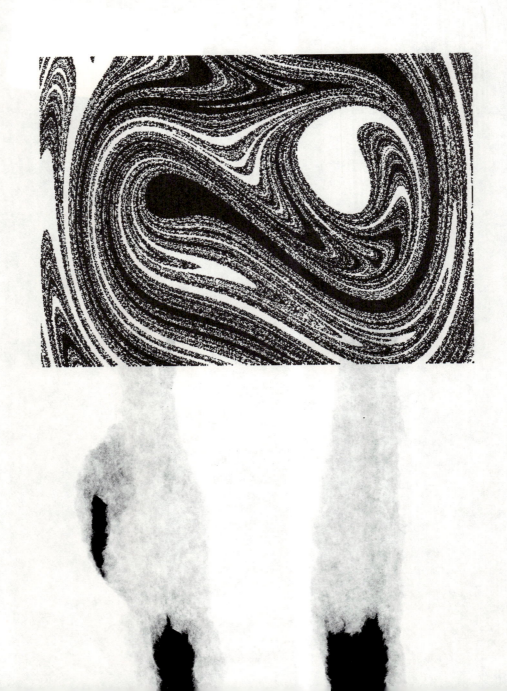

Chaos
and
Nonlinear Dynamics

An Introduction
for Scientists and Engineers

SECOND EDITION

Robert C. Hilborn

Department of Physics
Amherst College

OXFORD
UNIVERSITY PRESS

OXFORD

UNIVERSITY PRESS

Great Clarendon Street, Oxford OX2 6DP

Oxford University Press is a department of the University of Oxford.
It furthers the University's objective of excellence in research, scholarship,
and education by publishing worldwide in

Oxford New York

Auckland Bangkok Buenos Aires Cape Town Chennai
Dar es Salaam Delhi Hong Kong Istanbul Karachi Kolkata
Kuala Lumpur Madrid Melbourne Mexico City Mumbai Nairobi
São Paulo Shanghai Taipei Tokyo Toronto

Oxford is a registered trade mark of Oxford University Press
in the UK and in certain other countries

Published in the United States
by Oxford University Press Inc., New York

© Oxford University Press 1994, 2000

A catalogue record for this book is available from the British Library

Library of Congress Cataloging in Publication Data
(Data available)

ISBN 0 19 850723 2

Printed in Great Britain
on acid-free paper by
Biddles Ltd.,
Guildford and King's Lynn

First Edition Preface

In the last 10 years, the threads of chaos and nonlinear dynamics have spread across the scientific disciplines like a spider's intricate web. Chaos and nonlinear dynamics have provided new theoretical and conceptual tools that allow us to capture, understand, and link together the surprisingly complex behaviors of simple systems—the type of behavior called *chaos*—in essentially every field of contemporary science.

The universality of chaos is both intriguing and puzzling. What is it about the behavior of a convecting fluid, for example, that makes its transition from simple, regular behavior, to complex, chaotic behavior both qualitatively and quantitatively identical to the way an oscillating electrical circuit makes the same kind of transition? The theory of chaos clearly needs to be based on the fundamental laws of physics, chemistry, and biology. In a sense, however, the theory needs to transcend those laws to explain the *universality* of chaos. New ideas, new language, and new ways of reasoning about complex behavior are needed. These ideas, this language, and those modes of reasoning are what chaos theory in particular and nonlinear dynamics in general provide.

The study of nonlinear dynamics is by no means complete, but the field has now matured to the point that it makes sense to bring together in one book the essential elements. The foundations of the theory seem to be firmly in place, and the outlines of the final structure to be erected upon this foundation can now be discerned. This book provides an introduction to chaos and nonlinear dynamics for scientists and engineers who have little or no previous experience with the field. I have assumed no background other than some familiarity with introductory college-level physics and with calculus through elementary differential equations. After completing this book, the reader should be ready to grapple with the current literature in chaos.

Most of us who have been actively engaged in the study of chaos have learned the "tricks of the trade" in a rather piecemeal fashion: some abstract concepts from nonlinear mathematics, some stability theory from the engineers and most of our new ideas from the research literature. When I wanted to introduce my students to chaos, no book provided all of the needed background. Thus, I have written this book both to bring together the essentials of the field and to provide the background for those scientists, engineers, and students who want to discover what the excitement in chaos is all about.

Historically, the study of chaos is strongly rooted in the mathematical study of nonlinear dynamics, going back, at least, to the pioneering work of Henri Poincaré, the noted French mathematician (1854–1912). This heritage has bequeathed to chaos jargon that is (some would say excessively) mathematical in nature. In this book I have approached chaos from the point of view of a scientist who wants to describe and understand the complex behavior of real systems. Thus, I have chosen to introduce the concepts of chaos as descriptors of the behavior of actual systems rather than as abstract mathematical ideas. As these descriptions are refined, we recapture, I hope with more physical intuition and insight, the mathematical basis of those concepts.

Nonlinear dynamics and chaos, like most of contemporary physical science and engineering, is intimately tied to mathematics. To apply the concepts of nonlinear dynamics to her or his field, a scientist, engineer, economist, social scientist or physician must come to grips with at least some of the formalism and quantitative formulations of nonlinear dynamics. The concepts without the quantification are fruitless; likewise, quantification without the guide of concepts is blind number shuffling. Both aspects are necessary. Anyone wishing to make use of nonlinear dynamics must be willing to make an investment of time and energy to master some of the formalism. I have designed this book to provide an introduction to what I believe are the key parts of that formalism.

I believe that many scientists and their students are most comfortable with the traditional differential equations approach to the study of dynamics, and the standard undergraduate training in science and introductory calculus provides ample practice in solving (in closed form) and interpreting linear differential equations. Thus, I have begun the discussion from that point of view. However, the analytic results, which dominate the traditional approach to the study of linear systems, quickly become useless when dealing with nonlinear systems. Therefore, I gradually introduce the methods used to describe and think about these nonlinear systems. These methods generally require less formal mathematical manipulation than do the traditional analytic methods, but they force us to do more thinking.

I mentioned previously the background assumed of my readers: familiarity with college-level introductory physics and an acquaintance with calculus through elementary differential equations. Certain portions of the book do indeed stretch or even exceed those prerequisites. I have marked the title of those sections with an asterisk. The reader who feels that those sections exceed her or his mathematical fortitude or technical background can pick up the key ideas by reading the introductory paragraphs of those sections. There I have tried to give an elementary statement of what the more elaborate mathematics does with more rigor and generality. These more advanced sections are self-contained so that the reader who skips the mathematical treatment can still follow the flow of argument in the remainder of the book.

The results of research in science education agree with the reflections of experienced teachers: students, be they young people or experienced scientists venturing into new fields, must become actively engaged with the material they are attempting to learn. The readers of this book are no exception. To provide for that engagement, I have included some exercises in most sections that should be useful both for classroom instruction and for the reader who wants to tackle chaos directly. At the end of each section are references for further reading and some computer exercises. By working through these exercises, both with paper and pencil and with the computer, and by reading about the applications of nonlinear dynamics to various fields of science, the reader can begin to become engaged with the field.

The first two chapters introduce the key concepts, jargon, and important questions raised by chaos by looking at three simple systems that exhibit chaotic behavior: a simple electrical circuit, a model of biological population dynamics, and a set of differential equations modeling fluid convection. These examples were chosen because they show nearly the full spectrum of chaotic behavior, but they are sufficiently simple so that the basic science behind each of them can be easily understood. The surprise is that these simple systems exhibit exceedingly complex behavior. Simplicity of structure does not guarantee simplicity of behavior. By comparing the chaotic behavior of these systems, we recognize both qualitative and quantitative similarities. These similarities are quantified by the numbers first "discovered" by Feigenbaum, and I discuss carefully how well these numbers describe actual systems.

The remainder of the book then tackles the problem of building a theory of chaotic behavior. The key conceptual tool is the description of a system's behavior in state space, a geometrical construction similar to the phase space description familiar from classical mechanics and statistical mechanics. Poincaré sections further simplify the description of the dynamical behavior and allow conceptually simple, but analytically powerful means of classifying the types of dynamical behavior. We make contact with the mathematical scheme of iterated maps from these constructions. I spend some time developing the theory of these maps since they have been important in the historical development of the theory of chaos, and they elegantly and simply illustrate many of the fundamental types of chaotic behavior. Geometrical notions lead to a classification of the so-called routes to chaos and an understanding of how the system's behavior evolves as its environment, as described by suitable parameters, changes. I discuss in some detail the quasi-periodic, intermittency, and crisis routes to chaos.

Although the initial discussion focuses on dissipative systems—systems that "run down" unless provided with an external source of energy—since these systems are better models of most of the real world, I have included an introduction to the chaotic behavior of Hamiltonian systems—systems whose energy is conserved. Chaotic Hamiltonian systems are important theoretically and are crucial for an understanding of "quantum chaos," a subject I treat in the last chapter.

I then turn to the problem of describing chaos quantitatively. Introducing the notions of Lyapunov exponents, fractal dimensions, and various kinds of correlation exponents, I show how each of these quantifiers can be determined, at least in principle, from an analysis of a time series of sampled values of some dynamical variable of the system. All of these quantifiers are related, and some obey universal scaling laws, which tell how they vary as the system becomes more chaotic. I present a case history of the computation of the widely used correlation dimension to illustrate some of the pitfalls of quantifying chaos.

Recent research has emphasized that these simple descriptors are really only average quantities and that actual chaotic systems show a distribution or spectrum of values for each of these. For the case of fractal dimensions, these chaotic systems are described by what are called "multifractals." There is some indication that these distributions are themselves universal. An elegant thermodynamic formulation of chaotic behavior then leads naturally to a description of these distributions.

The penultimate chapter looks at systems with sufficient spatial extent to exhibit interesting patterns. In recent years there has been dramatic progress in understanding the physics of pattern formation and the related behavior called spatiotemporal chaos. The connection between pattern formation and chaotic dynamical behavior is outlined. I believe that this chapter is the only treatment of pattern formation at this level of presentation. Pattern formation is certainly one of the most important aspects of nonlinear science. Newcomers to the field should certainly be made aware of the fundamental issues.

Unfortunately, fluid dynamics and transport phenomena have all but disappeared from the standard undergraduate physics curriculum in the United States. Thus, I have included a brief introduction to these subjects in Chapter 11. The treatment is hardly exhaustive and provides just enough background so the reader can appreciate the fundamental issues in pattern formation and dissipative structures.

In the final chapter, I turn to a discussion of the problem of chaos and quantum mechanics. Quantum systems show peculiar behavior when their classical (non-quantum) counterparts display chaos. However, there are good reasons to believe that "pure" chaos cannot occur in quantum systems. What then is the connection between chaos and this peculiar behavior? Along with speculations about quantum chaos, I discuss the relationship between chaotic behavior and the more general notions of "complexity" as well as the import of chaos both for the technical development of science and its philosophical implications.

I have included several appendices, which gather together for convenient reference some of the technical background needed for understanding chaos and nonlinear dynamics. Fourier analysis and bifurcation theory are crucial in many aspects of nonlinear dynamics. Appendices A and B provide brief introductions to these subjects. In Appendix C, I present the details of the development of the now legendary Lorenz Model, starting from the fundamental equations of fluid flow and thermal energy diffusion. Appendix D gives an introduction to the scientific research literature on chaos. Appendix E contains the listings of some simple computer programs to illustrate the dynamics of the logistic map model. These programs can serve as useful examples to guide you in creating your own computer programs.

Let me describe the structure of the book. Each chapter is broken down into sections. Equations and exercises are numbered according to the section in which they occur. For example, Eq. (3.4-1) is the first equation in Chapter 3, Section 4. Section numbers are indicated on the top of even-numbered pages. Figures are numbered consecutively in each chapter. References to books are given with the author's name and the year of publication in square brackets, for example, [Gleick, 1987]. New concepts and terms of technical jargon are set in bold italics where they first appear. Double-line boxes set off important results and definitions. Technical comments and asides are indented.

References to papers and articles are cited by giving the initial letters of the family names of the first three authors, or the first three letters of a single author's family name, and the last two digits of the year of publication. For example, (HIL88) refers to a paper by Hilborn published in 1988. I trust that this citation method gives more information than just a numerical reference to a citation at the end of the chapter, without burdening the text with footnotes containing the full citation. The complete reference citations are given at the end of the first chapter in which those references appear. All references cited are gathered together in an alphabetical listing at the end of the book.

Throughout the book I have relied on simple (usually mathematical) models to illustrate new concepts. Some readers may find this approach frustrating. They may want more discussion of actual applications. I have described applications in several sections, but much of the narrative rides on these simple models. There are two reasons for this. First, these simple models provide us with well-controlled and well-defined "laboratories" for trying out and exploring the many unfamiliar and, in some cases, new concepts of nonlinear dynamics. We can ignore for the moment all the complexities and approximations associated with systems in the "real world." Second, many of the features associated with complex behavior in nonlinear dynamics are in fact independent of the details of the system being investigated. Hence, we can use what we learn about the dynamics of simple mathematical models to help us at least categorize and describe and often understand the behavior of lasers, heart beats, and convecting fluids. The use of these simple models is part and parcel of the methodology of nonlinear dynamics.

This book did not appear spontaneously out of (dare I say) the void of chaos. My thinking and writing about chaos and nonlinear dynamics have been influenced by many books. The text by Schuster—see the citation for [Schuster, 1988] at the end of Chapter 1—although it appeared in print after I had begun writing this book, shares much of the same conceptual strategy, but is written at the graduate level in physics.

I want to say a bit about what this book is not (reviewers take note). The history of nonlinear dynamics and chaos is not explored here except by a few references to early developments. These references (and all others in this book) are the ones I and my students have found useful. They do not necessarily point to the original creators or discoverers. Sorting out and understanding this history will take the skills of a disciplined historian of science. Also, this book is not a scholarly monograph on nonlinear dynamics and chaos. I have not proved many (or even most) of the theorems, nor I have explored all of the ramifications of the results stated here. My purpose is frankly (and, I believe, laudably) pedagogical. What I have tried to do is to provide an overview and a series of explanations of what the science of nonlinear dynamics and chaos is all about and what it does and what it (yet) cannot do.

Amherst, Mass.
October, 1992   R.C.H.

First Edition Acknowledgments

My writing of this book owes so much to so many that I hesitate to name a few for fear of leaving out many others. But particular credit must go to the undergraduate students at Oberlin College and at Amherst College, who have collectively over the years stimulated, challenged, and enriched my thinking and understanding of nonlinear dynamics. Among these students, special credit must go to Kenny Blum, Sharon Ross, Simon Kaplan, Clay Johnston, Ryan Wallach, Richard Rubenstein, Allen Rogel, and John Watson, who have worked with me on many different chaos projects. Jennifer Eden gave a careful and critical reading of the first five chapters.

Discussions and conversations with colleagues are too numerous to list in detail, but I must acknowledge particular debts to Jerry Gollub, Julio Ottino, Neal Abraham, John Delos, Bob Romer, Kannan Jagannathan, and Bob Krotkov. H. Eugene Stanley gave the entire manuscript a careful reading. His many suggestions for improvements are gratefully acknowledged. My agent Patricia Van der Leun and Oxford's Physical Sciences editor Jeffrey Robbins provided much needed encouragement. Copy editor Steve Bedney helped impose some order on my initial manuscript. Finally, without the patience of family and friends, this book would have been considerably more chaotic in both its writing and its final form.

Second Edition Preface

Preparing the second edition of a book is a unique combination of redemption (a chance to correct those typographical errors that slipped past the eyes of both the author and the copy editor), reflection (What is the best way to rephrase this argument?) and renewal (bringing references and examples up to date in a rapidly growing field). In reworking *Chaos and Nonlinear Dynamics* I decided, based on the positive feedback received from readers and reviewers of the first edition, to leave the basic structure intact and to focus on refinements and fine tuning.

Taking advantage of the tremendous improvements in computer graphics in the past ten years, I redid all of the figures. I moved a number of the more technical sections in Chapters 3 and 5 into separate appendices to allow the reader to move more rapidly through the first few chapters. The references have been updated selectively, again focusing on those references that have been most useful to me and my students. All claims for completeness are hereby denied.

In response to suggestion from readers of the first edition, I have reworked several arguments aiming for greater clarity. Many new references have been added on controlling and synchronizing chaotic systems and on stochastic resonance, three important areas of research that were just coming on the scene as I finished the first edition. I resisted the temptation to expand the sections of the text dealing with those topics because there are excellent review articles for readers interested in those issues and because those subjects go beyond the what I consider to be the basic essentials of the field. Several readers suggested adding a discussion of the Duffing oscillator, a useful model for many physical systems. That has been included as an appendix.

I have also added some web addresses for several nonlinear dynamics research groups. Those web sites contain useful information on contemporary research activities in nonlinear dynamics, tutorials on chaos, and many links to other nonlinear dynamics sites.

I invite my readers to send comments and suggestions. After all, this book has been written for you.

Amherst, Massachusetts
March, 2000   R. C. H.

Second Edition Acknowledgments

My education in nonlinear dynamics has continued at full pace since I finished the first edition of this book eight years ago. Particular credit must go to Rajarshi Roy, Kurt Wiesenfeld, and Bill Ditto who engaged me in many conversations about chaos during a sabbatical leave spent at the Georgia Institute of Technology in the spring of 1994. John Sommerer and Mingzhou Ding, both Georgia Tech colloquium speakers that year, stimulated some of my work on coupled chaotic systems and the use of the correlation dimension to analyze chaotic behavior.

Many readers of the first edition provided constructive feedback, most of which I have incorporated into the second edition. Many provided detailed lists of typographical errors and suggestions for improvements. I want to thank particularly Ben Gibson, Walter Becker, J. C. (Clint) Sprott, William J. Mullin, Robert Weller, Jessica Smith, Mark Wimbush, and Prof. Wishwamittar (Punjab University).

Katie Bilodeau and Lara Neel did an excellent job in converting the book's many equations to a more convenient format. My agent Patricia Van der Leun and Oxford's Physical Sciences editor Sonke Adlung provided much needed encouragement. I am grateful to the University of Nebraska-Lincoln for support in the 1999-2000 academic year during which time I prepared the second edition. Finally, thanks to my wife Diandra whose support, patience and wise counsel on things stylistic and scientific are indispensable in keeping chaos at bay.

Contents

I

THE PHENOMENOLOGY OF CHAOS

1

Three Chaotic Systems

"The very first of all, CHAOS came into being." Hesiod, *Theogeny* 116.

1.1 Prelude

Chaos is the term used to describe the *apparently* complex behavior of what we consider to be simple, well-behaved systems. Chaotic behavior, when looked at casually, looks erratic and almost random—almost like the behavior of a system strongly influenced by outside, random "noise" or the complicated behavior of a system with many, many degrees of freedom, each "doing its own thing."

The type of behavior, however, that in the last 20 years has come to be called *chaotic* arises in very simple systems (those with only a few active degrees of freedom), which are almost free of noise. In fact, these systems are essentially deterministic; that is, precise knowledge of the conditions of the system at one time allow us, at least in principle, to predict exactly the future behavior of that system. The problem of understanding chaos is to reconcile these apparently conflicting notions: randomness and determinism.

The key element in this understanding is the notion of *nonlinearity*. We can develop an intuitive idea of nonlinearity by characterizing the behavior of a system in terms of stimulus and response: If we give the system a "kick" and observe a certain response to that kick, then we can ask what happens if we kick the system twice as hard. If the response is twice as large, then the system's behavior is said to be linear (at least for the range of kick sizes we have used). If the response is not twice as large (it might be larger or smaller), then we say the system's behavior is nonlinear. In an acoustic system such as a record, tape, or compact disc player, nonlinearity manifests itself as a distortion in the sound being reproduced. In the next section, we will develop a more formal definition of nonlinearity. The study of nonlinear behavior is called *nonlinear dynamics*.

Why have scientists, engineers, and mathematicians become intrigued by chaos? The answer to that question has two parts: (1) The study of chaos has provided new conceptual and theoretical tools enabling us to categorize and understand complex behavior that had confounded previous theories; (2) chaotic behavior seems to be universal—it shows up in mechanical oscillators, electrical circuits, lasers, nonlinear optical systems, chemical reactions, nerve cells, heated fluids, and many other systems. Even more importantly, this chaotic behavior shows dramatic qualitative and quantitative universal features. These universal features are independent of the details of the particular system. This universality means that what we learn about chaotic behavior by studying, for example, simple

3

electrical circuits or simple mathematical models, can be applied immediately to understand the chaotic behavior of lasers and beating heart cells.

In this chapter, we will introduce some of the basic issues in the study of chaos by describing three quite different systems: a semiconductor diode circuit, a mathematical model of biological population growth, and a model of a convecting fluid. Our approach will be to describe how the behavior of these systems changes as some parameter that controls that behavior is changed. We will see that the behavior of these three systems is quite complex, but the complexities are similar, both qualitatively, and perhaps more surprisingly, quantitatively as well. In fact, this last sentence is a brief statement of the two themes that inform this book: (1) the theory of nonlinear dynamics allows us to describe and classify the complex behavior of these kinds of systems, and (2) the theory of chaos allows us to see an order and universality that underlies these complexities.

We would like to warn the reader that the results to be presented in this chapter represent only a narrow range of the vastly rich spectrum of behaviors exhibited by nonlinear systems. The theory of chaos and nonlinear dynamics teaches us to appreciate the richness of the world around us, even if we confine our attention to very simple systems. The reader is urged to postpone making too many generalizations about the uniqueness or pervasiveness of what we describe in these first three examples. Rather, the reader should play the role of the naturalist carefully observing and thinking about the flora and fauna of a new land.

1.2 Linear and Nonlinear Systems

Before we begin our journey into this new land, however, we should pause for a brief overview of the landscape of the strange territories we shall be visiting.

The word *chaos* is a piece of jargon used to describe a particularly complex type of behavior. Chaos *per se* is really only one type of behavior exhibited by nonlinear systems. The field of study is more properly (and more generally) called *nonlinear dynamics*, the study of the dynamical behavior (that is, the behavior in time) of a nonlinear system. But what is a nonlinear system? Here we will use the following rough definition, which we shall sharpen and clarify later:

> A nonlinear system is a system whose time evolution equations are nonlinear; that is, the dynamical variables describing the properties of the system (for example, position, velocity, acceleration, pressure, etc.) appear in the equations in a nonlinear form.

Let us illustrate this definition with two examples from elementary mechanics: one a linear system, the other nonlinear. In classical mechanics the behavior of a system consisting of a point particle with mass m and subject to a force F_x acting in the x direction and constrained to move in only the x direction is given by the following form of Newton's Second Law of Motion:

$$F_x(x,t) = ma = m\frac{d^2x}{dt^2} \tag{1.2-1}$$

Familiar from introductory physics is the case of a point mass subject to the force from an ideal spring, for which the force is given by

$$F_x(x) = -kx \tag{1.2-2}$$

Here x is the displacement of the spring from its equilibrium position (where $F_x = 0$), and k is the so-called spring constant, a measure of the stiffness of the spring. Combining Eqs. (1.2-1) and (1.2-2), we find the time evolution equation for the position of the particle:

$$\frac{d^2x}{dt^2} = -\frac{k}{m}x \tag{1.2-3}$$

This equation is linear in x and in the second derivative of x (the acceleration). Hence, we have described a linear system. If the mass is displaced from the equilibrium position and released, it will oscillate about the equilibrium position sinusoidally with an angular frequency

$$\omega = \sqrt{k/m} \tag{1.2-4}$$

For our second example, we give the force F_x a more complicated x dependence. For example, if $F = bx^2$, then the time evolution equation is

$$\frac{d^2x}{dt^2} = \frac{b}{m}x^2 \tag{1.2-5}$$

and the system is nonlinear, here, because the x position of the particle appears in the equation squared.

We say that a system is linear if, and only if, the following condition holds: Suppose that $g(x,t)$ and $h(x,t)$ are linearly independent solutions of the time evolution equation for the system; then $cg(x,t) + dh(x,t)$ is also a solution, where c and d are any numbers. Remark: $g(x,t)$ and $h(x,t)$ are linearly independent functions if (and only if) $\alpha g(x,t) + \beta h(x,t) = 0$ is true for all x and t implies that $\alpha = 0$ and $\beta = 0$.

We can also express the notion of nonlinearity in terms of the response of a system to a stimulus. Suppose $h(x,t)$ gives the response of the system to a particular stimulus $S(t)$. If we now change $S(t)$ to $2S(t)$, a linear system will have the response $2h(x,t)$. For a nonlinear system, the response will be larger or smaller than $2h(x,t)$.

Exercise 1.2-1. Add a stimulus term to the right-hand-side of Eq (1.2-3). Carry through the argument in the previous paragraph to show that the system described by Eq. (1.2-3) is linear if the stimulus is independent of x and its derivatives or if x or its derivatives appear in the stimulus term to the first power.

The Importance of Being Nonlinear

Nonlinear dynamics is concerned with the study of systems whose time evolution equations are nonlinear. What is the fuss over nonlinearity? The basic idea is the following: If a parameter that describes a linear system, such as the spring constant k in Eq. (1.2-3), is changed, then the frequency and amplitude of the resulting oscillations will change, but the qualitative nature of the behavior (simple harmonic oscillation in this example) remains the same. In fact, by appropriately rescaling our length and time axes, we can make the behavior for any value of k look just like that for some other value of k. As we shall see, for nonlinear systems, a small change in a parameter can lead to sudden and dramatic changes in both the qualitative and quantitative behavior of the system. For one value, the behavior might be periodic, for another value only slightly different from the first, the behavior might be completely aperiodic. (The question of why nonlinearity is crucial and how it leads to these sudden changes will be addressed in the later chapters of this book.) We should point out that almost all real systems are nonlinear at least to some extent. Hence, the type of behavior we study in this book is really much more common than the idealized linear behavior that has formed the core of instruction in physics for the last 300 years. To quote the famous mathematician Stanislaus Ulam, "Calling the subject nonlinear dynamics is like calling zoology 'nonelephant studies' " [Gleick, 1987]. Perhaps we should name this field of study "general dynamics" because the study of only linear systems restricts us to a rather narrow path through the vast territory of dynamics.

Nonlinearity and Chaos

Some sudden and dramatic changes in nonlinear systems may give rise to the complex behavior called chaos. The noun *chaos* and the adjective *chaotic* are used to describe the time behavior of a system when that behavior is aperiodic (it never exactly repeats) and is apparently random or "noisy." The key word here is *apparently*. Underlying this apparent chaotic randomness is an order determined, in some sense, by the equations describing the system. In fact, most of the systems that we shall be studying are completely deterministic. In general we need these three ingredients to determine the behavior of a system:

1. the time-evolution equations;
2. the values of the parameters describing the system;
3. the initial conditions.

A system is said to be *deterministic* if knowledge of the time-evolution equations, the parameters that describe the system, and the initial conditions [for example, the position x and the velocity dx/dt at $t = 0$ for Eqs. (1.2-3) and (1.2-5)], in principle completely determine the subsequent behavior of the system. The obvious problem is how to reconcile this underlying determinism with the overt (apparent) randomness.

Perhaps we can highlight this problem by playing the role of traditional (that is, "prechaos") scientists. If we see a system with complex, randomlike behavior, we might try to explain that behavior by either an argument based on the notion of "noise" or an argument based on "complexity." According to the noise argument, the complex behavior might be due to the influence of uncontrolled outside effects such as electrical pickup, mechanical vibrations, or temperature fluctuations. Because these outside influences are changing in uncontrolled (and perhaps, random) ways, the system's behavior appears random. According to the complexity argument, we recognize that most real systems in biology, chemistry, physics, and engineering are made of billions and billions of atoms and molecules. Since we cannot control (or even know) precisely the behavior of all these atoms and molecules (perhaps, the best we can do is to control their average behavior), it is not surprising that this lack of control leads to fluctuations and randomness in the overall behavior of the system. To be somewhat more technical, we could say that these complex systems have many *degrees of freedom*, and it is the activity of these many degrees of freedom that leads to the apparently random behavior. (In Chapter 3, we will define more precisely what is meant by a degree of freedom. For now, you may think of it as a variable needed to describe the behavior of the system.) Of course, in many cases, both noise and complexity might be contributing factors.

The crucial importance of chaos is that it provides an alternative explanation for this apparent randomness—one that depends on neither noise nor complexity. Chaotic behavior shows up in systems that are essentially free from noise and are also relatively simple—only a few degrees of freedom are active. Of course, if we believe that chaos does play a role in any given experiment, we need to establish that noise is not a major factor, and we ought to know the number of active degrees of freedom. Chaos theory provides us with the tools to carry out this analysis.

Thus far, we have emphasized the study of the *time* behavior of a system. Another important branch of nonlinear dynamics extends this study to those systems that have a significant spatial extent. Some of these systems, such as turbulent fluids, are of great practical importance. Most of them show the surprising ability to form intricate, and often beautiful spatial patterns almost spontaneously. After discussing time behavior in the first part of the book, we will take up the question of spatial patterns and their relation to chaos.

The Important Questions

Given this brief overview, we can now list some of the questions that the chaotician tries to answer:

1. What kinds of nonlinear systems exhibit chaos? (All chaotic systems are nonlinear, but not all nonlinear systems are chaotic.)
2. How does the behavior of a nonlinear system change if the parameters describing the system change?
3. How do we decide if a system is truly chaotic and how do we describe chaos quantitatively?

4. What are the universal features found in many nonlinear systems? Are these features truly universal or are there different kinds of chaos?
5. How do we understand this universality?
6. What does the study of chaos accomplish scientifically and technically?
7. What are the philosophical and methodological implications of chaos?

As we proceed through this book, we will refine and restate these questions and their answers in several different ways.

1.3 A Nonlinear Electrical System

Our first example of a chaotic system is a simple electrical circuit. We have chosen this circuit as our first example to emphasize two important points. First, chaotic behavior does occur in real systems; it is not an artifact limited to theoretical or numerical models. Second, chaotic behavior occurs in very simple systems, systems found in the everyday world around us. All of the components in this diode circuit are like those found in almost every radio, TV set, and VCR player. The circuit is a sinusoidally-driven series circuit consisting of a signal generator, an inductor, and a semiconductor diode. The circuit diagram is shown in Fig. 1.1.

How the Circuit Works

We will provide a brief, qualitative discussion of how the circuit operates. Although this discussion is not a necessary prerequisite for our first introduction to chaotic behavior, most readers should be more comfortable knowing something of the physics of the circuit behavior. Readers not interested in the circuit details may skip, without penalty, to the next section, *How the Circuit Behaves*.

We will begin with the diode. A diode is an electrical device whose essential property for our discussion is its ability to allow electrical current to flow easily in one direction and to impede greatly the flow of current in the other direction. The direction of easy current flow is indicated by the direction in which the large triangle in the diode's circuit symbol points (see Fig. 1.2). A simple analog can be used to understand the basic physics of the diode: a water pipe with a "flap valve" in it. The flap valve can swing up to the right to allow water to flow easily in one direction, the so-called forward-bias direction. When the water attempts to flow in the opposite direction, the flap valve closes against a stop in the pipe and halts water flow in the other direction, the so-called reverse-bias direction. (The valves in the human heart operate in an analogous way to regulate blood flow among the chambers of the heart.)

Three properties of this analog will be important to be able to understand chaos in the diode circuit. First, the flap valve needs a certain amount of time to close after being raised in the easy flow direction. Thus, before the valve can close, there will be a small surge of current in the reverse-bias direction. The time period for this closing of the valve (or diode) is called the *reverse-recovery time*. For the types of diodes used here the reverse-recovery time is a few microseconds. As we

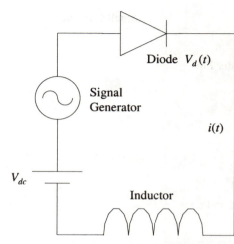

Fig. 1.1. The diode-inductor circuit. The signal generator and variable dc voltage source drive an electrical current $i(t)$ through the inductor and diode, connected in series. $V_d(t)$ is the (time-dependent) electrical potential difference across the diode.

shall see, interesting effects occur when the current in the circuit varies with a time scale on the order of the reverse-recovery time. Second, the closing time depends to some extent on the strength of the current that had been flowing in the forward direction. If only a small current had been flowing, then the valve can close quickly because it was not tilted up too far to the right. If the current had been larger, the valve would take longer to get to the closed position and a larger surge of current can flow temporarily in the reverse-bias direction.

A third property of the analog is the energy stored in the flap valve when it is pushed up by the flow of the water. This potential energy is returned to the water flow when the valve closes. This potential energy corresponds to the electrical energy stored in the diode's electrical capacitance. This electrical capacitance is a measure of how much electrical energy is stored (temporarily) in the configuration of electrical charge layers within the diode. For the diodes used here, the diode's capacitance value is about 100 picofarads, a value typical of small capacitors used in radio and TV circuitry.

Now let us turn to the inductor. The *inductor* is an electrical device, usually in the form of a coil of wire, that has the property of producing an electrical potential difference across it proportional to the <u>rate of change</u> of the electrical current. The proportionality constant is called the *inductance* for the inductor. Thus we say that the inductor acts to oppose <u>changes</u> in the current flow in the circuit. (The water flow analog is a turbine or paddle wheel whose mechanical rotational inertia opposes changes in water flow.) The inductor plays two related roles in the circuit. First, it provides an extra "degree of freedom" needed to make chaos possible in this circuit. Without the inductor the behavior of the current and potential differences in the circuit are so tightly linked together that chaos cannot occur. Second, in combination with the diode's electrical capacitance, it picks out a special frequency for oscillations of the circuit's current and voltage values. When the period of these oscillations (equal to the reciprocal of the oscillation frequency)

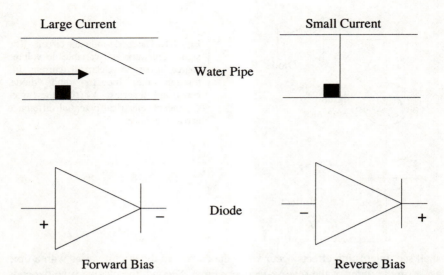

Fig. 1.2. The water pipe and valve analogy for the diode. The flap valve is free to pivot about its top connection. The flap closes against a stop in the pipe when water tries to flow from right to left (top right). When the diode is forward-biased (left-hand side), current flows easily. For an ideal diode there is no steady current flow in the reverse-bias direction (right side of the diagram).

is about the same as the reverse-recovery time, the current and voltage are changing rapidly enough that the nonlinear effects associated with the switching from forward-bias to reverse-bias become important and chaos becomes possible. For the circuit used here the inductance values are typically 50 millihenry. This is similar to inductance values found in many household electronic devices.

The signal generator provides a time-varying electrical voltage to drive the current in the circuit. For this circuit, we have used a generator that produces a voltage varying sinusoidally in time:

$$v(t) = V_o \sin 2\pi f t \qquad (1.3\text{-}1)$$

V_o is called the *amplitude* of the generator's voltage; f is the frequency of the signal, that is, the number of oscillations per second. For the circuit described here, f is typically in the range 20-70 kHz (20,000-70,000 cycles per second). The period of these oscillations is thus between 50 μsec and 14 μsec, a few times the reverse-recovery time of the diodes.

It has also been convenient, for the computer-control of the experiments, to add a steady, so-called dc bias voltage, V_{dc}, to the signal generator voltage. This bias voltage shifts the center voltage about which the signal generator's voltage oscillates.

How the Circuit Behaves: The Route to Chaos

To introduce the phenomenology of this driven nonlinear system and to present some of the jargon and terminology used to describe nonlinear systems and their chaotic behavior, we will describe the behavior of the circuit as one of the parameters of the system is varied. In this case we change the amplitude of the driving voltage. We will call this variable the ***control parameter***. In particular, we will observe the time dependence of the potential difference across the diode $V_d(t)$ and the current $i(t)$ flowing through the diode as we change the amplitude V_o of the sinusoidal signal generator voltage.

When V_o is approximately 0.5 volts (about equal to the maximum potential difference across the diode when it is conducting in the forward-bias direction), the diode acts like an ordinary diode. V_d exhibits the usual "half-wave rectified" waveform shown in the oscilloscope trace in Fig. 1.3. The term half-wave is used because the waveform looks somewhat like half of the sine wave being produced by the signal generator. Note that for aesthetic purposes the oscilloscope has been set so that the trace deflection is upward when the diode is *reverse-biased*. So far, the diode has behaved just as we might expect from elementary electronics: the diode signal has a periodicity that is the same as the periodicity of the signal generator.

When V_o reaches a value somewhere between 1 and 2 volts (depending on the detailed circuit conditions), we get our first surprise. At a well-defined value (denoted by V_1), the regular array of reverse-bias "pulses" splits into an alternating series.

The lower trace in Fig. 1.3 shows that the sequence of peaks in V_d is periodic with a period twice that of the driving voltage. We call this behavior "period-2" behavior. (Using this same terminology, the waveform in the upper trace in Fig. 1.3 would be called period-1.) This change from period-1 behavior to period-2 behavior is called a ***period-doubling bifurcation***.

Bifurcation means a splitting into two parts. The term *bifurcation* is commonly used in the study of nonlinear dynamics to describe any sudden change in the behavior of the system as some parameter is varied. The bifurcation then refers to the splitting of the behavior of the system into two regions: one above, the other below the particular parameter value at which the change occurs.

What causes this period-doubling? A rough physical explanation is the following: As the amplitude of the signal generator's voltage increases, the amount of current flowing in the forward-bias direction increases. When the amplitude is small the diode has time to stop the current flow completely in the reverse-bias direction. In the water flow analog, if the flap valve closes completely, the pressure drop across the valve (corresponding to the diode's electrical potential difference) in the reverse-bias direction will be largest, and no current will flow in the reverse-bias direction.

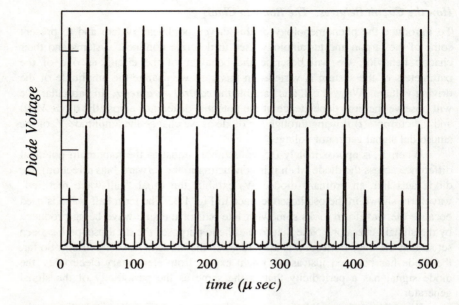

Fig. 1.3. Diode voltage as a function of time. Top trace: period-1. The voltage signal has the same period as the signal generator voltage. Bottom trace: period-2. The diode voltage signal's repetition period is now twice that of the signal generator. The signal generator frequency is about 40 kHz. The voltage scale is 0.2 volt per vertical division.

When the diode's voltage goes to the forward-bias polarity, in response to the signal generator's polarity change, the current can begin to flow immediately in that direction and the flow builds up to a large value, pushing the flap valve to a high position. When the polarity changes to produce reverse-bias, the flap valve, if it has been raised high enough, does not have time to close completely in the time between polarity reversals, and there is some flow in the reverse-bias direction during most of the reverse-bias time. Consequently, the pressure drop (electrical potential difference) in the reverse-bias direction is lower than it would have been if the valve had completely closed. When the polarity returns to the forward-bias direction, the reverse-bias direction flow must first stop before the forward-bias flow can begin. Hence the forward-bias current achieved is smaller and the flap valve does not open as far. This time, when the diode's potential switches to reverse-bias, the flap valve has time to close completely and the cycle can begin again. By way of contrast, for small amplitudes, the flap valve closes each time; so, each surge of forward-bias current is the same.

As V_o increases above V_1, one of the peaks in the sequence gets larger and the other smaller. When V_o reaches a voltage value V_2, the sequence changes again. Figure 1.4 shows that another period-doubling bifurcation has occurred. Now the signals consist of four distinct peak sizes with a repetition period four times that of the driving voltage. We call this a "period-4 sequence."

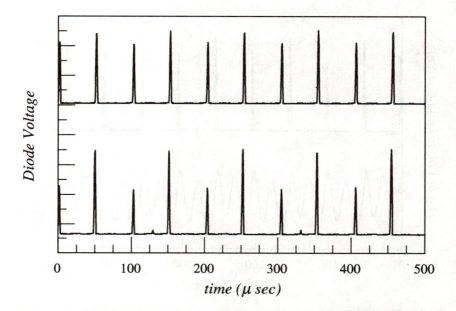

Fig. 1.4. Diode voltage as a function of time. Top trace: period-4. Bottom trace: period-8. Note that two of the period-4 "peaks " are too small to be seen on the scale of this figure. The period-8 behavior is discernible by paying attention to the smallest peaks.

In Fig. 1.4 some of the diode voltage "peaks" are difficult to see given the resolution of the diagram. The underlying problem is that the diode's voltage stays near 0.5 volts whenever the diode is forward-biased. Hence its "dynamic range" in the forward-bias direction is quite limited. However, the current has a wider dynamic range, and Fig. 1.5 shows that the same periodicity occurs for both the diode voltage and the diode current.

As we continue to increase V_o, we see further period-doubling bifurcations that lead to period-8 (shown in the lower trace of Fig. 1.4), period-16 and so on at ever smaller increments of V_o. We will eventually reach a voltage value, which we shall call V_∞, beyond which the sequence of peaks seems completely erratic. This is CHAOS! In Fig. 1.6, we have recorded two long sequences of V_d traces. The pattern seems to be completely aperiodic.

How do we know that we are seeing chaos and not just some "noisy" effect in our circuit? Several means of distinguishing chaos from noise have been developed and used successfully both in theoretical calculations and experimental studies. The one that seems to be the easiest to use experimentally is called ***divergence of nearby trajectories***. For the diode circuit, a trajectory is the time dependence of one of the variables, say, the voltage across the diode. This divergence effect can be observed easily with the diode circuit: With the circuit conditions set for chaotic behavior, we adjust the oscilloscope trigger-level so that the oscilloscope begins its trace just at the top of the largest V_d peak. We then adjust the time-base control of

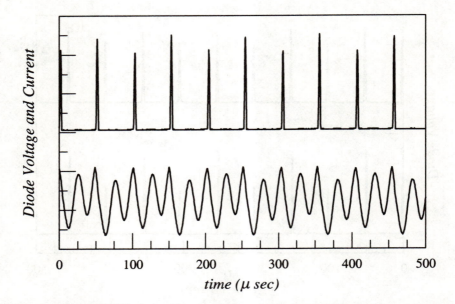

Fig. 1.5. Top trace: diode voltage as a function of time showing period-4 behavior. Bottom trace: a signal proportional to the circuit current for the same conditions. Here the four distinct peak sizes are more obvious. For the bottom trace, each vertical division corresponds to a current of about 0.1 mA.

the oscilloscope so that 10–15 peaks are displayed. In Fig. 1.6, you should note that the beginning (left-hand side) of each trace is nearly the same, but after a few cycles, the peaks differ by nearly the full range of peak heights.

What is happening is that the oscilloscope trigger circuitry starts the trace whenever the applied signal falls within a small range of values set by the trigger controls. Whenever the "trajectory" of voltage values $V_d(t)$ falls within the trigger window, the trace begins. Since the system is deterministic, if we picked out *exactly* the same value for each trigger, each trace would look exactly the same. However the trigger control picks out a small range of voltage values so we are looking at slightly different trajectories that pass close to each other. The signature of chaos is the divergence of nearby trajectories as evidenced here by the divergence of peak heights on the oscilloscope screen. Divergence of nearby trajectories distinguishes true chaos from "simple" noise, for which the trajectories would be smeared more or less the same for all times.

As V_o is increased further into the chaotic region, we find various *periodic windows* immersed in the chaos. For V_o between 2 and 3 volts (the exact value depends on the diode and the circuit conditions), there is a region of stable period-5 behavior. There is a region of stable period-3 behavior between 3 and 4 volts, as shown in Fig. 1.7. As V_o is increased still further, we see a period-doubling sequence of period-6, period-12, and so on, which culminates in another region of chaos.

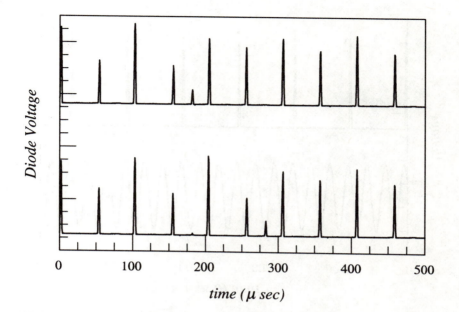

Fig. 1.6. Diode voltage as a function of time. Chaotic sequences in which the amplitudes of the "peaks" are not periodic.

Bifurcation Diagram

There is clearly a tremendous amount of information contained in this complex behavior. Several different means of summarizing that information have been developed.

One of the most useful of these is the so-called **bifurcation diagram**. To generate this kind of diagram, we record the value of, say, the peak heights of $V_d(t)$ or $i(t)$ as a function of the control parameter being varied. In practice, a computer samples the $i(t)$ signal at time intervals separated by the period of the signal generator. The sequence of sampled values is labeled I_1, I_2, and so on. The sampled values are then plotted as a function of the value of the control parameter. Two of these bifurcation diagrams are shown in Figs. 1.8 and 1.9, where we have recorded sampled values of $i(t)$ as a function of the dc bias voltage applied to the circuit (the control parameter) with V_o and the frequency f fixed. The values of 50 successive samples are stored in the computer and then plotted as a function of V_{dc}. (If V_o had been used as the control parameter, the bifurcation diagrams would be similar, but not identical.)

In these diagrams you can see period-1, where there is just a single value of $i(t)$ for a given V_{dc} value, period-2, where there are 2 values, and so on. (Because of the limited resolution of the computer graphics used to plot these data, some of the higher-order period-doublings are obscured.) When the behavior of the system is

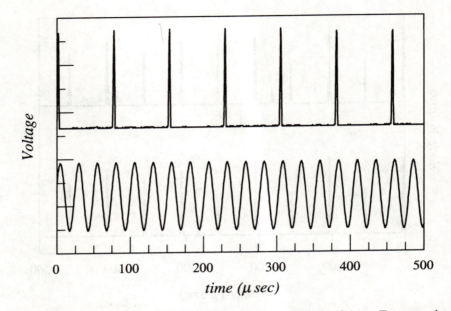

Fig. 1.7. Bottom trace: the signal generator voltage as a function of time. Top trace: the diode voltage for period-3 behavior.

chaotic, the sampled values seem to be smeared over the complete range of observed values.

Within the chaotic regions, several periodic windows can be seen. The period-3 window is usually the largest. Note that the period-3 window begins very abruptly. In moving from right to left in the bifurcation diagram, there is a very sudden transition from this periodic period-3 behavior to chaotic behavior. The lesson: not all transitions from periodic behavior to chaos involve period-doubling.

Let us pause a moment to summarize what we have seen. The key features of nonlinear behavior are the following:

1. Sudden changes in the qualitative behavior of the system as parameters are slowly changed. For example, period-1 changes suddenly to period-2.
2. Well-defined and reproducible changes from regular, periodic behavior to aperiodic, chaotic behavior. The time dependence of the behavior of the system may have little to do with the time dependence of the "forces" applied to the system.
3. Chaos can be distinguished from "noisy" behavior by looking at the divergence of nearby trajectories.

We will examine some of the *quantitative* features of this behavior in Chapter 2.

I_n

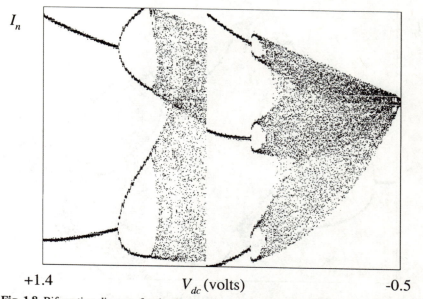

+1.4 V_{dc} (volts)   -0.5

Fig. 1.8. Bifurcation diagram for the diode circuit. The sampled values of the circuit current are plotted as a function of the dc bias voltage, the control parameter. Period-2 is seen at the far left. The broad, fuzzy bands are regions of chaotic behavior. Period-3 bifurcating to period-6 and eventually to more chaos is seen near the middle of the diagram. At the far right, the behavior returns to period-1.

1.4 A Mathematical Model of Biological Population Growth

For our second example of a chaotic system, we turn to a very simple mathematical model often used to describe the growth of biological populations. The mathematics of this model is important historically in the development of chaos theory. In the mid-1970's the biologist R.M. May wrote an influential (and highly readable) review article (MAY76), which described some of the bewildering complex behavior exhibited by this and other simple models. Shortly thereafter, Mitchell Feigenbaum (FEI78) discovered some of the universal quantitative features in numerical calculations based on this model. These universal features have become the hallmark of the contemporary study of chaos. Because of its mathematical simplicity, this model continues to be a useful test bed for new ideas in chaos.

The model can be built up in the following way. Let us consider a species, such as mayflies (no pun intended), whose individuals are born and die in the same season. We want to know how the number of mayflies 1 year hence, let us call that number N_1, is related to the original number of mayflies N_0. In the simplest situation we might guess that

I_n

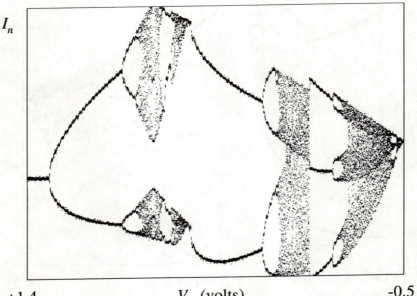

+1.4 V_{dc} (volts)   -0.5

Fig. 1.9. Another bifurcation diagram for a smaller signal generator amplitude.

$$N_1 = AN_0 \qquad (1.4\text{-}1)$$

where A is some number that depends on the conditions of the environment (food supply, water, general weather conditions, etc.). If $A > 1$, then the number of mayflies will increase. If $A < 1$, the number will decrease. If A remains the same for subsequent generations, the population would continue to increase, in the first case, leading to a Malthusian population explosion. In the second case, the population would decrease toward extinction.

We know, however, that if the population grows too much, there will not be enough food to support the larger population or perhaps that predators will have an easier time catching the flies. Hence, the population's growth will be limited. We can incorporate this limiting feature by adding another term to the model that by itself would be unimportant for small values of N, but becomes more important as N increases. One possible way of doing this is by introducing a term proportional to N^2, which then leads to

$$N_1 = AN_0 - BN_0^2 \qquad (1.4\text{-}2)$$

If $B \ll A$, then the second term in Eq. (1.4-2) will not be important until N gets sufficiently large. The minus sign means that the second term tends to *decrease* the population. We then use Eq. (1.4-2) repeatedly to find how N changes in subsequent years:

$$N_2 = AN_1 - BN_1^2$$
$$N_3 = AN_2 - BN_2^2$$
$$\vdots \qquad\qquad (1.4\text{-}3)$$

For our analysis it will be useful to modify Eq. (1.4-2) slightly. First, we note that according to Eq. (1.4-2), there is a maximum possible population number. In order to have $N_{n+1} > 0$, N_n cannot exceed

$$N^{max} = A / B \qquad\qquad (1.4\text{-}4)$$

Thus, we introduce a new variable

$$x_n = N_n / N^{max}$$

which gives the population as a fraction of the maximum possible population for our model. (To be meaningful in our model, x must lie in the range between 0 and 1.) Using this definition of x in Eq. (1.4-2) yields

$$x_{n+1} = Ax_n(1 - x_n) \equiv f_A(x) \qquad\qquad (1.4\text{-}5)$$

where x_n is the population (as a fraction of N^{max}) in the nth year. (We have put a double box around the previous equation because it has played such an important role in the development of the theory of chaos.)

> **Exercise 1.4-1.** Use the definition of x to verify that Eq. (1.4-5) follows from Eq. (1.4-2).

The function f_A is called the "iteration function" because we find the population fraction x in subsequent years by iterating (repeating) the mathematical operations indicated in Eq. (1.4-5). The function $f_A(x)$ is plotted in Fig. 1.10 for several values of the parameter A.

Now we want to find what this model tells us about the long term (after many seasons) value of the population fraction x and how that long-term value depends on A. Our linearly trained intuition would seem to tell us that we expect x to settle into some definite value since the environment, represented by the parameter A, remains constant. Further, we might expect that this value will change gradually if we change A gradually. The actual calculation runs as follows: Start with some value of x_0, compute x_1, then x_2, and so on:

$$x_1 = f_A(x_0), \quad x_2 = f_A(x_1), \quad x_3 = f_A(x_2),\ldots$$

We call this a sequence of iterations. The function $f_A(x)$ is sometimes called an *iterated map function*, since it maps one value of x, say x_0, in the range $0 \le x \le 1$ into another value of x, which we call x_1, in the same range if A is in the range $0 \le A \le 4$. For historical reasons, the function defined in Eq. (1.4-5) is called the *logistic map* function.

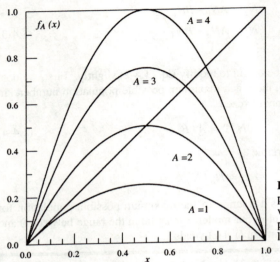

Fig. 1.10. The function $f_A(x)$ plotted as a function of x for various values of the parameter A. The diagonal line is a plot of $y = x$.

The sequence of x values generated by this iteration procedure will be called the **trajectory** or **orbit** in analogy to the sequence of position values for a planet or satellite taken at successive time intervals. Obviously, the first few points of a trajectory depend on the starting value of x. What may not be so obvious is that the eventual behavior of the trajectory is the same for almost all starting points between 0 and 1 for a given value of A.

However, some starting points are different from the others. For example, if we choose $x_0 = 0$, we see immediately that $f_A(x_0) = 0$, and the trajectory stays at $x = 0$ for all subsequent iterations. An x value, call it x^*, which gives

$$x_A^* = f_A(x_A^*) \tag{1.4-6}$$

is called, for obvious reasons, a **fixed point** of the iterated map. The subscript A indicates that x^* depends on the value of A. For the logistic map, Eq. (1.4-5), there are, in general, two fixed points

$$x_A^* = 0 \tag{1.4-7}$$

$$x_A^* = 1 - \frac{1}{A} \tag{1.4-8}$$

For $A < 1$, $x_A^* = 0$ is the only fixed point in the range of x that is of interest for our biological model. For $A > 1$, both fixed points fall in the range of interest between 0 and 1.

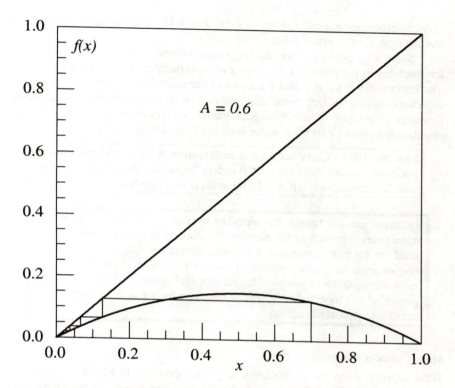

Fig. 1.11. A graphic representation of the iteration of Eq. (1.4-5) starting from $x_0 = 0.7$ with $A = 0.6$.

The Importance of Fixed Points

A simple geometric construction allows us to see why fixed points are important. In Fig. 1.10, we have plotted $y = f_A(x)$ for several values of A. We have also plotted the diagonal line $y = x$. Wherever the diagonal line crosses the curve for $f_A(x)$, the map function has a fixed point since there $x = f_A(x)$. From Fig. 1.10 it is easy to see that for $A < 1$, the only fixed point between 0 and 1 is $x = 0$, but for $A > 1$, there are two fixed points in the range of interest between 0 and 1.

Using Fig. 1.11, we can see how a trajectory that starts from some value of x different from 0 approaches 0 if $A < 1$. The procedure is: From the starting value of x_0 on the x axis, draw a line vertically to the f_A curve. This intersection value determines the value of x_1. Then draw a line from the intersection point parallel to the x axis to the diagonal $y = x$ line. Directly below this intersection point is x_1 on the x axis. From the intersection point on the diagonal line, draw a second vertical line to the f_A curve. The intersection point with the f_A curve determines x_2. As we continue this up-(or down)-and-over process, we are graphically carrying out the iteration procedure indicated in Eq. (1.4-5). Fig. 1.11 shows an iteration sequence starting from $x = 0.7$ with $A = 0.6$.

Since all trajectories (for starting values between 0 and 1 and for $A < 1$) approach the final value $x = 0$, the point $x = 0$ is called the ***attractor*** for those orbits. The interval $0 \le x \le 1$ is called the ***basin of attraction*** for that attractor since any trajectory starting in that range approaches $x = 0$ as the number of iterations increases. In terms of our biological model, we conclude that if $A < 1$, the population dies out ($x \rightarrow 0$) as n, the number of seasons, increases.

> **Exercise 1.4-2.** Carry out several such constructions for several starting values of x in the range between 0 and 1. Show that all these trajectories seem to approach $x = 0$ if $A < 1$. What happens if $x_0 = 0$ or 1?

> In more general terms, the ***attractor*** is that set of points to which trajectories approach as the number of iterations goes to infinity. As we shall see for more complicated systems, the system may have more than one attractor for a given parameter value. The ***basin of attraction*** for a particular attractor consists of that set of initial points $\{x_0\}$ each of which gives rise to a trajectory that approaches the attractor as n, the number of iterations, approaches infinity.

More Complex Behavior

What happens when the parameter A is greater than 1? In Fig. 1.12, we have plotted $f_A(x)$ with $A = 1.5$ along with the diagonal line $y = x$. If we follow our geometrical trajectory construction method, we see that a trajectory starting at $x = 0.10$, for example, now heads for the fixed point $x_A^* = 1 - 1/A = 1/3$ (for $A = 1.5$). In fact, any trajectory starting in the range $0 < x < 1$ approaches this same attractor.

At this point we may seem to have understood what our model tells us: Given any initial number x_0 lying between 0 and 1, the population fraction eventually approaches the attracting fixed point $x_A^* = 1 - 1/A$ if $A > 1$. For $A > 1$, $x^* = 0$ has become a "repelling fixed point" since trajectories that start near $x = 0$ move away from that value.

> **Exercise 1.4-3.** (a) What happens for $x_0 = 0$ and $x_0 = 1$? (b) Calculate numerically the trajectory sequences starting from $x = 0.25$ and $x = 0.75$ for $A = 1.5$. (c) How many iterations does it take for the trajectories to get within 0.001 of the final value $x = 0.3333$? (d) Explain, both graphically and numerically, why the two trajectories are the same after the first iteration.

There are surprises however in this simple model. The first surprise comes when A is just a bit greater than 3. Table 1.1 lists the trajectory values for the orbit starting at $x_0 = 0.25$ with $A = 3.1$. We see that the trajectory does not settle down to a single attracting value. In this case the trajectory values oscillate back and forth

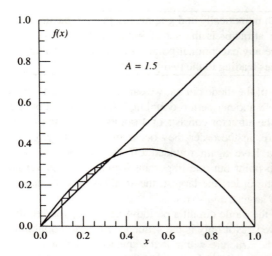

Fig. 1.12. Graphic representation of the iteration scheme with $A = 1.5$ and $x_0 = 0.10$. Note that the trajectory moves away from the fixed point at $x = 0$ and is attracted to the fixed point at $x = 1/3$.

between two values $x = 0.558$ . . . and $x = 0.764...$ In biological terms, the population fraction is high one year, low the next, then high again, then low again, and so on. Since the population fraction returns to the same value every 2 years, we call this, in analogy with the diode circuit example, period-2 behavior. We say that at $A = 3$, a period-doubling bifurcation occurs: For $A < 3$, the attractor consists of a single point $x = 1 - 1/A$. For A just greater than 3, the attractor consists of two points, whose values vary as A varies. (In Chapter 5, we will take up the mathematical theory of these maps in more detail. There we will see how to compute the A values for which these bifurcations occur.) Figure 1.13 shows the graphic construction of a trajectory that leads to this two-point attractor.

Table 1.1.
Trajectory values* for $A = 3.1$ and $x_0 = 0.250$.

n	x_n	n	x_n
0	0.250	11	0.561
1	0.581	12	0.764
2	0.755	13	0.559
3	0.574	14	0.764
4	0.758	15	0.559
5	0.569	16	0.764
6	0.760	17	0.559
7	0.565	18	0.764
8	0.762	19	0.558
9	0.562	20	0.764
10	0.763	21	0.558

*The x values have been rounded off to three significant figures.

Exercise 1.4-4. Verify both numerically and graphically that the basin of attraction of this two-point attractor is the set of initial x_0 in the range between 0 and 1. Are there any exceptional points between 0 and 1 that do not give rise to trajectories leading to the two-point attractor?

Based on our observations of the diode circuit, we can immediately anticipate what will happen. At $A = 3.44948$ another period-doubling bifurcation occurs. For A just greater than 3.44948..., the attractor consists of 4 points. For example, the attractor values for $A = 3.45$ are, in the order they occur in a trajectory, 0.852, 0.433, 0.847, 0.447, where we have again rounded off the numbers to three significant figures. We want to point out one important feature of these values. The values occur in the following order: the largest, the smallest, next to largest, next to smallest, then back to the largest, and so on.

If you refer to Fig. 1.5 for the diode circuit's period-4 behavior, you see that exactly the same order is followed. Not only is the sequence of period-doublings the same for both systems, but the structure within that sequence is the same. For the logistic map, further increases in A lead to period-8, period-16, and so on, occurring at ever smaller and smaller increments of A. For A just greater than 3.5699 . . ., the trajectory values never seem to repeat. The behavior is chaotic.

We can summarize the behavior of our model by plotting a bifurcation diagram; that is, for a given value of A, we compute the trajectory from some starting point and then plot, as a function of the parameter A, the attracting points for that trajectory. In practice, the trajectory is "close enough" to the final attractor after 20–50 iterations. (Of course, what "close enough" means numerically depends on the precision of the calculations.) Figure 1.14 was computed by picking a value of A, picking a starting point in the range between 0 and 1, iterating the map function 100 times to allow the trajectory to approach the attractor values and then plotting the next 100 values of x.

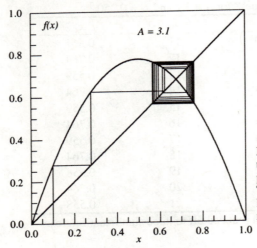

Fig. 1.13. Graphic representation of the iteration scheme leading to period-2 behavior. $A = 3.1$, $x_0 = 0.1$. The two attractor points lie at the top left and bottom right of the dark rectangular area.

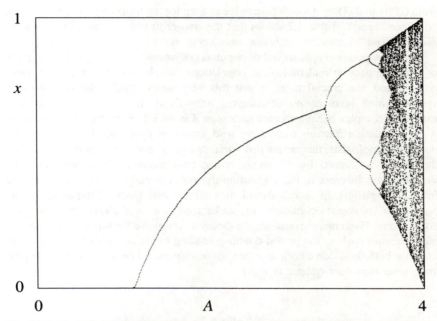

Fig. 1.14. The bifurcation diagram for the logistic map function. For $A < 1$, $x = 0$ is the attractor, for $1 < A < 3$, the attractor is the fixed point $x = 1 - 1/A$. At $A = 3$, period-2 begins. Periods higher than about 8 are obscured by the limited resolution of this plot. Chaos occurs in the bands where the dots seem to be smeared at random. Periodic windows for certain values of A within the chaotic bands can be seen as light vertical stripes.

The reader should immediately recognize the qualitative similarity between this bifurcation diagram and the ones in Figs. 1.8 and 1.9 for the diode circuit. Both show period-doublings leading to chaos and broad chaotic areas interrupted with periodic windows. However, there are differences. For example, the bifurcation diagram for the diode circuit "recollapses" back to period-1 behavior for large values of the parameter V_{dc}. The bifurcation diagram for the logistic map stops abruptly at $A = 4$. We will explore the reasons for these differences in Chapter 5.

Question: What happens to the logistic map trajectories for $A > 4$?

Is the "chaotic" behavior seen for the logistic map true chaos or is it some numerical artifact due to the iteration procedure? We can test for chaos by looking at two trajectories that start close to each other and checking for divergence of nearby trajectories. For example, Table 1.2 lists the orbits for three trajectories, each with $A = 3.99$. One trajectory starts from $x = 0.400$; a second from $x = 0.401$, and the third from $x = 0.4005$. We see that after only 10 iterations the first two trajectories are already 0.6 apart, a size about equal to the average value of the xs. Hence we conclude that the logistic map for $A = 3.99$ displays divergence of nearby trajectories. Suppose we "work twice as hard" and reduce the initial difference from 0.001 to 0.0005. Does it take twice as long for the trajectories to get the same "distance" apart? Table 1.2 shows that the answer to that last question is no. In fact we avoid "disaster" for only one more iteration.

Although this simple model of population dynamics is not a good description of what happens to real biological populations, which of course are much more complicated, the crucial point is that this very simple mathematical system—a system which involves no derivatives, integrals or fancy functions—exhibits extremely complex behavior as the parameter A is varied. Even for a fixed value of A the population fraction x can have wild variations from year to year. For the population biologists, this means that variations in the population need not (though they may) be caused by variations in the environment. In some sense, the variations are inherent in the population dynamics directly. The lessons learned from the logistic map model should then affect what kinds of explanations the population biologist considers to understand the variations in biological populations. Even more crucial for the theory of chaos are the universal features of this behavior, such as the period-doubling leading to chaos. This behavior is the same for both the diode circuit and the logistic map model (and, as we shall see, for many other nonlinear systems as well).

Table 1.2.

Trajectories for the Logistic Map with $A = 3.99$.

n	x_n	x_n	x_n
0	0.4000	0.4010	0.4005
1	0.9576	0.9584	0.9580
2	0.1620	0.1591	0.1605
3	0.5417	0.5338	0.5377
4	0.9906	0.9929	0.9918
5	0.0373	0.0280	0.0324
6	0.1432	0.1085	0.1250
7	0.4894	0.3860	0.4365
8	0.9971	0.9456	0.9814
9	0.0117	0.2052	0.0727
10	0.0462	0.6507	0.2691
11	0.1758	0.9069	0.7847
12	0.5781	0.3368	0.6740
13	0.9731	0.8912	0.8767
14	0.1043	0.3870	0.4314
15	0.3727	0.9465	0.9787

The values in the table have been rounded to four significant figures. The calculations were done with an accuracy of eight significant figures.

1.5 A Model of Convecting Fluids: The Lorenz Model

Our third example of a nonlinear system is a highly simplified model of a convecting fluid. The model was introduced in 1963 by the MIT meteorologist Edward Lorenz, who was interested in modeling convection in the atmosphere. What Lorenz demonstrated was that even a very simple set of equations may have solutions whose behavior is essentially unpredictable. Unfortunately for the development of the science of chaos, Lorenz published his results in the respectable but not widely read Journal of the Atmospheric Sciences, where they languished essentially unnoticed by mathematicians and scientists in other fields until the 1970s. Now that chaos is more widely appreciated, a minor industry studying the Lorenz model equations has developed (see, for example, [Sparrow, 1982]).

A detailed derivation of the Lorenz model equations is given in Appendix C. Here we will say just enough to give you a feeling for what the equations tell us. In simple physical terms, the Lorenz model treats the fluid system (say, the atmosphere) as a fluid layer that is heated at the bottom (due to the sun's heating the earth's surface, for example) and cooled at the top. The situation is illustrated in Fig. 1.15. The bottom of the fluid is maintained at a temperature T_w (the "warm" temperature), which is higher than the temperature T_c (the "cold" temperature) at the top. We will assume that the temperature difference $T_w - T_c$ is held fixed. (This type of system was studied experimentally by Bénard in 1900. Lord Rayleigh provided a theoretical understanding of some of the basic features in 1916. Hence, this configuration is now called a Rayleigh–Bénard cell.)

If the temperature difference $\delta T \equiv T_w - T_c$ is not too large, the fluid will remain stationary. Heat is transferred from bottom to top by means of thermal conduction. The tendency of the warm (less dense) fluid to rise is counterbalanced by the loss of heat from the warm fluid "packet" to the surrounding medium. The damping due to fluid viscosity prevents the packet from rising more rapidly than the time required for it to come to the same temperature as its neighbors. Under these conditions the temperature drops linearly with vertical position from T_w at the bottom of the layer to T_c at the top, as illustrated by the graph in Fig. 1.15. However, if the temperature difference increases sufficiently, the buoyant forces eventually become strong enough to overcome viscosity and steady circulating

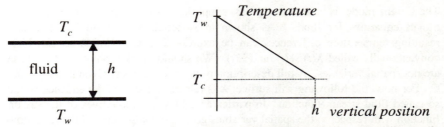

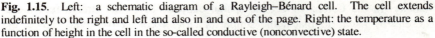

Fig. 1.15. Left: a schematic diagram of a Rayleigh–Bénard cell. The cell extends indefinitely to the right and left and also in and out of the page. Right: the temperature as a function of height in the cell in the so-called conductive (nonconvective) state.

currents develop. In this situation heat is transferred from the bottom to the top by the process of convection, the actual mass motion of the fluid. In simple terms, when the warm packet of fluid reaches the top of the layer, it loses heat to the cool region and then sinks to the bottom, where its temperature goes up again. The net result is a circulation pattern that is stable in time. The circulation pattern is shown schematically in Fig. 1.16.

With a further increase in the temperature difference δT, the circulating currents and the resulting temperature differences within the fluid start to vary in time. This is an example of another typical nonlinear feature: Although the fluid environment is perfectly stable in time (recall that we assume that the temperatures at the top and bottom are maintained at fixed values), the system "spontaneously" develops time-dependent behavior. This never occurs for a linear system. If a linear system is subject to steady "forces," its response (after initial transients die away) will be steady in time. (Strictly speaking, this last statement is true only if the system has dissipation or "friction." Otherwise the transients never die away.) If the forces themselves vary in time in a certain way, then the linear system response will eventually settle into the same time variation.

In more abstract terms, we say that the nonlinear system can spontaneously break the *time translation symmetry* of the equations and the environment. Time translation symmetry means that the equations and the environment are exactly the same for any value of the time variable t. That is, the conditions at t and at $t + \tau$ are the same for any values of t and τ. However, once the system's behavior becomes time dependent, that time translation symmetry is broken. For a linear system, the symmetry is broken only if the external conditions imposed on the system break the symmetry. A nonlinear system can break the symmetry spontaneously. In Chapter 11 we will see similar symmetry breaking in the formation of spatial patterns as well. In fact, the circulating currents in a Rayleigh–Bénard cell are a simple example of spatial-symmetry breaking.

The Lorenz Equations

The Lorenz model is based on a (gross) simplification of the fundamental Navier–Stokes equations for fluids. As shown in Appendix C, the fluid motion and resulting temperature differences can be expressed in terms of three variables, conventionally called $X(t)$, $Y(t)$, and $Z(t)$. We should quickly point out that these are not spatial variables. A full description of these variables is given in Appendix C. For now, the following will suffice: X is related to the time-dependence of the so-called fluid stream function. In particular, taking the derivatives of the stream function with respect to the spatial variables gives the components of the fluid flow velocity. In the Lorenz model the spatial dependence of the stream function is chosen "by hand" to match the simple pattern of convective rolls. Hence, the

T_c

T_w

Fig. 1.16. Cross-section view of the circulating convection "rolls" in a Rayleigh–Bénard cell.

Lorenz model cannot be expected to apply to fluids that develop more complex spatial patterns.

The variables Y and Z are related to the time dependence of the temperature deviations away from the linear temperature drop from bottom to top, which obtains for the nonconvective steady-state situation. In particular, Y is proportional to the temperature difference between the rising and falling parts of the fluid at a given height, while Z is proportional to the deviation from temperature linearity as a function of vertical position.

Using these variables, we may write the Lorenz model equations as three coupled differential equations:

$$\dot{X} = p(Y - X)$$
$$\dot{Y} = -XZ + rX - Y \qquad (1.5\text{-}1)$$
$$\dot{Z} = XY - bZ$$

Here $\dot{X}$ indicates, as usual, the derivative with respect to time

$$\dot{X} \equiv \frac{dX}{dt} \qquad (1.5\text{-}2)$$

p, r, and b are adjustable parameters: p is the so-called Prandtl number, which is defined to be the ratio of the kinetic viscosity of the fluid to its thermal diffusion coefficient. In rough terms, the Prandtl number compares the rate of energy loss from a small "packet" of fluid due to viscosity (friction) to the rate of energy loss from the packet due to thermal conduction. r is proportional to the **Rayleigh number**, which is a dimensionless measure of the temperature difference between the bottom and top of the fluid layer. As the temperature difference increases, the Rayleigh number increases. The final parameter b is related to the ratio of the vertical height h of the fluid layer to the horizontal size of the convection rolls. It turns out that for $b = 8/3$, the convection begins for the smallest value of the Rayleigh number, that is, for the smallest value of the temperature difference δT.

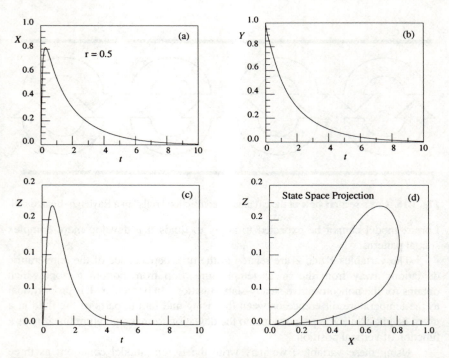

Fig. 1.17. In (a), (b), and (c), X, Y, and Z are plotted as functions of time for the Lorenz model with $r = 0.5$, $p = 10$, and $b = 8/3$. In (d), the trajectory is shown as a projection onto the ZX plane of state space. In all cases the trajectory started at the initial point $X = 0$, $Y = 1$, $Z = 0$.

This is the value usually chosen for the study of the Lorenz model. p is then chosen for the particular fluid under study. Lorenz (LOR63) used the value $p = 10$ (which corresponds roughly to cold water), a value that had been used in a previous study of Rayleigh–Bénard convection by Saltzman (SAL62). We let r, the Rayleigh number, be the adjustable control parameter.

The Lorenz model, although based on what appears to be a very simple set of differential equations, exhibits very complex behavior. The equations look so simple that one is led to guess that it would be easy to write down their solutions, that is, to give X, Y, and Z as functions of time. In fact, as we shall discuss later, it is now believed that it is in principle impossible to give the solutions in analytic form, that is, to write down a formula that would give X, Y, and Z for any instant of time. Thus, we must solve the equations numerically, which, in practice, means that a computer does the numerical integration for us. Here, we will describe just a few results of such an integration. The analytic underpinnings for these results will be discussed later.

Behavior of Solutions to the Lorenz Equations

For small values of the parameter r, that is, for small temperature differences δT, the model predicts that the stationary, nonconvecting state is the stable condition. In terms of the variables X, Y, and Z, this state is described by the values $X = 0$, $Y = 0$, and $Z = 0$. For values of r just greater than 1, steady convection sets in. There are two possible convective states: one corresponding to clockwise rotation, the other to counterclockwise for a given convective roll. As we shall see, some initial conditions lead to one state, other conditions to the other state. Lord Rayleigh showed that if $p > b + 1$, then this steady convection is unstable for large enough r and gives way to more complex behavior. As r increases, the behavior has regions of chaotic behavior intermixed with regions of periodicity and regions of "intermittency," which cycle back and forth, apparently randomly, between chaotic and periodic behavior.

To illustrate some of this behavior, let us start our examination of the Lorenz model by looking at the behavior of the system for values of r less than 1. Rayleigh's analysis predicts that the system should settle into the steady, nonconvective state indicated by $X = 0$, $Y = 0$, $Z = 0$. Figure 1.17 shows the results of a numerical integration of the Lorenz equations starting from the initial state $X = 0$, $Y = 1$, $Z = 0$; that is, we have started the system with a small amount of circulation and slight temperature deviations. As time goes on, however, the system relaxes to the steady nonconvective state at $X = 0$, $Y = 0$, $Z = 0$.

It will be useful to look at this behavior in two complementary graphic presentations. One graph plots the variables X, Y, and Z as functions of time, as in Fig. 1.17(a–c). The other graphs display the evolution of the system by following the motion of a point in XYZ space. Since the variables X, Y, and Z specify the state of the system for the Lorenz model, we call this space the ***state space*** for the system. For the Lorenz model, the state space is three-dimensional. We will usually follow the system with a two-dimensional projection, say on the XY or ZX planes of this state space. As time goes on, the point in state space will follow a path, which we shall call a trajectory. Figure 1.17(d) shows a ZX plane projection of the trajectory in state space. From Fig. 1.17, we see that the trajectory "relaxes" to the condition $X = 0$, $Y = 0$, and $Z = 0$ corresponding to the nonconvecting state illustrated in Fig. 1.15.

Exercise 1.5-1. Use the X, Y and Z versus time graphs to determine how the trajectory point circulates around the loop in the state space projection of Fig. 1.17.

Exercise 1.5-2. Show that the Lorenz model equations have a total of three fixed points, one at $X = 0$, $Y = 0$, and $Z = 0$; the other two at $X = Y = \pm\sqrt{b(r-1)}$ with $Z = r - 1$. (Obviously, the latter two fixed points are of interest only for $r > 1$.)

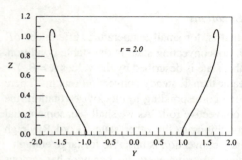

Fig. 1.18. State space projections onto the *YZ* plane for trajectories in the Lorenz model with *r* = 2. One attractor corresponds to clockwise rotation, the other to counterclockwise rotation of the fluid at a particular spatial location.

Note that at the point $X = 0$, $Y = 0$, $Z = 0$, all of the time derivatives in the Lorenz equations are 0. We call such a point a *fixed point* of the state space. We shall see that these fixed points play a crucial role in the dynamics of nonlinear systems.

> For a system described by a set of first-order differential equations, such as Eq. (1.5-1) for the Lorenz model, a point in the state space for which all of the time derivatives of the state space variables are 0 is said to be a *fixed point* for that system. (What we call fixed points for systems of differential equations are also called *equilibrium points*, or *critical points*, or *singular points* by other authors.) If the system starts at one of these fixed points, it stays at that fixed point for all time. Since the time derivatives of the state space variables are 0 at the fixed point, those variables cannot change in time.

For *r* values less than 1, all trajectories, no matter what their initial conditions, eventually end up approaching the fixed point at the origin of our *XYZ* state space. To use the language introduced for the logistic map, we can say that for *r* < 1, all of the *XYZ* space is the *basin of attraction* for the *attractor* at the origin.

For *r* > 1, we have three fixed points. The one at the origin turns out to be a repelling fixed point in the sense that trajectories starting near it tend to move away from it. The other two fixed points are attracting fixed points if *r* is not too large. Some initial conditions give rise to trajectories that approach one of the fixed points; other initial conditions give rise to trajectories that approach the other fixed point. (In Chapter 4, we will see more quantitatively what is different about these fixed points.) For *r* just greater than 1, the other two fixed points become the attractors in the state space. Thus, we say that *r* = 1 is a bifurcation point for the Lorenz model. Figure 1.18 illustrates the behavior of two trajectories starting from different initial points.

Let us describe this behavior in more physical terms. If *r* increases to a value just greater than 1 (recall that this means that we have increased the temperature difference between the bottom and top of the fluid layer), the fixed point at the origin becomes a repelling fixed point. This tells us that the so-called conductive state (the state with no fluid convection) has become unstable. The slightest

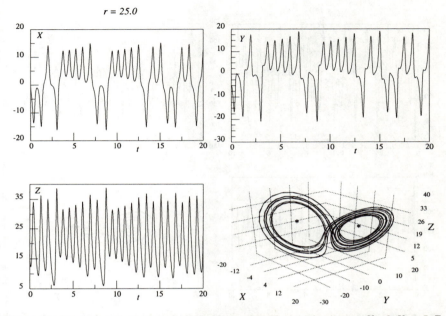

Fig. 1.19. Solutions to the Lorenz equations for $r = 25$. The initial point was $X = 0$, $Y = -5$, $Z = 15$. In the state space diagram in the lower right panel, the two off-origin fixed points at $Z = 24$, $X = 8$ and $Y = 8$ are indicated by asterisks.

deviation from the conditions $X = 0$, $Y = 0$, $Z = 0$ sends the state space trajectory away from the origin. For r just greater than 1, the trajectories are attracted to one or the other of the other two fixed points at $X = Y = \pm\sqrt{b(r-1)}$. Those two fixed points correspond to steady (time-independent) convection, one with clockwise rotation, the other counterclockwise. Some initial conditions give rise to trajectories that head toward one fixed point; other initial conditions lead to the other fixed point. The left-hand side of Fig. 1.18 shows the YZ plane state space projection for a trajectory starting from $X = 0$, $Y = -1$, $Z = 0$. The right-hand side of Fig. 1.18 shows a trajectory starting from a different set of initial values: $X = 0$, $Y = +1$, $Z = 0$. Note, in particular, that the system settles into a state with nonzero values of Y and Z, that is, the fluid is circulating.

An interesting question to ask for any dynamical system is the following: What region of initial conditions in XYZ space leads to trajectories that go to each of the fixed points? In other words, what are the basins of attraction for the attracting fixed points? How do these regions change as the parameters describing the system change? We shall see later that these regions can be quite complicated geometrically. In fact, in order to describe them, we need to use the relatively new geometrical concept of *fractals*. All of this, however, will be taken up in due course. Let us continue to increase the temperature difference for our fluid layer.

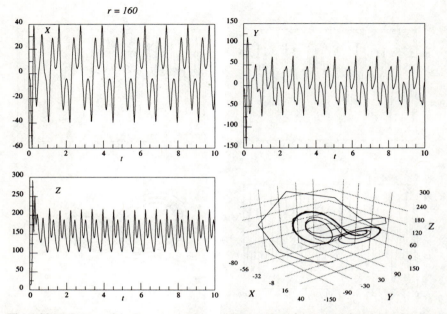

Fig. 1.20. Solutions of the Lorenz equations with $r = 160$. After an initial transient that lasts until about $t = 3$, the solutions are periodic (but not sinusoidal). The jaggedness of the transient trajectory in the XYZ state space plot (lower right) is a graphing artifact. The calculations were actually carried out with much smaller time steps.

Nothing dramatic happens until r reaches about 13.93 where we find that repelling regions develop around the off-origin fixed points. There are still small basins of attraction surrounding the two off-origin fixed points, which give rise to trajectories attracted to those two points. Trajectories starting outside these small regions, however, are repelled from the vicinity of the fixed points. If we examine the graphs of $X(t)$, $Y(t)$, and $Z(t)$ shown in Fig. 1.19, then we see that the new conditions correspond to time dependent variations in the fluid flow and the corresponding temperature differences. The corresponding state space diagram is shown in the lower right of Fig. 1.19.

This observation should be the cause of some reflection. Here, we have a system for which the externally controlled "forcing" (that is, the imposed temperature difference) is independent of time. Yet the system has developed spontaneously a nontrivial time dependence. As we mentioned before, a nonlinear system can break the time-translation symmetry of its fundamental equations and external environment. (The period-2, period-4, and so on, variations of populations in the logistic map model are also examples of the spontaneous breaking of time-translation symmetry.)

The time behavior in this region of r values is quite complex. So let us move on to examine another region near $r = 160$. Figure 1.20 shows the time dependence of X, Y, and Z for $r = 160$. The behavior is not simple harmonic (that is, it is not

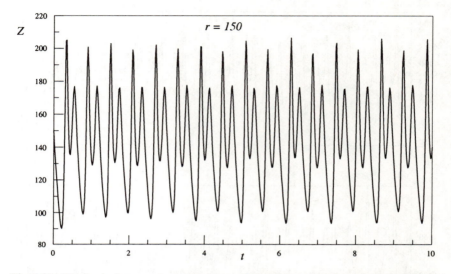

Fig. 1.21. $Z(t)$ for the Lorenz equations with $r = 150$. After an initial transient, the behavior is periodic with a period twice that seen in Fig. 1.20. Notice the alternating heights of the largest peaks in this figure.

sinusoidal), but it is periodic. We can understand the physical nature of the system's behavior by looking at the graphs of X and Y as functions of time. We see that X oscillates symmetrically around the value $X = 0$. This tells us that the fluid is convecting first in the clockwise direction, then counterclockwise, continually reversing as time goes on. The temperature difference between up flow and down flow, represented by the variable Y, also oscillates symmetrically around 0. Note that the Z variable, on the other hand, oscillates around a nonzero value (approximately 160 for the case displayed in Fig. 1.20).

Exercise 1.5-3. Show explicitly that the Lorenz equations are unchanged under time translation, that is, if t is replaced by $t + \tau$ then the equations are the same. Thus, we say that the Lorenz equations have time translation symmetry.

When the Rayleigh number is <u>decreased</u> to about $r = 150$, we find that the periodic behavior of Z suddenly changes. Figure 1.21 indicates that we now have period-2 behavior. (Please note a complication: The fundamental period is also slightly changed from Fig. 1.20. The Lorenz model does not have any external periodic forcing, as did the diode circuit, to determine the fundamental period.) The period-2 behavior is most easily recognized by looking at the largest upward "peaks" or downward "troughs" in Fig. 1.21. We see that a period-doubling bifurcation has occurred.

At $r \approx 146$, we find that $Z(t)$ bifurcates again, now with a period four times the original. [Similar, but less dramatic changes occur in $Y(t)$ and $X(t)$.] As r

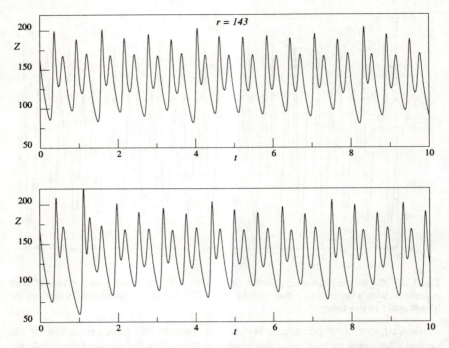

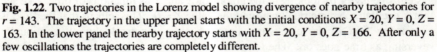

Fig. 1.22. Two trajectories in the Lorenz model showing divergence of nearby trajectories for $r = 143$. The trajectory in the upper panel starts with the initial conditions $X = 20$, $Y = 0$, $Z = 163$. In the lower panel the nearby trajectory starts with $X = 20$, $Y = 0$, $Z = 166$. After only a few oscillations the trajectories are completely different.

decreases below about 144, the behavior of all the variables becomes completely aperiodic. We have seen yet another period-doubling route to chaos. (We have not generated a complete bifurcation diagram for this range of r values because of the amount of computation time involved.)

Some comments are in order: In the case of the Lorenz model, the period-doubling progresses as the parameter r is <u>decreased</u>. In both of our previous examples, the doublings occurred as a parameter was increased. This distinction is really an unimportant one as we shall see in Chapter 2. In addition, we should again emphasize that the Lorenz model shows a vast range of complex behavior. We have described only a very limited part of that behavior, carefully selected, of course, to match the kind of behavior exhibited by our other two examples.

Divergence of Trajectories in the Lorenz Model

We now want to address the crucial question in deciding whether or not the Lorenz model equations exhibit chaotic behavior for some range of r values: Do nearby trajectories diverge for that range of r values? Figure 1.22 shows two trajectories for $r = 143$ with slightly different initial conditions. We see that after only a few oscillations the trajectories are completely different. Although this result does not prove the existence of the divergence of nearby trajectories on the average, it does suggest that the Lorenz model displays chaotic behavior for $r = 143$.

1.6 Determinism, Unpredictability, and Divergence of Trajectories

What is the importance of the divergence of nearby trajectories? We have claimed that this property is a signature of the kind of behavior we want to call chaotic and that this property allows us to distinguish between aperiodic behavior due to chaos and that due to external noise. The theoretical details will be taken up in later chapters. Here we want to discuss this behavior qualitatively.

The importance of divergence of nearby trajectories is the following: If a system, like the Lorenz model, displays divergence of nearby trajectories for some range of its parameter values, then the behavior of that system becomes essentially *unpredictable*. The system is still deterministic in the sense that if we knew the initial conditions of a trajectory exactly, then we could predict the future behavior of that trajectory by integrating the time-evolution equations for the system. If we make the smallest change in those initial conditions, however, the trajectory quickly follows a completely different path. Since there is always some imprecision in specifying initial conditions in any real experiment or real numerical calculation, we see that the actual future behavior is in fact unpredictable for a chaotic system. To make this point more forcefully, we like to say that the future of a chaotic system is *indeterminable* even though the system is *deterministic*.

This unpredictability is related to the fact that we cannot write down a closed-form solution for the nonlinear equations used to describe the system. A closed-form solution is a "formula" $X(t) = X_0 \tanh \pi (a\ t^2)$, for example, or a series solution, perhaps with an infinite number of terms, $X(t) = a_1(t) + a_2(t) + a_3(t)\ldots$ If such a closed-form solution could be found, then we could predict the future behavior of the system simply by evaluating the formula for some value of t corresponding to a future time. For a slightly different set of initial conditions, we would just evaluate the formula for those new initial conditions. Since the formula is presumably continuous in its dependence on parameters and initial conditions, small changes in those parameters and initial conditions would lead to small changes in $X(t)$. So, the large changes in $X(t)$ that occur for a chaotic system when we make small changes in the initial conditions cannot be represented by a closed-form solution. For a chaotic system, we must integrate the equations step-by-step to find the future behavior. (In essence we have to let the "experiment" run to find out what will happen.) The divergence of nearby trajectories means that any small error in specifying the initial conditions will be "magnified" as we integrate the equations. Thus, a small change in initial conditions leads to grossly different long-term behavior of the system, and we cannot in practice predict that long-term behavior in detail.

The unpredictability problem in nonlinear systems can be even worse than we imagine. For example, we might think that even though we cannot predict the detailed behavior of a trajectory, at least we know that the trajectory will end up within a particular attracting region in state space and will remain within that region. Unfortunately, many nonlinear systems have multiple attractors for a given set of parameter values. Trajectories starting at different state space points will end up on different attractors. Each attractor has its own basin of attraction. In some cases these basins have relatively simple geometric structures, and we can easily determine which initial conditions will lead to motion on the different attractors. In other cases, the basins can be so intertwined (forming so-called *riddled basins*) that even the smallest change in initial conditions can lead to a trajectory that ends up on a different attractor. In that case we lose even the modest ability to predict which attractor the trajectory will end up on (SOO93a) (LAW94)(LAI99).

The effect of the divergence of nearby trajectories on the behavior of nonlinear systems has been expressed in an elegant metaphor known as the ***butterfly effect***. This metaphor first appeared in the title of a talk given by E. N. Lorenz[1] at the December 1972 meeting of the American Association for the Advancement of Science in Washington, D.C.: "Predictability: Does the Flap of a Butterfly's Wings in Brazil set off a Tornado in Texas." Earlier, Lorenz had used a seagull for this metaphor, but the name took an interesting Nabokovian twist with this paper's title. Lorenz's point was that if the atmosphere displays chaotic behavior with divergence of nearby trajectories or sensitive dependence on initial conditions, then even a small effect, such as the flapping of a butterfly's (or other avian creature's) wings would render our long-term predictions of the atmosphere (that is, the weather) completely useless.

The conflict between determinism and its (purported) opposite, free will, has been a long-standing problem in philosophy. Newtonian mechanics appears to present us with a picture of a deterministic, clockwork universe in which all of the future is determined from the force laws, and the "initial conditions" of the objects that make up that world. From this point of view, all of our actions are completely determined, and there is no free will. This determinism was dramatized by the great French mathematician Pierre Simon Laplace (1749–1827), who in the introduction to his book *Theory of Probability* [Laplace, 1812] wrote:

> Let us imagine an Intelligence who would know at a given instant of time all forces acting in nature and the position of all things of which the world consists; let us assume, further, that this Intelligence would be capable of subjecting all these data to mathematical analysis. Then it could derive a result that would embrace in one and the same formula the motion of the largest bodies in the universe and of the lightest atoms. Nothing would be uncertain for this Intelligence. The past and the future would be present to its eyes.

---

[1] We thank Prof. Lorenz for some useful private correspondence on the historical origin of this metaphor.

From Laplace's point of view, the initial state of an isolated system is the "cause" from which the subsequent behavior of the system flows. Physical laws, expressed as differential equations, for example, provide the connecting link between the cause and its effects. The (assumed) existence of physical laws provides a deterministic link, according to this view, between the past and the future. There is no room for chance or free will[2]. There is even some historical evidence that Isaac Newton himself, the creator of Newtonian mechanics, drew back from the deterministic picture that seemed to emerge from his discoveries. Newton wanted to leave room for the active participation of God in the evolution of the universe.

As we have seen, however, nonlinear systems, and chaotic systems in particular, make the implementation of Laplace's calculating Intelligence impossible. Even the smallest imprecision in the specification of the initial conditions presented to that Intelligence would make the predictions for the future behavior of a chaotic system (of which the universe has an abundance) impossible. Thus, even God must allow these chaotic systems to evolve to see what will happen in the future. There is no short cut to prediction for chaotic systems.

We should also point out that the twentieth-century development of quantum mechanics, with its inherent probabilities and uncertainty relations, has undermined Laplace's scheme from an entirely different direction. The relationship between chaos and quantum mechanics is discussed in Chapter 12.

1.7 Summary and Conclusions

In this chapter we have examined the behavior of three simple systems that exhibit chaotic behavior: a real experimental system, a simple algebraic iterative model, and a (relatively) simple set of differential equations. We hope that you are impressed with the complexity of the behavior of these simple systems. Although the systems are all quite different, there are important similarities in their behavior as the parameters describing the systems change. These nonlinear systems show sudden and dramatic changes (bifurcations) in their behavior even for small changes in their parameters. Under certain conditions their behavior becomes aperiodic with the appearance of randomness. This chaotic behavior can be distinguished from noisy behavior by looking at the divergence of nearby trajectories. More importantly, we see that we do not need either external noise or complexity to produce complex, randomlike behavior. We have seen that the period-doubling route to chaos occurs in at least three very different kinds of systems (but other routes to chaos are possible). This observation suggests that there may be various universal features of the approach to chaos and perhaps for chaos itself. In the next chapter we shall examine some of the universal *quantitative* features of chaos.

---

[2] For a discussion of the philosophical problems of causality and determinism, see Phillip Frank, *Philosophy of Science* (Prentice-Hall, Englewood Cliffs, NJ, 1957).

1.8 Further Reading

Popularizations

James Gleick, *Chaos, Making a New Science* (Viking, New York, 1987). Now almost a "cult" book. Rather journalistic and hyperbolic in its writing, but it does try to give a scientifically honest picture of what chaos is all about. Includes good biographical sketches of many of the major players in the development of chaos.

John Briggs and F. David Peat, *Turbulent Mirror* (Harper & Row, New York, 1989). Aimed at the lay audience, this book tosses in a lot of (generally unexplained) technical jargon. Sometimes overly cute. For example, the chapter numbers run backward from the middle of the book to the end.

Ian Stewart, *Does God Play Dice? The Mathematics of Chaos* (Blackwell, New York, 1989). For the scientifically literate reader. Emphasizes the mathematical approach to chaos, but does pay some attention to experiments.

Peter Smith, *Explaining Chaos* (Cambridge University Press, Cambridge, 1998). An excellent philosophical look at chaos theory and fractals. A close examination of what chaos is about and what it can legitimately claim to explain.

Collections of Reprints

Hao Bai-Lin, *Chaos* (World Scientific, Singapore, Vol I. 1984, Vol II, 1989). A wide-ranging collection of reprints of important papers both theoretical and experimental. It includes a brief (and very dense) introduction to chaos by the editor.

Pedrag Cvitanovic, *Universality in Chaos*, 2nd ed. (Adam Hilger, Bristol, 1989). The editor has also included his views on the universality issues in chaos.

E. Ott, T. Sauer, and J. A. Yorke, *Coping with Chaos* (Wiley, New York, 1994). A collection of reprints dealing with the analysis of experimental data for chaotic systems.

Robert C. Hilborn and Nicholas B. Tufillaro, *Chaos and Nonlinear Dynamics* (American Association of Physics Teachers, College Park, MD, 1999). Contains an extensive bibliography and 22 reprinted articles, selected particularly for undergraduate students and for faculty new to the field.

Introductory Books for Scientists and Engineers

The following are listed more or less in order of increasing demands on your mathematical sophistication.

Larry S. Liebovitch, *Fractals and Chaos Simplified for the Life Sciences* (Oxford University Press, New York, 1998). A brief introduction with lots of references to applications in biology and medicine.

David Peak and Michael Frame, *Chaos Under Control: The Art and Science of Complexity* (W. H. Freeman, New York, 1994). A delightful book intended for a course for first-year college students.

Daniel Kaplan and Leon Glass, *Understanding Nonlinear Dynamics*, (Springer-Verlag, New York, Berlin, Heidelberg, 1995). Biological and medical orientation.

Garnett P. Williams, *Chaos Theory Tamed* (National Academy Press, Washington, DC, 1997). A nice introduction that assumes little background. Detailed sections on Fourier analysis, time series analysis, and so on.

Gregory Baker and Jerry Gollub, *Chaotic Dynamics, An Introduction*, 2nd ed. (Cambridge University Press, New York, 1996). A rather brief (255 pp.) introduction to chaos. Emphasizes the driven, damped pendulum and the use of the personal computer. The authors claim that this book is aimed at physics and math majors at the second and third year undergraduate level. This book covers only a few topics and does not give a comprehensive introduction to nonlinear dynamics. However, it is quite clearly written and does give some feeling for the key concepts.

Francis C. Moon, *Chaotic and Fractal Dynamics, An Introduction for Applied Scientists and Engineers* (Wiley, New York, 1992). Requires some background in applied science terminology. Tells how the ideas of chaos help understand important engineering phenomena. The formal development is rather brief, but the book provides a good overview. Includes a section on chaos gadgets: devices you can build to demonstrate chaotic dynamics.

P. Bergé, Y. Pomeau, and C. Vidal, *Order within Chaos* (Wiley, New York, 1986). A somewhat dated introduction (the French original was published in 1984) assuming roughly a first-level graduate student background in physics. Particularly good discussion of quasi-periodicity and intermittency.

J. M. T. Thompson and H. B. Stewart, *Nonlinear Dynamics and Chaos* (Wiley, New York, 1986). Two books in one! Apparently each author wrote half of the book, so there is some repetition and an abrupt change in style half-way through. Not very careful about defining terms before they are used freely in the text. Covers a wide range of "classical "nonlinear dynamics problems; but there is not much on modern methods such as time-series analysis, generalized dimensions, and so on.

Steven H. Strogatz, *Nonlinear Dynamics and Chaos: With Applications in Physics, Biology, Chemistry and Engineering* (Addison–Wesley, Reading, MA, 1994). An excellent book for an introductory applied-mathematics course at the advanced undergraduate level.

H. G. Schuster, *Deterministic Chaos, An Introduction*, 3rd revised ed. (Wiley, New York, 1995). A rather compact (319 pp.) introduction at roughly the graduate-level in physics. The arguments are very dense. Schuster does give a wide-ranging overview, and he does try to provide at least outline proofs of many important results.

E. Atlee Jackson, *Perspectives of Nonlinear Dynamics*, Vol. 1 and Vol. 2 (Cambridge University Press, New York, 1989, 1991). A very thoughtful and engaging book. A careful look at mathematical assumptions. Gradually builds up complexity of results, but rather heavy emphasis on analytical methods (perturbation methods, averaging methods, etc.). A good follow-up after reading this book.

N. Tufillaro, T. Abbott, and J. Reilly, *An Experimental Approach to Nonlinear Dynamics and Chaos* (Addison–Wesley, Reading, MA, 1992). This book, at the upper-undergraduate and graduate physics level, treats nonlinear dynamics by focusing on several experimental systems and using several computer-based models. The book has particularly good discussions of the analysis of experimental data.

E. Ott, *Chaos in Dynamical Systems* (Cambridge University Press, Cambridge, 1993). An insightful introduction to chaos at the beginning graduate level.

T. Kapitaniak and S. R. Bishop, *The Illustrated Dictionary of Nonlinear Dynamics and Chaos* (Wiley, Chicester and New York, 1999). This book contains an extensive set of definitions of the terms used in nonlinear dynamics, complete with illustrations in many cases. Advanced undergraduate, beginning graduate level.

For the Mathematically Inclined Reader

R. L. Devaney, *Chaos, Fractals, and Dynamics, Computer Experiments in Mathematics* (Addison–Wesley, Reading, MA, 1990). An introduction (without proofs) to some of the fascinating mathematics of iterated maps, Julia sets, and so on. Accessible to the good secondary school student and most college undergraduates.

R. L. Devaney, *A First Course in Chaotic Dynamical Systems* (Addison–Wesley, Reading, MA, 1992). A comprehensive introduction accessible to readers with at least a year of calculus.

Brian Davies, *Exploring Chaos: Theory and Experiment* (Perseus Books, Reading, MA, 1999). An introduction to dynamical systems examining the mathematics of iterated map functions and some simple ordinary differential equation models. Includes exercises using the software *Chaos for Java*. (See the software listings at the end of Chapter 2.)

R. L. Devaney, *An Introduction to Chaotic Dynamical Systems* (Benjamin–Cummings, Menlo Park, CA, 1986). Here are the proofs for the fascinating mathematics of iterated maps. This is definitely a mathematics book, but, with a modest amount of work and attention to detail, quite accessible to the nonmathematician.

Kathleen Alligood, Timothy Sauer and James A. Yorke, *Chaos, An Introduction to Dynamic Systems* (Springer-Verlag, New York, 1997). Contains lots of computer experiments for undergraduates who have completed calculus and differential equations.

D. Gulick, *Encounters with Chaos* (McGraw–Hill, New York, 1992). This book provides a very readable introduction to the mathematics of one- and two-dimensional iterated map functions with many nice proofs, examples, and exercises. Briefly covers fractals and systems of differential equations.

R. H. Abraham and C. D. Shaw, *Dynamics: The Geometry of Behavior* (Addison–Wesley, Reading, MA, 1992). F. D. Abraham, R. H. Abraham, and C. D. Shaw, *Dynamical Systems: A Visual Introduction* (Science Frontier Express, 1996). The picture books of chaos! Outstanding diagrams of heteroclinic and homoclinic tangles, and the like. These books are most useful after you have had some general introduction to chaos. The texts dodge careful definitions (the authors' philosophy is to give a visual introduction to dynamical systems).

J. Hale and H. Koçak, *Dynamics and Bifurcations* (Springer-Verlag, New York, 1991). This book provides a well-thought-out introduction to the mathematics of dynamical systems and bifurcations with many examples. Easily accessible to the advanced undergraduate.

J. Guckenheimer and P. Holmes, *Nonlinear Oscillations, Dynamical Systems, and Bifurcations of Vector Fields*, 3rd ed. (Springer-Verlag, New York, 1990). A classic in the field, but you need to know your diffeomorphisms from your homeomorphisms. If you are serious about the study of chaos, you will eventually come to this book.

Other Books, Collections of Essays, etc.

A. V. Holden (ed.), *Chaos* (Princeton University Press, Princeton, NJ, 1986). A collection of 15 essays by active researchers in the field. Many useful ideas and perspectives, but you should be familiar with basic issues before tackling this book.

L. Glass and M. C. Mackey, *From Clocks to Chaos, The Rhythms of Life* (Princeton University Press, Princeton, NJ, 1988). A well-written book showing how the ideas of nonlinear dynamics and chaos can be applied to the analysis of rhythmic effects in physiology. Most of the physiological discussions are accessible to the nonspecialist. Reflecting the general mathematical level in the biological sciences, this book is quite a bit less sophisticated mathematically compared to most books on chaos.

Harvey Gould and Jan Tobochnik, *An Introduction to Computer Simulation Methods, Applications to Physical Systems*, Part I (Addison–Wesley, Reading, Mass., 1987). A marvelous book on computer methods; very readable and pedagogical. Chapter 7 leads the reader through a numerical study of the logistic map and its surprisingly complex behavior.

D. Ruelle, *Chaotic Evolution and Strange Attractors* (Cambridge University Press, New York, 1989). Ruelle is one of the masters of nonlinear dynamics. In this short book (96 pp.) he presents what is really an extended essay (with mathematics) on what he considers to be the important issues in the statistical analysis (via time series) of chaotic systems.

Some Introductory Journal Articles

R. May, "Simple Mathematical Models with Very Complicated Dynamics," *Nature* **261**, 459–467 (1976). A stimulating introduction to iterated maps ante Feigenbaum.

J. P. Crutchfield, J. D. Farmer, N. H. Packard, and R. S. Shaw, "Chaos," *Scientific American* **255** (6), 46–57 (December, 1986). A good overview of the field of chaos and its implications for science. Emphasizes the ideas of state space and attractors.

R. V. Jensen, "Classical Chaos," *American Scientist* **75**, 168–81 (1987). A well-written, detailed treatment of the major issues in the current study of chaos but with an emphasis on mathematics and theory.

D. R. Hofstadter, "Metamagical Themas," *Scientific American* **245**, (5) 22–43 (1981). A popular level discussion of the mathematics of iterated maps.

N. B. Abraham, J. P. Gollub, and H. L. Swinney, "Testing Nonlinear Dynamics," *Physica D* **11**, 252–64 (1984). Summary of a 1983 conference; gives a good idea of the range of experimental and theoretical activity in chaos. Unfortunately, already somewhat dated.

R. Van Buskirk and C. Jeffries, "Observation of Chaotic Dynamics of Coupled Nonlinear Oscillators," *Phys. Rev. A* **31**, 3332–57 (1985). A detailed description of how one can study chaos using simple semiconductor diodes. Lots of pictures and diagrams.

J.-P. Eckmann, "Roads to Turbulence in Dissipative Dynamical Systems," *Rev. Mod. Phys.* **53**, 643–54 (1981). An early essay, but still a useful survey.

E. Ott, "Strange Attractors and Chaotic Motions of Dynamical Systems," *Rev. Mod. Phys.* **53**, 655–72 (1981). This survey appears in the same issue as the Eckmann article cited earlier. Again, somewhat dated, but useful.

M. F. Doherty and J. M Ottino, "Chaos in Deterministic Systems: Strange Attractors, Turbulence, and Applications in Chemical Engineering," *Chemical Engineering Science* **43**, 139–83 (1988). A wide-ranging and thoughtful survey of chaos in both dissipative and conservative systems, with an eye on engineering applications, this article is written at roughly the beginning graduate student level.

Max Dresden, "Chaos: A New Scientific Paradigm—or Science by Public Relations," *The Physics Teacher* **30**, 10–14 and 74–80 (1992). [Reprinted in Hilborn and Tufillaro, 1999]. An engaging introduction to the fundamental issues of chaos.

The issue of predictability in Newtonian mechanics is discussed in J. Lighthill, "The Recently Recognized Failure of Predictability in Newtonian Dynamics," *Proc. Roy. Soc. Lond. A* **407**, 35–50 (1986).

P. Holmes, "Poincaré, Celestial Mechanics, Dynamical-Systems Theory and 'Chaos'," *Physics Reports* **193**, 137–63 (1990). This wide-ranging essay provides insight into the historical development of nonlinear dynamics at a moderately sophisticated mathematical level.

D. Ruelle, "Where can one hope to profitably apply the ideas of chaos," *Physics Today* **47** (7), 24–30 (1994).

Robert C. Hilborn and Nicholas B. Tufillaro, "Resource Letter: ND-1: Nonlinear Dynamics," *Am. J. Phys.* **82** (9), 822–834 (1997). An extensive bibliography with comments.

The Lorenz Model

E. N. Lorenz, "Deterministic Nonperiodic Flow," *J. Atmos. Sci.* **20**, 130–41 (1963). (Reprinted in [Cvitanovic, 1984]).

B. Saltzman, "Finite Amplitude Free Convection as an Initial Value Problem-I," *J. Atmos. Sci.* **19**, 329–41 (1962).

C. T. Sparrow, *The Lorenz Equations: Bifurcations, Chaos, and Strange Attractors* (Springer-Verlag, New York, Heidelberg, Berlin, 1982).

Prediction and Basins of Attraction

J. C. Sommerer and E. Ott, "A physical system with qualitatively uncertain dynamics," *Nature* **365**, 136–140 (1993)

Y.-C. Lai and R. L. Winslow, "Riddled Parameter Space in Spatiotemporal Chaotic Dynamical Systems," *Phys. Rev. Lett.* **72**, 1640–43 (1994).

Y.-C. Lai and C. Grebogi, "Riddling of Chaotic Sets in Periodic Windows," *Phys. Rev. Lett.* **83**, 2926–29 (1999).

Novels and plays

Chaos and nonlinear dynamics play major roles in the following:

Michael Crichton, *Jurassic Park* (Ballentine Books, New York, 1990).

Kate Wilhelm, *Death Qualified, A Mystery of Chaos* (Fawcett Crest, New York, 1991).

Tom Stoppard, *Arcadia* (Faber and Faber, London and Boston, 1993).

World Wide Web Sites

All of the major centers of nonlinear science research have extensive Web sites. A few bookmarks to get you started include:

Chaos at the University of Maryland (http://www-chaos.umd.edu). In particular see their "Chaos Database," which has a nice search engine for an extensive bibliographic data base.

The Institute of Nonlinear Science at the University of California at San Diego (http://inls.ucsd.edu).

The Center for Nonlinear Dynamics at University of Texas at Austin (http://chaos.ph.utexas.edu).

The Center for Nonlinear Science at Los Alamos National Lab (http://cnls.lanl.gov). In particular, see the CNLS Nonlinear Science e-print archive (http://cnls.lanl.gov/pbb.announce.html), and the Nonlinear Dynamics Archive (http://cnls.lanl.gov/nbt/intro.html).

G. Chen, "Control and synchronization of chaotic systems (a bibliography)" is available by anonymous ftp from (uhoop.egr.uh.edu/pub/TeX/chaos.tex).

"Nonlinear Dynamics Bibliography" maintained by the University of Mainz (http:/www.uni-mainz.de/FB/Physik/Chaos/chaosbib.html).

The Applied Chaos Laboratory at the Georgia Institute of Technology (http://www.physics.gatech.edu/chaos/).

2

The Universality of Chaos

Chaos is come again. Shakespeare, *Othello*, III, iii

2.1 Introduction

In the preface and in the previous chapter, we mentioned that nonlinear systems display many universal quantitative features, both in their approach to chaos and in their chaotic behavior itself. In Chapter 1, we pointed out some of the qualitative features common to many nonlinear systems. In this chapter, we will give a brief introduction to some of the quantitative universal features. In a sense, the remainder of the book is devoted to developing an understanding of those quantitative features.

It is difficult to overstate the importance of these universal features. If each nonlinear system did "its own thing," in its own way, then the study of dynamical systems would have languished as a branch of applied science, important certainly in applications, but providing no new general principles. After all, it is those general principles that lead to advances in the fundamentals of science. What has surprised almost everyone is the vast number of common features discovered in the behavior of nonlinear systems. These features include the sequences of bifurcations connecting regular, periodic behavior to chaotic behavior. In this chapter, we will explore some of the completely unexpected quantitative aspects of these bifurcations. These features seem to be largely independent of the physical, chemical or biological details of the system under investigation. This universality has made nonlinear dynamics a truly interdisciplinary field of study.

The approach in this chapter will again be descriptive rather than deductive in order to provide an overview without the burden of detailed mathematical proofs. The mathematical details providing support (if not proof) for these universal features will be taken up in later chapters.

2.2 The Feigenbaum Numbers

In Chapter 1 we saw that three quite different systems followed the period-doubling route to chaos, at least for some range of their control parameters. Of course, as we have also seen, there are other ways for the system to change from periodic to chaotic behavior besides the period-doubling route. It was in the study of period-doubling in the logistic map, however, that theoretical physicist Mitchell Feigenbaum discovered (FEI78) the first of these universal quantitative features, and the theory of these universal quantitative features is more highly developed for

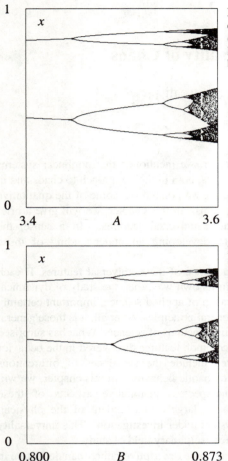

Fig. 2.1. A portion of the bifurcation diagram for the logistic map function, Eq. (2.2-1) showing the period-doublings leading to chaos indicated by the fuzzy bands at the right of the diagram.

Fig. 2.2. Part of the bifurcation diagram for the sine map, Eq. (2.2-2).

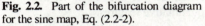

the period-doubling behavior than for some of the other routes to chaos, which we shall discuss in later chapters.

Feigenbaum's first clue that there might be some universality underlying chaos came from the observation that several different functions, when used as iterated maps, lead to the same convergence in the bifurcation diagram as simple, periodic behavior changed to chaotic behavior through a sequence of period-doublings. To illustrate what Feigenbaum discovered, we have plotted in Fig. 2.1 a portion of the bifurcation diagram for the logistic map,

$$x_{n+1} = Ax_n(1 - x_n) \tag{2.2-1}$$

Figure 2.2 uses a map based on the trigonometric sine function

$$x_{n+1} = B\sin(\pi x_n) \tag{2.2-2}$$

In both cases we see the (now) obvious period-doubling route to chaos. Feigenbaum was the first to realize that the "rate of convergence" of the two sequences was the same for both maps.

For anyone trained in mathematical physics, the instinctive reaction, when faced with the kind of convergence seen in Fig. 2.1 and Fig. 2.2, is to look for a geometric convergence ratio. If the convergence is geometric, then the ratio of differences of parameter values at which successive period-doublings occur should be the same for all the splittings.

The calculation runs as follows: Let A_1 be the parameter value where period-1 gives birth to period-2 (see Fig. 2.3). A_2 is the value where period-2 changes to period-4, and so on. In general we denote A_n as the parameter value at which period-2^n is "born." We then examine the ratio

$$\delta_n = \frac{A_n - A_{n-1}}{A_{n+1} - A_n} \qquad (2.2\text{-}3)$$

Feigenbaum found that indeed the ratio was approximately the same for all values of n and that, more importantly and surprisingly, for large values of n the ratio approached a number that was the same for both map functions! This value of δ has now been named the "Feigenbaum δ (delta)":

$$\boxed{\delta \equiv \lim_{n \to \infty} \delta_n = 4.66920161... \qquad (2.2\text{-}4)}$$

The number 4.669... is destined, we believe, to take its place along side the fine structure constant (1/137) (which tells us the ratio of the strength of electromagnetic forces to that of nuclear forces) in the (sparsely populated) Pantheon of universal numbers in physics.

> **Research Problem**: Relate the Feigenbaum δ value, 4.669..., to some other fundamental numbers, for example, π, e (2.718...), the golden mean ratio $(\sqrt{5} - 1)/2$, and so on (no known solution).

Of course, seeing two (nearly) identical ratios is no proof of the universality of these numbers. Later Feigenbaum was able to establish that any iterated map function that has a parabolic shape near its maximum value (and which satisfies some other conditions to be discussed in chapter 5) will have the same convergence ratio as the order of the bifurcation, labeled by n, approaches infinity.

The theory that Feigenbaum developed to explain the universality of δ actually used a slightly different definition of the parameter values in calculating the ratios. Rather than referring to the parameter value at which a bifurcation occurs, the parameter values chosen are those for which the abscissa of the maximum of the map function x_{max} is part of the "orbit" for a particular periodicity. (For the logistic map of Eq. (2.2-1), $x_{max} = 0.5$.) These orbits are called *supercycles* for reasons we shall discuss in Chapter 5. A^S_n is then the value of A for which x_{max} is part of the period 2^n orbit. It seems reasonable geometrically that both

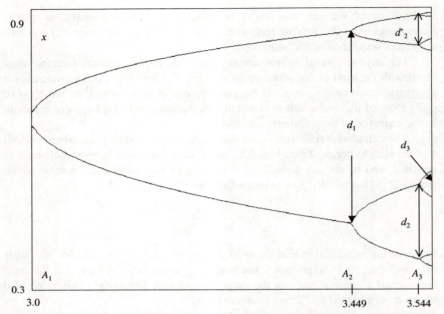

Fig. 2.3. A portion of the bifurcation diagram for the logistic map. The As indicate the parameter values at which period-doubling bifurcations occur. The ds indicate the relative sizes of the bifurcation patterns at the bifurcation points.

definitions give rise to the same ratio in the limit of high-order bifurcations. In terms of numerical computations, the supercycle definition is somewhat easier to implement. However, for experiments (see the next section), only the bifurcation point values can be determined since the (equivalent) map function and its maximum value are not known, in general.

Exercise 2.2-1. Table 2.1 lists the A_n values for the logistic map (Eq. 2.2-1). Also listed are the supercycle values A^S_n, the parameter values for which x_{max} is part of the trajectory for period 2^n. (For reasons that will be explained in Chapter 5, the supercycle values are somewhat easier to compute to high precision. They are also important in the theoretical development discussed in Chapter 5.) The B^S_ns are the supercycle parameter values for the sine map (Eq. 2.2-2). Compute δ_n for these parameter values and compare your results to the value given in Eq. (2.2-4). (In a computer exercise at the end of this chapter, we describe how to calculate these numbers to high accuracy. Obviously, reading them off the bifurcation diagram cannot be done with high precision.)

Table 2.1.
Bifurcation and Supercycle Values
for the Logistic Map and the Sine Map

n	A_n	A^s_n	B^s_n
1	3.00000	3.23607	0.77734
2	3.4931	3.49856	0.84638
3	3.54402	3.55463	0.86145
4	3.56437	3.56667	0.86469
5	3.56875	3.56924	0.86539
∞	3.569946…	—	—

Note that δ is defined in such a way that it is independent of the units in which the parameter is measured (that is, we can multiply all the parameter values by a common factor and still get the same δ value). It is also independent of a shift in the zero of the scale on which the parameter values are measured (that is, we can add to or subtract from the parameter values a common number and still get the same δ value).

2.3 Convergence Ratio for Real Systems

In Chapter 1, we saw that at least one real experiment (the diode circuit) showed a period-doubling route to chaos. The same scenario has been seen in fluid convection, modulated lasers, acoustic waves, chemical reactions, mechanical oscillators and many more systems. The obvious question to ask is the following: Does Feigenbaum's δ describe what happens in the experiments? One difficulty is immediately apparent: Feigenbaum's argument requires that the sequence of values $\delta_n \to 4.669\ldots$ only for large values of n, but as n gets large, the successive bifurcations get closer together and more difficult to locate precisely in an actual experiment. In fact, in only a few experiments (such as the diode circuit) have period-16 and period-32 been seen unambiguously. Hence, the experimenter can make precise measurements of the bifurcation parameter values only for low-order bifurcations; the theory makes a rigorous prediction only for high-order bifurcations. (As an aside, we should point out that it is rather fortuitous for the historical development of chaos theory that the logistic map and the sine-function map have ratios that converge very quickly to 4.669; Feigenbaum could "see" convergence to a universal number after computing only a few δ_n values. Other map functions, some of which will be described in Chapter 5, are not so well-behaved in their rate of convergence.)

Table 2.2.
Voltage values for bifurcations in the diode circuit.

n	V_n (30 kHz)	V_n (85 kHz)
1	−3.470(8)	−0.946(1)
2	−2.505(8)	−0.438(1)
3	−2.234(7)	−0.330(1)

With these words of caution in mind, let us compare the experimentally determined ratios (for low-order bifurcations) with the Feigenbaum δ value. Table 2.2 lists the values of the dc bias voltages at which successive bifurcations occur for the diode circuit. (For this experiment it was easier to use the dc bias voltage, rather than the signal voltage amplitude, as the control parameter. Like many other nonlinear systems, the diode circuit has a wide range of behavior. To generate the data in Table 2.2, we purposely chose circuit conditions that led to a bifurcation diagram closely resembling that for the logistic map.)

Using the values listed in Table 2.2, we find that $\delta = 3.57(10)$ for the 30 kHz data and $\delta = 4.7(1)$ for the 85 kHz data. The figures in parentheses give the uncertainty in the previous significant figure. For example, 4.66(2) means 4.66 ± 0.02.

Two comments are in order:

1. It is important (and, in fact, crucial) to take into account experimental uncertainties in calculating these ratios. We see that for the higher-order bifurcations, the experimental uncertainties become as large as the parameter differences, and the resulting relative uncertainty in δ becomes quite large.

2. The experimentally determined values for δ_n are close to but often do not agree with the Feigenbaum value of 4.669 even within the range of experimental uncertainties.

Table 2.3.
Feigenbaum δ Values Determined by Experiment

Experiment	Ref.	Max $n+1$ observed	Value of δ_n	Difference From 4.669
Fluid convection	GMP81	4	4.3(8)	−0.3(8)
Diode circuit	TPJ82	5	4.3(1)	−0.3(1)
Optical bistability	HKG82	3	4.3(3)	−0.3(3)
Acoustic waves in helium	STF82	3	4.8(6)	+0.2(6)
Josephson-junction analog	YEK82	3	4.4(3)	−0.2(3)

The latter statement might seem to be a cause for despair. After all, one of the usual criteria for the validity of a scientific theory is its ability to predict phenomena quantitatively. We must keep in mind, however, that period-doubling experiments and the Feigenbaum theory operate at opposite ends of the period-doubling sequence. Given that fact, contrary to our proposed feeling of despair, we ought to have a sense of exhilaration that the numbers agree as well as they do. Table 2.3 lists some representative experimentally determined values for δ for a variety of physical systems. In all cases the measured values are within 20% or so of the Feigenbaum value.

Some Preliminary Reflections on Explaining Universality

We should reflect on this "agreement" between theory and experiment for a moment. It is not at all obvious what a semiconductor diode, a convecting fluid, or a modulated laser have in common with the logistic map. We should feel a sense of awe and wonder that there is some quantitative link between these experimental results and a simple mathematical iterated map.

In a traditional physics setting, if we see some common quantitative and qualitative features in the behavior of different systems (such as the well-known free-fall acceleration near the surface of the Earth, if we are able to ignore air resistance), we look for some common underlying physical cause. It should be clear, however, that this kind of explanation is not appropriate here. To explain this new kind of universality, we need to adopt a different level of explanation. This we shall do in the following chapters where we will begin to see that this universality can be explained as a result of common geometries in an abstract state space description of the dynamics of these systems.

2.4 Using δ to Make Predictions

The numerical agreement between the values of δ found in experiments and the value found by Feigenbaum from a mathematical model points to an underlying unity, which we shall explore in Chapter 5. At a more practical level, the existence of a universal number such as δ allows us to make quantitative predictions about the behavior of a nonlinear system, even if we cannot solve the equations describing the system. More importantly, this is true even if we do not know what the fundamental equations for the system are, as is often the case. For example, if we observe that a particular system undergoes a period-doubling bifurcation from period-1 to period-2 at a parameter value A_1, and from period-2 to period-4 at a value A_2, then we can use δ to predict that the system will make a transition from period-4 to period-8 at an A_3 value given by

$$A_3 = \frac{A_2 - A_1}{\delta} + A_2 \tag{2.4-1}$$

As we shall see later, however, observing the first two period-doublings produces no guarantee that a third will occur, but if it does occur, Eq. (2.4-1) gives us a reasonable prediction of the parameter value near which we should look to see the transition.

We can also use δ to predict the parameter value to which the period-doubling sequence converges and at which point chaos begins. To see how this works, we first write an expression for A_4 in terms of A_3 and A_2, in analogy with Eq. (2.4-1). (We are, of course, assuming that the same δ value describes each ratio. This is not exact, in general, but it does allow us to make a reasonable prediction.)

$$A_4 = \frac{A_3 - A_2}{\delta} + A_3 \qquad (2.4-2)$$

We now use Eq. (2.4-1) in Eq. (2.4-2) to obtain

$$A_4 = (A_2 - A_1)\left(\frac{1}{\delta} + \frac{1}{\delta^2}\right) + A_2 \qquad (2.4-3)$$

If we continue to use this procedure to calculate A_5, A_6, and so on, we just get more terms involving powers of $(1/\delta)$ in the sum. We recognize this sum as a geometric series. We can sum the series to obtain the result

$$A_\infty = (A_2 - A_1)\left(\frac{1}{\delta - 1}\right) + A_2 \qquad (2.4-4)$$

After we have observed the first two period-doublings in a system, we can make a prediction of the parameter value at which chaos should appear. However, we do not expect this prediction to be exact, first, since it is based on experimentally determined numbers A_2 and A_1, and more importantly because we have assumed that all the bifurcation ratios are described by the same value of δ. Nevertheless, this prediction does usually get us reasonably close to the region in which chaos begins.

We can use the results of this extrapolation to predict the parameter values for the onset of chaos for the logistic map and for the diode circuit. Using the results given in Table 2.1, we find that for the logistic map, the onset of chaos is predicted to occur at $A = 3.572$ if we use the bifurcation values of A and at $A = 3.563$ if we use the supercycle values of A. The actual value is $A = 3.5699\ldots$ For the diode circuit, using the voltage values in Table 2.2 in Eq. (2.4-4) predicts chaos at $-2.24(1)$ volts while the experiment shows chaos beginning at $-2.26(1)$ volts for the 30 kHz data. The predicted value is $-0.299(9)$ volts and the observed value is $-0.286(8)$ volts for the 85 kHz data. We see that the agreement is not exact, but nevertheless, considering that we have done no detailed calculations of the dynamics of these systems, we have surprisingly good agreement.

Exercise 2.4-1. (a) Verify the calculation leading from Eq. (2.4-3) to Eq. (2.4-4). (b) Prove that

$$\left(A_\infty - A_n\right)\delta^n = \left(A_2 - A_1\right)\frac{\delta^2}{\delta - 1}$$

Note that the right-hand-side is independent of n. N.B. This result will be used several times in later chapters.

2.5 Feigenbaum Size Scaling

As part of his numerical investigation of simple mapping functions such as the logistic map, Feigenbaum recognized that each successive period-doubling bifurcation is just a smaller replica, with more branches, of course, of the bifurcation just before it. This observation suggested that there might be a universal size-scaling in the period-doubling sequence. Figure 2.3 illustrates the definition of the "size" ratio, now designated as the Feigenbaum α (alpha) value:

$$\alpha = \lim_{n \to \infty} \frac{d_n}{d_{n+1}} = 2.5029\ldots \qquad (2.5\text{-}1)$$

where d_n is the "size" of the bifurcation pattern of period 2^n just before it gives birth to period-2^{n+1}. The ratio involves the ds for the corresponding parts of the bifurcation pattern. For example, the size of the largest of the period-4 segments is compared to the size of the largest period-8 segment. Feigenbaum also argued that the size d_2 of the larger of the period-4 "pitchforks" ought to be α times the size of the smaller of the period-4 pitchforks d_2' as shown in Fig. 2.3.

Feigenbaum actually used a slightly different definition of d_n in his determination of α. As in the definition of δ, the ds refer to distances in the bifurcation diagram when the point x_{max} is part of the trajectory. We shall make use of this choice of distance in Chapter 5 and Appendix F. We intuitively expect that the two definitions give the same numerical values for high-order bifurcations.

Applying this ratio to the description of experimental data carries the same caveats we mentioned for the convergence ratio δ: The theory leading to the number 2.5029... applies only in the limit of high-order bifurcations, while experiments are constrained to look at relatively low-order bifurcations. However, in those few cases in which α has been determined experimentally, we find reasonable but not exact agreement between the measured values and the prediction of Eq. (2.5-1). We should be elated to have any prediction at all. We see a priori little reason for the size scaling in the logistic map to be related to the size scaling in the experiments.

Exercise 2.5-1. Use a ruler to measure the appropriate lengths in Fig. 2.3 to determine the Feigenbaum α value. Estimate your measurement uncertainties.

Exercise 2.5-2. Use numerically generated values from the logistic map to estimate α. Estimate numerical uncertainties due to the finite precision of your computer's arithmetic.

Something to think about: It is rather curious that the Feigenbaum δ is just about the "right" size. This means that if δ were an order of magnitude larger (say, about 40 or 50), then period-doublings would occur so quickly as a function of parameter values that it would be very difficult to see them experimentally. For example, if δ were about equal to 50, the difference $A_3 - A_2$ would be about 2 per cent (1/50) of the difference $A_2 - A_1$ and a modest amount of experimental care would be needed to observe it. [As a rough rule of thumb, it is easy to carry out measurements with 10 per cent experimental uncertainty; 1 per cent is not too difficult; 0.1 per cent requires considerable care and effort.] But the next difference $A_4 - A_3$ would be only 0.0004 of the difference $A_2 - A_1$ and observing it would be very difficult.

On the other hand, if δ were too small, then the period-doubling sequence would be so spread out as a function of the parameter that it would be difficult to change the parameter in question over the range required to see the full period-doubling sequence. (Note that δ must be greater than 1 in order to have the sequence converge at all. Therefore, "too small" means too close to 1.) This latter result would seem to be less constraining than the first because, you might think, we could then concentrate our attention on the higher-order bifurcations more easily. However, given the wide parameter range over which the sequence would occur, we might never see those higher-order bifurcations at all, or we might not recognize them as part of a sequence.

Exercise 2.5-3. Carry through the same kind of argument to show that the Feigenbaum α is also about the "right size."

Question: Is there a connection between the values for δ and α? The answer is given in Chapter 5 and Appendix F.

2.6 Self-Similarity

The two Feigenbaum numbers tell us something very important about the period-doubling sequence: Different pieces of our bifurcation diagram are just smaller replicas of other pieces. To be specific, we can note that the lower section of Fig. 2.3 between A_3 and A_2 looks just like the region between A_1 and A_2 if we expand the parameter axis between A_3 and A_2 by the factor δ and also expand the vertical axis for the same region by the factor α. (The upper portion requires a magnification of α^2.) [Again, we should remind ourselves that this replication is exact only in the limit of high-order bifurcations. However, it is nearly exact for low-order bifurcations as well.]

A geometric structure that has this replicating behavior under the appropriate magnifications is said to be *self-similar*: each subpart when appropriately magnified looks just like a larger part. As we shall see in Chapter 5, this self-similarity plays a key role in the theory of the universality of δ and α. In later chapters, we will see this notion of self-similarity in many other aspects of nonlinear dynamics as well. Such self-similar objects are called *fractals* because their geometric dimension (suitably defined) is often a fraction, not an integer.

Why is self-similarity so important? The basic idea is the following: If a geometric structure exhibits self-similarity, then it has no *inherent* size scale. That is, if we look at some subsection of this structure, at some level of magnification, it looks like any other subsection at some other level of magnification. There is no way for us to tell by looking at one subsection what length scale we are seeing. This remarkable feature means, as we shall see many times in this book, that many features of the geometric structure must be independent of the details of the model that gave rise to it.

A dramatic illustration of this self-similarity is shown in Figs. 2.4 and 2.5. Figure 2.4 shows an expanded portion of the bifurcation diagram for the logistic map in the region containing a period-5 window. Figure 2.5 is an expanded version of the boxed area of Fig. 2.4. This shows a period-doubling sequence converging to chaos, and inside that chaotic area is yet another period-5 window. At a slightly larger value of A is a period-3 window, which period-doubles to yet more chaos. Many of the features seen in the full bifurcation diagram are apparently reproduced again on a much smaller scale. As we shall see in Chapter 5, this self-replication allows us to make very powerful, quantitative statements about the behavior of the system that gives rise to such a bifurcation diagram. We find that these quantitative statements take the form of scaling laws, much like the famous scaling laws for thermodynamic properties of a system near a phase transition.

2.7 Other Universal Features

The Feigenbaum numbers do not, by any means, exhaust the range of universality that has been discovered in nonlinear systems. In later chapters, we will discuss other predictions of universality concerning the power spectra of chaotic systems, the influence of noise on these systems, and various measures of the "amount of chaos." It is safe to predict that as our study of nonlinear systems develops, still more universal features will be found.

The cautious reader should ask: Do these quantitative features hold for *all* nonlinear systems? The answer is No. There seem to be, however, classes of

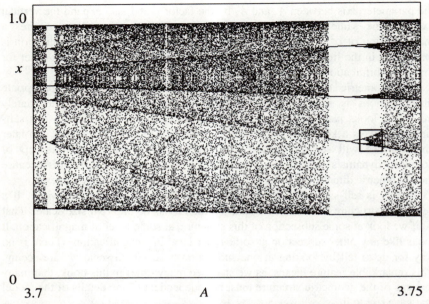

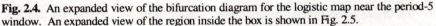

Fig. 2.4. An expanded view of the bifurcation diagram for the logistic map near the period-5 window. An expanded view of the region inside the box is shown in Fig. 2.5.

nonlinear systems for which the Feigenbaum numbers (and various generalizations of them) do provide good quantitative descriptions. A complete methodology for deciding in advance which "universality class" is appropriate for which nonlinear system (either theoretical models or real experimental systems) is lacking so far. Nevertheless it is useful to know that apparently there are only a small number of such classes. (At least, we have so far recognized only a small number of these classes.) For example, nonlinear systems whose dynamics reduce to what we shall call "one-dimensional dynamics," a notion that will be made more precise in Chapter 5, often show the period-doubling route to chaos, and the Feigenbaum numbers apply to those systems.

2.8 Models and the Universality of Chaos

Now that we have had at least a brief overview of the landscape of chaos, we would like to raise some fundamental questions about chaos and what it holds for scientific methodology. In short, we want to raise the question (and several issues surrounding it): What do we get out of chaos?

A crucial issue in understanding the significance of chaos in scientific methodology is the problem of the use of models in science. You may have noticed this terminology in Chapter 1, where we discussed a model of biological population growth and a model (the Lorenz model) for a convecting fluid. What does *model* mean here? First, we point out that most "real-life" systems are far too complicated

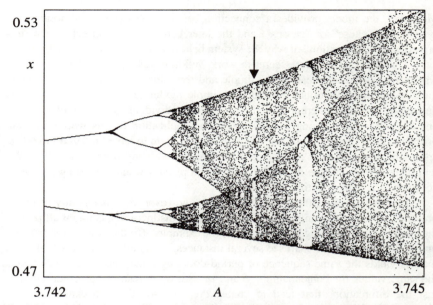

Fig. 2.5. An expanded view of the region inside the box in Fig. 2.4. The self-similarity of the bifurcation diagram is indicated by the existence of yet another period-5 window, indicated by the arrow, inside the chaotic band.

to be described directly by fundamental laws, which for the sake of discussion here, we will take to be the microscopic laws of physics, such as Schrödinger's equation in quantum mechanics, Maxwell's equations in electromagnetic theory, and so on. For example, the diode and the convecting fluid system described in Chapter 1 each have trillions of atoms and there is no practical way of applying the fundamental microscopic laws to these systems. Instead, we extract what we believe are the most important features of the phenomena being studied (the current and voltages in the diode circuit; temperature and fluid velocity in the fluid experiment, for example), and then we build theoretical models (mathematical descriptions) of the behavior of those selected features. In the process, we hope that the features we have neglected in our model are not significant for the phenomena under consideration. For example, we did not specify the color of the fluid in the Lorenz model. We assumed that the nature of the fluid could be specified in terms of its density, thermal conductivity, and its viscosity and that the color is not important. If the model yields predictions (usually quantitative predictions) that are in agreement with the observed phenomena, then we generally feel that we are justified in claiming that our model is, in some sense, a correct description of the phenomena. Presumably, other models, which pay attention to other features or which describe the behavior of the selected features in a different way, would not yield the same predictions. Hence, we use the criterion of successful (quantitative) prediction as a means of selecting the correct model from a set of possible models.

Moreover, the model provides a connection between the initial conditions of the system (the "cause" or "causes") and the later behavior (the "effects") and thus provides an "explanation" of why the system behaves the way it does.

This general procedure seems to work well in physics and often in chemistry, where the systems are reasonably simple and reproducible. However, in biology there is considerable doubt whether the simple modeling procedure can work as desired because the inherent diversity among the individuals of a particular species and the differences among species makes the application of this whole program problematic. (These difficulties, however, do not seem to have hindered biologists and social scientists from embracing mathematical modeling whole-heartedly.) We will therefore restrict the discussion to cases in physics where modeling is usually accepted as a valid scientific procedure.

Even within the simple cases of physics, however, the universality of chaos presents us with a dilemma: Several (apparently) distinct models might all predict the same period-doubling route to chaos, for example. (In the case histories taken up in later chapters, we shall see several instances of this dilemma.) Each of these models yields the same sequence of period-doublings, and, moreover, each yields the same values for Feigenbaum's δ and α. So, if we restrict our observation to just the bifurcations that lead to chaos, then we see that we cannot use the observation of chaos or even the quantitative values of δ and α to help us decide which model is correct. Of course, if we observe period-doublings and the model does not predict period-doubling, then we can discard that model. In general, however, there will be several models with distinctly different physical descriptions for a given system, all of which predict the same transition to chaos.

There are two ways to proceed. First, one could look at finer details of the models' predictions to choose the correct one. For example, for the diode circuit of Chapter 1 we could ask for the best agreement with exact voltage values for the bifurcations or the exact shape of the $i(t)$ curves. If we do that, however, the model becomes so specific to the case at hand that it is not useful for other diodes in different circuits; therefore, this criterion is too strict[1].

Another possibility is to look for those features that are common to all the "successful" models of the diode and to point to those common features as providing an explanation of the diode's behavior. If we do this, what we find is disturbing to many scientists. We find that the common features are not physical features of the systems we are studying. Let us illustrate this by an example where there is a common physical feature. Simple harmonic (sinusoidal) oscillation is observed in many physical systems. When we try to understand the ubiquity of these oscillations, we find that the common "cause" is the fact that most forces, for small disturbances from "equilibrium," are reasonably described by the same kind of force law: The force is proportional to the displacement (in general terms) from

---

[1] This tension between detailed prediction and general explanation has been explored, for example, in Nancy Cartwright's *How the Laws of Physics Lie* (Oxford University Press, Oxford and New York, 1983).

equilibrium and directed in such a way as to pull the system back toward equilibrium. Such a force law always gives rise to simple harmonic oscillation. In this case we have a physical explanation, in terms of the behavior of forces, for the common feature of simple harmonic motion. When we consider the case of chaos, there does not seem to be a common physical explanation. What we do find in common is the behavior of the models in an abstract state space, the details of which we shall introduce in Chapters 3 and 4. The explanation of the diode's behavior, at least in terms of understanding its transition to chaos, therefore lies in understanding what goes on in this state space as control parameters are varied.

Given this situation, we are often asked the questions "What do we learn from chaos?" and "Why bother with chaos?" Lying behind these questions is the prejudice that we study the behavior of systems only to learn more about the fundamental, microscopic structure of that system. As we have argued earlier, chaos and the transitions to chaos do not seem to help us in that endeavor. Then why study chaos? To answer this question, we have to cast off the blinders that most twentieth-century physicists have worn. The blinders have kept our attention focused on learning more about the microscopic world that underlies the phenomena we observe. We cannot deny that this has been an immensely successful enterprise both in terms of what we have learned about the fundamental structure of matter and in the application of those fundamental ideas to the practical needs of society. In this drive toward the microscopic, however, many scientists have lost sight of the complexity of phenomena outside the tightly controlled domains of laboratory experiments. In some sense we expect that this complexity follows from the fundamental microscopic laws and is, in some way, embodied in those laws. However, the fundamental laws do not seem to give us the means to talk about and understand this complexity. If we are to understand and explain the universality of chaos, for example, we need to go beyond the specific predictions made by the fundamental laws for specific systems. We must approach this complexity at a different "level of explanation." Instead of seeing chaotic behavior as yet another tool to help us probe the microscopic world, we should think of this complexity as an essential part of the world around us, and science should attempt to understand it. Nonlinear dynamics and the theory of chaos are our first (perhaps rather feeble) attempts to come to grips with this dynamical complexity.

2.9 Computers and Chaos

While we are in a reflective mood, we want to raise the question of why chaos was not "discovered" much sooner. As we shall see in this book, almost all of the theory of chaos and nonlinear dynamics can be understood with only a moderate background in mathematics. Most of the phenomena of chaos occur in physical systems whose basic mechanisms could be understood in terms of physics that is a least a century old. Why, then, did the study of chaos suddenly explode in the 1970s and 1980s?

Even a cursory reading of the history of chaos, the definitive version of which is yet to be written, shows that Poincaré knew about, at least in a rough way, most of the crucial ideas of nonlinear dynamics and chaos. What Poincaré and the rest of the scientific world lacked until recently is a way of coming to grips with these ideas and exploring their consequences. The high-speed digital computer and, particularly, computer-driven graphics are the key tools that have made much of the progress in chaos and nonlinear dynamics possible. As we argued in Chapter 1, we need computers to generate the numerical solutions to nonlinear equations. Without some way of understanding that numerical output, however, little progress can be made. Computer graphics provides a way of visualizing the behavior of these nonlinear systems and allowing us to build intuition about the solutions and how they change as parameters of the system change. Of course, we also need some theoretical concepts to provide some organization for what we see and to guide us through the maze of complex behavior of nonlinear systems. It is safe to say, however, that if Poincaré had had a Macintosh or IBM personal computer, then the field of nonlinear dynamics would be much further along in its development than it is today.

Another important issue arises in numerical computations: If we take into account the combined influence of round-off errors in numerical computations and the property of divergence of nearby trajectories for chaotic behavior, how can we trust numerical computations of trajectories to give us reliable results? (As an aside, we should note that the same problem arises in experimental measurements in which "noise" plays the role of round-off errors.) If the system's behavior is chaotic, then even small numerical errors are amplified exponentially in time. Perhaps all of our results for chaotic systems are artifacts of the numerical computation procedure. Even if they are not artifacts, perhaps the numerical values of the properties depend critically on the computational procedures. If that is true, how general are our results?

Although it is difficult to answer these questions once and for all, it is comforting to know that while it is true that the details of a particular trajectory do depend on the round-off errors in the numerical computation, the trajectory actually calculated does follow very closely <u>some</u> trajectory of the system. That is, the trajectory you calculate might not be the one you thought you were going to calculate, but it is very close to one of the other possible trajectories of the system. In more technical terms, we say that the computed trajectory *shadows* some possible trajectory of the system. (A proof of this shadowing property for chaotic systems is given in GHY90 and SGY97.) As we will see in later chapters, we are most often interested in properties that are averaged over a trajectory; in many cases those average values are independent of the particular trajectory we follow. So, as long as we follow some possible trajectory for the system, we can have confidence that our results are a good characterization of the system's behavior. We note that they are special cases involving coupled chaotic systems (discussed in Chapter 11) for which the shadowing theorem may fail (LAG99).

There are cases, however, in which even shadowing becomes problematic. This issue is discussed in DGS94. Furthermore, in more complex systems (systems with many degrees of freedom), there may be situations in which deterministic modeling with computers can fail to give meaningful results (LAG99). The lesson is that some degree of skepticism is always appropriate when using computers to model nonlinear systems.

2.10 Further Reading

The Feigenbaum Numbers

M. J. Feigenbaum, "Universal Behavior in Nonlinear Systems," *Los Alamos Science* **1**, 4–27 (1980). Reprinted in [Cvitanovic, 1984]. The first sections of this paper give a quite readable introduction to some of the universal features of one-dimensional iterated map functions.

R. M. May, "Simple Mathematical Models with Very Complicated Dynamics," *Nature* **261**, 459–67 (1976). Reprinted in [Cvitanovic, 1984] and in [Hao, 1984]. A very influential and quite interesting look at the behavior of iterated map functions, written before the "discovery" of Feigenbaum universality.

G. B. Lubkin, "Period-Doubling Route to Chaos Shows Universality," *Physics Today* **34**, 17–19 (1981). An account of some of the history leading up to Feigenbaum's discovery and its early impact on physics.

P. J. Myrberg, "Sur l'Itération des Polynomes Réels Quadratiques," *J. Math. Pure Appl.* **41**, 339–51 (1962). Apparently the first recognition of the infinite cascade of period-doublings.

Experimental Measurements on Period-doubling and the Feigenbaum Numbers

M. Giglio, S. Musazzi, V. Perini, "Transition to Chaotic Behavior via a Reproducible Sequence of Period-Doubling Bifurcations," *Phys. Rev. Lett.* **47**, 243–46 (1981).

J. Testa, J. Perez, and C. Jeffries, "Evidence for Universal Chaotic Behavior of a Driven Nonlinear Oscillator," *Phys. Rev. Lett.* **48**, 714–17 (1982).

F. A. Hopf, D. L. Kaplan, H. M. Gibbs, and R. L. Shoemaker, "Bifurcations to Chaos in Optical Bistability," *Phys. Rev. A* **25**, 2172–82 (1982).

C. W. Smith, M. J. Tejwani, and D. A. Farris, "Bifurcation Universality for First-Sound Subharmonic Generation in Superfluid Helium-4," *Phys. Rev. Lett.* **48**, 492–94 (1982).

W. J. Yeh and Y. H. Kao, "Universal Scaling and Chaotic Behavior of a Josephson-junction Analog," *Phys. Rev. Lett.* **49**, 1888–91 (1982).

Modeling and Shadowing

P. J. Denning, "Modeling Reality," *American Scientist* **78**, 495–98 (1990). A thoughtful discussion of modeling and its implications in science.

C. Grebogi, S. M. Hammel, J. A. Yorke, and T. Sauer, "Shadowing of Physical Trajectories in Chaotic Dynamics: Containment and Refinement," *Phys. Rev. Lett.* **65**, 1527–30 (1990). The shadowing question for chaotic trajectories is studied in some detail.

T. Sauer, C. Grebogi, J. A. Yorke, "How Long Do Numerical Chaotic Solutions Remain Valid?" *Phys. Rev. Lett.* **79**, 59–62 (1997).

Y.-C. Lai and C. Grebogi, "Modeling of Coupled Oscillators," *Phys. Rev. Lett.* **82**, 4803–06 (1999).

2.11 Computer Exercises

As we discussed briefly in Section 2.9, the computer and computer graphics have played a crucial role in the development of the theory of chaos and nonlinear dynamics. If you want to use these ideas in your own work or even if you just want to come to grips with the basic concepts, you need to use a computer to allow you to explore changes in dynamics, to calculate Feigenbaum numbers, and, as we shall see later, to evaluate quantities that allow us to make quantitative and predictive statements about chaotic behavior.

For better or worse, most of us do not want to spend much time doing detailed computer programming. Fortunately, there are now available several software packages that permit us to carry out fairly sophisticated calculations in nonlinear dynamics with no programming required.

To encourage you to use a computer to help in developing your intuition for nonlinear systems, we have included three types of computer exercises. The first uses some very simple computer programs for the logistic map function, the listings for which are included in Appendix E. These programs are sufficiently simple that they can be readily adapted to run on almost any personal computer with little programming effort. The listings for these programs are good illustrations of the iteration algorithms common to many studies in nonlinear dynamics. The second category of exercises uses commercially available software packages. We have chosen to give exercises based primarily on two packages described shortly because they are readily available at relatively low cost. We also describe several other software packages you may find useful. The third category of exercises is for the experienced computer programmer. Those exercises require you to write your own programs. Useful information and suggestions for writing your own software are given in [Devaney, 1990] and in [Baker and Gollub, 1996] and in [Gould and Tobochnik, 1996].

We have listed some software packages for nonlinear dynamics (the prices listed are the ones current at this time of writing and, of course, may change). All of the programs require that your computer have some sort of graphics display (for example, VGA or SVGA). You should contact each publisher for detailed hardware requirements.

1. *Chaos Demonstrations 3*, J. C. Sprott and G. Rowlands (Physics Academic Software, Box 8202, North Carolina State University, Raleigh, NC 27695-8202), $90. For IBM and compatible personal computers. A set of programs covering a variety of iterated maps, systems described by differential equations, Julia sets, Mandlebrot set, and fractals. Can be run in a purely demonstration mode. You can also adjust some parameters. Has several 3-d animations. "3-d glasses" included. Physics Academic Software information is available at the web site http://www.aip.org/pas/catalog.html.

2. *Chaotic Dynamics Workbench*, R. W. Rollins (Physics Academic Software, Box 8202, North Carolina State University, Raleigh, NC 27695-8202, 1990), $90. For IBM and compatible personal computers. Focuses on systems described by a few ordinary differential equations. Can generate state space diagrams, Poincaré sections, calculate Lyapunov exponents. Parameters and initial conditions can be changed at will, even while the program is integrating the equations so you can see the effects of transients. You have a great deal of flexibility in choosing methods of integration, integration step sizes, and what is to be plotted on the graphs. The program can also store a sequence of Poincaré sections; so you can make a "movie" illustrating the system's behavior on the attractor in state space.

3. *Chaotic Mapper*, J. B. Harold. (Physics Academic Software, Box 8202, North Carolina State University, Raleigh, NC 27695-8202, 1993), $60. For IBM and compatible personal computers. Covers one– and two–dimensional iterated maps, systems described by differential equations, Julia sets, and the chaos game map. Calculates state space trajectories, Poincaré sections, and Lyapunov exponents. You can also enter your own map equations or sets of differential equations.

4. *Phaser*, H. Koçak. This is a set of programs accompanying the author's book *Differential and Difference Equations through Computer Experiments*, 2nd ed. (Springer-Verlag, New York, Berlin, Heidelberg, Tokyo, 1989). Diskettes available separately for $39. For IBM and compatible personal computers. Covers a wide range of iterated maps (difference equations) and differential equations systems. A more complex program with very flexible graphics. [Hale and Kocak, 1991] makes use of exercises in *Phaser*.

5. Helena Nusse and James A. Yorke, *Dynamics*: *Numerical Explorations*, 2nd ed. (Springer, New York, 1998). The accompanying program allows you to produce bifurcation diagrams, basins of attraction, and so on for a variety of iterated map and differential equation systems. Highly recommended.

6. *Strange Attractors, Creating Patterns in Chaos*, Julien C. Sprott (M&T Books, New York, 1993). A book with accompanying software that allows you to produce a wide variety of computer graphics of strange attractors.

7. *Chaos, a Program Collection for the PC*, H. J. Korsch and H.-J. Jodl (Springer-Verlag, New York, Berlin, Heidelberg, Tokyo, 1994). Includes diskettes for the programs.

8. *Chaos for Java*, Brian Davies. Free software available over the internet from Australian National University (http:/sunsite.anu.edu.au/education/chaos). Produces nice bifurcation diagrams (with an excellent zoom feature), graphical iteration procedures, and orbits for ordinary differential equation models. Used as exercises in [Davies, 1999].

Powerful computer-aided mathematics packages such as MAPLE and MATHEMATICA can be quite useful for studying nonlinear systems. See, for example

R. H. Enns and G. C. McGuire, *Nonlinear Physics with Maple for Scientists and Engineers* (Birkhäuser, Boston, 1997) and *Nonlinear Physics with Maple for Scientists and Engineers: A Laboratory Manual* (Birkhäuser, Boston, 1997).

In referring to various commercial software packages, we will assume that the reader has used the introductory material that comes with each package to become familiar with how the package operates. Although we refer to *Chaos Demonstrations* and *Chaotic Dynamics Workbench* explicitly because we have found them useful and relatively easy to use, other software packages could be used as well.

Exercises

CE2-1. Graphic Iteration Method. Use *Chaos Demonstrations'* Logistic Map $x(n+1)$ vs. $x(n)$ section [access via the V (View) key] or the program Graphit in Appendix E to carry out the graphical iteration method for the logistic map. The set of programs *Chaos for Java* are also useful here. (a) Start with parameter value $A = 2.9$. (A is the same as L in *Chaos Demonstrations*.) Show that all initial values of x between 0 and 1 lead to sequences ending up on the fixed point $x = 1 - 1/A$. (b) Show that for $A = 3.0$, the convergence to the fixed point is very slow. (c) Show that you get period-2 behavior for $A = 3.2$. (d) Find period-4 and, perhaps, period-8 bifurcation values. (e) Try other parameter values to observe chaotic trajectories.

CE2-2. Bifurcation Diagrams. Use *Chaos Demonstrations'* Logistic Map bifurcation diagram section or the program Bifur in Appendix E or *Chaos for Java* to generate a bifurcation diagram for the logistic map with the parameter A ranging from 2.9 to 4.0. (a) Identify period-doubling bifurcations, chaotic regions, and periodic windows. (b) Restrict the range of the parameter A to get a magnified view of various parts of the bifurcation diagram. (In Bifur, you can also restrict the range of x values to get magnification in the vertical direction.)

CE2-3. Divergence of Nearby Trajectories. One of *Chaos Demonstrations'* Views (V) for the Logistic Map plots the difference in trajectory values for two slightly different initial conditions. Use this program to show that the differences tend to 0 when the trajectories are periodic but diverge when the trajectories are chaotic (parameter $A > 3.5699 \dots$).

CE2-4. The Lorenz Model. Use *Chaos Demonstrations'* Lorenz Attractor section to explore the Lorenz model. The program plots the *X-Y* state space projection of the trajectories for particular values of the parameters *p*, *r*, and *b* of Eq. (1.5-1). (These are labeled *a*, *d*, and *b*, respectively, in *Chaos Demonstrations*.) (a) Show that for $a = 10$, $b = 8/3$, and $d = 20$ the trajectories end up on one or the other of the off-axis fixed points. (b) Verify that the locations of the off-axis fixed points are in agreement with the coordinate values given in Exercise 1.5-2. (c) Explore the behavior of the trajectories as a function of the parameter *d*. (d) Observe the behavior of the trajectories near the origin of the *X-Y* plane. Explain their behavior in terms of the (unstable) fixed point at $X = Y = Z = 0$. (e) Explore other combinations of parameter values. Try to find some periodic orbits.

CE2-5. The Lorenz Model and period-doubling. (a) Use *Chaotic Dynamics Workbench* to study the Lorenz model trajectories both as functions of time and in state space. The *X-Z* projection for $r > 24$ leads to the famous "butterfly attractor." (b) For $r < 24$ observe the chaotic transients that occur before the trajectories settle onto one of the off-origin fixed points. (c) Find the period-doubling sequence that occurs, as described in Chapter 1, near $r = 160$.

CE2-6. A Cautionary Tale. Use *Chaotic Dynamics Workbench* to vary the integration step size in integrating the equations for the Lorenz model. Show that if the step size becomes too large the periodic or chaotic character of the trajectories for a particular set of parameter values can apparently change. The moral here is that in numerical integration it is important to verify, by using smaller step sizes (and, consequently, more computer time) that your results are not an artifact of the particular integration step size you have chosen.

CE2-7. Feigenbaum Numbers. Write a computer program for the logistic map that allows you to find the supercycle parameter values, that is, those parameter values for which the attracting trajectory of a particular iterate of the map function contains the value $x = 0.5$. Find these values for period-2, period-4, period-8, and period-16. Use your program to calculate the Feigenbaum δ and α values. Hints: Have the program search for the *A* values for which $f^{(n)}(0.5) = 0.5$. Use the known value of δ to estimate where the *A*s occur for higher values of *n* once you have found them for lower values. Question: Given the arithmetic precision of the computer language you are using, to what value of *n* are you limited unless you are very clever in your programming?

II

TOWARD A THEORY OF

NONLINEAR DYNAMICS AND CHAOS

3

Dynamics in State Space: One and Two Dimensions

Before the beginning of great brilliance, there must be Chaos. *I Ching*, Image # 3

3.1 Introduction

In this chapter we will begin to build up the theoretical framework needed to describe more formally the kinds of complex behavior that we learned about in Chapters 1 and 2. We will develop the formalism slowly and in simple steps to see the essential features. We will try to avoid unnecessary mathematical jargon as much as possible until we have built a firm conceptual understanding of the framework.

The key theoretical tool in this description is a *state space* or *phase space* description of the behavior of the system. This type of description goes back to the French mathematician Henri Poincaré in the 1800s and has been widely used in statistical mechanics since the time of the American physicist J. Willard Gibbs (about 1900) even for systems that are linear and not chaotic [Gibbs, 1902]. Of course, we are most interested in the application of state space ideas to nonlinear systems; the behavior of linear systems emerges as a special case.

The notion of *fixed points* (also called *equilibrium points* or *stationary points* or *critical points* or *singular points*) in state space plays a key role in understanding the dynamics of nonlinear systems. Much of this chapter will be spent cataloging fixed points. In addition, we will meet *limit cycles*, which describe periodic behavior that can occur only in state spaces with two or more dimensions. For both fixed points and limit cycles the notions of stability and instability are crucial to understand how trajectories behave in the neighborhood of the fixed point or limit cycle. We will describe how to determine the nature of that stability or instability mathematically. Finally, we will introduce bifurcation theory to describe how fixed points and limit cycles change their stability and how fixed points and limit cycles are born and die as the control parameters of the system are changed.

This is a rather long chapter, but it builds a substantial foundation for all of our subsequent work. Throughout this chapter, we shall emphasize how the dimensionality of the state space limits the kind of behavior that can occur for deterministic systems. Armed with these theoretical tools, we will be ready for the study of chaotic behavior in Chapter 4.

Appendices G, I, and J treat three extended examples: the Duffing oscillator, the van der Pol oscillator, and a model of laser relaxation oscillations. These examples provide nice illustrations of the basic ideas of this chapter but may be skipped on a first reading.

3.2 State Space

In Chapter 1, we introduced rather casually the notion of a state space description of the behavior of a dynamical system. Now we want to develop this notion more carefully and in more detail. Let us start with a very simple example: the motion of a point mass on an ideal (Hooke's Law) spring, oscillating along the x axis. For this system, Newton's Second Law ($\vec{F} = m\vec{a}$) tells us that

$$F_x = m\frac{d^2 x}{dt^2} = -kx \tag{3.2-1}$$

where (as in Chapter 1) k is the spring constant, and m is the particle's mass. The motion of this system is determined for all time by specifying its position coordinate x at some time and its velocity

$$\dot{x} = \frac{dx}{dt} \tag{3.2-2}$$

at some time. Traditionally, we choose $t = 0$ for that time, and $x(t = 0)$ and $dx/dt(t = 0) \equiv \dot{x}_0$ are the "initial conditions" for the system. The motion, in fact, is given by the equation

$$x(t) = x_0 \cos\omega t + \frac{\dot{x}_0}{\omega}\sin\omega t \tag{3.2-3}$$

where $\omega = \sqrt{k/m}$ is the (angular) frequency of the oscillations. By differentiating Eq. (3.2-3) with respect to time, we find the equation for the velocity

$$\dot{x}(t) = -\omega x_0 \sin\omega t + \dot{x}_0 \cos\omega t \tag{3.2-4}$$

Exercise 3.2-1. Differentiate Eq. (3.2-4) with respect to time to find the acceleration and show that it is consistent with Eq. (3.2-1).

Since knowledge of $x(t)$ and $\dot{x}(t)$ completely specifies the behavior of this system, we say that the system has "two degrees of freedom." (See the comment on terminology on the next page.) At any instant of time we can completely specify the state of the system by indicating a point in an $\dot{x}$ versus x plot. This plot is then what we call the *state space* for this system. In this case the state space is two-dimensional as shown in Fig. 3.1.

Note that the dimensionality of the state space is generally not the same as the spatial dimensionality of the system. The state space dimensionality is determined by the number of variables needed to specify the dynamical state of the system. Our oscillator moves (by construction) in just one spatial dimension, but the state

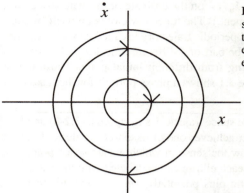

Fig. 3.1. Phase portrait for the mass on a spring. The ellipses are state space trajectories for the system. The larger the ellipse, the larger the total mechanical energy associated with the trajectory.

space is two-dimensional. Later we shall see examples of systems that "live" in three spatial dimensions, but whose state spaces have an infinite number of dimensions.

Two notes on terminology:
1. In the literature on dynamical systems and chaos, the terms *phase space* and *state space* are often used interchangeably. The term phase space was borrowed from Josiah Willard Gibbs in his treatment of statistical mechanics. The use of this notion in dynamical systems and chaos is somewhat more general than that used by Poincaré and Gibbs; so, we prefer (and will use) the term *state space*.
2. There is also some ambiguity about the use of the term *degree of freedom*. In the classical mechanics of point particles a degree of freedom refers to a pair of variables, such as the position coordinate along the x axis and the corresponding component of the linear momentum p_x. In this usage, our simple mass on a spring has one degree of freedom. (We shall use this definition in Chapter 8.) In dynamical systems theory, the number of degrees of freedom is usually defined as the number of independent variables needed to specify the dynamical state of the system (or alternately, but equivalently, as the number of independent initial conditions that can be specified for the system). We will use the latter definition of degree of freedom (except in Chapter 8). In the first sense of "degrees of freedom," the corresponding phase space must always have an even number (2, 4, 6, …) of dimensions. However, in the theory of dynamical systems and chaos, it will often be useful to have state spaces with an odd number of dimensions. The Lorenz model of Chapter 1 is one such example.

As time evolves, the initial state point in state space follows a trajectory, which, in the case of the mass on a spring, is just an ellipse. (The ellipse can be transformed into a circle by plotting $\dot{x}_0 / \omega$ on the ordinate of the state space plot, but that is simply a geometric refinement.) The trajectory closes on itself because the motion is periodic. Such a closed periodic trajectory is called a *cycle*. Another initial point (not on that ellipse) will be part of a different trajectory. A collection of several such trajectories originating from different initial points constitutes a *phase portrait* for the system. Figure 3.1 shows a phase portrait for the mass on a spring system.

Exercise 3.2-2. Plot several state space trajectories using Eqs. (3.2-3) and (3.2-4). Show that $\dot{x} = 0$ when x achieves one of its extreme values. Put arrows on the trajectories to show the sense in which a state space point traverses the ellipse. Show that each ellipse can be labeled with the value of the mechanical energy (kinetic plus potential) $\frac{1}{2}mv^2 + \frac{1}{2}kx^2$ for that trajectory.

A state space and a rule for following the evolution of trajectories starting at various initial conditions constitute what is called a *dynamical system*. The mathematical theory of such systems is called *dynamical systems theory*. This theory has a long and venerable history quite independent of the more recent theory of chaos and was particularly well developed by Russian mathematicians (see, for example, [Arnold, 1983]). Because of the extensive groundwork done by mathematicians studying dynamical systems, scientists and mathematicians investigating chaos have been able to make relatively rapid progress in recent years.

3.3 Systems Described by First-Order Differential Equations

Our theoretical treatment will at first be limited to a special (but rather broad) class of systems for which the equations giving the time-dependence of the state space variables can be expressed as a set of coupled first-order differential equations. To be specific, let us consider a system that has three degrees of freedom (in the second sense described in Section 3.2). Hence, we need three state variables, say, u, v, and w, to describe the state of the system. We will assume that the dynamics of the system can be expressed as a set of three first-order differential equations. That is, the equations involve only the first derivatives of u, v, and w with respect to time:

$$\dot{u} = f(u,v,w)$$
$$\dot{v} = g(u,v,w) \qquad\qquad (3.3\text{-}1)$$
$$\dot{w} = h(u,v,w)$$

The functions f, g, and h depend on the variables u, v, and w (but not their time derivatives) and also on one or more control parameters, not denoted explicitly. In general u, v, and w occur in all three of f, g, and h, and we say we have a set of "coupled differential equations." Time itself does not appear in the functions f, g, and h. In such a case the system is said to be *autonomous*. The Lorenz model equations of Chapter 1 are of this form. The time behavior of the system can be tracked by following the motion of a point whose coordinates are $u(t)$, $v(t)$, $w(t)$ in a three-dimensional *uvw* state space.

You might note, however, that the mass-on-a-spring model discussed earlier was not of this form. In particular, Eq. (3.2-1) has a second-order time derivative, rather than just a first-order time derivative. However, we can transform Eq. (3.2-1) into the standard form by introducing a new variable, say, v such that

$$v = \frac{d^2 x}{dt^2} \qquad (3.3-2)$$

Using Eq. (3.3-2) and Eq. (3.2-1), we can write the time evolution equations for the spring system as

$$v = -\frac{k}{m} x \qquad (3.3-3)$$

$$\dot{x} = v \qquad (3.3-4)$$

Exercise 3.3-1. Use new variables, analogous to the one introduced in Eq. (3.3-2), to convert the following differential equations into the standard form given by Eq. (3.3-1).

(a) $\dfrac{d^2 x}{dt^2} = -kx + \gamma \dot{x}$

(b) $\dfrac{d^3 x}{dt^3} = bx^2$

(c) $\dfrac{d^2 x}{dt^2} == kx + b \sin \omega t$

Hint for (c): See the following paragraph.

We can broaden considerably the class of systems to which Eq. (3.3-1) applies by the following "trick." Suppose that after applying the usual reduction procedure, the functions on the right-hand side of Eq. (3.3-1) still involve the time variable. (In that case, we say the system is **nonautonomous**.) This case most often arises when the system is subject to an externally applied time-dependent "force." For a two-degree-of-freedom system, the standard equations will be of the form

$$\dot{u} = f(u, v, t) \qquad (3.3-5)$$

$$\dot{v} = g(u, v, t) \qquad (3.3-6)$$

We can change these equations to a set of autonomous equations by introducing a new variable w whose time derivative is given by

$$\dot{w} \equiv \frac{dt}{dt} = 1 \qquad (3.3\text{-}7)$$

The dynamical equations for the system then become

$$\dot{u} = f(u,v,w)$$
$$\dot{v} = g(u,v,w) \qquad (3.3\text{-}8)$$
$$\dot{w} = 1$$

We have essentially enlarged the number of dimensions of the state space by 1 to include time as one of the state space variables. The advantage of this trick is that it allows us to treat nonautonomous systems (those with an imposed time dependence) on the same footing as autonomous systems. The price we pay is the difficulty of treating one more dimension in state space.

Why do we use this standard form (first-order differential equations) for the dynamical equations? The basic reason is that this form allows a ready identification of the fixed points of the system, and (as mentioned earlier) the fixed points play a crucial role in the dynamics of these systems. Recall that the *fixed points* are defined as the points in state space for which all of the time derivatives of the state variables are 0. Thus, with our standard form equations the fixed points are determined by requiring that

$$f(u,v,w) = 0$$
$$g(u,v,w) = 0 \qquad (3.3\text{-}9)$$
$$h(u,v,w) = 0$$

Thus, we find the fixed points by solving the three (for our three-dimensional example) coupled *algebraic* equations.

Question: What happens for a nonautonomous system? If we have used our "trick" to write the dynamical equations as suggested above, it should be clear that we can never have a fixed point because the derivative for the time variable is never zero. (The time variable never stops changing!) Thus, we will need special techniques to handle such systems. See Section 3.16.

An important question: Can the dynamical equations for all systems be reduced to the form of Eq. (3.3-1)? The answer is yes <u>if</u> (and this is an important *if*) we are willing to deal with an infinite number of degrees of freedom. For example, systems that are described by partial differential equations (that is, equations with partial derivatives rather than ordinary derivatives) or systems described by integral-differential equations (with both integrals and derivatives occurring in essential ways) or by systems with time-delay equations (where the state of the system at time t is determined not only by what is happening at that time but also by what happened earlier), all can be reduced to a set of first-order ordinary differential equations, but with an infinite number of equations coupled together. The state space then has an infinitely large number of dimensions, clearly a situation difficult to think about as well as to draw. In later chapters we shall look at some examples of this sort of system. Fortunately, in at least some cases, experience indicates that only a few of the infinitely many degrees of freedom are "active," and we can model the system with a finite number of equations. (This is in fact how the Lorenz model equations are derived.) For now, we will restrict ourselves to a finite number of coupled equations.

3.4 The No-Intersection Theorem

Before beginning the analysis of the types of trajectories and fixed points that can occur in state space, we state a fundamental and important theorem:

> **The No-Intersection Theorem**: Two distinct state space trajectories cannot intersect (in a finite period of time). Nor can a single trajectory cross itself at a later time.

By *distinct* trajectories, we mean that one of the trajectories does not begin on one of the points of the other trajectory. The parenthetical comment about a *finite* period of time is meant to exclude those cases for which distinct trajectories approach the same point as $t \rightarrow \infty$. (In the excluded case, we say the trajectories approach the point *asymptotically*.)

The basic physical content of this theorem is a statement of *determinism*. We have already mentioned that the state of a dynamical system is specified by its location in state space. Furthermore, if the system is described by equations of the form of Eq. (3.3-1), then the time derivatives of the state variables are also determined by the location in state space. Hence, how the system evolves into the future is determined solely by where it is now in state space. Hence, we cannot have two trajectories intersect in state space. If two trajectories *did* cross at some point, then the two trajectories would have the same values of their state variables and the same values of their time derivatives, yet they would evolve in different ways. This is impossible if their time evolution is described by equations like Eq. (3.3-1). As we shall see, the No-Intersection Theorem highly constrains the behavior of trajectories in state space.

The No-Intersection Theorem can also be based mathematically on uniqueness theorems for the solutions of differential equations. For example, if the functions f, g, and h on the right-hand-side of Eq. (3.3-1) are continuous functions of their arguments, then only one solution of the equations can pass through a given point in state space. [The more specific mathematical requirement is that these functions be continuous and at least once differentiable. This is the so-called Lipschitz condition (see, for example, [Hassani, 1991], pp. 570–71).

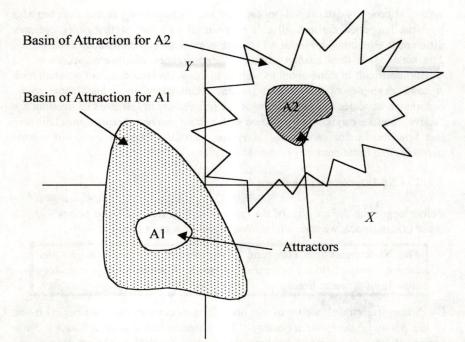

Fig. 3.2. A sketch of attractors A1 and A2 and basins of attraction in state space. Trajectories starting inside the dotted basin eventually end up in the attractor region inside the dotted region. Trajectories starting in the other basin head for the other attractor. For starting points outside these two basins, the trajectories may go toward a third attractor (not shown). The line bounding a basin of attraction forms a separatrix.

We shall see two apparent violations of this theorem. The first occurs for those asymptotic "intersections" mentioned earlier. The second occurs when we project the trajectory onto a two–dimensional plane for the sake of illustration. For example, Fig. 1.19 shows a *YZ* plane projection of a trajectory for the Lorenz model. The trajectory seems to cross itself several times. However, this crossing occurs only in the two–dimensional projection. In the full three–dimensional state space the trajectories do not cross.

3.5 Dissipative Systems and Attractors

In our current discussion of state space and its trajectories, we will limit our discussion to the case of dissipative systems. (Systems for which dissipation is unimportant will be discussed in Chapter 8.) As mentioned in Chapter 1, a dissipative system displays the nice feature that the long-term behavior of the system is largely independent of how we "start up" the system. We will elaborate this point in Section 3.9. (Recall, however, that there may be more than one possible "final state" for the system.) Thus, for dissipative systems, we generally ignore the transient behavior associated with the start up of the system and focus our attention on the system's long-term behavior.

As the dissipative system evolves in time, the trajectory in state space will head for some final state space point, curve, area, and so on. We call this final point or curve (or whatever geometric object it is) the *attractor* for the system since a number of distinct trajectories will approach (be attracted to) this set of points in state space. For dissipative systems, the properties of these attractors determine the dynamical properties of the system's long-term behavior. However, we will also be interested in how the trajectories approach the attractor.

The set of initial conditions giving rise to trajectories that approach a given attractor is called the *basin of attraction* for that attractor. If more than one attractor exists for a system with a given set of parameter values, there will be some initial conditions that lie on the border between the two (or more) basins of attraction. See Fig. 3.2. These special initial conditions form what is called a *separatrix* since they separate different basins of attraction.

The geometric properties of basins of attraction can often be complicated. In some cases the boundaries are highly irregular, forming what are called *fractal basin boundaries* (GMO83, MGO85). In other cases, the basins of attraction can be highly intertwined, forming what are called *riddled basins of attraction* (SOO93a): any point in one basin is close to another point in another basin of attraction. As we mentioned in Chapter 1, the existence of such complicated structures means that our ability to predict even which attractor a system will evolve to is severely compromised.

In the next sections we will describe the kinds of trajectories and attractors that can occur in state spaces of different dimensions. The dimensionality of the state space is important because the dimensionality and the No-Intersection Theorem together highly constrain the types of trajectories that can occur. In fact, we shall see that we need at least three state space dimensions in order to have a chaotic trajectory. We will, however, begin the cataloging with one and two dimensions to develop the necessary mathematical and conceptual background.

3.6 One-Dimensional State Space

A one-dimensional system, in the sense of dimension we are using here, has only one state variable, which we shall call X. This is, as we shall see, a rather uninteresting system in terms of its dynamical possibilities; however, it will be useful for developing our ideas about trajectories and state space. For this one dimensional state space, the dynamical equation is

$$\dot{X} = f(X) \tag{3.6-1}$$

The state space is just a line: the X axis.

First let us consider the fixed points for such a system, that is the values of X for which $\dot{X} = 0$. Why are the fixed points important? If a trajectory happens to get to a fixed point, then the trajectory stays there. Thus, the fixed points divide the X axis up into a number of "noninteracting" regions. We say the regions are noninteracting because a trajectory that starts from some initial X value in a region located between two fixed points can never leave that region.

Exercise 3.6-1. Provide the details of the proof of the last statement in the previous paragraph.

Now we want to investigate what happens to trajectories that are near a fixed point. For a one-dimensional state space, there are three types of fixed points:
1. **Nodes** (**sinks**): fixed points that attract nearby trajectories.
2. **Repellors** (**sources**): fixed points that repel nearby trajectories.
3. **Saddle points**: fixed points that attract trajectories on one side but repel them on the other. (The origin of the term *saddle point* will become obvious when we get to the two-dimensional case.)

A node is said to be a **stable fixed point** in the sense that trajectories that start near it are drawn toward it much like a ball rolling to a point of stable equilibrium under the action of gravity. A repellor is an example of an **unstable fixed point** in analogy with a ball rolling off the top of a hill. The top of the hill is an equilibrium point, but the situation is unstable: The slightest nudge to the side will cause the ball to roll away from the top of the hill. A saddle point attracts trajectories in one direction while repelling them in the other direction.

How do we determine what kind of fixed point we have? The argument goes as follows: Let X_o be the location of the fixed point in question. By definition, we have

$$\dot{X}\Big|_{X=X_o} = f(X_o) = 0 \qquad (3.6\text{-}2)$$

Now consider a trajectory that has arrived at a point just to the right of X_o. Let us call that point $X = X_o + x$ (see Fig. 3.3). We shall assume that x is small and positive. If $f(X_o + x)$ is positive (for x positive), then $\dot{X}$ is positive and hence the trajectory point will move <u>away from</u> X_o (toward more positive X values). On the other hand, if $f(X_o + x)$ is negative (for x positive), then $\dot{X}$ is negative and the trajectory moves to the left <u>toward</u> the fixed point X_o. Conversely, if we start to the left of X_o along the X axis, then we need $f(X_o - x)$ positive to move toward X_o and $f(X_o - x)$ negative to move away from X_o. These two cases are illustrated in Fig. 3.3. When trajectories on both sides of X_o move away from X_o, the fixed point is a repellor. When trajectories on both sides of X_o move toward X_o, the fixed point is a node.

Both of these cases can be summarized by noting that the <u>derivative</u> of $f(X)$ with respect to X evaluated at X_o is <u>negative</u> for a node and <u>positive</u> for a repellor. The value of this derivative at the fixed point is called the **characteristic value** or **eigenvalue** (from the German *eigen* = characteristic) of that fixed point. We call the characteristic value λ.

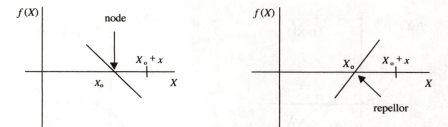

Fig. 3.3. On the left, $f(X)$ in the neighborhood of a node located at X_o. On the right, $f(X)$ in the neighborhood of a repellor located at X_o.

$$\lambda = \left. \frac{df(X)}{dX} \right|_{X=X_o} \tag{3.6-3}$$

We summarize these results in Table 3.1. The crucial and important lesson here is that we can determine the character of the fixed point and consequently the behavior of the trajectories near that fixed point by evaluating the derivative of the function $f(X)$ at that fixed point.

What happens when the characteristic value is equal to 0? The fixed point might be a node or a repellor or a saddle point. To find out which is the case we need to look at the <u>second</u> derivative of f with respect to X as well as the first derivative. For a saddle point, the second derivative has the same sign on both sides of X_o (see Fig. 3.4). Thus, we see that for a saddle point the trajectory is attracted toward the fixed point on one side, but repelled from the saddle point on the other.

For the node and repellor with characteristic value equal to 0, the second derivative changes sign as X passes through X_o (it is positive on the left and negative on the right for the node and negative on the left and positive on the right for the repellor). These kinds of "flat" nodes and repellors attract and repel trajectories more slowly than the nodes and repellors with nonzero characteristic values. For the type I saddle point, trajectories are attracted from the left but repelled on the right. The attraction and repulsion are reversed for the type II saddle points.

We will not discuss these types of "flat" fixed points further because, in a sense, they are relatively rare. They are rare because they require both the function $f(X)$ and its first derivative to be 0. If we have only one control parameter to adjust, then it is "unlikely" that we can satisfy both conditions simultaneously for some range of parameter values. In more formal terms, we talk about the ***structural***

Table 3.1
Characteristic Values

$\lambda < 0$	fixed point is a node
$\lambda > 0$	fixed point is a repellor

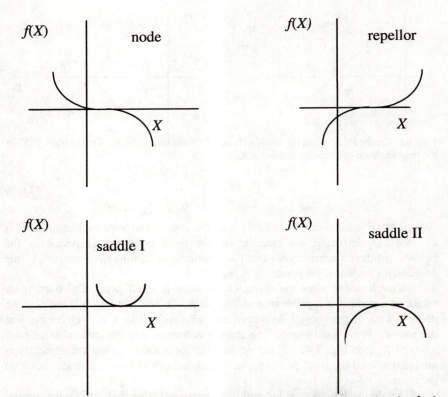

Fig. 3.4. Four possible types of fixed points in one-dimension with characteristic value $\lambda = 0$. These fixed points are structurally unstable.

stability of the fixed point. If the fixed point keeps the same character when the shape or position of the function changes slightly (for example, as a control parameter is adjusted), then we say that the fixed point is structurally stable. If the fixed point changes character or disappears completely under such changes, then we say it is structurally unstable. For example, the nodes and repellors shown in Fig. 3.3 are structurally stable because shifting the function $f(X)$ up and down slightly or changing its shape slightly does not alter the character of the fixed point. However, the fixed points shown in Fig. 3.4 are structurally unstable. For example, a small change in the function, say, shifting it up or down by a small amount, will cause a saddle point to either disappear completely or change into a node-repellor pair (see Fig. 3.5).

To examine in detail what constitutes a small change in the function $f(X)$ and how to decide whether a particular structure is stable or unstable would lead us rather far afield (see [Guckenheimer and Holmes, 1990]). Most of the work in nonlinear dynamics focuses on structurally stable state space portraits because in any real experiment the only properties that we can observe are those that exist for

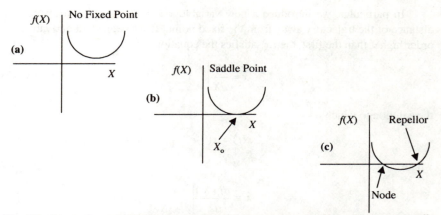

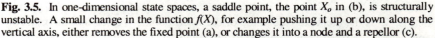

Fig. 3.5. In one-dimensional state spaces, a saddle point, the point X_o in (b), is structurally unstable. A small change in the function $f(X)$, for example pushing it up or down along the vertical axis, either removes the fixed point (a), or changes it into a node and a repellor (c).

some finite range of parameter values. We can never set the experimental conditions absolutely precisely, and "noise" always smears out parameter values. However, as we shall see, structurally unstable conditions are still important: In many cases they mark the border between two different types of behavior for the system. We will return to this issue at the end of this chapter in the discussion of bifurcations.

3.7 Taylor Series Linearization Near Fixed Points

The formal discussion of the nature of fixed points can be summarized very compactly using the mathematical notion of a *Taylor series expansion* of the function $f(X)$ for X values in the neighborhood of the fixed point X_o:

$$f(X) = f(X_o) + (X - X_o)\frac{df}{dX}$$
$$+ \frac{1}{2}(X - X_o)^2 \frac{d^2 f}{dX^2} \tag{3.7-1}$$
$$+ \frac{1}{6}(X - X_o)^3 \frac{d^3 f}{dX^3} + \ldots$$

where all the derivatives are evaluated at $X = X_o$. At a fixed point for a dynamical system, the first term on the right-hand side of Eq. (3.7-1) is 0, by the definition of fixed point. The Taylor series expansion tells us that the function $f(X)$ near X_o is determined by the values of the derivatives of f evaluated at X_o and the difference between X and X_o. This information together with the dynamical equation (3.6-1) is sufficient to predict the behavior of the system near the fixed point.

In particular, we introduce a new variable $x = X - X_o$, which measures the distance of the trajectory away from the fixed point. If we neglect all derivatives of order higher than the first, then x satisfies the equation

$$\dot{x} = \left.\frac{df}{dX}\right|_{X_o} x \tag{3.7-2}$$

the solution to which is

$$x(t) = x(0)e^{\lambda t} \tag{3.7-3}$$

where

$$\lambda \equiv \left.\frac{df(X)}{dX}\right|_{X_o} \tag{3.7-4}$$

that is, λ is the characteristic value of the fixed point. We see that the trajectory approaches the fixed point (a node) exponentially if $\lambda < 0$ and is repelled from the fixed point (a repellor) exponentially if $\lambda > 0$. λ is also called the *Lyapunov exponent* for the region around the fixed point. We should emphasize that these results hold only in the immediate neighborhood of the fixed point where the Taylor series expansion Eq. (3.7-1), keeping only the first derivative term, is a good description of the function $f(X)$.

Definition: The Lyapunov exponent for a region of one-dimensional state space near a fixed point is the characteristic value λ of that fixed point.

$$\lambda \equiv \left.\frac{df(X)}{dX}\right|_{X_o}$$

3.8 Trajectories in a One-Dimensional State Space

What kinds of trajectories can we have in a one-dimensional state space? First, we

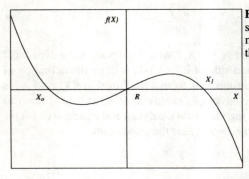

Fig. 3.6. In a one-dimensional state space, two nodes (here labeled X_0 and X_1) must have a repellor R located between them.

should note that our analysis thus far simply tells us what happens in the neighborhood of fixed points or what we might call the *local* behavior of the system. As we have seen, this local behavior is determined by the nature of the derivatives of the time evolution function evaluated at the fixed point. To obtain a larger-scale picture of the trajectories (a so-called *global* picture or a *global phase portrait*), we need to consider the relationship between the positions of different kinds of fixed points.For the one-dimensional case, the possible relationships among fixed points are highly constrained by the property of the continuity of the function f. More explicitly, if we assume that f must vary smoothly as a function of X, with no sudden jumps or steps (the physics behind this statement is the assumption that the "velocity" $\dot{X}$ must vary smoothly), then the two neighboring fixed points cannot be nodes, nor could both be repellors. Nor could one be a type I and the other a type II saddle point. Under the assumption of a continuous function f, we can have only certain patterns of fixed points. We can "prove" this last result simply by graphing $f(X)$ and trying to connect various types of fixed points with a smooth (continuous) curve.

As an example, consider Fig. 3.6 in which we have nodes at X_0 and X_1. In order for $f(X)$ to go through both nodes and to be continuous in the interval between the two nodes, it must pass through a repellor, here at R.

Exercise 3.8-1. Prove the following statements for a one-dimensional state space. (a) Two repellors must have a node between them. (b) A type I saddle point and a type II saddle point must have either a repellor or a node between them. Distinguish those two cases. (c) A type I saddle point can have another type I saddle point as a neighbor. (d) A type II saddle point can have another type II saddle point as a neighbor.

We can establish one further restriction: If the trajectories for our system are to be *bounded* (that is, trajectories stay within some finite region of state space for all time), then the "outermost" fixed points along the X axis must be either nodes or a type I saddle point on the left or a type II saddle point on the right. If the system has a saddle point, then there must be a node further along the X axis on the repelling side of the saddle point if the trajectories are to remain bounded.

The lesson of this section is that given the pattern of fixed points for a particular system, we can piece together a global phase portrait, a picture of how trajectories must travel through the state space. For a one-dimensional system this is quite easy if we are given the pattern of fixed points. We draw trajectory arrows that point toward nodes and the attracting side of saddle points. We draw trajectory arrows that point away from repellors and away from the repelling side of saddle points. From this picture we can then give a qualitative description of how trajectories move in the state space.

Exercise 3.8-2. Is it possible to have trajectories that represent oscillatory (repetitive) motion for a system with a one-dimensional state space?

Exercise 3.8-3. The logistic differential equation. The following differential equation has a "force" term that is identical to the logistic map function introduced in Chapter 1

$$\dot{X} = AX(1 - X)$$

(a) Find the fixed points for this differential equation.
(b) Determine the characteristic value and type of each of the fixed points.

Exercise 3.8-4. (a) Show that the solution of the differential equation given in Ex. 3.8-3 is

$$X(t) = \frac{X_o}{X_o - (X_o - 1)e^{-At}}$$

where X_o is the initial ($t = 0$) value of X. (b) Sketch the solution $X(t)$ as a function of time for several values of X_o and relate the behavior of the solutions to the nature of the fixed points. (c) Why are the solutions to the logistic differential equation relatively simple while the behavior of the trajectories for the logistic map equation in Chapter 1 are very complicated?

3.9 Dissipation Revisited

Earlier in this chapter, we mentioned that we would be interested primarily in dissipative systems. How do we know if a particular system, here represented by a particular function $f(X)$, is dissipative or not? If we are modeling a real physical system, the dissipation is due to friction (in the generalized sense), viscosity, and so on, and usually we can decide on physical grounds whether or not dissipation is important. However, it would be useful to have a mathematical tool that we could use to recognize a dissipative system directly from its dynamical equations. Given this tool we could check to see if a mathematical model we have developed (or which someone has given to us) includes dissipation or not.

To assess dissipation, we will use an important conceptual tool: a "cluster" of initial conditions. In the one-dimensional case, the cluster of initial conditions is some (relatively) small segment of the X axis. (We exclude segments that contain fixed points for what will become obvious reasons.) Let us suppose that this line segment runs from X_A to X_B (with $X_B > X_A$). See Fig. 3.7. The length of the segment is $X_B - X_A$. We want to examine what happens to the length of this line segment as time evolves and the trajectory points in that segment move through the state space. The time rate of change of the length of this segment is given by

Fig. 3.7. A "cluster of initial conditions," indicated by the heavy line, along the X axis.

$$\frac{d}{dt}(X_B - X_A) = \dot{X}_B - \dot{X}_A = f(X_B) - f(X_A) \qquad (3.9\text{-}1)$$

Thus, if $f(X_B) < f(X_A)$, the length of the segment will shrink as time goes on. If the line segment is sufficiently short, we can use the Taylor series expansion

$$f(X_B) = f(X_A) + \frac{df}{dX}\bigg|_{X_A}(X_B - X_A) + \dots \qquad (3.9\text{-}2)$$

to relate $f(X_B)$ to $f(X_A)$. If we let $L = X_B - X_A$, and keep only the first derivative term in Eq. (3.9-2), then we can write Eq. (3.9-1) in the form

$$\frac{1}{L}\frac{dL}{dt} = \frac{1}{L}[f(X_B) - f(X_A)] = \frac{df(X)}{dX} \qquad (3.9\text{-}3)$$

From Eq. (3.9-3), we see that the length of the segment of initial conditions will decrease if $f(X_B) < f(X_A)$ or, equivalently, if df/dX is negative. This condition will be satisfied if the trajectories are approaching a node, since the derivative of f is negative at a node and, by continuity, in the neighborhood of a node. (We are excluding the structurally unstable fixed points from our consideration.)

The previous analysis concentrated on the behavior near a single fixed point. More generally, we can ask for the "average" behavior over the history of some trajectory. It may turn out that a cluster of initial conditions first expands, as it leaves the region around a repellor, and then later contracts as it approaches a node. On the average, the cluster of trajectory points must experience contraction for a bounded dissipative system.

The readers with substantial mathematical experience will recognize the last equation as a pedestrian version of the "divergence theorem." In Section 3.13 we shall develop that theorem more formally for the two-dimensional case.

3.10 Two-Dimensional State Space

We now extend our discussion of state space to two-dimensional systems, where we shall see that the greater freedom provided by the higher dimensionality increases significantly the variety of behaviors and at the same time lifts some, but not all, of the geometrical constraints on the pattern of fixed points. Also, we shall see that a new type of attractor, a *limit cycle*, must be introduced to describe some of these new types of behavior.

Our discussion for two-dimensional state spaces will proceed along the same lines as the discussion of one-dimensional systems. We assume that the equations describing the dynamics of the system can be written as a pair of coupled, first-order differential equations for the state variables, which we shall label X_1 and X_2. (Occasionally, we will use x and y as the independent variables, but we want to emphasize that in general the state space variables are not spatial coordinate variables.) The time evolution equations are

$$\dot{X}_1 = f_1(X_1, X_2)$$
$$\dot{X}_2 = f_2(X_1, X_2)$$

(3.10-1)

The behavior of the system is followed by looking at trajectories in an X_1-X_2 state space. Just as in one-dimension, the fixed points of Eq. (3.10-1) play a major role in the dynamics of the system. The fixed points, of course, are those points (X_{1o}, X_{2o}) satisfying

$$f_1(X_{1o}, X_{2o}) = 0$$
$$f_2(X_{1o}, X_{2o}) = 0$$

(3.10-2)

You have probably already anticipated the next step: The character of the fixed point and the behavior of trajectories in the neighborhood of the fixed point are determined by the derivatives of the functions f_1 and f_2 evaluated at the fixed point; however, since f_1 and f_2 generally depend on both X_1 and X_2, there are four partial derivatives to consider

$$\frac{\partial f_1}{\partial X_1}, \quad \frac{\partial f_1}{\partial X_2}, \quad \frac{\partial f_2}{\partial X_1}, \quad \frac{\partial f_2}{\partial X_2}$$

(3.10-3)

The question then is how the characteristics of the fixed point depend on those four partial derivatives.

A Special Case

Before considering the general problem of fixed point characteristics in two dimensions, let us first look at a particularly simple case—the case for which only two of the four derivatives are not equal to 0. In particular, let us assume that at the fixed point (X_{1o}, X_{2o}) the derivatives have the following values:

$$\frac{\partial f_1}{\partial X_1} = \lambda_1 \qquad \frac{\partial f_1}{\partial X_2} = 0$$

(3.10-4)

$$\frac{\partial f_2}{\partial X_1} = 0 \qquad \frac{\partial f_2}{\partial X_2} = \lambda_2$$

In this special case, what happens along the X_1 direction in the neighborhood of the fixed point depends only on λ_1, and what happens along the X_2 direction depends only on λ_2. For this case, we say that the X_1 and X_2 axes are the **characteristic directions** with the associated characteristic values λ_1 and λ_2. (Please keep in mind that this independence of the X_1 and X_2 motions holds only in this special case and only in the vicinity of this fixed point.)

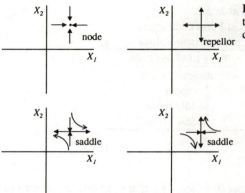

Fig. 3.8. Sample trajectories near each of the four types of fixed points with real characteristic values in two dimensions.

Types of Fixed Points in Two Dimensions

We can now begin to construct the catalog of types of fixed points in two dimensions by fitting together the possible types of one-dimensional behavior. We shall soon see, however, that there are new types of behavior possible in two dimensions. In the simplest case, λ_1 and λ_2 are both real numbers and both are nonzero. (When a characteristic value equals 0, then we need a more complicated analysis, just as we did in one-dimension.) By using arguments like those leading up to Eq. (3.6-3), we can see that there are four possible fixed points as listed in Table 3.2. In Fig. 3.8, sample trajectories are shown in the neighborhood of those fixed points.

We are now in a position to understand why a saddle point is called a saddle point. The behavior of trajectories near a saddle point is analogous to the behavior of a ball rolling under the influence of gravity on a saddle-shape surface as shown in Fig. 3.9. In that picture, a ball rolling along the x axis will be attracted to the saddle point at $(0,0)$. A ball rolling along the y axis will roll away from (be "repelled by") the saddle point.

In more formal terms the connection is made by defining a function $g(x,y)$ (to use the variables indicated in Fig. 3.9) such that

$$f_1(x, y) = -\frac{\partial g(x, y)}{\partial x}$$
$$f_2(x, y) = -\frac{\partial g(X,Y)}{\partial y}$$

(3.10-5)

The "force functions" f_1 and f_2 are given by the negative gradients of the "potential function" $g(x,y)$. Then at a fixed point of the f_1, f_2 system the function g has an extremum (a local maximum or minimum). At a saddle point, the function g, as shown in Fig. 3.9, has a minimum while moving along the x axis but a maximum while moving along the y axis. For a mechanical system the function $g(x,y)$ might represent the potential energy function for the system.

Table 3.2.
Possible Fixed Point Characters
with Real Characteristic Values

λ_1	λ_2	Type of Fixed Point
< 0	< 0	attracting node
> 0	> 0	repellor
> 0	< 0	saddle point
< 0	> 0	saddle point

Some Terminology

Saddle points, and in particular the special trajectories that head directly toward or directly away from a saddle point, play an important role, as we shall see, in organizing the behavior of all possible trajectories in state space. Because of this role, special terminology has been developed to talk about these trajectories.

The sets of points that form the trajectories heading directly to (approaching the saddle point as $t \to \infty$) or directly away from a saddle point are sometimes called the ***invariant manifolds*** associated with that saddle point. More specifically, the trajectories heading directly toward the saddle point form what is called the ***stable manifold*** (because the characteristic value $\lambda < 0$ along those trajectories), while the trajectories heading directly away from the saddle point form what is called the ***unstable manifold***. Other authors (e.g. [Abraham and Shaw, 1984] and [Thomson and Stewart, 1986]) call these same manifolds ***insets*** and ***outsets*** respectively. We prefer to call them ***in-sets*** and ***out-sets*** to avoid possible confusion with the usual English meanings of the words inset and outset.

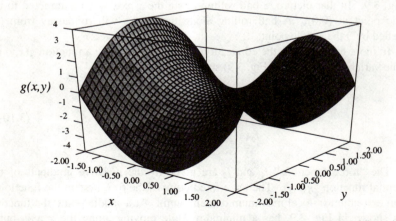

Fig. 3.9. A saddle-type surface for a two-dimensional state space. The saddle point is located at $(x,y) = (0,0)$.

The Importance of Saddle Points

To get a feeling for the importance of saddle points and their in-sets and out-sets, let us consider a system that has only one fixed point. If that fixed point is a saddle point, and if the characteristic values are not equal to zero, then the in-sets and out-sets of that saddle point divide the state space up into four "quadrants." A trajectory that is not an in-set or an out-set is confined to the quadrant in which it starts as illustrated in the lower half of Fig. 3.8. In that sense, the in-sets and out-sets "organize" the state space. The out-sets and in-sets are part of the separatrices (if there are any) for the state space.

For this kind of saddle point (for which neither of the characteristic values is 0), the trajectories near the saddle point but not on either the in-set or out-set look like sections of hyperbolas. Hence, this kind of saddle point is called a *hyperbolic point*. In fact, the term *hyperbolic* is applied to any fixed point whose characteristic values are not equal to 0. (In the general case to be discussed later, the real parts of the characteristic values are not 0.) In this language, the one-dimensional saddle points discussed in the previous section, which we called structurally unstable, are *nonhyperbolic* because the associated characteristic value is 0.

3.11 Two-dimensional State Space: The General Case

In the most general case in two dimensions, all four of the derivatives in Eq. (3.10-3) are nonzero. How do we characterize the fixed point in that situation? It turns out that in this case there are still just two characteristic values associated with the fixed point, but the associated characteristic directions are no longer the X_1 and X_2 directions, in general.

At this point a specific example will help illustrate these ideas. We will describe the equations used to model a certain set of chemical reactions [Nicolis and Prigogine, 1989], called the Brusselator model because its originators worked in Brussels. The equations are

$$\dot{X} = A - (B+1)X + X^2Y$$
$$\dot{Y} = BX - X^2Y$$

$$(3.11\text{-}1)$$

A and B are positive numbers that represent the control parameters, and X and Y are variables proportional to the concentrations of some of the intermediate products in the chemical reaction. One can imagine monitoring these concentrations as functions of time with some appropriate electrodes or with some optical absorption measurements that are sensitive to those chemical concentrations.

First let us find the fixed points for this set of equations. By setting the time derivatives equal to 0, we find that the fixed points occur at the values X,Y that satisfy

$$A - (B+1)X + X^2Y = 0 \qquad\qquad (3.11\text{-}2)$$

$$BX - X^2Y = 0 \qquad (3.11\text{-}3)$$

We see that there is just one point (X,Y) which satisfies these equations, and the coordinates of that fixed point are $X_o = A$, $Y_o = B/A$.

What is the character of that fixed point? To see how we find these characteristic values, let us return to our general two-dimensional state space and make use of a Taylor series expansion of Eq. (3.10-1) in the neighborhood of the fixed points (X_{1o}, X_{2o}):

$$\dot{X}_1 = f_1(X_1, X_2) = (X_1 - X_{1o})\frac{\partial f_1}{\partial X_1} + (X_2 - X_{2o})\frac{\partial f_1}{\partial X_2} + \dots \qquad (3.11\text{-}4a)$$

$$\dot{X}_2 = f_2(X_1, X_2) = (X_1 - X_{1o})\frac{\partial f_2}{\partial X_1} + (X_2 - X_{2o})\frac{\partial f_2}{\partial X_2} + \dots \qquad (3.11\text{-}4b)$$

In Eq. (3.11-4), we have evaluated the derivatives at the fixed point (X_{1o}, X_{2o}), and the ellipsis indicates all derivatives higher than the first, which we are ignoring. (Note that we use partial derivatives in the Taylor series expansion because the functions depend on both X_1 and X_2.) It is useful to introduce new variables $x_1 = (X_1 - X_{1o})$ and $x_2 = (X_2 - X_{2o})$, which indicate the deviation away from the fixed point. Noting that

$$\dot{x}_1 = \dot{X}_1 \text{ and } \dot{x}_2 = \dot{X}_2 \qquad (3.11\text{-}5)$$

and ignoring all the higher-order derivative terms, we may write Eq. (3.11-4) as

$$\dot{x}_1 = \frac{\partial f_1}{\partial x_1}x_1 + \frac{\partial f_1}{\partial x_2}x_2$$
$$\dot{x}_2 = \frac{\partial f_2}{\partial x_1}x_1 + \frac{\partial f_2}{\partial x_2}x_2 \qquad (3.11\text{-}6)$$

Please note that Eqs. (3.11-6) are *linear*, first-order differential equations with constant coefficients (the factors multiplying x_1 and x_2 are independent of time) for the new state variables x_1 and x_2. There are many standard techniques for solving such differential equations. We shall use a method that gets us to the desired results as quickly as possible.

To simplify the notation, we shall write

$$f_{ij} = \frac{\partial f_i}{\partial x_j} \qquad (3.11\text{-}7)$$

where i and $j = 1$ or 2. First, we find a differential equation for x_1 alone by differentiating the first equation in Eq. (3.11-6) with respect to time and then eliminating $\dot{x}_2$ by the use of the second equation in Eq. (3.11-6):

$$\ddot{x}_1 = f_{11}\dot{x}_1 + f_{12}\dot{x}_2$$
$$= f_{11}\dot{x}_1 + f_{12}(f_{21}x_1 + f_{22}x_2) \tag{3.11-8}$$

We now use the first of Eq. (3.11-6) again to eliminate x_2:

$$\ddot{x}_1 = (f_{11} + f_{22})\dot{x}_1 + (f_{12}f_{21} - f_{11}f_{22})x_1 \tag{3.11-9}$$

To solve Eq. (3.11-9), let us assume that the solution can be written in the form

$$x_1(t) = Ce^{\lambda t} \tag{3.11-10}$$

where λ is a constant to be determined, and C is a constant (independent of time) to be determined from the initial ($t = 0$) conditions. Let us pause a second to note that if λ is positive (and real) then the trajectory will be repelled by the fixed point; that is, we have an unstable fixed point. If λ is negative (and real), then the trajectory approaches the fixed point; that is, we have a stable fixed point. As we shall see later, λ may also be a complex number.

Let us return to our solution. If we use Eq. (3.11-10) in Eq. (3.11-9), then we find that

$$\lambda^2 - (f_{11} + f_{22})\lambda + (f_{11}f_{22} - f_{12}f_{21}) = 0 \tag{3.11-11}$$

We call Eq. (3.11-11) the ***characteristic equation*** for λ, whose value depends only on the derivatives of the time evolution functions evaluated at the fixed point. Eq. (3.11-11) is a quadratic equation for λ and in general has two solutions, which we can write down from the standard quadratic formula:

$$\lambda_{\pm} = \frac{f_{11} + f_{22} \pm \sqrt{(f_{11} + f_{22})^2 - 4(f_{11}f_{22} - f_{12}f_{21})}}{2} \tag{3.11-12}$$

We have denoted λ_+ as the result obtained with the + sign in front of the square root in Eq. (3.11-12) and λ_- the result obtained with the − sign. Obviously, the characteristic values will be real numbers if the argument under the square root sign in Eq. (3.11-12) is positive. They will be complex numbers if the argument is negative.

The most general solution of Eq. (3.11-9) can then be written as

$$x_1(t) = Ce^{(\lambda_+)t} + De^{(\lambda_-)t} \tag{3.11-13}$$

where C and D are constants that can be found from the initial conditions $x_1(t = 0)$ and $x_2(t = 0)$.

Before interpreting the general solutions, we should note that we have considerable freedom in the choice of coordinates for the state space. We started out using x_1 and x_2 as the coordinates, but we could equally well use x_1 and $\dot{x}_1$ since, from Eq. (3.11-6), if we know x_1 and $\dot{x}_1$, then we know x_2. We could, however, also use x_2 and $\dot{x}_1$ or other pairings of variables. The general geometric behavior of the trajectories is not influenced by the choice of state space coordinates.

Exercise 3.11-1. Show that Eq. (3.11-12) reduces to Eq. (3.10-4) in the special case $f_{12} = 0$ and $f_{21} = 0$.

Exercise 3.11-2. Start with Eq. (3.11-13) and find C and D in terms of the initial conditions $x_1(0)$ and $x_2(0)$. Hint: differentiate Eq. (3.11-13) and use Eq. (3.11-9) to get a second condition on C and D. This exercise requires a modest amount of algebra, and the final results are not particularly pretty.

Exercise 3.11-3. Show that x_2 satisfies a differential equation that is exactly the same as Eq. (3.11-9). Combine these results to specify the complete dynamics of the system represented by Eq. (3.11-6).

3.12 Dynamics and Complex Characteristic Values

What are the dynamics of the system when the characteristic values are not real, but are complex numbers? This situation occurs when the argument of the square root in Eq. (3.11-12) for the characteristic values is negative. We shall find that this case describes behavior in which trajectories spiral in toward or away from the fixed point, as illustrated in Fig. 3.10.

When the argument of the square root in Eq. (3.11-12) is negative, we may write the characteristic values as

$$\lambda_\pm = R \pm i\Omega \qquad (3.12\text{-}1)$$

where

$$i = \sqrt{-1}$$

$$R = \frac{1}{2}(f_{11} + f_{22})$$

$$\Omega = \frac{1}{2}\sqrt{\left|(f_{11} + f_{22})^2 - 4(f_{11}f_{22} - f_{12}f_{21})\right|} \qquad (3.12\text{-}2)$$

Using the standard mathematical language of complex numbers, we say that R is the "real part" and Ω is the "imaginary part" of these complex numbers. The two eigenvalues λ_+ and λ_- form a complex conjugate pair: λ_+ is the complex conjugate of λ_- and vice versa. To see what the trajectory behavior is like in this case, we use these characteristic values in the equation for $x_1(t)$:

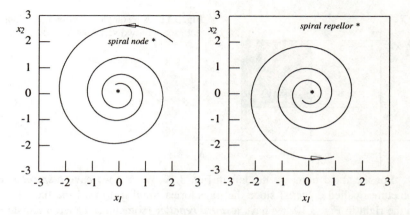

Fig. 3.10. A spiral node (left) and a spiral repellor (right) occur when the characteristic values of a fixed point are complex numbers.

$$x_1(t) = Ce^{(\lambda_+)t} + De^{(\lambda_-)t}$$
$$= e^{Rt}\left[Ce^{i\Omega t} + De^{-i\Omega t}\right] \tag{3.12-3}$$

To see what is going on, let us consider the special case $x_1(0) = 0$, which tells us that $C = -D$. We now use the famous Euler formula

$$e^{i\theta} = \cos\theta + i\sin\theta \tag{3.12-4}$$

to write

$$x_1(t) = Fe^{Rt}\sin(\Omega t) \tag{3.12-5}$$

where F is a constant that depends on $x_2(0)$. From this result we see that x_1 oscillates in time with an angular frequency Ω while the amplitude of the oscillation increases or decreases exponentially (depending on whether $R > 0$ or $R < 0$). x_2 undergoes similar behavior. The corresponding state space behavior is shown schematically (with different initial conditions) in Fig. 3.10. For more general initial conditions, the state space behavior is still the same: oscillations with exponentially increasing or decreasing amplitude.

Exercise 3.12-1. Show that the constant F in Eq. (3.12-5) is given by
$$F = f_{12}x_2(0)/\Omega$$

Exercise 3.12-2. For the reader with a bit more algebraic fortitude: Work out the most general solution of Eq. (3.12-3) in terms of $x_1(0)$ and $x_2(0)$ and show that the behavior is as described at the end of the previous paragraph.

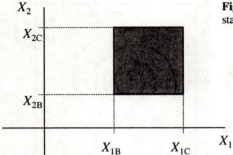

Fig. 3.11. A rectangle of initial conditions in state space of two variables X_1 and X_2.

For the fixed point on the left in Fig. 3.10, we say we have a ***spiral node*** (sometimes called a ***focus***) since the trajectories spiral in toward the fixed point. On the right in Fig. 3.10, we have a ***spiral repellor*** (sometimes called an ***unstable focus***). In the special case when $R = 0$, the trajectory forms a closed loop around the fixed point. This closed loop trajectory is called a ***cycle***. If trajectories in the neighborhood of this cycle are attracted toward it as time goes on, then the cycle is called a ***limit cycle***. We need a more detailed analysis to see if this cycle is itself stable or unstable. An analysis of cycle behavior will be taken up in Section 3.16.

It is important to realize that the spiral type behavior shown in Fig. 3.10 and the cycle type behavior discussed in the Section 3.16 are possible only in state spaces of two (or higher) dimensions. They cannot occur in a one-dimensional state space because of the No-Intersection Theorem (recall Exercise 3.8-2).

3.13 Dissipation and the Divergence Theorem

Now we show how we can test for dissipation in two-dimensional state space. We shall then see that, in principle, the generalization to many dimensions is easy. In two dimensions, we start with a cluster of initial conditions of the two variables X_1 and X_2 in some (small) area delimited by the coordinates (X_{1C}, X_{2C}) and (X_{1B}, X_{2B}) as shown in Fig. 3.11.

Again we compute the rate of change of that area

$$A = (X_{1C} - X_{1B})(X_{2C} - X_{2B}) \tag{3.13-1}$$

$$\frac{dA}{dt} = (X_{1C} - X_{1B})\{f_2(X_{1B}, X_{2C}) - f_2(X_{1B}, X_{2B})\}$$
$$+ \{f_1(X_{1C}, X_{2B}) - f_2(X_{1B}, X_{2B})\}(X_{2C} - X_{2B}) \tag{3.13-2}$$

where we have used the time-evolution equations

$$\dot{X}_1 = f_1(X_1, X_2)$$
$$\dot{X}_2 = f_2(X_1, X_2)$$
(3.13-3)

We make use of a Taylor series expansion

$$f_1(X_{1C}, X_{2B}) = f_1(X_{1B}, X_{2B})$$
$$+(X_{1C} - X_{1B}) \frac{\partial f_1}{\partial X_1}\bigg|_{X_{1B}, X_{2B}} + \dots$$
(3.13-4)

with a similar expression for f_2. When these expansions are substituted into Eq. (3.13-2), we obtain, after dividing through by A

$$\frac{1}{A} \frac{dA}{dt} = \frac{\partial f_1}{\partial X_1} + \frac{\partial f_2}{\partial X_2}$$
(3.13-5)

Once again we see that the relative growth or shrinkage of the area containing the set of initial conditions is determined by the derivatives (here partial derivatives) of the time evolution functions. If the right-hand side of Eq. (3.13-5) is negative, then the initial phase space area shrinks to 0, and we say that the system is dissipative. The trajectories all collapse to an attractor whose geometric dimension is less than that of the original state space. For two state space dimensions, the attractor could be a point (a node) or a curve (a limit cycle). It should be (almost) obvious that for N dimensions, the evolution of an N-dimensional volume V of initial conditions in state space is given by

$$\frac{1}{V} \frac{dV}{dt} = \sum_{i=1}^{N} \frac{\partial f_i}{\partial x_i} \equiv div(f)$$
(3.13-6)

where the right-hand equality defines what is called the ***divergence*** of the set of functions f_i. If $div(f) < 0$ on the average over state space, we know that the initial volume of initial conditions will collapse onto a geometric region whose dimensionality is less than that of the original state space, and we know that the state space has at least one attractor.

> **Exercise 3.13-1.** Evaluate $div(f)$ for the Lorenz model equations introduced in Chapter 1 for parameter values $r = 0.5$, $p = 10$, and $b = 8/3$. Is the Lorenz system dissipative throughout its state space?

3.14 The Jacobian Matrix for Characteristic Values

We would now like to introduce a more elegant and general method of finding the characteristic equation for a fixed point. This method makes use of the so-called ***Jacobian matrix*** of the derivatives of the time evolution functions. Once we see how this procedure works, it will be easy to generalize the method, at least in principle, to find characteristic values for fixed points in state spaces of any dimension. The Jacobian matrix for the system is defined to be the following square array of the derivatives:

$$J = \begin{pmatrix} f_{11} & f_{12} \\ f_{21} & f_{22} \end{pmatrix} \tag{3.14-1}$$

where the derivatives are evaluated at the fixed point. We subtract λ from each of the principal diagonal (upper left to lower right) elements and set the determinant of the matrix equal to 0:

$$\begin{vmatrix} f_{11} - \lambda & f_{12} \\ f_{21} & f_{22} - \lambda \end{vmatrix} = 0 \tag{3.14-2}$$

Multiplying out the determinant in the usual way then yields the characteristic equation (3.11-11). The Jacobian matrix method is obviously easily extended to d-dimensions by writing down the d-by-d matrix of derivatives of the d time-evolution functions f_n, forming the corresponding determinant, and then (at least in principle) solving the resulting dth order equation for the characteristic values.

We now introduce some terminology from linear algebra to make some very general and very powerful statements about the characteristic values for a given fixed point. First, the **trace** of a matrix, such as the Jacobian matrix (3.14-1), is defined to be the sum of the principal diagonal elements. For Eq. (3.14-1) this is explicitly

$$TrJ = f_{11} + f_{22} \tag{3.14-3}$$

If we look at the solution for the characteristic values given in Eq. (3.11-12), we see that the sum of the diagonal elements is in fact equal to the sum of the characteristic values.

$$\lambda_+ + \lambda_- = f_{11} + f_{22} = TrJ \tag{3.14-4}$$

According to Eq. (3.13-5), however, this is just the combination of derivatives needed to test whether or not the system's trajectories collapse toward an attractor. To make a connection with the previous section, we note that $TrJ = 2R$, so that we see that the <u>sign</u> of TrJ determines whether the fixed point is a node or a repellor.

Linear algebra also tells us how to find the directions to be associated with the characteristic values. For a saddle point, these directions will be the directions for the in-sets and out-sets in the immediate neighborhood of the saddle point. The basic idea is that by transforming the coordinate system, (in general the new coordinates are linear combinations of the original coordinates), we can bring the Jacobian matrix to the so-called diagonal form in which only the principal diagonal elements are non-zero. In that case the matrix has the form (for a two-dimensional state space)

Table 3.3
Fixed Points for Two-dimensional State Space

	$TrJ < 0$	$TrJ > 0$
$\Delta > (1/4)(TrJ)^2$	spiral node	spiral repellor
$0 < \Delta < (1/4)(TrJ)^2$	node	repellor
$\Delta < 0$	saddle point	saddle point

$$\begin{pmatrix} \lambda_1 & 0 \\ 0 & \lambda_2 \end{pmatrix} \qquad (3.14\text{-}5)$$

In linear algebra this procedure is called "finding the eigenvalues and eigenvectors of the matrix." For our purposes, the eigenvalues are the characteristic values of the fixed point and the eigenvectors give the associated characteristic directions. However, we will not need these eigenvectors for most of our purposes. The interested reader is referred to the books on linear algebra listed at the end of the chapter.

We now introduce one more symbol:

$$\Delta = f_{11}f_{22} - f_{21}f_{12} \qquad (3.14\text{-}6)$$

Δ is called the ***determinant*** of that matrix. Then we may show that the nature of the fixed point is determined by TrJ and Δ as listed in Table 3.3.

> **Exercise 3.14-1.** Use the definitions in Eqs. (3.14-3) and (3.14-6) and Eq. (3.11-12) to verify the entries in Table 3.3.

Summary of Fixed Point Analysis for Two-dimensional State Space

1. Write the time evolution equations in the first-order time derivative form of Eq. (3.10-1).

$$\dot{X}_1 = f_1(X_1, X_2)$$
$$\dot{X}_2 = f_2(X_1, X_2) \qquad (3.10\text{-}1)$$

2. Find the fixed points of the evolution by finding those points that satisfy

$$f_1(X_1, X_2) = 0$$
$$f_2(X_1, X_2) = 0$$

3. At the fixed points, evaluate the partial derivatives of the time evolution functions to set up the Jacobian matrix

$$J \equiv \begin{pmatrix} f_{11} & f_{12} \\ f_{21} & f_{22} \end{pmatrix} \qquad (3.14\text{-}1)$$

4.   Evaluate the trace and determinant of the Jacobian matrix at the fixed point and
     use Table 3.3 to find the type of fixed point.
5.   Use Eq. (3.11-12) to find the numerical values of the characteristic values and
     to specify the behavior of the state-space trajectories near the fixed point with
     Eq. (3.11-13).

Example: The Brusselator Model

As an illustration of our techniques, let us return to the Brusselator Model given in
Eq. (3.11-1). The Jacobian matrix for that set of equations is

$$J = \begin{pmatrix} (B-1) & A^2 \\ -B & -A^2 \end{pmatrix} \qquad (3.14\text{-}7)$$

Following the Jacobian determinant method outlined earlier, we find the
characteristic values:

$$\lambda_\pm = \frac{1}{2}\left[(B-1) - A^2\right]$$
$$\pm \frac{1}{2}\sqrt{\left(A^2 - (B-1)\right)^2 - 4A^2} \qquad (3.14\text{-}8)$$

In the discussion of this model, it is traditional to set $A = 1$ and let B be the
control parameter. Let us follow that tradition. We see that with $B < 2$, both
characteristic values have negative real parts and the fixed point is a spiral node.
This result tells us that the chemical concentrations tend toward the fixed point
values $X_o = A = 1$, $Y_o = B$ as time goes on. They oscillate, however, with the
frequency $\Omega = |B(B-4)|^{1/2}$ as they head toward the attractor. For $2 < B < 4$, the
fixed point becomes a spiral repellor. However, our analysis cannot tell us what
happens to the trajectories as they spiral away from the fixed point. As we shall
learn in the next section, they tend to a limit cycle as shown in Fig. I.1 in Section I
(for a different model).

> **Exercise 3.14-2.** Characterize the fixed point of the Brusselator model for
> $B > 4$.

3.15 Limit Cycles

In state spaces with two or more dimensions, it is possible to have cyclic or periodic
behavior. This very important kind of behavior is represented by closed loop
trajectories in the state space. A trajectory point on one of these loops continues to
cycle around that loop for all time. These loops are called *limit cycles* if the cycle is
isolated, that is if trajectories nearby either approach or are repelled from the limit
cycle. The discussion in the previous section indicated that motion on a limit cycle

in state space represents oscillatory, repeating motion of the system. The oscillatory behavior is of crucial importance in many practical applications, ranging from radios to brain waves.

We shall formulate the analysis in answer to two questions: (1) When do limit cycles occur? and (2) When is a limit cycle stable or unstable? The first question is answered for a two-dimension state space by the famous **Poincaré–Bendixson Theorem**. The theorem can be formulated in the following way:

1. Suppose the long-term motion of a state point in a two-dimensional state space is limited to some finite-size region; that is, the system doesn't wander off to infinity.
2. Suppose that this region (call it R) is such that any trajectory starting within R stays within R for all time. [R is called an "invariant set" for that system.]
3. Consider a particular trajectory starting in R. The Poincaré–Bendixson Theorem states that there are only two possibilities for that trajectory:

   a. The trajectory approaches a fixed point of the system as $t \rightarrow \infty$.
   b. The trajectory approaches a limit cycle as $t \rightarrow \infty$.

A proof of this theorem is beyond the scope of this book. The interested reader is referred to [Hirsch and Smale, 1974]. We can see, however, that the results are entirely reasonable if we take into account the No-Intersection Theorem and the assumption of a bounded region of state space in which the trajectories live. The reader is urged to draw some pictures of state space trajectories in two dimensions to see that these two principles guarantee that the only two possibilities are fixed points and limit cycles.

> It is important to note that the Poincaré–Bendixson Theorem works only in two dimensions because only in two dimensions does a closed curve separate the space into a region "inside" the curve and a region "outside." Thus a trajectory starting inside the limit cycle can never get out and a trajectory starting outside can never get in. This argument is an excellent example of the power of topological arguments in the study of dynamical systems. Further, from the Poincaré–Bendixson Theorem we arrive at an important result: Chaotic trajectories (in a bounded system) cannot occur in a state space of two dimensions. *For systems described by differential equations, we need at least three state-space dimensions for chaos.*

The Brussellator model displays the typical situation in which a limit cycle develops. An invariant region R contains a repelling fixed point. Trajectories starting near the repelling fixed point are pushed away and (if there is no attracting fixed point in R) must head toward a limit cycle (which can be proved to enclose the repellor).

As an aside, we should point out that if the differential equation describing the system is more complicated than the type considered here, then limit cycles and even chaotic behavior can occur even for what appear to be one-dimensional systems. For example, the following differential equation is a so-called ***delay-differential equation***:

$$\frac{du(t)}{dt} = f(t) + \lambda g(t - T) \qquad (3.15\text{-}1)$$

Note that the second term on the right-hand side of this equation depends on the behavior of the function g evaluated at an earlier time $t - T$. If we try to use the "trick" introduced in Section 3.3 to reduce Eq. (3.15-1) to a set of autonomous first-order differential equations, we find we need an infinite number of them! From a state space point of view, the system described by Eq. (3.15-1) is certainly not a one-dimensional system. Having limit cycle or chaotic behavior for such a system does not violate the No-Intersection Theorem or the Poincaré–Bendixson Theorem.

Finally, we point out that some powerful theorems in topology link the numbers of nodes and saddle points that can occur as a function of state space dimensionality and topology (for example, whether the state space has "holes" or not). For two-dimensional state spaces, the relevant theorem is called the ***Poincaré Index Theorem***. For an introductory exposition of the Poincaré Index Theorem, see [Kaplan and Glass, 1995]. For extension of these ideas to higher-dimensional state spaces, see GLA75.

3.16 Poincaré Sections and the Stability of Limit Cycles

We have seen that in state spaces of two (or more) dimensions, a new type of behavior can arise: motion on a limit cycle. The obvious question is the following: Is the motion on the limit cycle stable? That is, if we push the system slightly away from the limit cycle, does it return to the limit cycle (at least asymptotically) or is it repelled from the limit cycle? As we shall see, both possibilities occur in actual systems.

You might expect that we would proceed much as we did for nodes and repellors, by calculating characteristic values involving derivatives of the functions describing the state space evolution. In principle, one could do this, but Poincaré showed that an algebraically and conceptually much simpler method suffices. This method uses what is called a ***Poincaré section*** of the limit cycle. The Poincaré section is closely related to the stroboscopic portraits used in Chapter 1 to discuss the behavior of the diode circuit.

For a two-dimensional state space, the Poincaré section is constructed as follows. In the two-dimensional state space, we draw a line segment that cuts through the limit cycle as shown in Fig. 3.12 (a). This line can be any line segment, but in some cases one might wish to choose the X_1 or X_2 axes. Let us call the point at which the limit cycle crosses the line segment going, say, from left to

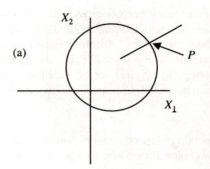

(a)

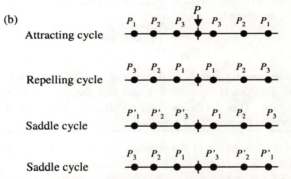

(b)

Fig. 3.12. (a) The Poincaré line segment intersects the limit cycle at point P. (b) The four possibilities for sequences of Poincaré intersection points for trajectories near a limit cycle in two dimensions.

right, point P. (We need to specify right-to-left or left-to-right to pick out just one of the two possible crossing points.)

If we now start a trajectory in the state space at a point that is close to, but not on, the limit cycle, then that trajectory will cross the Poincaré section line segment at a point other than P. Let's call the first crossing point P_1. As the trajectory evolves, it will cross the Poincaré line segment again at points P_2, P_3, and so on. If the sequence of points approaches P as time goes on for any starting point in the neighborhood of the limit cycle, we say that we have an *attracting limit cycle* or, equivalently, a *stable limit cycle*. In other words, the limit cycle is an attractor for the system. If the sequence of intersection points moves away from P (for any trajectory starting near the limit cycle), we say we have a *repelling limit cycle* or, equivalently, an *unstable limit cycle*. Another possibility is that the points are attracted on one side and repelled on the other: In that case we say that we have a *saddle cycle* (in analogy with a saddle point). These possibilities are shown graphically in Fig. 3.12 (b).

How do we describe these properties quantitatively? We use what is called a *Poincaré map function* (or *Poincaré map*, for short). The essential idea is that given a point P_1, where a trajectory crosses the Poincaré line segment, we can in principle determine the next crossing point P_2 by integrating the time-evolution equations describing the system. So, there must be some mathematical function, call it F, that relates P_1 to P_2: $P_2 = F(P_1)$. (Of course, finding this function F is equivalent to solving the original set of equations and that may be difficult or impossible in actual practice.) In general, we may write

$$P_{n+1} = F(P_n) \tag{3.16-1}$$

In general the function F depends not only on the original equations describing the system, but on the choice of the Poincaré line segment as well.

Exercise 3.16-1. For the simple harmonic oscillator, the state space coordinates $x(t)$, the position coordinate, and $v(t)$, the velocity of the oscillator, are given by

$$x(t) = \omega_o \cos(\omega t)$$
$$v(t) = -\omega x_o \sin(\omega t)$$

if the initial conditions are $x(t = 0) = x_0$ and $v(t = 0) = 0$. Take the positive x axis as the Poincaré section line and find the corresponding Poincaré map function. Then do the same for a Poincaré section line at an angle θ with respect to the positive x axis.

To analyze the nature of the limit cycle, we can analyze the nature of the function F and its derivatives. Two points are important to notice:

1. The Poincaré section reduces the original two-dimensional problem to a one-dimensional problem.
2. The Poincaré map function states an iterative (finite-size time step) relation rather than a differential (infinitesimal time step) relation.

The last point is important because F gives P_{n+1} in terms of P_n. The time interval between these points is roughly the time to go around the limit cycle once, a relatively big jump in time. On the other hand, a one-dimensional differential equation $\dot{x} = f(x)$ tells us how x changes over an infinitesimal time interval. The function F is sometimes called an *iterated map function* (or *iterated map*, for short). (Because of the importance of iterated maps in nonlinear dynamics, we shall devote Chapter 5 to a study of their properties.)

Let us note that the point P on the limit cycle satisfies $P = F(P)$. Any point P^* that satisfies $P^* = F(P^*)$ is called a *fixed point* of the map function. If a trajectory crosses the line segment exactly at P^*, it returns to P^* on every cycle. In analogy with our discussion of fixed points for differential equations, we can ask what happens to a point P_1 close to P^*. In particular, we ask what happens to the distance between P_1 and P^* as the system evolves. Formally, we look at

$$P_2 - P^* = F(P_1) - F(P^*) \tag{3.16-2}$$

and use a Taylor series expansion about the point P^* to write

$$P_2 - P^* = F(P^*) + \frac{dF}{dP}\bigg|_{P^*} (P_1 - P^*) + \ldots - F(P^*) \qquad (3.16\text{-}3)$$

If we define $d_i = (P_i - P^*)$, we see that

$$d_2 = \frac{dF}{dP}\bigg|_{P^*} d_1 \qquad (3.16\text{-}4)$$

We now define the ***characteristic multiplier*** M for the Poincaré map:

$$M = \frac{dF}{dP}\bigg|_{P^*} \qquad (3.16\text{-}5)$$

M is also called the ***Floquet multipler*** or the ***Lyapunov multiplier***. In terms of M, we can write Eq. (3.16-4)

$$d_2 = M d_1 \qquad (3.16\text{-}6)$$

We find in general

$$d_{n+1} = M^n d_1 \qquad (3.16\text{-}7)$$

We see that if $M < 1$, then $d_2 < d_1$, $d_3 < d_2$, and so on: The intersection points approach the fixed point P. In that case the cycle is an attracting limit cycle. If $M > 1$, then the distances grow with repeated iterations, and the limit cycle is a repelling cycle. For saddle cycles, M is equal to 1 but the derivative of the map function is greater than 1 on one side of the cycle and less than 1 on the other side. However, based on our discussion of saddle points for one-dimensional state spaces, we expect that saddle cycles are rare in two-dimensional state spaces. Table 3.4 lists the possibilities.

Exercise 3.16-2. In a state space with two dimensions, M cannot be negative. Show that this must be the case to avoid violating the No-Intersection Theorem. Hint: If M is negative, then the intersection points must oscillate from one side of P^* to the other on subsequent iterations.

We can also define a ***characteristic exponent*** associated with the cycle by the equation

$$M \equiv e^{\lambda} \qquad (3.16\text{-}8)$$

or

$$\lambda \equiv \ln(M) \qquad (3.16\text{-}9)$$

Table 3.4.
The Possible Limit Cycles and Their Characteristic
Multipliers for Two–Dimensional State Space

Characteristic Multiplier	Type of Cycle
M < 1	Attracting Cycle
M > 1	Repelling Cycle
M = 1	Saddle Cycle
	(rare in two-dimensions)

The idea is that the characteristic exponent plays the role of the Lyapunov exponent but the time unit is taken to be the time from one crossing of the Poincaré section to the next.

Let us summarize: The Poincaré section method allows us to characterize the possible types of limit cycles and to recognize the kinds of changes that take place in those limit cycles. However, in most cases, we cannot find the mapping function F explicitly; therefore, our ability to predict the kinds of limit cycles that occur for a given system is limited.

Appendix I provides a discussion of these ideas for a model system called the van der Pol oscillator. Appendix J looks a some simple models of laser dynamics in which these ideas are also useful.

3.17 Bifurcation Theory

We have seen that the characteristic values associated with a fixed point depend on the various parameters used to describe the system. As the parameters change, for example as we adjust a voltage in a circuit or the concentration of chemicals in a reactor, the nature of the characteristic values and hence the character of the fixed point may change. For example, an attracting node may become a repellor or a saddle point. The study of how the character of fixed points (and other types of state space attractors) change as parameters of the system change is called *bifurcation theory*. (Recall that the term *bifurcation* is used to describe any sudden change in the dynamics of the system. When a fixed point changes character as parameter values change, the behavior of trajectories in the neighborhood of that fixed point will change. Hence the term bifurcation is appropriate here.) Being able to classify and understand the various possible bifurcations is an important part of the study of nonlinear dynamics. However, the theory, as it is presently developed, is rather limited in its ability to predict the kinds of bifurcations that will occur and the parameter values at which the bifurcations take place for a particular system. Description, however, is the first step toward comprehension and understanding.

In this section we will give a brief introduction to bifurcation theory as applied to state spaces with one and two dimensions. Furthermore, we will restrict the discussion to those bifurcations that occur when just <u>one</u> control parameter of the system is changed. A more extensive discussion of bifurcations is given in Appendix B. We should also emphasize that simple bifurcation theory treats only

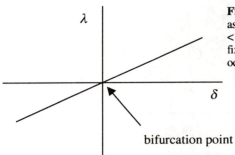

Fig. 3.13. The characteristic value plotted as a function of the control parameter. For δ < 0 the fixed point is a node. For δ > 0 the fixed point is a repellor. A bifurcation occurs at $\delta = 0$.

bifurcation point

the changes in stability of a particular attractor (or, as we shall see in Chapter 4, a particular basin of attraction). Since in general a system may have, for fixed parameter values, several attractors in different parts of state space, we often need to consider the overall dynamical system (that is, its "global" properties) to see what happens to trajectories when a bifurcation occurs.

To keep track of what is happening as the control parameter is varied, we will use two types of diagrams. One type, which we have seen before, is the bifurcation diagram, in which we plot the location of the fixed point (or points) as a function of the control parameter. In the second type of diagram, we plot the characteristic values of the fixed point as a function of the control parameter.

To see how this kind of analysis proceeds, let us begin with the one-dimensional state space case. In a one-dimensional state space, a fixed point has just one characteristic value λ. The crucial assumption in the analysis is that λ varies smoothly (continuously) as some parameter, call it μ, varies. For example, if $\lambda(\mu) < 0$ for some value of μ, then the fixed point is a node. As μ changes, λ might increase (become less negative), going through zero, and then become positive. The node then changes to a repellor when $\lambda > 0$.

Let us consider a specific example:

$$\dot{x} = (\mu - 1)(x - a) \qquad (3.17\text{-}1)$$

The fixed point is located at $x = +a$. The characteristic value is $\mu - 1$. Thus, for $\mu < 1$, the fixed point is a node. For $\mu > 1$, the fixed point is a repellor. We say that at $\mu = 1$ there is a bifurcation and the node (a stable fixed point) changes to a repellor (an unstable fixed point).

Exercise 3.17-1. What kind of fixed point does this example have at $\mu = 1$?

In the literature on nonlinear dynamics, it is traditional to redefine the control parameter and the independent variable such that the bifurcation occurs when the parameter equals 0 and, sometimes, when the independent variable equals 0. Thus, for our previous example, we define a new parameter $\delta = \mu - 1$ and a new variable $y = x - a$. Then the dynamical equation becomes

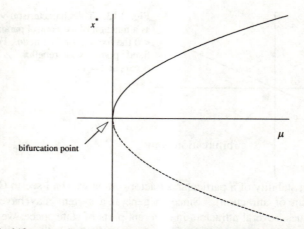

Fig. 3.14. The bifurcation diagram for the repellor-node (saddle-node) bifurcation. The solid line indicates the x value for the node as a function of the parameter value. The dashed line is for the repellor. Note that there is no fixed point at all for $\mu < 0$.

$$\dot{y} = \delta y \qquad (3.17\text{-}2)$$

We then plot the characteristic value as a function of the parameter as shown in Fig. 3.13.

The previous example was rather artificial in many ways. In particular, at the bifurcation value of $\mu = 1$, every value of x is a fixed point. Let's consider a different one-dimensional model:

$$\dot{x} = \mu - x^2 \qquad (3.17\text{-}3)$$

For μ positive, there are two fixed points: one at $x = +\sqrt{\mu}$, the other at $x = -\sqrt{\mu}$. For μ negative there are no fixed points (assuming, of course, that x is a real number). If we use Eq. (3.6-3), which defines the characteristic value for a fixed point, to find the characteristic value of the two fixed points (for $\mu > 0$), we see that the fixed point at $x = -\sqrt{\mu}$ is a repellor, while the fixed point at $x = +\sqrt{\mu}$ is a node.

If we start with $\mu < 0$ and let it increase, we find that a bifurcation takes place at $\mu = 0$. At that value of the parameter we have a saddle point, which then changes into a repellor-node pair as μ becomes positive. We say that we have a ***repellor-node bifurcation*** at $\mu = 0$.

If we shift the function shown in Fig. 3.5 so that the fixed point occurs at $x = 0$, then Fig. 3.5 shows how the function $f(x)$ changes as we pass through the bifurcation. Figure 3.14 shows the bifurcation diagram for the repellor-node bifurcation. Note that at the repellor-node bifurcation point, the fixed point of the system is structurally unstable in the sense discussed in Section 3.6. Structurally unstable points are important because their existence indicates a possible bifurcation.

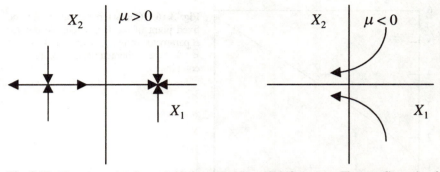

Fig. 3.15. Phase portraits above and below a saddle-node bifurcation. The one-dimensional trajectories have been lifted to a two-dimensional state space as described in the text. For $\mu >$ 0, there is a saddle point at $X_1 = -\sqrt{\mu}$ and a node at $X_1 = +\sqrt{\mu}$. There is no fixed point for $\mu <$ 0.

In the nonlinear dynamics literature, the bifurcation just described is usually called a *saddle-node bifurcation*, *tangent bifurcation*, or a *fold bifurcation*. The origin of these names will become apparent when we see analogous bifurcations in higher-dimensional state spaces. For example, if we imagine the curves in Fig. 3.14 as being the cross section of a piece of paper extending into and out of the plane of the page, then the bifurcation point represents a "fold" in the piece of paper. Also, Fig. 3.5 shows how the function in question becomes tangent to the x axis at the bifurcation point.

To visualize the trajectories, it is customary to add another dimension to the state space. Trajectories moving along the added X_2 direction are assumed to be attracted toward the X_1 axis (the original x axis). The one-dimensional repellor then becomes a two-dimensional saddle point. Thus in this *lifted* or *suspended state space*, the bifurcation involves the interaction of a saddle point and a node. Hence, we call this event a saddle-node bifurcation. Figure 3.15 gives a sketch of the system's (lifted) phase portrait above and below the bifurcation value $\mu = 0$.

There are other types of bifurcations possible in a one-dimensional state space, but the repellor-node bifurcation is the most common. Appendix B contains a more elaborate treatment of bifurcation theory and a list of references to the appropriate literature.

Bifurcations in Two Dimensions

Let us begin with a consideration of fixed points. We will then take up the question of bifurcations of limit cycles. In a two-dimensional state space, a fixed point can have either a single characteristic value (an exceptional case) or, more generally, two real characteristic values or a pair of complex-valued characteristic values. (Recall that the two complex values form a complex conjugate pair.) If the two characteristic values turn out to be just real values, then we can plot those values as a function of parameter value. For example, in the Brusselator model, the

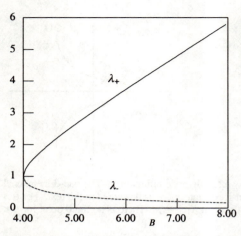

Fig. 3.16. Characteristic values for the fixed point of the Brusselator model for B parameter values greater than 4. For $B < 4$, the characteristic values are complex.

characteristic values are given by Eq. (3.14-8). For the parameter $B > 4$, the characteristic values are real as shown in Fig. 3.16.

For $0 \leq B \leq 4$, the characteristic values are complex, with both real and imaginary parts. Figure 3.17 shows the real and imaginary parts of the characteristic values for the Brusselator model.

Corresponding to the repellor-node bifurcation in one dimension, the saddle-node bifurcation is very common in two-dimensional state space systems. The behavior of the trajectories near the fixed points is often modeled with the so-called **normal form equations** (see Appendix B for more details):

$$\dot{x}_1 = \mu - x_1^2$$
$$\dot{x}_2 = -x_2$$

(3.17-4)

To emphasize that these equations hold only in the neighborhood of a fixed point and close to the bifurcation value of the parameter, we have used lower case letters

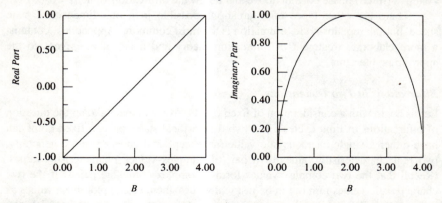

Fig. 3.17. The real and imaginary parts of the characteristic value λ_+ for the Brusselator model are shown as a function of the parameter B. For λ_-, the imaginary part is negative.

for the variables.

From Eq. (3.17-4) we see that there are two fixed points (when $\mu > 0$); one at $(x_1, x_2) = (+\sqrt{\mu}, 0)$; the other at $(-\sqrt{\mu}, 0)$. The one at $(+\sqrt{\mu}, 0)$ is a node; the one at $(-\sqrt{\mu}, 0)$ is a saddle point. For μ negative, there is no fixed point. A saddle-node bifurcation occurs at $\mu = 0$ where the saddle-node pair is born. Figure 3.15 shows the corresponding phase portraits.

Limit Cycle Bifurcations

As we saw earlier, a fixed point in a two-dimensional state space may also have complex-valued characteristic values for which the trajectories have spiral-type behavior. A bifurcation occurs when the characteristic values move from the left-hand side of the complex plane to the right-hand side; that is, the bifurcation occurs when the real part of the characteristic value goes to 0.

We can also have limit cycle behavior in two-dimensional systems. The birth and death of a limit cycle are bifurcation events. The birth of a stable limit cycle is called a *Hopf bifurcation* (named after the mathematician E. Hopf). (Although this type of bifurcation was known and understood by Poincaré and later studied by the Russian mathematician A. D. Andronov in the 1930s, Hopf was the first to extend these ideas to higher-dimensional state spaces.) Since we can use a Poincaré section to study a limit cycle and since for a two-dimensional state space, the Poincaré section is just a line segment, the bifurcations of limit cycles can be studied by the same methods used for studying bifurcations of one-dimensional dynamical systems.

A Hopf bifurcation can be modeled using the following normal form equations:

$$\dot{x}_1 = -x_2 + x_1\{\mu - (x_1^2 + x_2^2)\} \tag{3.17-5a}$$

$$\dot{x}_2 = +x_1 + x_2\{\mu - (x_1^2 + x_2^2)\} \tag{3.17-5b}$$

The geometric form of the trajectories is clearer if we change from (x_1, x_2) coordinates to polar coordinates (r, θ) defined in the following equations and illustrated in Fig. 3.18.

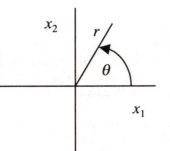

Fig. 3.18. The definition of polar coordinates. r is the length of the radius vector from the origin. θ is the angle between the radius vector and the positive x_1 axis.

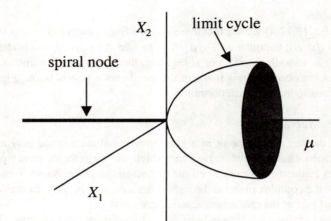

Fig. 3.19. The bifurcation diagram for the Hopf bifurcation. For $\mu < 0$, the fixed point is a spiral node. For $\mu > 0$, the fixed point at the origin is a spiral repellor and the attractor for the system is the limit cycle.

$$r = \sqrt{(x_1^2 + x_2^2)}$$
$$\tan\theta = \frac{x_2}{x_1}$$
(3.17-6)

Using these polar coordinates, we write Eqs. (3.17-5) as

$$\dot{r} = r\{\mu - r^2\} \equiv f(r)$$
(3.17-7a)

$$\dot{\theta} = 1$$
(3.17-7b)

Exercise 3.17-2. Use the relations in Eq. (3.17-6) to show that Eqs. (3.17-7) follow from Eq. (3.17-5)

Now let us interpret the geometric nature of the trajectories that follow from Eqs. (3.17-7). The solution to Eq. (3.17-7b) is simply

$$\theta(t) = \theta_o + t$$
(3.17-8)

that is, the angle continues to increase with time as the trajectory spirals around the origin. For $\mu < 0$, there is just one fixed point for r, namely $r = 0$. By evaluating the derivative of $f(r)$ with respect to r at $r = 0$, we see that the characteristic value is equal to μ. Thus, for $\mu < 0$, that derivative is negative, and the fixed point is stable. In fact, it is a spiral node.

For $\mu > 0$, the fixed point at the origin is a spiral repellor; it is unstable; trajectories starting near the origin spiral away from it. There is, however, another fixed point for r, namely, $r = \sqrt{\mu}$. This fixed point for r corresponds to a limit cycle with a period of 2π [in the time units of Eqs. (3.17-7)]. We say that the limit cycle is born at the bifurcation value $\mu = 0$. Fig. 3.19 shows the bifurcation diagram for the Hopf bifurcation.

In a two-dimensional state space the possible types of bifurcations are also limited. As explained in Appendix B, the saddle-node bifurcation and the Hopf bifurcation are the most "common" two-dimensional bifurcations for models with one control parameter. As we shall see in the next chapter, once we move to a state space with three or more dimensions the number of common bifurcations increases tremendously.

Appendix J contains a detailed discussion of a case history of bifurcations in which a simplified model of laser dynamics illustrates several of the features discussed in this chapter.

3.18 Summary

In this chapter we have developed much of the mathematical machinery needed to discuss the behavior of dynamical systems. We have seen that fixed points and their characteristic values (determined by derivatives of the functions describing the dynamics of the system) are crucial for understanding the dynamics. We have also seen that the dimensionality of the state space plays a major role in determining the kinds of trajectories that can occur for bounded systems.

Moreover, as the control parameters of a system change, the character of fixed points and the nature of trajectories near them can change dramatically at bifurcation points. Bifurcation diagrams are used to describe the change in behavior near bifurcation points. We again saw that the dimensionality of the state space limits the kinds of bifurcations that can commonly occur.

In state spaces with two or more dimensions, limit cycles, describing periodic behavior, can appear. The stability of a limit cycle can be discussed by means of a Poincaré section and the characteristic multiplier determined by the derivative of the corresponding Poincaré map function. A limit cycle may be born via a Hopf bifurcation.

The growth and behavior of these limit cycles form an important paradigm in nonlinear dynamics and illustrate several crucial features that can occur in nonlinear systems but not in linear systems. In Chapter 1, we discussed qualitatively the novelty of "spontaneous" generation of time-dependent behavior of a system "living" in an environment that is completely steady in time. No such behavior is possible for a linear system. A linear system can maintain time-dependent behavior only in the idealized (and unrealistic) case in which dissipative forces (such as friction) are completely absent. If we include dissipation, a linear system must eventually relax to a time dependence determined by the time dependence of the external forces (in the most general sense of force) applied to it. On the other hand, nonlinear systems can spontaneously break the time symmetry of the environment, at least for some range of control parameter values. For a two-dimensional state space system, this time-dependent behavior must be a limit cycle if the system remains bounded.

3.19 Further Reading

Books on Phase Space and State Space

J. B. Marion and S. T. Thornton, *Classical Dynamics of Particles and Systems,* 4th ed. (Saunders, Fort Worth, 1995). This text, aimed at intermediate level undergraduate physics majors, gives a nice introduction to nonlinear oscillations, phase space, and so on in Chapter 4.

J. W. Gibbs, *Elementary Principles in Statistical Mechanics* (C. Scribner's Sons, New York, 1902). A classic and still worth reading for its careful consideration of questions of fundamentals.

V. I. Arnold, *Geometric Methods in the Theory of Ordinary Differential Equations* (Springer-Verlag, New York, Heidelberg, Berlin, 1983). One of the best introductions to the use of geometric methods for differential equations.

A. P. Pippard, *Response and Stability* (Cambridge University Press, Cambridge, 1985). A delightful book, filled with physical insight. This extended essay treats driven oscillators, nonlinear oscillators, bifurcations, catastrophes, phase transitions, and broken symmetries at the advanced undergraduate level.

P. Hagedorn, *Nonlinear Oscillations* (Clarendon Press, Oxford, 1981). A more advanced treatise on analytic methods of dealing with non-linear oscillators. Good discussion of stability criteria and limit cycles.

M. W. Hirsch and S. Smale, *Differential Equations, Dynamical Systems, and Linear Algebra* (Academic Press, New York, 1974). Contains a proof of the Poincaré–Bendixson theorem.

J. A. Sanders and F. Verhulst, *Averaging Methods in Nonlinear Dynamical Systems* (Springer-Verlag, New York, Berlin, Heidelberg, 1984). Provides a systematic introduction to many of the analytic tools used to get approximate descriptions of the behavior of nonlinear systems.

Linear Alegebra and Mathematical Methods

Mary L. Boas, *Mathematical Methods in the Physical Sciences*, 2nd ed. (John Wiley and Sons, New York, 1983).

S. Hassani, *Foundations of Mathematical Physics* (Allyn and Bacon, Boston, 1991).

Seymour Lipschutz, *Linear Algebra* (Schaum's Outline Series)(McGraw–Hill, New York, 1968). An inexpensive book with hundreds of problems and worked examples.

D. Gulick, *Encounters with Chaos* (McGraw–Hill, New York, 1992). A mathematician's introduction to iterated maps and systems of differential equations as dynamical systems.   Chapter 3 has a particularly nice introduction to the linear algebra of matrices.   Many good examples and exercises.

Fractal Basin Boundaries and Riddled Basins of Attraction

Fractal basin boundaries are discussed in C. Grebogi, S. W. McDonald, E. Ott, and J. A. Yorke, "Final State Sensitivity: An obstruction to predictability," *Phys. Lett. A* **99**, 415–418 (1983) and S. W. McDonald, C. Grebogi, E. Ott, and J. A. Yorke, "Fractal Basin Boundaries," *Physica D* **17**, 125–153 (1985).

The driven damped pendulum exhibits fractal basin boundaries as discussed in E. G. Gwinn and R. M. Westervelt, "Fractal Basin Boundaries and Intermittency in the Driven Damped Pendulum," *Phys. Rev. A* **33**, 4143–55 (1986).

J. C. Alexander, J. A. Yorke, Z. You, and I. Kan, "Riddled Basins," *Int. J. Bifur. and Chaos* **2**, 795–80 (1992).   This paper introduced the notion of riddled basins of attraction.

J. C. Sommerer and E. Ott, "A physical system with qualitatively uncertain dynamics," *Nature* **365**, 136–140 (1993).   Gives a nice example of riddled basins of attraction.

J. F. Hagey, T. L. Carroll, and L. M. Pecora, "Experimental and Numerical Evidence for Riddled Basins in Coupled Chaotic Systems," *Phys. Rev. Lett.* **73**, 3528–31 (1994).

State Space Topological Considerations

See [Kaplan and Glass, 1995], pp. 253 ff.

L. Glass, "Combinatorial and topological methods in nonlinear chemical kinetics," *J. Chem. Phys.* **63**, 1325–35 (1975).

[Guckenheimer and Holmes, 1983], pp. 50–51.   A discussion of the Poincaré Index Theorem.

The Brusselator Model

G. Nicolis and I. Prigogine, *Exploring Complexity* (W. H. Freeman, San Francisco, 1989).   This wide-ranging book contains a detailed discussion of the Brusselator Model and its dynamics.

Poincaré Sections

Most of the more formal books listed at the end of Chapter 1 have extensive discussions of Poincaré sections.

The Poincaré section technique can be generalized in various ways.   For example, in cases where a system is driven by "forces" with two incommensurate frequencies, taking a Poincaré section within a Poincaré section is useful.   See F. C. Moon and W. T. Holmes "Double Poincaré Sections of A Quasi-Periodically Forced, Chaotic Attractor," *Physics Lett. A* **111**, 157–60 (1985) and [Moon, 1992].

3.20 Computer Exercises

CE3-1. Use *Chaos Demonstrations* to study the van der Pol equation limit cycles in state space. (See Appendix I for a discussion of the van der Pol model.) Vary the parameter *h* (equivalent to the parameter *R* used in Appendix I) to see how the oscillations change from simple harmonic (for small values) to relaxation oscillations for larger values.

CE3-2. Use *Chaotic Dynamics Workbench* to study the Shaw–Van der Pol Oscillator with the force term set to 0 (to make the state space two-dimensional). Observe the time dependence of the dynamical variables and the state space diagrams as the coefficient *A* (corresponding to *R* in Appendix I) increases.

4

Three-Dimensional State Space and Chaos

*The chaos is come of the organized disorder, The consistently inappropriate
and the simple wrong. George Barker, The First American Ode.*

4.1 Overview

In the previous chapter, we introduced some of the standard methods for analyzing
dynamical systems described by systems of ordinary differential equations, but we
limited the discussion to state spaces with one or two dimensions. We are now
ready to take the important step to three dimensions. This is a crucial step, not
because we live in a three-dimensional world (remember that we are talking about
state space, not physical space), but because in three dimensions dynamical systems
can behave in ways that are not possible in one or two dimensions. Foremost
among these new possibilities is chaos.

First we will give a hand-waving argument (we could call it heuristic if we
wanted to sound more sophisticated) that shows why chaotic behavior may occur in
three dimensions. We will then discuss, in parallel with the treatment of the
previous chapter, a classification of the types of fixed points that occur in three
dimensions. However, we gradually wean ourselves from the standard analytic
techniques and begin to rely more and more on graphic and geometrical
(topological) arguments. This change reflects the flavor of current developments in
dynamical systems theory. In fact, the main goal of this chapter is to develop
geometrical pictures of trajectories, attractors, and bifurcations in three-dimensional
state spaces.

Next, we will discuss the types of attractors that can occur for dissipative
dynamical systems with three-dimensional state spaces. Two new possibilities
emerge: (1) quasi-periodic attractors and (2) chaotic attractors.

When the parameters of a system are changed, chaotic behavior may appear
and disappear in several different ways, even for the same dynamical system: We
may have several *routes to chaos*. These routes can be put into two broad
categories with several subdivisions within each category. One category includes
sequences of bifurcations involving limit cycles (or equivalently, fixed points of the
associated Poincaré map). (The period-doubling sequence in Chapter 1 belongs to
this category.) The other category involves changes in trajectories associated with
several fixed points or limit cycles. Since these changes involve the properties of
trajectories ranging over a significant volume of state space, these changes are
called "global" bifurcations (in contrast to "local" bifurcations associated with
changes in individual fixed points).

Rather sudden changes from regular to chaotic behavior, such as we have seen with the Lorenz model in Chapter 1, are characteristic of these global bifurcations. Although the nature of the long-time attractor changes suddenly as a parameter is varied, these sudden changes are often heralded by ***chaotic transients***. In a chaotic transient, the system's trajectory wanders through state space, in an apparently chaotic fashion. Eventually, the trajectory approaches a regular, periodic attractor. As the control parameter is changed, the chaotic transient lasts longer and longer until finally the asymptotic behavior is itself chaotic.

The questions we want to address are the following: How does this complicated chaotic behavior develop? How does the system evolve from regular, periodic behavior to chaotic behavior? What changes in the fixed points and in trajectories in state space give rise to these changes in behavior?

Of course, it is impossible to give a single, simple answer to any of these questions. Different dynamical systems seem to behave quite differently depending on both parameter values and initial conditions. These complications arise because, in general, a dynamical system may have several attractors (like the two fixed points of the Lorenz model) that "coexist" for a given range of parameter values. The system can change its behavior because the attractors change their characters (for example, a limit cycle becomes unstable and is replaced by a period-doubled limit cycle) or the basins of attraction can interact (both with each other and with in-sets and out-sets of saddle points and saddle cycles) in such a way as to give rise to chaotic dynamics. We shall discuss each of these possibilities in later sections of this chapter.

4.2 Heuristics

We will describe, in a rather loose way, why three (or more) state space dimensions are needed to have chaotic behavior. First, we should remind ourselves that we are dealing with dissipative systems whose trajectories eventually approach an attractor. For the moment we are concerned only with the trajectories that have settled into the attracting region of state space. When we write about the divergence of nearby trajectories, we are concerned with the behavior of trajectories within the attracting region of state space.

In a somewhat different context we will need to consider sensitive dependence on initial conditions. Initial conditions that are <u>not</u>, in general, part of an attractor can lead to very different long-term behaviors on different attractors. Those behaviors, determined by the nature of the attractor (or attractors), might be time-independent or periodic or chaotic.

As we saw in Chapter 1, chaotic behavior is characterized by the divergence of nearby trajectories in state space. As a function of time, the "separation" (suitably defined) between two nearby trajectories increases exponentially, at least for short times. The last restriction is necessary because we are concerned with systems whose trajectories stay within some bounded region of state space. The system does not "blow up." There are three requirements for chaotic behavior in such a situation:

1. no intersection of different trajectories;
2. bounded trajectories;
3. exponential divergence of nearby trajectories.

These conditions cannot be satisfied simultaneously in one- or two-dimensional state spaces. You should convince yourself that this is true by sketching some trajectories in a two-dimensional state space on a sheet of paper. However, in three dimensions, initially nearby trajectories can continue to diverge by wrapping over and under each other. Obviously sketching three-dimensional trajectories is more difficult. You might try using some relatively stiff wire to form some trajectories in three dimensions to show that all three requirements for chaotic behavior can be met. You should quickly discover that these requirements lead to trajectories that initially diverge, then curve back through the state space, forming in the process an intricate layered structure. Figure 4.1 is a sketch of diverging trajectories in a three-dimensional state space.

The crucial feature of state space with three or more dimensions that permits chaotic behavior is the ability of trajectories to remain within some bounded region by intertwining and wrapping around each other (without intersecting!) and without repeating themselves exactly. Clearly the geometry associated with such trajectories is going to be strange. In fact, such attractors are now called *strange attractors*. In Chapter 9, we will give a more precise definition of a strange

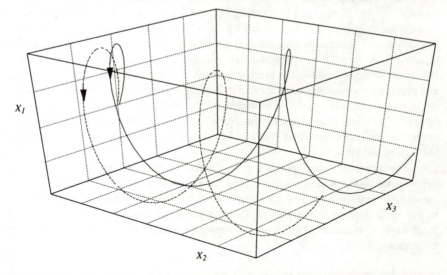

Fig. 4.1. A sketch of trajectories in a three–dimensional state space. Notice how two nearby trajectories can continue to behave quite differently from each other yet remain bounded by weaving in and out and over and under each other.

attractor in terms of the notion of fractal dimension. If the behavior on the attractor is chaotic, that is, if the trajectories on the attractor display exponential divergence of nearby trajectories (on the average), then we say the attractor is chaotic. Many authors use the terms *strange attractor* and **chaotic attractor** interchangeably, but in principle they are distinct [GOP84].

The notion of exponential divergence of nearby trajectories is made formal by introducing the **Lyapunov exponent**. If two nearby trajectories on a chaotic attractor start off with a separation d_0 at time $t = 0$, then the trajectories diverge so that their separation at time t, denoted by $d(t)$, satisfies the expression

$$d(t) = d_0 e^{\lambda t} \qquad (4.2\text{-}1)$$

The parameter λ in Eq. (4.2-1) is called the Lyapunov exponent for the trajectories. If λ is positive, then we say the behavior is chaotic. (Section 4.13 takes up the question of Lyapunov exponents in more detail.) From this definition of chaotic behavior, we see that chaos is a property of a <u>collection</u> of trajectories.

Chaos, however, also appears in the behavior of a single trajectory. As the trajectory wanders through the (chaotic) attractor in state space, it will eventually return near some point it previously visited. (Of course, it cannot return exactly to that point. If it did, then the trajectory would be periodic.) If the trajectories exhibit exponential divergence, then the trajectory on its second visit to a particular neighborhood will have subsequent behavior, quite different from its behavior on the first visit. Thus, the impression of the time record of this behavior will be one of nonreproducibility, nonperiodicity, in short, of chaos.

To illustrate some of the issues involved here, let us consider a counterexample. Imagine a ball perched precariously at the (unstable) equilibrium point at the top of a hill surrounded by an infinite plane surface. This situation displays sensitive dependence on initial conditions: The path the ball takes depends sensitively on how it is disturbed and pushed away from the top of the hill. However, if the hill and plane are frictionless, then the ball keeps rolling forever, and there are no bounded trajectories. If friction is present, then the ball eventually stops rolling at some point determined by the direction and size of the initial "push." Each final state is associated with a particular initial condition. There is no attracting region of the state space, which pulls in trajectories from some finite basin of attraction.

The point of these remarks is to remind us that our notions of sensitive dependence on initial conditions and divergence of nearby trajectories are meaningful and useful only for those systems that are bounded and have attractors in the sense defined in Chapter 3. (In Chapter 8, we shall see how to generalize these ideas to bounded Hamiltonian systems for which there is no dissipation and no attractor.)

4.3 Routes to Chaos

The study of chaos has brought two surprises: (1) the ubiquity of chaotic behavior and (2) the universality of the routes to chaos. Why are these surprising? Although mathematicians such as Poincaré long ago recognized the <u>possibility</u> of what we now call chaotic behavior, the general impression was that this type of behavior was in some sense unusual and pathological; in any case, its detailed character would be particular to the nonlinear system being studied. The standard general methods developed to solve linear differential equations fail for nonlinear differential equations. Solutions to only a few nonlinear differential equations were known and each of these needed to be developed by methods particular to the case at hand. Mathematicians reasoned, we might imagine, that since no general solution methods exist, there could be no general character to the solutions. What we are now just beginning to appreciate is the universality that does exist among the solutions to these nonlinear equations, at least, in the case of systems with just a few degrees of freedom. What may be even more surprising is that this universality seems to be exhibited by the actual physical, chemical, and biological systems that we model with these equations.

The universality we want to describe is the universality of the routes or transitions to chaos. How does a (nonlinear) system change its behavior from regular (either stationary or periodic) to chaotic (or vice versa) as the control parameters of the system are (slowly) changed? (The parenthetical "slowly" is introduced to remind us that we are concerned with the long-term behavior of the system after transients have died out. In an experiment, we change a parameter's value, let the transients die out, and then look at the "asymptotic" behavior.) What we have learned in the past twenty years or so is that the transitions to chaos exhibited by many experimental systems and the equations used to describe them can be grouped into just a few broad categories. Once we find which type of transition a system exhibits, we can then make many qualitative and quantitative predictions about how that transition will proceed.

Two facts about these transitions are important to note: First, a given system may exhibit several different types of transitions to chaos for different ranges of parameter values (recall our examples in Chapter 1). Second is a point of humility: The theory of nonlinear systems is not yet sufficiently developed to allow us to tell in advance what type(s) of transition will occur for what range of parameters for a given system. More "routes to chaos" will undoubtedly be recognized (discovered), particularly as we learn to deal with systems with larger numbers of degrees of freedom. However, at present the recognized transitions to chaos can be gathered under two large headings as shown in Table 4.1.

In the first category of transitions (via local bifurcations), a limit cycle occurs for a range of parameter values. As some control parameter of the system is changed, the limit cycle behavior "disappears" and chaotic behavior appears. In the second category (via global bifurcations), the long-term behavior of the system is influenced by unstable fixed points or cycles as well as by an attractor (or several

Table 4.1

Transitions to Chaos

I. Via Local Bifurcations

    A. Period-doubling
    B. Quasi-periodicity
    C. Intermittency
        1. Type I (tangent bifurcation intermittency)
        2. Type II (Hopf bifurcation intermittency)
        3. Type III (period-doubling intermittency)
        4. On-off intermittency

II. Via Global Bifurcations

    A. Chaotic transients
    B. Crises

---

attractors). As a parameter is changed, the transient trajectories, which would eventually end up approaching the fixed point (or cycle) become more and more complicated, producing what we call *chaotic transients*. These chaotic transients eventually last forever and the long-term behavior of the system is chaotic.

The two categories of transitions are also distinguished to some extent by the mathematical and geometrical tools needed to analyze them. In the local bifurcation category, the Poincaré section technique allows us to understand and describe how the transition takes place. The three subcategories indicate the ways the fixed points of the Poincaré section evolve as parameters are changed. In the global bifurcation category, the analysis requires following trajectories over a significant range of state space and seeing how these trajectories are influenced by various attractors, fixed points and cycles that may coexist in different parts of the state space. Poincaré sections are still used, but information about individual fixed points and limit cycles ("local information") is no longer sufficient to determine the nature of the long-term trajectories. We need a broader, more global view of the structure of state space.

We shall also see that the two categories of transitions are not as distinct as a first reading might suggest. In fact, to understand the dynamics of intermittency, for example, we need to track trajectories through a significant region of state space. Conversely, global bifurcations are influenced by the locations and properties of (local) fixed points. However, the two categories do serve as a useful starting point for organizing our thoughts about and our understanding of the development and emergence of chaotic behavior.

4.4 Three-Dimensional Dynamical Systems

We will now introduce some of the formalism for the description of a dynamical system with three state variables. We call a dynamical system three-dimensional if it has three independent dynamical variables, the values of which at a given instant of time uniquely specify the state of the system. We assume that we can write the time-evolution equations for the system in the form of the standard set of first-order ordinary differential equations. (Dynamical systems modeled by iterated map functions will be discussed in Chapter 5.) Here we will use x with a subscript 1, 2, or 3 to identify the variables. This formalism can then easily be generalized to any number of dimensions simply by increasing the numerical range of the subscripts. The differential equations take the form

$$\dot{x}_1 = f_1(x_1, x_2, x_3)$$
$$\dot{x}_2 = f_2(x_1, x_2, x_3) \quad\quad (4.4\text{-}1)$$
$$\dot{x}_3 = f_3(x_1, x_2, x_3)$$

The Lorenz model equations of Chapter 1 are of this form. Note that the three functions f_1, f_2, and f_3 do not involve time explicitly; again, we say that the system is autonomous.

As an aside, we note that some authors like to use a symbolic "vector" form to write the system of equations:

$$\dot{\vec{x}} = \vec{f}(\vec{x}) \quad\quad (4.4\text{-}2)$$

Here $\vec{x}$ stands for the three symbols x_1, x_2, x_3, and $\vec{f}$ stands for the three functions on the right-hand side of Eqs. (4.4-1).

The differential equations describing two-dimensional systems subject to a time-dependent "force" (and hence nonautonomous) can also be written in the form of Eq. (4.4-1) by making use of the "trick" introduced in Chapter 3: Suppose that the two-dimensional system is described by equations of the form

$$\dot{x}_1 = f_1(x_1, x_2, t)$$
$$\dot{x}_2 = f_2(x_1, x_2, t) \quad\quad (4.4\text{-}3)$$

The trick is to introduce a third variable, $x_3 = t$. The three "autonomous" equations then become

$$\dot{x}_1 = f_1(x_1, x_2, x_3)$$
$$\dot{x}_2 = f_2(x_1, x_2, x_3) \quad\quad (4.4\text{-}4)$$
$$\dot{x}_3 = 1$$

which are of the same form as Eq. (4.4-1). As we shall see, this trick is particularly useful when the time-dependent term is periodic in time.

Exercise 4.4-1. The "forced" van der Pol equation is used to describe an electronic triode tube circuit subject to a periodic electrical signal. The equation for $q(t)$, the charge oscillating in the circuit, can be put in the form

$$\frac{d^2q}{dt^2} + \gamma(q)\frac{dq}{dt} + q(t) = g\sin\omega t$$

Use the trick introduced earlier to write this equation in the standard form of Eq. (4.4-1).

4.5 Fixed Points in Three Dimensions

The fixed points of the system of Eqs. (4.4-1) are found, of course, by setting the three time derivatives equal to 0. [Two-dimensional forced systems, even if written in the three-dimensional form (4.4-4), do not have any fixed points because, as the last of Eqs. (4.4-4) shows, we never have $\dot{x}_3 = t = 0$. Thus, we will need other techniques to deal with them.] The nature of each of the fixed points is determined by the three characteristic values of the Jacobian matrix of partial derivatives evaluated at the fixed point in question. The Jacobian matrix is

$$J = \begin{pmatrix} \dfrac{\partial f_1}{\partial x_1} & \dfrac{\partial f_1}{\partial x_2} & \dfrac{\partial f_1}{\partial x_3} \\[2mm] \dfrac{\partial f_2}{\partial x_1} & \dfrac{\partial f_2}{\partial x_2} & \dfrac{\partial f_2}{\partial x_3} \\[2mm] \dfrac{\partial f_3}{\partial x_1} & \dfrac{\partial f_3}{\partial x_2} & \dfrac{\partial f_3}{\partial x_3} \end{pmatrix} \qquad (4.5\text{-}1)$$

In finding the characteristic values of this matrix, we will generally have a cubic equation, whose roots will be the three characteristic values labeled $\lambda_1, \lambda_2, \lambda_3$.

Some mathematical details: The standard theory of cubic equations tells us that a cubic equation of the form

$$\lambda^3 + p\lambda^2 + q\lambda + r = 0 \qquad (4.5\text{-}2)$$

can be changed to the "standard" form

$$x^3 + ax + b = 0 \qquad (4.5\text{-}3)$$

by the use of the substitutions

$$x = \lambda + p/3$$

$$a = \frac{1}{3}(3q - p^2) \qquad (4.5\text{-}4)$$

$$b = \frac{1}{27}(2p^3 - 9qp + 27r)$$

If we now introduce

$$s = \left(\frac{b^2}{4} + \frac{a^3}{27}\right)$$

$$A = (-b/2 + \sqrt{s})^{\frac{1}{3}} \qquad (4.5\text{-}5)$$

$$B = (-b/2 - \sqrt{s})^{\frac{1}{3}}$$

the three roots of the x equation can be written as

$$A + B$$

$$-\left(\frac{A+B}{2}\right) + \left(\frac{A-B}{2}\right)\sqrt{-3} \qquad (4.5\text{-}6)$$

$$-\left(\frac{A+B}{2}\right) - \left(\frac{A-B}{2}\right)\sqrt{-3}$$

from which the characteristic values for the matrix can be found by working back through the set of substitutions. Most readers will be greatly relieved to know that we will not make explicit use of these equations. But it is important to know the <u>form</u> of the solutions.

There are three cases to consider:

1. The three characteristic values are real and unequal ($s < 0$).
2. The three characteristic values are real and at least two are equal ($s = 0$).
3. There is one real characteristic value and two complex conjugate values ($s > 0$).

Case 2 is just a borderline case and need not be treated separately. Just as we did in the previous chapter, we can classify the types of fixed points according to the nature of their characteristic values. We are about to list the fixed points that can occur in three-dimensional state space. In Figs. 4.2 and 4.3, we have drawn sketches of trajectories in state space to indicate the behavior near the fixed point. We have also shown a graph of the complex plane indicating the real and imaginary parts of the corresponding characteristic values.

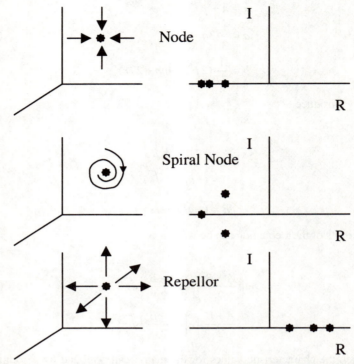

Fig. 4.2. On the left, sketches of trajectories for fixed points in a three–dimensional state space. The location of the fixed point is indicated by the asterisk. On the right, the characteristic values for the fixed points are indicated in the complex plane. The imaginary part is plotted on the vertical axis; the real part on the horizontal axis.

For state spaces with three or more dimensions, it is common to specify the so-called *index* of a fixed point.

> The **index** of a fixed point is defined to be the number of characteristic values of that fixed point whose real parts are positive.

In more geometric terms, the index is equal to the spatial dimension of the out-set of that fixed point. For a node (which does not have an out-set), the index is equal to 0. For a repellor, the index is equal to 3 for a three-dimensional state space. A saddle point can have either an index of 1, if the out-set is a curve, or an index of 2, if the out-set is a surface as shown in Fig. 4.3.

The four basic types of fixed points for a three-dimensional state space are:

1. **Node**. All the characteristic values are real and negative. All trajectories in the neighborhood of the node are attracted toward the fixed point without looping around the fixed point.

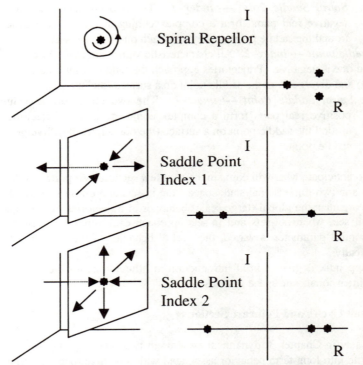

Fig. 4.3. More sketches of trajectories for fixed points in a three–dimensional state space. Not shown are possible spiral versions of the two types of saddle points. On the right, the characteristic values for the fixed points are indicated in the complex plane. For a spiral index-1 saddle point, trajectories spiral toward the fixed point on the in-set surface. For a spiral index-2 saddle point, the trajectories spiral away from the fixed point on the out-set surface.

    1s. ***Spiral Node***. All the characteristic values have negative real parts but two of them have nonzero imaginary parts (and in fact form a complex conjugate pair). The trajectories spiral around the node on a "surface" as they approach the node.

2.   ***Repellor***. All the characteristic values are real and positive. All trajectories in the neighborhood of the repellor diverge from the repellor.

    2s. ***Spiral Repellor***. All the characteristic values have positive real parts, but two of them have nonzero imaginary parts (and in fact form a complex conjugate pair). Trajectories spiral around the repellor (on a "surface") as they are repelled from the fixed point.

3.   ***Saddle point — index–1***. All characteristic values are real. One is positive and two are negative. Trajectories approach the saddle point on a surface (the in-set) and diverge along a curve (the out-set).

3s. ***Spiral Saddle Point — index–1***. The two characteristic values with negative real parts form a complex conjugate pair. Trajectories spiral around the saddle point as they approach on the in-set surface.

4. ***Saddle point — index–2***. All characteristic values are real. Two are positive and one is negative. Trajectories approach the saddle point on a curve (the in-set) and diverge from the saddle point on a surface (the out-set).

4s. ***Spiral Saddle Point — index–2***. The two characteristic values with positive real parts form a complex conjugate pair. Trajectories spiral around the saddle point on a surface (the out-set) as they diverge from the saddle point.

To anticipate what will come in the next few sections, we point out that just as in one- and two-dimensional state spaces, the in-sets and out-sets of saddle points tend to organize the global (large scale) behavior of trajectories in state space. As we shall see, when out-sets and in-sets approach each other (in loose terms) as some control parameter is varied, the overall behavior of the system can change dramatically.

Appendix B gives a brief introduction to bifurcations for the fixed points in three-dimensional state space systems.

4.6 Limit Cycles and Poincaré Sections

As we saw in Chapter 3, dynamical systems in two (and higher) dimensions can also settle into long-term behavior associated with repetitive, periodic limit cycles. We also learned that the Poincaré section technique can be used to reduce the dimensionality of the description of these limit cycles and to make their analysis simpler.

First, we focus on the construction of a Poincaré section for the system. For a three-dimensional state space, the Poincaré section is generated by choosing a ***Poincaré plane*** (a two-dimensional surface) and recording on that surface the points at which a given trajectory cuts through that surface. (In most cases the choice of plane is not crucial as long as the trajectories cut the surface ***transversely***, that is, the trajectories do not run parallel or almost parallel to the surface as they pass through; see Fig. 4.4.) For autonomous systems, such as the Lorenz model equations, we choose some convenient plane in the state space, say, the XY plane for the Lorenz equations. When a trajectory crosses that plane passing from, for example, negative Z values to positive Z values, we record that crossing point.

If the system has a natural period associated with it, say the period of a (periodic) "force" applied to the system, then the Poincaré plane could be a surface corresponding to a definite (but arbitrarily chosen) phase of that force. In the latter case the Poincaré section is analogous to a "stroboscopic portrait" of a mechanical system recorded with a flash (or strobe) lamp fired once every period of the motion. In that sense we can say that the Poincaré section "freezes the motion" of the dynamical system.

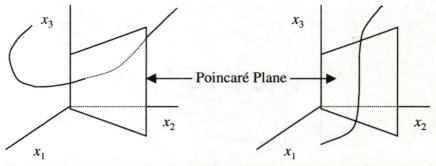

Fig. 4.4. A Poincaré section for a three–dimensional state space. On the left the trajectory crosses the Poincaré plane transversely. On the right the intersection is not transverse because the trajectory runs parallel to the plane for some distance.

For the sake of concreteness, let us consider the latter case with periodic forcing with independent variables X_0 and X_1. See Fig. 4.5. (All of our results will apply equally well to the autonomous case.) The Poincaré section is formed by recording the values of X_0 and X_1 whenever the phase of the periodic force reaches some definite value. (As usual, we restrict the phase to be between 0 and 2π.) Therefore, we actually have a three-dimensional state space generated geometrically by rotating the X_0X_1 plane about some axis (see Fig. 4.5).

If after transients have died away, the asymptotic behavior of the system is periodic with a period equal to the period of the force $g(t)$, then the Poincaré section record will consist of a single point whose coordinates in the plane we label x_1^* and x_2^*. On the other hand, if the long-term behavior is a subharmonic of the periodic force, say with period $T = N\,T_f$, (where N is a positive integer) then, in general, the Poincaré section record will consist of N points whose coordinates can be labeled $(x_1^*, x_2^*)_i$, $i = 1,\ldots, N$.

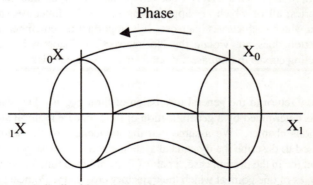

Fig. 4.5. For a periodically driven system, the state space is a three–dimensional cylinder generated by rotating the X_1–X_0 plane about an axis. An elliptical trajectory in the X_0–X_1 plane resides on the surface of a torus in the three–dimensional state space. Here the torus is shown in a cross sectional view. (For this diagram, the coordinates are labeled X_1 and X_0 so the rotation of the X_0–X_1 plane can be made more obvious typographically.) The phase of the driving force indicates the circumferential location of a trajectory point.

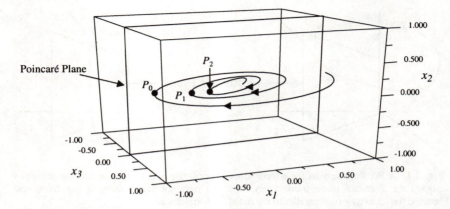

Fig. 4.6. The sequence of points P_0, P_1, P_2, . . . is the record of successive intersections of a single trajectory with the Poincaré plane (the plane with $x_3 = 0$) as the trajectory goes from $x_3 > 0$ to $x_3 < 0$.

In later discussions, it will be useful to indicate on the Poincaré section the record of trajectory intersections with the plane as trajectories approach or diverge from the periodic points. For example, Fig. 4.6 shows a sequence of points P_0, P_1, P_2, . . . as a trajectory approaches an attracting limit cycle in a three-dimensional state space. (Compare Fig. 4.6 with Fig. 3.13.) The reader should be warned that in some diagrams found in the literature this series of dots will be connected with a smooth curve intersecting (x_1^*, x_2^*). It is important to remember that this curve is not a trajectory. In fact the Poincaré intersection of any single trajectory is just a sequence of points as shown in Fig. 4.6. If a smooth curve is drawn on this kind of diagram, it represents the intersection points of an infinite family of trajectories, all of which are approaching (x_1^*, x_2^*). Later we shall see cases in which such curves intersect. It is important to remember that this intersection does not violate the No-Intersection Theorem because the intersecting curves in this case are not themselves trajectories.

We now return to the general discussion of limit cycles. The stability of the limit cycle is determined by a generalization of the Poincaré multipliers introduced in the previous chapter. We assume that the uniqueness of the solutions to the equations used to describe the dynamical system entails the existence of a Poincaré map function (or in the present case, a pair of Poincaré map functions), which relate the coordinates of one point at which the trajectory crosses the Poincaré plane to the coordinates of the next (in time) crossing point. (Again we assume we have chosen a definite crossing sense; e.g., from top to bottom, or from left to right.) These functions take the form

$$x_1^{(n+1)} = F_1(x_1^{(n)}, x_2^{(n)})$$
$$x_2^{(n+1)} = F_2(x_1^{(n)}, x_2^{(n)})$$

(4.6-1)

where the parenthetical superscript indicates the crossing point number.

Here these Poincaré map functions have arisen from the consideration of a Poincaré section for trajectories arising from a set of differential equations. In Chapter 5, we shall consider such map functions as interesting models in their own right, independent of this particular heritage.

The fixed points of the Poincaré section are those points that satisfy

$$x_1^* = F_1(x_1^*, x_2^*)$$
$$x_2^* = F_2(x_1^*, x_2^*)$$

(4.6-2)

Each fixed point in the Poincaré section corresponds to a limit cycle in the full three-dimensional state space.

We can characterize the stability of these fixed points by finding the characteristic values of the associated Jacobian matrix of derivatives [sometimes called the *Floquet matrix*, after Gaston Floquet (1847–1920), a French mathematician who studied, among other things, the properties of differential equations with periodic terms]. This matrix is analogous to the Jacobian matrix used to determine the characteristic values of a fixed point in the full state space. The Jacobian matrix *JM* is given by

$$JM = \begin{pmatrix} \dfrac{\partial F_1}{\partial x_1} & \dfrac{\partial F_1}{\partial x_2} \\ \dfrac{\partial F_2}{\partial x_1} & \dfrac{\partial F_2}{\partial x_2} \end{pmatrix}$$

(4.6-3)

where the matrix is to be evaluated at the Poincaré map fixed point in question. The characteristic values of this matrix determine the stability of the limit cycle. A stable limit cycle attracts nearby trajectories, while an unstable limit cycle repels nearby trajectories. In principle, we can use the mathematical methods given in Chapter 3 to find these characteristic values. In practice, however, we most often cannot find these characteristic values explicitly, since, to do that, we would need to know the exact form of the Poincaré map function, and in most cases, we do not know that function. [In Chapter 5, we will examine some models that do give us the map function directly. However, for systems described by differential equations in state spaces of three (or more) dimensions, it is in general impossible to find the map functions.]

Since the Jacobian matrix is a 2×2 matrix for a Poincaré section in a three-dimensional state space, the fixed point has two characteristic values. Hence, we have the same set of stability cases here that we had for fixed points in a two-dimensional state space, with one addition: The intersection points may alternate from one side of the fixed point to the other. (Recall that this alternation was not possible in two dimensions because the trajectory would have to cross itself. In three dimensions the trajectory can wind over and under itself to give the alternation without intersection.)

Dissipation

For a 2×2 matrix, there are two characteristic values. We denote the characteristic values as M_1 and M_2 since we use them as *Floquet multipliers* in determining how trajectories approach or diverge from the Poincaré intersection point of the limit cycle. Just as for Poincaré sections in a two-dimensional state space, the criterion for dissipation can be formulated in terms of the multipliers since dissipation is linked to the contraction of clusters of initial conditions. Because M_1, the first multiplier, determines the expansion in the x_1 direction and M_2 the expansion in the x_2 direction, we see that the product $M_1 M_2$ determines the expansion or contraction of *areas* in the Poincaré plane. For a dissipative system, we must have $M_1 M_2 < 1$ on the average (not only near the fixed points). In Chapter 8, we shall consider model map systems that preserve state-space area. They have $M_1 M_2 = 1$.

Stability of Limit Cycles

As we saw in two-dimensional systems, if the fixed point is to be stable and have trajectories in its neighborhood attracted to it, then the *absolute value* of each multiplier must be less than 1. [In state spaces with three or more dimensions, we can have $M < 0$; so the stability criterion is formulated using the absolute value of the multipliers.]

The types of limit cycles are

    I.   **Stable limit cycle** (node for the Poincaré map)
    II.   **Repelling limit cycle** (repellor for the Poincaré map)
    III.   **Saddle cycle** (saddle point for the Poincaré map)

Table 4.2 lists the categories of characteristic multipliers, the associated Poincaré plane fixed points and the corresponding limit cycles for three-dimensional state spaces. (Compare this table to Table 3.4 for limit cycles in two-dimensional state spaces.)

Table 4.2
Characteristic Multipliers for Poincaré Sections
of Three-Dimensional State Spaces

Type of Fixed Point	Characteristic Multiplier	Corresponding Cycle
Node	$\lvert M_1 \rvert, \lvert M_2 \rvert < 1$	Limit Cycle
Repellor	$\lvert M_1 \rvert, \lvert M_2 \rvert > 1$	Repelling Cycle
Saddle	$\lvert M_1 \rvert < 1, \lvert M_2 \rvert > 1$	Saddle Cycle

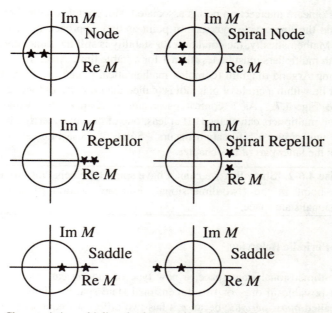

Fig. 4.7. Characteristic multipliers in the complex plane. If both multipliers lie within a circle of unit radius (the unit circle), then the corresponding limit cycle is stable. If one (or both) of the multipliers lies outside the unit circle, then the limit cycle is unstable.

Exercise 4.6-1. Let Tr be the trace of the Jacobian matrix in Eq. 4.6-3 and Δ its determinant.

(a) Show that the two characteristic multipliers can be expressed as

$$M_{1,2} = \frac{1}{2}[Tr \pm \sqrt{Tr^2 - 4\Delta}]$$

(b) Show that the expansion (contraction) factor for state space areas is given by

$$M_1 M_2 = \Delta$$

(c) In Chapter 3, we found that the <u>trace</u> of the Jacobian matrix is the signature of dissipation for a dynamical system modeled by a set of differential equations. Here, we use the <u>determinant</u> of the Jacobian matrix. Explain the difference.

Of course, the characteristic multipliers could also be complex numbers. Just as we saw for fixed points in a two-dimensional state space, the complex multipliers will form a complex-conjugate pair. In more graphic terms, the successive Poincaré intersection points associated with complex-valued multipliers rotate around the limit cycle intersection point as they approach or diverge from that point. Mathematically, the condition for stability is still the same: the absolute value of both multipliers must be less than 1 for a stable limit cycle. In terms of the corresponding Argand diagram (complex mathematical plane), both characteristic values must lie within a circle of unit radius (called the ***unit circle***) for a stable limit cycle. See Fig. 4.7. As a control parameter is changed the values of the characteristic multipliers can change. If at least one of the characteristic multipliers crosses the unit circle, a bifurcation occurs. Some of these bifurcations will be discussed in the latter part of this chapter.

> **Exercise 4.6-2.** Show that we cannot have spiral type behavior around a saddle point in the (two-dimensional) Poincaré section of a three-dimensional state space.

4.7 Quasi-Periodic Behavior

For a three-dimensional state space, a new type of motion can occur, a type of motion not possible in one- or two-dimensional state spaces. This new type of motion is called *quasi-periodic* because it has two different frequencies associated with it; that is, it can be analyzed into two independent, periodic motions. For quasi-periodic motion, the trajectories are constrained to the surface of a torus in the three-dimensional state space. A mathematical description of this kind of motion is given by:

$$x_1 = (R + r\sin\omega_r t)\cos\omega_R t$$
$$x_2 = r\cos\omega_r t \qquad\qquad (4.7\text{-}1)$$
$$x_3 = (R + r\sin\omega_r t)\sin\omega_R t$$

where the two angular frequencies are denoted by ω_R and ω_r. Geometrically, Eqs. (4.7-1) describe motion on the surface of a torus (with the center of the torus at the origin), whose large radius is R and whose cross-sectional radius is r. In general the torus (or doughnut-shape or the shape of the inner tube of a bicycle tire) will look something like Fig. 4.8. The frequency ω_R corresponds to the rate of rotation around the large circumference with a period $T_R = 2\pi/\omega_R$, while the frequency ω_R corresponds to the rate of rotation about the cross section with $T_r = 2\pi/\omega_r$. A general torus might have elliptical cross sections, but the ellipses can be made into circles by suitably rescaling the coordinate axes.

The Poincaré section for this motion is generated by using a Poincaré plane that cuts through the torus. What the pattern of Poincaré map points looks like depends on the numerical relationship between the two frequencies as illustrated in Fig. 4.9. If the ratio of the two frequencies can be expressed as the ratio of two integers (that is, as a "rational fraction," 14/17, for example), then the Poincaré section will consist of a finite number of points. This type of motion is often called

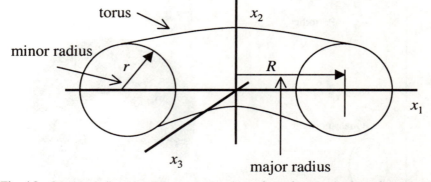

Fig. 4.8. Quasi-periodic trajectories roam over the surface of a torus in three–dimensional state space. Illustrated here is the special case of a torus with circular cross sections. r is the minor radius of the cross section. R is the major radius of the torus. A periodic trajectory on the surface of the torus would close on itself. On the right. a perspective view of a torus and a Poincaré plane.

frequency-locked motion because one of the frequencies is locked, often over a finite control parameter range, so that an integer multiple of one frequency is equal to another integer multiple of the other. (The terms *phase-locking* and *mode-locking* are also used to describe this behavior.)

If the ratio of frequencies cannot be expressed as a ratio of integers, then the ratio is called "irrational" (in the mathematical, not the psychological sense). For the irrational case, the Poincaré map points will eventually fill in a continuous curve in the Poincaré plane, and the motion is said to be *quasi-periodic* because the motion never exactly repeats itself. (Russian mathematicians call this *conditionally periodic*. See, for example, [Arnold, 1983]. The term *almost periodic* is also used in the mathematical literature.)

In the quasi-periodic case the motion, strictly speaking, never exactly repeats itself (hence, the modifier *quasi*), but the motion is not chaotic; it is composed of two (or more) periodic components, whose presence could be made known by measuring the frequency spectrum (Fourier power spectrum) of the motion. We should point out that detecting the difference between quasi-periodic motion and motion with a rational ratio of frequencies, when the integers are large, is a delicate question. Whether a given experiment can distinguish the two cases depends on the resolution of the experimental equipment. As we shall see later, the behavior of the system can switch abruptly back and forth between the two cases as a parameter of the system is varied. The important point is that the attractor for the system is a two-dimensional <u>surface</u> of the torus for quasi-periodic behavior.

This notion can be generalized to higher-dimensional state spaces. For example, quasi-periodic motion in a four-dimensional state space may be characterized by three frequencies, none of which are related by a rational ratio to any of the others. The trajectories then wander completely over the "surface" of a three-dimensional torus. If there are only two frequencies, the motion would of course be restricted to the two-dimensional surface of a torus.

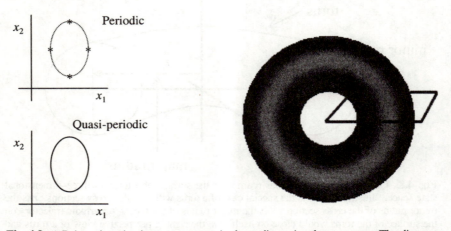

Fig. 4.9. A Poincaré section intersects a torus in three–dimensional state space. The diagram on the upper left shows the Poincaré map points for a two-frequency periodic system with a rational ratio of frequencies. The intersection points are indicated by asterisks. The diagram on the lower left is for quasi-periodic behavior. The ratio of frequencies is irrational, and eventually the intersection points fill in a curve (sometimes called a "drift ring") in the Poincaré plane.

We have now seen the full panoply of regular (nonchaotic) attractors: fixed points (dimension 0), limit cycles (dimension 1), and quasi-periodic attractors (dimension 2 or more). We are ready to begin the discussion of how these attractors can change into chaotic attractors.

We will give only a brief discussion of the period-doubling, quasi-periodic, and intermittency routes. These will be discussed in detail in Chapters 5, 6, and 7, respectively. A discussion of crises will be found in Chapter 7. As we shall see, the chaotic transient route is more complicated to describe because it requires a knowledge of what trajectories are doing over a range of state space. We can no longer focus our attention locally on just a single fixed point or limit cycle.

4.8 The Routes to Chaos I: Period-Doubling

As we discussed earlier, the period-doubling route begins with limit cycle behavior of the system. This limit cycle, of course, may have been "born" from a bifurcation involving a node or other fixed point, but we need not worry about that now. As some control parameter changes, this limit cycle becomes unstable. Again this event is best viewed in the corresponding Poincaré section. Let us assume that the periodic limit cycle generates a single point in the Poincaré section. If the limit cycle becomes unstable by having one of its characteristic multipliers become more negative than -1 (which, of course, means $|M| > 1$), then, in many situations, the new motion remains periodic but has a period twice as long as the period of the original motion. In the Poincaré section, this new limit cycle exhibits two points, one on each side of the original Poincaré section point (see Fig. 4.10).

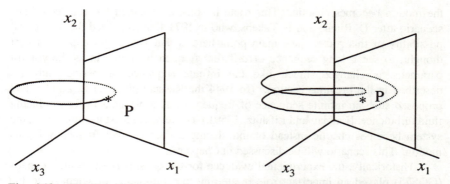

Fig. 4.10. The Poincaré section of a trajectory that has undergone a period-doubling bifurcation. On the left is the original periodic trajectory, which intersects the Poincaré plane in one point. On the right is the period-doubled trajectory, which intersects the Poincaré plane in two points, one on each side of the original intersection point.

This alternation of intersection points is related to the characteristic multiplier associated with the <u>original</u> limit cycle, which has gotten more negative than −1. Since $|M| > 1$, the trajectory's map points are now being "repelled" by the original map point. The minus sign tells us that they alternate from one side to the other, as we can see formally from Eq. (3.16-6). This type of bifurcation is also called a *flip bifurcation* because the newly born trajectory flips back and forth from one side of the original trajectory to the other.

> **Question**: Why don't we see period-tripling, quadrupling, etc.? Is there a simple explanation? See Chapter 5.

As the control parameter is changed further, this period-two limit cycle may become unstable and give birth to a period-four cycle with four Poincaré intersection points. Chapter 5 will examine in detail how, when, and where this sequence occurs. The period-doubling process may continue until the period becomes infinite; that is, the trajectory never repeats itself. The trajectory is then chaotic.

4.9 The Routes to Chaos II: Quasi-Periodicity

In the quasi-periodic scenario, the system begins again with a limit cycle trajectory. As a control parameter is changed, a second periodicity appears in the behavior of the system. This bifurcation event is a generalization of the *Hopf bifurcation* discussed in Chapter 3; so, it is also called a Hopf bifurcation. In terms of the characteristic multipliers, the Hopf bifurcation is marked by having the two complex-conjugate multipliers cross the unit circle simultaneously.

If the ratio of the period of the second type of motion to the period of the first is not a rational ratio, then we say, as described previously, that the motion is quasi-periodic. Under some circumstances, if the control parameter is changed further, the motion becomes chaotic. This route is sometimes called the Ruelle–Takens scenario after D. Ruelle and F. Takens, who in 1971 first suggested the theoretical possibility of this route. The main point here is that you might expect, at first thought, to see a long sequence of different frequencies come in as the control parameter is changed, much like the infinite sequence of period-doublings described in the previous section. (In 1944 the Russian physicist L. Landau had proposed such an infinite sequence of frequencies as a mechanism for producing fluid turbulence. [Landau and Lifshitz, 1959].) However, at least in some cases, the system becomes chaotic instead of introducing a third distinct frequency for its motion. This scenario will be discussed in Chapter 6.

Historically, the experimental evidence for the quasi-periodic route to chaos (GOS75) played an important role in alerting the community of scientists to the utility of many of the newly emerging ideas in nonlinear dynamics. During the late 1970s and early 1980s there were many theoretical conjectures about the necessity of the transition from two-frequency quasi-periodic behavior to chaos. More recent work (see for example, BAT88) has shown that systems with significant spatial extent and with more degrees of freedom can have quasi-periodic behavior with three or more frequencies before becoming chaotic.

4.10 The Routes to Chaos III: Intermittency and Crises

Chapter 7 contains a detailed discussion of intermittency and crises; so, we will give only the briefest description here. The intermittency route to chaos is characterized by dynamics with irregularly occurring bursts of chaotic behavior interspersed with intervals of apparently periodic behavior. As some control parameter of the system is changed, the chaotic bursts become longer and occur more frequently until, eventually, the entire time record is chaotic.

A crisis is a bifurcation event in which a chaotic attractor and its basin of attraction suddenly disappear or suddenly change in size as some control parameter is adjusted. Alternatively, if the parameter is changed in the opposite direction, the chaotic attractor can suddenly appear "out of the blue" or the size of the attractor can suddenly be reduced. As we shall see in Chapter 7, a crisis event involves the interaction between a chaotic attractor and an unstable fixed point or an unstable limit cycle.

4.11 The Routes to Chaos IV: Chaotic Transients and Homoclinic Orbits

In our second broad category of routes to chaos, the global bifurcation category, the chaotic transient route is the most important for systems modeled by sets of ordinary differential equations. Although the chaotic transient route to chaos was one of the first to be recognized in a model of a physical system (the Lorenz model), the theory of this scenario is, in some ways, less well developed than the theory for period-doubling, quasi-periodicity, and intermittency. This lack of development is due to the fact that this transition to chaos is not (usually) marked

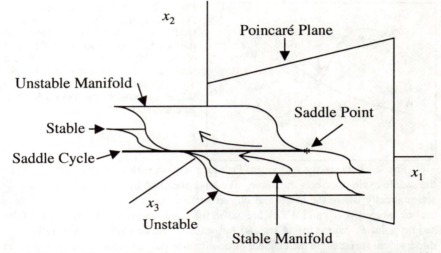

Fig. 4.11. A saddle cycle in a three–dimensional state space. The stable and unstable manifolds are surfaces that intersect at the saddle cycle. Where the saddle cycle intersects a Poincaré plane we have a saddle point for the Poincaré map function. A portion of one trajectory, repelled by the saddle cycle, is shown on the unstable manifold, and a portion of another trajectory, approaching the saddle cycle, is shown on the stable manifold.

by any change in the fixed points of the system or the fixed points of a Poincaré section. The transition is due to the interaction of trajectories with various unstable fixed points and cycles in the state space. The common features are the so-called **homoclinic** orbits and their cousins, **heteroclinic** orbits. These special orbits may suddenly appear as a control parameter is changed. More importantly, these orbits strongly influence the nature of other trajectories passing near them.

What is a homoclinic orbit? To answer this question, we need to consider *saddle cycles* in a three-dimensional state space. (These ideas carry over in a straightforward fashion to higher-dimensional state spaces.) You should recall that saddle points and saddle cycles and, in particular, their in-sets and out-sets serve to organize the state space. That is, the in-sets and out-sets serve as "boundaries" between different parts of the state space and all trajectories must respect those boundaries. We will focus our attention on a saddle point in the Poincaré section of the state space. This saddle point corresponds to a saddle cycle in the original three-dimensional state space (see Fig. 4.11). We can consider the saddle cycle to be the intersection between two surfaces: One surface is the in-set (that is, the *stable manifold*) associated with the cycle. The other surface is the out-set (*unstable manifold*) associated with the cycle. Trajectories on the in-set approach the saddle cycle as time goes on. Trajectories on the out-set diverge from the cycle as time goes on. Trajectories that are near to, but not on, the in-set will first approach the cycle and then be repelled roughly along the out-set (unstable manifold).

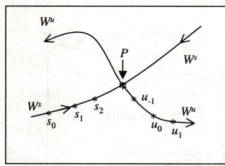

Fig. 4.12. Point P is a saddle point in a Poincaré section. It corresponds to a saddle cycle in the full three–dimensional state space. The intersection of the in-set of the saddle cycle with the Poincaré plane generates the curve labeled W^s. The intersection of the out-set of the saddle cycle with the Poincaré plane generates curve W^u.

Fig. 4.12 shows the equivalent Poincaré section with a saddle point P where the saddle cycle intersects the plane. We have sketched in some curves to indicate schematically where the in-set and out-set surfaces cut the Poincaré plane. These curves, labeled $W^s(P)$ and $W^u(P)$, are called the *stable* and *unstable manifolds* of the saddle point P. Since the in-set and out-set of a saddle cycle are generally two-dimensional surfaces (in the original three-dimensional state space), the intersection of one of these surfaces with the Poincaré plane forms a curve. It is crucial to realize that these curves are <u>not</u> trajectories. For example, if we pick a point s_0 on $W^s(P)$, a point at which some trajectory intersects the Poincaré plane, then the Poincaré map function F gives us s_1, the coordinates of the point at which the trajectory next intersects the plane. From s_1, we can find s_2, and so on. The sequence of points lies along the curve labeled $W^s(P)$ and approaches P as $n \to \infty$. Similarly, if u_0 is a point along $W^u(P)$, then $F(u_0) = u_1$, $F(u_1) = u_2$, and so on, generates a series of points that diverges from P along $W^u(P)$. If we apply the inverse of the Poincaré map function $F^{(-1)}$ to u_0, we generate a sequence of points u_{-1}, u_{-2}, and so on, that approaches P as $n \to -\infty$. The curves drawn in Fig. 4.12 represent the totality of such sequences of points taken with infinitely many starting points. For any one starting point, however, the sequence jumps along W^s or W^u, it does <u>not</u> move smoothly like a point on a trajectory in the original state space.

As a control parameter is changed, it is possible for $W^s(P)$ and $W^u(P)$ to approach each other and in fact to intersect, say, at some point q. If this intersection occurs, we say that we have a ***homoclinic intersection*** at q, and the point q is called a ***homoclinic (intersection) point***. It is also possible for the unstable manifold of one saddle point to intersect the stable manifold from some other saddle point. In that case we say we have a ***heteroclinic intersection***. Other heteroclinic combinations are possible. For example, we could have the intersection of the unstable manifold surface of an index-2 saddle point and the stable manifold of a saddle cycle. (For a nice visual catalog of the possible kinds of intersections, see [Abraham and Shaw, 1992][Abraham, Abraham, and Shaw, 1996].) For now we will concentrate on homoclinic intersections.

We now come to an important and crucial theorem:

If the in-sets and out-sets of a saddle point in the Poincaré section of a dynamical system intersect at one homoclinic intersection point q_0, then there must be an infinite number of homoclinic intersections.

To prove this statement, we consider the result of applying the mapping function F to q_0. We get another point q_1. Since q_0 belongs to both W^s and W^u, so must q_1, since we have argued in the previous paragraphs that applying F to a point on W^s or W^u generates another point on W^s or W^u. By continuing to apply F to this sequence of points, we generate an infinite number of homoclinic points. Fig. 4.13 shows part of the resulting *homoclinic tangle*, which must result from the homoclinic intersections.

Please note that the smooth curves drawn in Fig. 4.13 are <u>not</u> individual trajectories. (We cannot violate the No-Intersection Theorem!) The smooth curves are generated by taking infinitely many starting points on W^s and W^u. Only those trajectories that hit one of the homoclinic points will hit (some of) the other homoclinic points.

What is the dynamical significance of a homoclinic point and the related homoclinic tangle? If we now shift our attention back to the original three-dimensional state space, we see that a homoclinic point in the Poincaré section corresponds to a continuous trajectory in the original state space. When a homoclinic intersection occurs, one trajectory on the unstable manifold joins another trajectory on the stable manifold to form a <u>single</u> new trajectory whose Poincaré intersection points are the homoclinic points described earlier. (To help visualize this process, recall that the in-set and out-set of a saddle cycle in the three-dimensional state space are, in general, two-dimensional surfaces.) This new trajectory connects the saddle point to itself and hence is called a ***homoclinic trajectory*** or ***homoclinic orbit*** and is said to form a ***homoclinic connection***. As our previous theorem states, this homoclinic trajectory must intersect the Poincaré plane an infinite number of times.

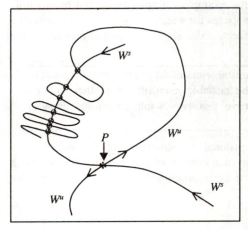

Fig. 4.13. A homoclinic tangle results from the homoclinic intersection of the unstable manifold $W^u(P)$ with the stable manifold $W^s(P)$ of the saddle point P. Each of the circled points is a homoclinic (intersection) point. For clarity's sake, only a portion of the tangle is shown.

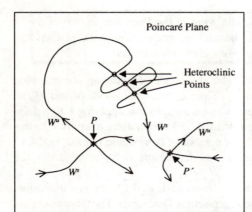

Fig. 4.14. The Poincaré section of a heteroclinic connection. Two saddle cycles intersect the plane at P and P' respectively. A heteroclinic orbit links together two saddle cycles forming a heteroclinic connection. For the sake of clarity only the part of the tangle involving the unstable manifold of P intersecting the stable manifold of P' is shown

How does a homoclinic orbit lead to chaotic behavior? To understand this, we need to consider other trajectories that come near the saddle point of the Poincaré section. Generally speaking, the trajectories approach the saddle point close to (but not on) the in-set (stable manifold), but they are then forced away from the saddle point near the out-set (unstable manifold). After a homoclinic tangle has developed, a trajectory will be pushed away from the saddle point by the out-set part of the tangle, but it will be pulled back by the in-set part. It is easy to see that the homoclinic tangle can lead to trajectories that seem to wander randomly around the state space region near the saddle point.

The same general type of behavior can result from a *heteroclinic orbit*, which connects one saddle point (or saddle cycle in the original state space) to another. A second heteroclinic orbit takes us from the second saddle point back to the first. When such a combined trajectory exists, we say we have a *heteroclinic connection* between the two saddle cycles. Figure 4.14 shows schematically a part of a heteroclinic connection in a Poincaré section of a three-dimensional state space. It is also possible to have heteroclinic orbits that link together sequentially three or more saddle cycles. Figure 4.15 shows examples of homoclinic and heteroclinic connections in three-dimensional state spaces for which the in-sets and out-sets of a saddle cycle are two-dimensional surfaces. Also shown are partial pictures of the resulting Poincaré sections.

> **Exercise 4.11-1.** Prove that the unstable manifold of one saddle point (or saddle cycle) cannot intersect the unstable manifold of another saddle point (or cycle). Similarly, prove that two stable manifolds cannot intersect.

Figure 4.16 shows three-dimensional constructions of homoclinic and heteroclinic tangles resulting from the intersections of the in-sets and out-sets of saddle cycles. Partial diagrams of the corresponding Poincaré sections are also shown.

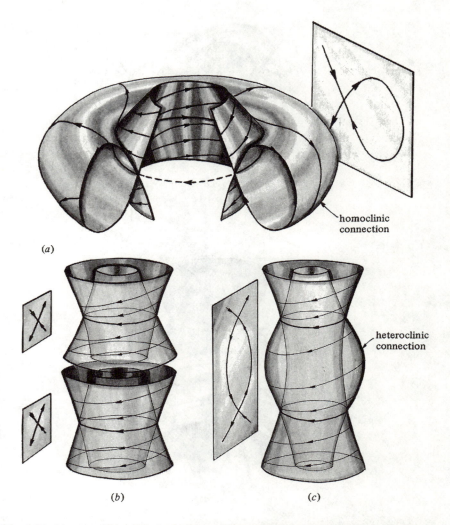

Fig. 4.15. Diagrams illustrating the formation of homoclinic and heteroclinic connections in a three–dimensional state spaces. In (a) the in-set and out-set of a saddle cycle join smoothly to form a homoclinic connection. In (b) and (c), the out-set of one saddle cycle joins the in-set of a second saddle cycle to form a heteroclinic connection. In both cases, examples of two–dimensional Poincaré sections are shown. (From [Ottino, 1989])

As an example of how homoclinic and heteroclinic connections affect dynamics, let us return to the Lorenz model introduced in Chapter 1. This model provides a nice example of chaotic transients due to homoclinic and heteroclinic connections eventually leading to chaotic behavior. A homoclinic connection is formed when the parameter r is near 13.93 (with $b = 8/3$ and $p = 10$, as in Chapter 1). At that parameter value, the one-dimensional out-set of the fixed point at the

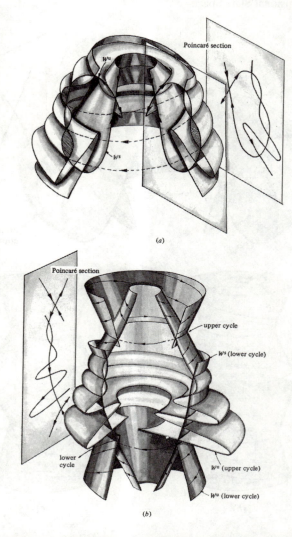

(a)

(b)

Fig. 4.16. A three–dimensional representation of the homoclinic (top) and the heteroclinic (bottom) tangles resulting from the intersections of the in-set and out-set surfaces associated with saddle cycles. Also shown are sketches of parts of the two–dimensional Poincaré sections resulting from the tangles. (From [Ottino,1989])

origin, which is now a saddle point, touches, over its entire length, the two-dimensional in-set of that same saddle point. Actually, a double homoclinic connection is formed because there are two branches of the one-dimensional out-set, one leading to one of the off-origin fixed points, the other leading to the other off-origin fixed point. Trajectories passing near the homoclinic connections are successively repelled by and attracted to the saddle point at the origin many times while wandering back and forth between regions around the two off-origin fixed

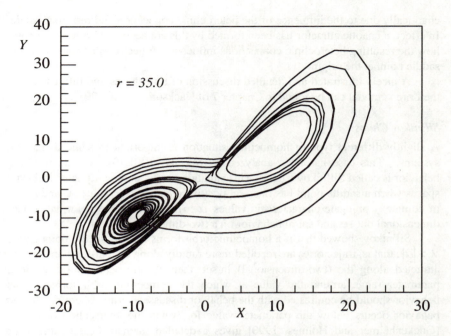

Fig. 4.17. An illustration of heteroclinic behavior in the Lorenz model at $r = 35.0$. The XY plane projection of the trajectory is shown. The trajectory starts near the saddle point at the origin. The out-set of the origin connects to the in-set of one of the saddle cycles near the fixed point at $X = Y = +9.5$. The trajectory then leaves that saddle cycle on its out-set which connects to the in-set of the other saddle cycle. After orbiting near the fixed point at $X = Y = -9.5$, the trajectory heads back toward the origin on the out-set of the saddle cycle. Near the origin, it is again repelled.

points before finally settling into one of the off-origin fixed points. Such behavior looks chaotic, but because it is really only transient behavior, it is called *transient chaos*. For more information about transient chaos, a fascinating topic in its own right, see the references at the end of this chapter.

When this homoclinic connection occurs, two saddle cycles are also created. These saddle cycles play an essential role in the development of the chaotic attractor in the Lorenz model. As r increases beyond 13.93, the saddle cycles, which surround the off-origin fixed points (which themselves are spiral nodes), begin to decrease in size and contract around the spiral nodes. At $r = 24.74$, the real part of the complex eigenvalues of the spiral nodes goes to 0, and the saddle cycles collapse onto the nodes. Before this, however, at $r < 24.06$, the out-set of the saddle point at the origin falls outside the saddle cycles. For $r > 24.06$ the out-set falls inside the saddle cycles; therefore, at $r = 24.06$ (approximately), the out-set of the saddle point at the origin must intersect the saddle cycles to form a heteroclinic connection. For $r < 24.74$, there are still two (small) basins of attraction near the two spiral nodes, but trajectories starting outside these two small basins wander chaotically due to the influence of the heteroclinic connection and resulting tangle. In effect, a chaotic attractor has been formed by this connection. Figure 4.17 shows how the resulting heteroclinic connections influence a trajectory that starts near the saddle point at the origin.

A nicely illustrated and detailed discussion of trajectories and bifurcations in the Lorenz model can be found in Chapter 7 of [Jackson, Vol. 2, 1991].

Sil'nikov Chaos

A slightly different type of homoclinic situation is important in some dynamical systems. This case has been analyzed by Sil'nikov (SIL70), and the resulting behavior is called Sil'nikov Chaos. This situation occurs in three-dimensional state spaces when a saddle point has one real positive characteristic value λ and a pair of complex conjugate characteristic values $\alpha \pm i\beta$: The saddle point has a one-dimensional out-set and spiral behavior on a two-dimensional in-set.

Sil'nikov showed that if a homoclinic orbit forms for this saddle point and if $\lambda > |\alpha|$, that is, trajectories are repelled more rapidly along the out-set than they are attracted along the (two-dimensional) in-set, then chaotic behavior occurs in a parameter range around the value at which the homoclinic orbit forms. This behavior should be contrasted with the behavior discussed earlier where no chaotic behavior occurs below the parameter value for which the intersection develops. [Guckenheimer and Holmes, 1990] gives a detailed mathematical treatment of Sil'nikov chaos.

4.12 Homoclinic Tangles and Horseshoes

A very elegant and useful geometric model of the effect of homoclinic and heteroclinic tangles on state-space orbits is the Smale horseshoe. This equestrian metaphor was introduced by the mathematician Stephen Smale (SMA67) to capture the essence of the effects of homoclinic tangles on dynamical systems. The horseshoe construction has the additional benefit of providing a scheme that allows mathematical proofs of many important aspects of the dynamics of the system. We shall introduce the basic horseshoe idea here. In Chapter 5, we will take up the mathematical results from this construction.

To understand Smale's construction, let us consider a small rectangle of initial points surrounding a saddle point in the Poincaré section of a dynamical system. As the system evolves, this rectangle of points will tend to be stretched out along the unstable manifold direction W^u and compressed along the W^s direction. The rectangle will eventually reach the tangled region of W^u shown in Fig. 4.13, and its shape will resemble a horseshoe. As the system evolves further, this horseshoe will in fact eventually overlap with the original rectangle. Smale constructed a mapping function, now known as the Smale horseshoe map, which captures the essence of this process.

In the Smale horseshoe map, a square of initial points is first stretched in one direction and compressed in the orthogonal direction. The now elongated rectangle

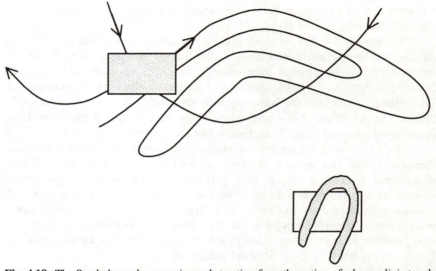

Fig. 4.18. The Smale horseshoe map is an abstraction from the action of a homoclinic tangle on a rectangle of initial conditions. In the upper part of the figure a rectangle of initial conditions is shown superposed on part of a homoclinic tangle. Under the evolution of the system that rectangle will be stretched along the unstable manifold direction and compressed along the stable manifold direction. In the lower part of the figure is Smale's abstraction of that effect in the shape of a horseshoe superposed on the original rectangle.

is folded and overlaid on the initial square (see Fig. 4.18). The process is iterated, and one looks for those points that remain within the area of the initial square as the number of iterations goes to infinity. This stretching in one direction, compressing in another, combined with the folding, mimics the effect of the homoclinic tangle on trajectories in the dynamical system. The famous Smale–Birkhoff Homoclinic Theorem proves [Guckenheimer and Holmes, 1990] that having a homoclinic tangle guarantees that the system will have "horseshoe dynamics."

> **Smale–Birkoff Theorem**: If a dynamical system described by a set of ordinary differential equations has a homoclinic intersection, then the system has intersections with a Poincaré plane whose behavior is described by a horseshoe map.

Stretching, Compression and Folding

Although the original Smale horseshoe map does not have an attractor and hence cannot be a model for the chaotic behavior of a dissipative system, many authors have decided to equate chaotic behavior with horseshoe dynamics because the stretching (in at least one state space direction) gives rise to exponential divergence of nearby initial conditions. Certainly in the general sense of requiring stretching in one direction, compression in another, combined with folding to keep the system in a finite region of state space, horseshoe dynamics must be a general feature of all chaotic behavior.

As mentioned earlier, in many systems the stretching and folding is actually effected by heteroclinic connections, which link the unstable manifold of one saddle cycle (in the original state space, a saddle point in the corresponding Poincaré section) to the stable manifold of another saddle cycle or saddle point. The unstable manifold of the latter may then reconnect back to the original saddle cycle. The net effect of this heteroclinic cycle on a cluster of initial condition points is the same topologically as the effect of a homoclinic connection.

The effect of a homoclinic or heteroclinic tangle on a cluster of initial condition points can be seen elegantly in fluid mixing experiments. A two-dimensional fluid flow subject to a periodic perturbation may show chaotic trajectories for tracer particles suspended in the fluid. (We will explore this connection in more detail in Chapter 11.) Tracer particles injected near a saddle point (called a *hyperbolic point* in the fluid dynamics literature) show horseshoe type behavior with stretching, folding, and reinjection near the saddle point (see [Ottino, 1989] and OTT89 for beautiful pictures of these effects).

4.13 Lyapunov Exponents and Chaos

Our discussion of chaotic behavior has so far been qualitative. Now we want to introduce one method of quantifying chaotic behavior. There are two motivations here. First, we want some quantitative test for chaotic behavior; something that can, at least in principle, distinguish chaotic behavior from noisy behavior due to random, external influences. Second, we would like to have some quantitative measure of the degree of chaoticity; so, we can see how chaotic behavior changes as the system's parameters are changed. In this section, we will introduce Lyapunov exponents as one possible quantitative measure of chaos. In Chapter 5, we will describe how to find Lyapunov exponents for iterated maps. In Chapters 9 and 10, we will describe how to find Lyapunov exponents, as well as other quantifiers of chaos, from experimental data. In this section we will focus attention on dynamical systems described by a set of ordinary differential equations.

As we have seen in Section 3.7, a Lyapunov exponent is a measure of the rate of attraction to or repulsion from a fixed point in state space. In Section 4.2, we indicated that we could also apply this notion to the divergence of nearby trajectories in general at any point in state space. For a one-dimensional state space, let x_0 be one initial point and x a nearby initial point. Let $x_0(t)$ be the trajectory that arises from that initial point, while $x(t)$ is the trajectory arising from the other initial point. Then, if we follow the line of reasoning leading to Eq. (3.7-3), we can show that the "distance" s between the two trajectories, $s = x(t) - x_0(t)$ grows or contracts exponentially in time. Let us work through the details.

The time development equation is assumed to be

$$\dot{x}(t) = f(x) \tag{4.13-1}$$

Since we assume that x is close to x_0, we can use a Taylor series expansion to write

$$f(x) = f(x_0) + \frac{df(x)}{dx}\bigg|_{x_0} (x - x_0) + \dots \qquad (4.13\text{-}2)$$

We now find that the rate of change of distance between the two trajectories is given by

$$
\begin{aligned}
\dot{s} &= \dot{x} - \dot{x}_0 \\
&= f(x) - f(x_0) \qquad\qquad\qquad (4.13\text{-}3) \\
&= \frac{df}{dx}\bigg|_{x_0} (x - x_0)
\end{aligned}
$$

where we have kept only the first derivative term in the Taylor series expansion of $f(x)$. Since we expect the distance to change exponentially in time, we introduce the Lyapunov exponent λ as the quantity that satisfies

$$s(t) = s(t = 0)e^{\lambda t} \qquad (4.13\text{-}4)$$

If we differentiate Eq. (4.13-4) with respect to time, we find

$$
\begin{aligned}
\dot{s} &= \lambda s(t = 0)e^{\lambda t} \\
&= \lambda s
\end{aligned}
\qquad (4.13\text{-}5)
$$

Comparing Eq. (4.13-5) and Eq. (4.13-3) yields

$$\lambda = \frac{df(x)}{dx}\bigg|_{x_0} \qquad (4.13\text{-}6)$$

Thus we see that if λ is positive, then the two trajectories diverge; if λ is negative, the two trajectories converge.

In state spaces with two (or more) dimensions, we can associate a (local) Lyapunov exponent with the rate of expansion or contraction of trajectories for each of the directions in the state space. In particular, for three dimensions, we may define three Lyapunov exponents, which are the eigenvalues of the Jacobian matrix evaluated at the state space point in question. In the special case for which the Jacobian matrix has zeroes everywhere except for the principal (upper-left to lower-right) diagonal, the three eigenvalues (and hence the three local Lyapunov exponents) are given by

$$\lambda_1 = \frac{\partial f_1}{\partial x_1} \qquad (4.13\text{-}7\text{a})$$

$$\lambda_2 = \frac{\partial f_2}{\partial x_2} \qquad (4.13\text{-}7\text{b})$$

$$\lambda_3 = \frac{\partial f_3}{\partial x_3} \qquad (4.13\text{-}7c)$$

where the partial derivatives are evaluated at the state space point in question.

In practice, we know that the derivative of the time evolution function generally varies with x; therefore, we want to find an average of λ over the history of a trajectory. If we know the time evolution function, we simply evaluate the derivative of the time evolution function along the trajectory and find the average value. (For a dissipative, one-dimensional system, we know that this average Lyapunov exponent must be negative.)

We define a chaotic system to be a system which has at least one positive **average** Lyapunov exponent.

Behavior of a Cluster of Initial Conditions

To develop a geometric interpretation of the Lyapunov exponents, we consider a small rectangular volume of initial conditions with sides s_1, s_2, and s_3 surrounding this point with the sides oriented along the three state space axes. That volume will evolve in time as:

$$V(t) = s_1 s_2 s_3 e^{(\lambda_1 + \lambda_2 + \lambda_3)t} \qquad (4.13\text{-}8)$$

If we compare Eq. (4.13-8) with Eq. (3.13-6), we see that the sum of the three Lyapunov exponents gives us the mathematical *divergence* of the set of time evolution functions. Again, in practice, we are interested in the average of these Lyapunov exponents over the history of a trajectory. For a dissipative system, the average of the sum of the exponents must be negative.

For a three-dimensional state space system described by a set of three first-order differential equations, one of the average Lyapunov exponents must be 0 unless the attractor is a fixed point (HAK83). (The 0 value for a Lyapunov exponent corresponds to the negligible attraction or repulsion of trajectories starting from nearby points that could be carried into each other by following the same trajectory for a short time.) Thus, for a dissipative system, at least one of the remaining average Lyapunov exponents must be negative. If the system is chaotic, one of the Lyapunov exponents is positive for a three-dimensional state space.

In state spaces with four (or more) dimensions, we might have more than one positive average Lyapunov exponent. In those cases, we say we have *hyperchaos*. One possible route from periodic behavior to hyperchaotic behavior is discussed in HAL99.

It may be helpful to visualize what is happening with a more pictorial construction. For a dissipative system, we pick an initial point and let the resulting trajectory evolve until it is on the attracting region in state space. Then, we pick a trajectory point and construct a small sphere around it. Next, we follow the evolution of trajectories starting from initial points inside that sphere (some of

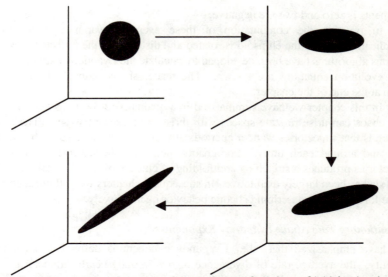

Fig. 4.19. A schematic representation of the evolution of a sphere of initial points in state space. The sphere starts in the upper-left-hand side. Time increases as we go clockwise around the figure. For a dissipative system, the volume associated with the set of initial points must go to 0. A chaotic system will exhibit exponential stretching of the sphere in at least one direction.

which may not be on the attractor). In general, the sphere may be stretched in one direction and contracted in others as time goes on as shown schematically in Fig. 4.19. The sphere will be distorted into an ellipsoid. (For a dissipative system, the volume must contract to 0 if we follow the system long enough.)

If we evaluate the (assumed) exponential rates of stretching and contraction for the different axes of the ellipsoid, we can find the Lyapunov exponents for that region of the attractor. Repeating this procedure along the trajectory allows us to calculate the set of average Lyapunov exponents for the system. This set of average Lyapunov exponents is called the *spectrum of Lyapunov exponents*. If at least one of the average Lyapunov exponents is positive, then we conclude that the system displays (on the average) divergence of nearby trajectories and is "truly" chaotic. Table 4.3 summarizes the relationship between the spectrum of Lyapunov exponents and the type of attractor. $(0, -, -,)$ means that one of the Lyapunov

Table 4.3

Spectra of Lyapunov Exponents and Associated Attractors
Three–dimensional State Space

Signs of λs	Type of Attractor
$(-,-,-)$	Fixed Point
$(0,-,-)$	Limit Cycle
$(0,0,-)$	Quasi-Periodic Torus
$(+,0,-)$	Chaotic

exponents is zero and two are negative.

In practice the computation of these average Lyapunov exponents is complicated because the ellipsoid is rotated and distorted as the trajectories evolve. Various algorithms have been developed to calculate the Lyapunov exponents if the time evolution equations are known. The reader should consult the references given at the end of the chapter.

In this chapter we have summarized in a qualitative way the general ideas of how chaos can arise in state spaces with three or more dimensions. The crucial feature is that trajectories wander aperiodically through the state space by winding over and around each other. For chaotic behavior, the divergence of nearby trajectories produces a stretching and folding of clusters of initial conditions. This analysis has been largely qualitative. In subsequent chapters we will build up more of the formalism for describing chaotic behavior quantitatively.

A Cautionary Tale About Lyapunov Exponents

We have emphasized that "the" Lyapunov exponent is defined as an average quantity: the average rate of divergence of two initially nearby trajectories. We can legitimately conclude that the behavior of the system is chaotic only if this average Lyapunov exponent is positive.

It is possible to find exponential divergence of nearby trajectories even for systems that are not chaotic in the strict sense of our definition. This pseudochaos occurs if the trajectories start off in the neighborhood of the out-set of a saddle point in state space. On the out-set of the saddle point, the characteristic exponent is positive. Trajectories that are near the out-set will have behaviors that are close to the behavior on the out-set since the characteristic values change smoothly through state space. Thus, two initially nearby trajectories in the neighborhood of the out-set of the saddle point will diverge exponentially with a local Lyapunov exponent close to the positive characteristic value associated with the out-set of the saddle point. The danger lies in assuming that the average Lyapunov exponent is positive based on following the trajectories for only a short time.

As an example of this kind of behavior, let us consider the simple pendulum, well known from introductory physics courses. The position of the pendulum is given by the angle θ between the pendulum rod and a vertical line. By convention, we choose $\theta = 0$ when the pendulum is hanging straight down. We describe the motion of the system by following trajectories in a state space whose coordinates are the angular velocity ($\dot{\theta}$) and the angle (θ). Using $u = \dot{\theta}$, we write the state space equations for the system as

$$\dot{u} = -\frac{g}{L}\sin\theta$$

$$\dot{\theta} = u$$

(4.13-9)

where, as usual, g is the local gravitational field strength (9.8 N/kg near Earth's surface) and L is the length of the pendulum.

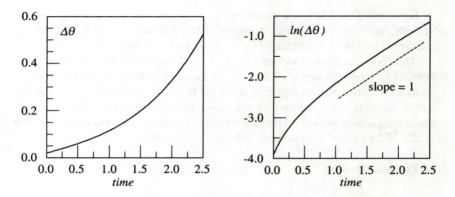

Fig. 4.20. On the left the difference between the angular positions of the simple pendulum (with $g/L = 1$) for two initially nearby trajectories is plotted as a function of time. One trajectory starts at $\theta_0 = 3.00$ radians and $d\theta/dt = 0.0724$ and the other at $\theta_0 = 3.02$ radians and $d\theta/dt = 0$. The natural log of those differences is plotted on the right. For exponential divergence, the plot on the right should be a straight line whose slope is the local Lyapunov exponent. The dashed line on the right has a slope of 1.

By using the methods of Chapter 3 for this two-dimensional state space, we see that the system has fixed points at $\dot{\theta} = 0$ and $\theta = 0, \pi$. The fixed point at $\theta = \pi$ is a saddle point with characteristic values

$$\lambda_+ = +\sqrt{\frac{g}{L}}$$

$$\lambda_- = -\sqrt{\frac{g}{L}}$$

(4.13-10)

Exercise 4.13-1. Verify the statements about fixed points and characteristic values for the simple pendulum.

Thus, we see that if two trajectories start in the neighborhood of the out-set of the saddle-point, they will be pulled toward the unstable manifold and then diverge (until they leave the neighborhood of the saddle point) with a characteristic exponent of $\sqrt{g/L}$. In Fig. 4.20, we have plotted the difference between the angular positions for two trajectories, one with initial angle $\theta_0 = 3.00$ radians, the other with $\theta_0 = 3.02$ radians. The angular velocity in the first case is 0.0724 radians per second while it is 0 in the second case. (This choice gives the pendulum the same initial energy in both cases.) (We have set $\sqrt{g/L} = 1$.) On the right of Fig. 4.20 is plotted the natural logarithm of the difference between the two angular positions as a function of time. You can see that after an initial transient, the divergence is exponential with a characteristic value of 1 (as expected) up to about $t = 2.5$.

However, if we follow the trajectories for a longer time, we would see that the difference between the angular positions later decreases, giving an average Lyapunov exponent of 0 as required for our model with no friction (no dissipation). If we had followed the trajectories only to $t = 2.5$, we might have (erroneously) concluded that the behavior was chaotic. To avoid this pitfall, we must follow the trajectories a sufficiently long time to allow the trajectories to wander over the full extent of the state space region they will visit. (In practice, it may not be easy to tell how long is long enough.) Then the (average) Lyapunov exponent will be a true indicator of the chaotic or nonchaotic behavior of the system.

4.14 Further Reading

Routes to Chaos

Routes to chaos are discussed in most of the introductory books listed at the end of Chapter 1.

[Abraham and Shaw, 1992] and F. D. Abraham, R. H. Abraham, and C. D. Shaw, *Dynamical Systems: A Visual Introduction*, (Science Frontier Express, 1996). These books provide lavishly illustrated examples of homoclinic and heteroclinic tangles, including their effects on the Lorenz attractor. Various bifurcation events are depicted graphically.

J. M. Ottino, *The Kinematics of Mixing: Stretching, Chaos, and Transport* (Cambridge University Press, Cambridge, 1989) pp.111–15. Provides an excellent, illustrated introduction to homoclinic and heteroclinic tangles.

[Guckenheimer and Holmes, 1990]. The authors discuss homoclinic and heteroclinic orbits throughout their book.

M. A. Harrison and Y.-C. Lai, "Route to high-dimensional chaos," *Phys. Rev. A* **59**, R3799–R3802 (1999).

Strange but not Chaotic Attractors

C. Grebogi, E. Ott, S. Pelikan, and J. A. Yorke, "Strange Attractors that are not Chaotic," *Physica D* **13**, 261–68 (1984).

Quasi-Periodicity

L. D. Landau and E. M. Lifshitz, *Fluid Mechanics*, (Pergamon, London, 1959). Discusses the conjecture of an infinite number of periods required to explain turbulence.

J. P. Gollub and H. L. Swinney, "Onset of Turbulence in a Rotating Fluid," *Phys. Rev. Lett.* **35**, 927–30 (1975). The first experimental evidence for the Ruelle-Takens quasi-periodic route to chaos.

P. M. Battelino, "Persistence of Three-Frequency Quasiperiodicity under Large Perturbation," *Phys. Rev. A* **38**, 1495–502 (1988). This paper shows that in some cases more than three incommensurate frequencies may coexist before chaos begins.

Transient Chaos

H. Krantz and P. Grassberger, "Repellers, Semi-attractors, and Long-lived Chaotic Transients," *Physica D* **17**, 75–86 (1985).

T. Tél, "Transient Chaos," in [Hao, 1990].

Sil'nikov Chaos

L. P. Sil'nikov, "A contribution to the problem of the structure of an extended neighborhood of a rough equilibrium state of saddle-focus type," *Math. USSR Sbornik* **10**, 91–102 (1970).

[Hale and Kocak, 1991]. Section 17.3 gives a brief introduction to Sil'nikov chaos.

[Guckenheimer and Holmes, 1990]. Gives an extended mathematical treatment of Sil'nikov chaos.

The Horseshoe Map

S. Smale, "Differentiable Dynamical Systems," *Bull. Amer. Math. Soc.* **73**, 747–817 (1967).

[Guckenheimer and Holmes, 1991].

[Gulick, 1992].

[Ottino, 1989] and J. M. Ottino, "The Mixing of Fluids," *Scientific American*, **260** (1), 56–67, (January, 1989). Illustrates the connection between horseshoe dynamics and fluid tracer experiments. Beautiful examples of stretching, compression, and folding.

Lyapunov Exponents

H. Haken, "At Least One Lyapunov Exponent Vanishes if the Trajectory of an Attractor does not Contain a Fixed Point," *Phys. Lett. A* **94**, 71–74 (1983).

Various methods for finding Lyapunov exponents are discussed in:

A. Wolf, J. B. Swift, H. L. Swinney, and J. A. Vasano, "Determining Lyapunov Exponents from a Time Series," *Physica D* **7**, 285–317 (1985).

A. Wolf "Quantifying Chaos with Lyapunov Exponents," in *Chaos*, A. V. Holden, ed. (Princeton University Press, Princeton, 1986).

S. Neil Rasband, *Chaotic Dynamics of Nonlinear Systems* (Wiley, New York, 1990).

4.15 Computer Exercises

CE4-1. Use *Chaotic Dynamics Workbench* or *Chaos for Java* (see the information at the end of Chapter 2) to explore the behavior of the Lorenz model near the parameter values $r = 13.96$ and $r = 24$-25. Use Poincaré sections in *Chaotic Dynamics Workbench* to see the effects of the repelling fixed point at the origin and the repelling cycles near the off-origin fixed points.

CE4-2.   Use *Chaotic Dynamics Workbench* to calculate the Lyapunov exponents for the Lorenz model. Verify that the sum of the exponents is negative for the Lorenz model. Verify that one Lyapunov exponent is (close to) 0. Verify that for a chaotic orbit, one Lyapunov exponent is positive.

CE4-3. Consider a system described by the simple differential equation

$$\dot{x} + x + x^2 = 0$$

(a) First, write this system as a pair of first-order differential equations and then find the locations of the fixed points in the $\dot{x} - x$ two-dimensional state space. Show that there is a saddle point at $(-1,0)$ and a repellor at $(0,0)$. (b) This system has a homoclinic orbit associated with the saddle point. Write a computer program to display the trajectories in the neighborhood of the saddle point and the repellor. Try to find the homoclinic orbit. [Hint: The homoclinic orbit passes through the point $(0, 1/\sqrt{3})$. In Chapter 8, we will see why that is the case.]

CE4-4.   The Rössler model (ROS76) is a simplified version of the Lorenz equations introduced in Chapter 1. The model's equations take the following form:

$$\dot{x} = -y - z$$
$$\dot{y} = x + ay$$
$$\dot{z} = b + z(x - c)$$

a, b, and c are parameters while x, y, and z are the dynamical variables. Find the fixed points of the system. Then use *Chaos for Java* to explore the behavior of this system with $a = 0.2$ and $b = 0.2$, using c as the variable parameter. (Start with c in the range between 2.5 and 6.) View the behavior with *xyz* state-space plots, a "return map" using x_{max}, the (local) maximum value of x plotted as a function of c, and a Poincaré section. See [Strogatz, 1994], pp. 376–379 for some graphical results for this system. Write your own program to generate a bifurcation diagram, plotting x_{max} as a function of c.

5

Iterated Maps

What she's doing is, every time she works out a value for *y*, she using *that* as her next value for *x*. And so on. Like a feedback. She's feeding the solution back into the equation, and then solving it again. Iteration, you see. Tom Stoppard, *Arcadia* (p. 44).

5.1 Introduction

In this chapter we will develop the theory and analysis of iterated map functions viewed as dynamical systems. One motivation for studying such maps is their origin in the description of intersections of state space trajectories with Poincaré sections as described in the previous chapter. These maps, however, have a mathematical life of their own. Furthermore, in many cases, we can use the maps as models for physical systems even if we do not know the underlying differential equation model. This approach to modeling has its pitfalls, some of which we will discuss later, but nevertheless it can give us useful insights for the dynamics of complex systems.

The initial discussion will focus on those systems whose dynamics can be described by so-called one-dimensional iterated maps. A one-dimensional iterated map is based on a function of a single (real) variable and takes the form

$$x_{n+1} = f(x_n) \tag{5.1-1}$$

Section 5.2 provides some arguments about the conditions under which we might expect such one-dimensional iterated maps to be good models for systems whose "natural" description would be in the form of differential equations.

The reader should be warned that the theory of nonlinear dynamics does not (yet) provide us with the tools needed to say in advance under what circumstances, if any, we can use the results developed in this chapter to describe the behavior of a particular system. We can say, however, that many systems, both theoretical and experimental, do seem to be well described by such a scheme, at least for some range of control parameter values. We believe that the arguments given in Section 5.2 should make this fact reasonable.

In any case, the mathematical theory of one-dimensional iterated maps has played an important role, both historically and conceptually, in the development of chaos theory, and these maps can be studied quite fruitfully in their own right. We shall present a pedestrian view of some of the theory in the remaining sections of the chapter. However, one of the themes of this book is that nonlinear dynamics is a scientific study of the "real world." So, we must focus on those aspects of the mathematical development that are most closely related to the description of actual systems. Much of the discussion, however, will concentrate on the mathematics. Applications will be taken up in later chapters. The references at the end of the chapter will satisfy the needs of the mathematically inclined and of those readers whose appetites have been whetted for a more detailed mathematical treatment.

5.2 Poincaré Sections and Iterated Maps

As we discussed in the previous chapter, the Poincaré section of the state space dynamics simplifies the geometric description of the dynamics by removing one of the state space dimensions. The key point is that this simplified geometry nevertheless contains the "essential" information about the periodicity, quasi-periodicity, chaoticity, and bifurcations of the system's dynamics.

For a three-dimensional state space, the Poincaré section is a two-dimensional plane chosen so that the trajectories intersect the plane transversely (in the sense defined in Section 4.6). Figure 5.1 shows such a Poincaré plane, where we have set up a Cartesian coordinate system with coordinates u and v (to distinguish them from state space coordinates x_1, x_2, and x_3).

The assumed uniqueness and determinism of the solutions of the differential equations describing the dynamics of the system imply the existence of a Poincaré map function, which relates a trajectory intersection point to the next intersection point. Suppose a trajectory intersects the Poincaré plane at the point (u_1, v_1). Then, after "circling" around state space, the trajectory next crosses the plane at (u_2, v_2). In essence, we assume that there exists a pair of Poincaré map functions that relate (u_2, v_2) to (u_1, v_1):

$$u_2 = P_u(u_1, v_1)$$
$$v_2 = P_v(u_1, v_1)$$

$$(5.2-1)$$

From the pair (u_2, v_2), we can find (u_3, v_3) and so on. Hence, if the map functions P_u and P_v are known, we have essentially all the information we need to characterize the dynamics of the system. We want to emphasize that we need not restrict ourselves to the long-term behavior of the system, that is, we need not restrict ourselves to what we have called the attractor for dissipative systems. However, most of the applications of the Poincaré section technique will focus on the attractors and how they evolve as parameters are changed.

In a (perhaps) surprisingly large number of cases, the Poincaré map functions reduce to a one-dimensional iterated map. To see what this collapse means, let us start with a three-dimensional state space and a two-dimensional Poincaré plane. In general, we need to know both coordinates (say, u and v) of one trajectory intersection point on the plane, as well as the map functions, to be able to predict the location of the next intersection point. However, in some cases, the system behaves in such a way that less information is required. For example, we may need

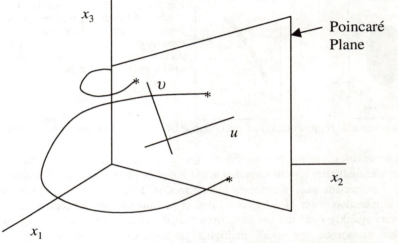

Fig. 5.1. A Poincaré plane, which is the Poincaré section for trajectories in a three–dimensional state space. The coordinates u and υ label the location of the intersection points (indicated by asterisks) in the plane.

to know only u_n in order to be able to predict υ_n. In this case the mapping functions can be written as

$$\upsilon_n = P_\upsilon(u_n)$$
$$u_{n+1} = P_u(u_n, P(u_n)) \tag{5.2-2}$$
$$\equiv f_A(u_n)$$

where A is some control parameter.

Let us assume that we are dealing with a dissipative system You will recall from our discussion in Chapters 2 and 3 that dissipative systems (by definition) are those for which a cluster of initial points, occupying some finite (nonzero) volume in state space, collapses into a cluster of 0 volume as the system evolves in time. (The points may still be spread over a finite area or along a finite-length curve, but the <u>volume</u> tends to 0.) That 0 volume cluster is part of the state space attractor for the system. From the point of view of a Poincaré section, this cluster of initial points corresponds to (for a three-dimensional state space) a cluster of points occupying a finite area of the Poincaré surface of section. (We could think of building our volume of initial points around the Poincaré plane and then recording the intersection points as the subsequent trajectories cross the plane.) As time evolves, the collapse of the volume of points in the three-dimensional state space means that the Poincaré points will collapse onto a "curve" (perhaps of very complicated geometry) on the Poincaré plane. (In the special case in which the attractor is a periodic limit cycle in the three-dimensional state space, the Poincaré section will consist of a finite number of points, rather than a curve.) Figure 5.2

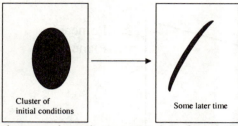

Fig. 5.2. A sketch of the evolution of a cluster of initial condition points viewed on the Poincaré plane of a three–dimensional dissipative system. As time goes on, the cluster collapses to a curve or a series of points.

shows a schematic representation of the collapse of the cluster of initial points into a curve.

In some cases, the attractor for the dynamical system under study will have a characteristic multiplier for one direction in the Poincaré plane that is much smaller than the characteristic multiplier for the other direction. This small multiplier means that an intersection point of a trajectory that is just outside the attractor will be pulled very quickly onto the attractor from that direction. (In the language of Lyapunov exponents, the small multiplier corresponds to a large negative Lyapunov exponent.) The net result of this rapid collapse is that after a very short time (short compared to the overall time evolution of the system), we need concern ourselves only with the evolution of the system along this curve in the Poincaré plane. If this curve has sufficiently simple geometry, then we say that the evolution of the system is essentially one-dimensional corresponding to evolution along the curve. We make this statement more formally later.

Although the theory of nonlinear dynamics does not yet provide a complete analysis of when this kind of collapse to one-dimensional dynamics will occur, the general working rule-of-thumb is that this one-dimensional behavior occurs when the system is "sufficiently" dissipative. (We put quotation marks around sufficiently because there is no general rule for telling us how much dissipation is sufficient.) For example, in the diode circuit described in Chapter 1, we get sufficient dissipation by increasing the amount of electrical resistance in the circuit. In a fluid experiment, we could (in principle) increase the viscosity of the fluid. In many such cases, we do find that the Poincaré map function becomes effectively one-dimensional as the amount of dissipation increases, even for systems with state spaces of more than three dimensions. This last remark indicates that our analysis is more general than one might conclude from our considerations of three-dimensional state spaces.

We can recognize this one-dimensional behavior by recording measured values of some variable, say u, taken at successive intervals of time and then plotting u_{n+1} as a function of u_n. If that graph results in a single-valued functional relationship

$$u_{n+1} = f_A(u_n) \tag{5.2-3}$$

then we say the evolution is essentially one-dimensional.

Figure 5.3 shows some data taken with the diode circuit described in Chapter 1. The value of the current flowing through the diode was recorded at a fixed phase

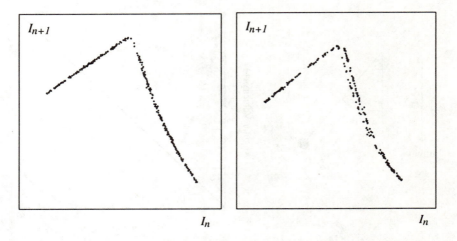

Fig. 5.3. A return map for the diode circuit described in Chapter 1. The $(n+1)$th sample of the current I is plotted as a function of the nth sample. On the left, the circuit dissipation is relatively high and the evolution is essentially one–dimensional. On the right, with lower relative dissipation, the double-valuedness (loop) seen on the far right indicates higher–dimensional behavior. (A technical point: the relative dissipation of the circuit was varied by changing the drive frequency. On the left the drive frequency is 30 kHz. On the right, it is 50 kHz. In other words, the Q ("quality factor") for the circuit is lower on the left and higher on the right. Lower Q corresponds to higher relative dissipation.)

point of the sinusoidal voltage driving the circuit, thereby forming a stroboscopic portrait. We have plotted I_{n+1} as a function of I_n. On the left, the dissipation in the circuit was relatively high and the displayed points indicate that knowing I_n allows us to predict I_{n+1}. Thus, we say the behavior is one-dimensional. On the right, the amount of dissipation was weaker. Toward the far right of the diagram, a loop is formed that indicates that there are two possible I_{n+1} values for a certain range of I_n values. In this region the behavior is not one-dimensional.

Of course, tracking one variable does not tell us the details of the behavior of other variables. If period-doubling occurs for one variable, however, then it generally occurs for the others as well. Hence, following the behavior of one variable allows us to determine the overall periodicity and bifurcations for the system.

In a few cases, effective one-dimensional behavior has been demonstrated for models based on differential equations. A popular example for physicists is the model of a pendulum driven periodically by an external torque and subject to frictional damping. The pendulum is taken to consist of a mass M attached to the end of a rigid rod of length L as shown in Fig. 5.4. (We assume the rod has 0 mass). The other end of the rod is fixed at the pivot point. As usual the angular displacement from the vertical direction is indicated by the angle θ. For this model the time evolution equation is

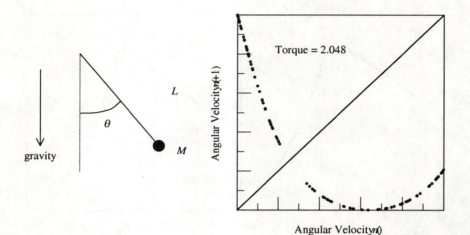

Fig. 5.4. On the left, a schematic diagram of the pendulum. On the right is a plot of the $(n+1)$th sample of the angular velocity variable as a function of the nth for the driven damped pendulum. The angular velocity was sampled when the external torque was at phase 0. The amplitude of the external torque = 2.048 MgL; the frequency ratio is 2/3 and the damping factor is 1.0. Note that the functional dependence is essentially one–dimensional. The $y = x$ line is drawn for reference.

$$\frac{d^2\theta}{dt^2} = -\frac{g}{L}\sin\theta - \gamma\frac{d\theta}{dt} + K\cos(\omega t) \qquad (5.2\text{-}4)$$

On the right of the previous equation, the first term represents the torque due to gravity; the second gives the damping due to friction, and the third describes the external torque, which is assumed to vary periodically in time.

    If we now introduce some dimensionless variables, then the equations are more suited for numerical work. The natural oscillation frequency of the pendulum in the absence of damping and external torques is $\omega_0 = \sqrt{g/L}$ for small amplitude motion; therefore, we choose to measure time in units of $1/\omega_0$. The damping is proportional to $1/Q$ (the reciprocal of the quality factor). Small Q means a large amount of damping per oscillation. (Q is defined as the ratio of the total mechanical energy of the system E to the energy lost to dissipation per oscillation cycle.) The torque is measured in units of the product MgL, which is the torque required to hold the pendulum at rest, 90° from the vertical position. Using those variables, the time evolution equation becomes

$$\frac{d^2\theta}{ds^2} = -\sin\theta - b\frac{d\theta}{ds} + F\cos(Ds) \qquad (5.2\text{-}5)$$

Here b measures the amount of damping; s is the dimensionless time variable; F is the amount of torque relative to MgL, and D is the ratio of the drive frequency to the natural oscillation frequency. Our standard procedures show that this system has a three-dimensional state space, where the variables can be taken to be the angular displacement, the angular velocity, and the phase of the external torque.

Figure 5.4 shows the return map plotted for the angular velocity of the driven damped pendulum with $D = 2/3$, $b = 1.0$, and $F = 2.048$, which yields chaotic behavior following a period-doubling sequence of bifurcations. (The values were sampled when the phase of the external torque was at 0.) The return map shows that for this set of parameter values, the behavior of the angular velocity can be modeled by a one-dimensional iterated map.

Much theoretical work has gone into analyzing how Poincaré map functions for the Lorenz model, introduced in Chapter 1, reduce to one-dimensional iterated maps. For further information, see the references at the end of this chapter.

5.3 One-Dimensional Iterated Maps

The discussion in the previous section should have convinced you that using one-dimensional iterated maps as models for the dynamics in a state space with three (or more) dimensions is at least plausible. We will now turn to a discussion of such map functions as examples of dynamical systems in their own right. Later, we will return to the question of applying what we have learned to the full dynamics of the system.

The mathematical literature on one-dimensional iterated maps is vast and sophisticated. In the limited space available in this book we cannot do justice to the elegance of the many mathematical results that are known. We hope, however, to bring the readers to the point where they will be ready to tackle the more detailed treatments given in the references at the end of the chapter.

Let us begin with a brief review of some of the notions introduced in Chapter 1. We will now call the independent variable x; the iterated map function will be $f(x)$. In general the map function depends on some parameter p, which we will not usually show explicitly. When we want to emphasize the parameter dependence, we will display the parameter p as a subscript $f_p(x)$.

The iteration scheme begins with an initial value of x, call it x_0, and we generate a trajectory (or orbit) by successive applications of the map function:

$$x_1 = f(x_0), \quad x_2 = f(x_1),\dots \qquad (5.3\text{-}1)$$

and so on.

> We should note that one-dimensional iterated maps show a much greater range of dynamical behavior than do one-dimensional differential equation systems because the iterated maps are free from the constraints of continuity. By that we mean that x for the iterated map can jump from one value to another without passing through the intermediate values of x. As we saw in Chapter 3, the constraint of continuity on the solutions of a one-dimensional differential equation severely restricts the kinds of behavior that can occur.

It will be helpful to have a specific example in front of us. We will use the logistic map from Chapter 1 as our model system. (We will discuss several other map functions later.) Recall that the logistic map function is given by

$$x_{n+1} = Ax_n(1 - x_n) \qquad (5.3\text{-}2)$$

where A is the control parameter, taken to be a positive, real number.

We are most interested in maps that keep the values of x within some finite interval along the x axis. If the function $f(x)$ always produces x values within that interval when we start with an x_0 within that interval, we say that the function f is "a map of the interval onto itself." (It is traditional to scale the x values so that interval is either $0 \le x \le 1$ or $-1 \le x \le +1$.) For example, for the logistic map, the relevant interval is between 0 and 1.

Let us also review the notion of fixed points. In analogy with fixed points for dynamical systems described by differential equations, we are often interested in the fixed points of the map function. x^* is a fixed point of the map function $f(x)$ if it satisfies

$$x^* = f(x^*) \qquad (5.3\text{-}3)$$

If Eq. (5.3-3) is satisfied, then repeated application of the function f to x^* yields the same value. It is important to note that a given map function may have several fixed points for a given control parameter value. For example, the logistic map has fixed points at $x^* = 0$ and $x^* = 1 - 1/A$.

What happens to trajectories that start near a fixed point; that is, what is the *stability* of the fixed point? If those trajectories approach x^* as the iteration process proceeds ($n \to \infty$), we say that x^* is an *attracting fixed point* or (equivalently) a *stable fixed point*. (The phrase *asymptotically stable* is also used.) If the trajectories move away from x^*, then we say that x^* is a *repelling fixed point* or (equivalently) an *unstable fixed point*.

As we saw in the previous chapters, we can investigate the stability of a fixed point by finding the derivative of the map function evaluated at the fixed point. In geometric terms, we are looking at the slope of the map function at the fixed point. Figure 5.5 shows a graphical representation of the iteration process introduced in Chapter 1. In the immediate neighborhood of a fixed point, the function can be approximated as a straight line.

It should be obvious from those graphs that the following criteria hold for the stability of the fixed point (the subscript on the derivative reminds us that we are evaluating the derivative at the fixed point):

$$x^* \text{ is a stable fixed point if } |df/dx|_* < 1$$

$$x^* \text{ is an unstable fixed point if } |df/dx|_* > 1$$

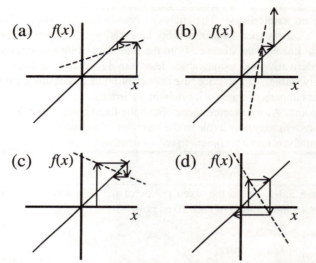

Fig. 5.5. A graphical representation of the effect of repeated use of map function $f(x)$. The map function is plotted as a dashed line. The $f(x) = x$ line is solid. (a) $0 < df/dx < 1$ gives a stable fixed point. (b) $df/dx > 1$ gives an unstable fixed point. (c) $-1 < df/dx < 0$ gives a stable fixed point. (d) $df/dx < -1$ gives an unstable fixed point. Note that when $df/dx < 0$, the successive iteration points alternate from one side of the fixed point to the other.

In Chapter 3, we used a Taylor series expansion near x^* to arrive at the same results analytically. For a trajectory starting at x_0 near x^*, we can write

$$x_1 = f(x_0) = f(x^*) + \left.\frac{df}{dx}\right|_{x^*} (x_0 - x^*) + \dots$$

$$= x^* + \left.\frac{df}{dx}\right|_{x^*} (x_0 - x^*) + \dots \tag{5.3-4}$$

Keeping only the first-derivative term in the expansion, we write this result in terms of the difference between x_n and x^*:

$$(x_n - x^*) = \left.\frac{df}{dx}\right|_{x^*} (x_{n-1} - x^*) \tag{5.3-5}$$

If the magnitude of the derivative evaluated at the fixed point is < 1, then the distance decreases with subsequent iteration, and the trajectory approaches the fixed point. If the magnitude of the derivative is greater than 1, then the distance increases, and the trajectory moves away from the fixed point. Note that if the derivative is negative, then the sign of the difference changes with each iteration, and the trajectory points alternate being on the left or the right of the fixed point.

A note on terminology: The analysis given in terms of the Taylor series above is often called "linear stability analysis" since we are keeping only the term linear in the distance from the trajectory point to the fixed point. If the derivative has a magnitude less than 1, we say the fixed point is "linearly stable." In practice, the important point is that this analysis tells us what happens to trajectories only in the immediate neighborhood of the fixed point. As we venture further from the fixed point, other fixed points of the system may play a role in the behavior of trajectories and the Taylor series analysis is not sufficient to tell us what will happen.

Exercise 5.3-1. Find the fixed points of the following map functions. Determine the stability of each fixed point.

$$x_{n+1} = f(x_n) = x_n^2$$
$$x_{n+1} = f(x_n) = \sin x_n$$

Exercise 5.3-2. Explore the logistic map model with the parameter $A < 0$. Find the fixed points and study their stability as A changes.

5.4 Bifurcations in Iterated Maps: Period-Doubling, Chaos, and Lyapunov Exponents

As we have seen with other dynamical systems, the nature of the fixed points of iterated maps can change as the control parameters of the system change. In this section we will examine bifurcations in the logistic map as a model of the kinds of bifurcations that can occur for one-dimensional iterated maps. In particular we will focus on the sequence of period-doublings that leads to chaotic behavior.

First let us examine the stability of the two fixed points $x^* = 0$ and $x^* = 1 - 1/A$. We evaluate df/dx at each of those fixed points:

$$\left.\frac{df}{dx}\right|_{x=0} = A$$

$$\left.\frac{df}{dx}\right|_{x=1-1/A} = 2 - A \tag{5.4-1}$$

We see from Eq. (5.4-1) that the fixed point at $x = 0$ is an attracting (stable) fixed point for $A < 1$ and that the fixed point at $x = 1 - 1/A$ is a repelling (unstable) fixed point for $A < 1$. Thus, for $A < 1$, we expect all trajectories with $0 < x_0 < 1$ to approach the fixed point at $x = 0$ (except for the trajectory that starts exactly at $x_0 = 1 - 1/A$).

Exercise 5.4-1. What happens to logistic map trajectories with $x_0 < 0$ and $x_0 > 1$ for $A < 1$? Can you explain that behavior in terms of a fixed point of the map function $f(x)$?

We also see from Eq. (5.4-1) that for $1 < A < 3$, the two fixed points exchange stability. That is, for $A > 1$ the fixed point at $x = 0$ is an unstable fixed point. For $1 < A < 3$, the fixed point at $x = 1 - 1/A$ becomes a stable fixed point.

Exercise 5.4-2. Use the graphic iteration technique and also a computer calculation of trajectory values to show that all trajectories starting between $x = 0$ and $x = 1$ approach $x^* = 1 - 1/A$ for the logistic map when $1 < A < 3$.

The trajectory behavior becomes more interesting for $A > 3$. As we saw in Chapter 1, for A values just greater than 3, the trajectories settle into a pattern of alternation between two points, which we shall label x_1^* and x_2^*. These values satisfy the equations

$$x_2^* = f(x_1^*)$$
$$x_1^* = f(x_2^*)$$
(5.4-2)

These two points are attracting fixed points of a "two-cycle." Thus, we say that at $A = 3$, the logistic map trajectories undergo a *period-doubling bifurcation*. Just below $A = 3$, the trajectories converge to a single value of x. Just above $A = 3$, the trajectories tend to this alternation between two values of x.

To describe what happens at $A = 3$, we introduce what is called the *second-iterate of f*. The second-iterate of the map function is defined to be

$$f^{(2)}(x) \equiv f(f(x))$$
(5.4-3)

that is, the second-iterate of f is what we get by applying the function f twice, first to the value of x, then to the result of the first application. (We will use the parentheses around the superscript to remind us that we are concerned with the second-iterate, not the square of the function.) The two-cycle points x_1^* and x_2^* are fixed points of the second-iterate function:

$$x_1^* = f^{(2)}(x_1^*)$$
$$x_2^* = f^{(2)}(x_2^*)$$
(5.4-4)

Thus, we conclude that for A just greater than 3, these two fixed points of the second-iterate function become stable fixed points.

Let us see how the derivatives of the map function and of the second-iterate function change at the bifurcation value, which we shall label as $A = A_1 = 3$. Eq. (5.4-1) tells us that df/dx (the derivative of the map function) passes through the value -1 as A increases through 3. What happens to the derivative of $f^{(2)}$? We can evaluate the derivative of the second-iterate function by using the ***chain rule of differentiation***:

$$\frac{df^{(2)}(x)}{dx} = \frac{df(f(x))}{dx}$$

$$= \frac{df}{dx}\bigg|_{f(x)} \frac{df}{dx}\bigg|_{x} \qquad (5.4\text{-}5)$$

If we now evaluate the derivative at one of the fixed points, say, x_1^*, we find

$$\frac{df^{(2)}(x)}{dx}\bigg|_{x_1^*} = \frac{df}{dx}\bigg|_{x_2^*} \frac{df}{dx}\bigg|_{x_1^*}$$

$$= \frac{df^{(2)}(x)}{dx}\bigg|_{x_2^*} \qquad (5.4\text{-}6)$$

In arriving at the last result, we made use of $x_2^* = f(x_1^*)$ for the two fixed points. Eq. (5.4-6) states a rather surprising and important result: The derivatives of $f^{(2)}$ are the same at both the fixed points that are part of the two-cycle! This result tells us that both of these fixed points are stable or both are unstable and that they have the same "degree" of stability or instability.

We gain insight about what happens at the bifurcation value $A = 3$ by realizing that for $A = 3$, the fixed point $1 - 1/A$ for $f(x)$ coincides with the two fixed points for $f^{(2)}(x)$. Since the derivative of $f(x)$ is equal to -1 for that value of A, Eq. (5.4-6) tells us that the derivative of $f^{(2)}$ is equal to $+1$ for that value of A. As A increases further, the derivative of $f^{(2)}$ decreases and the two fixed points become stable. Figure 5.6 shows a graph of $f^{(2)}(x)$ for a value of A just below 3 and a value just above 3. For A just greater than 3, we see that the slope of $f^{(2)}$ at those two fixed points is less than 1 and that hence they are stable fixed points of $f^{(2)}$. From the perspective of $f(x)$, a trajectory jumps back and forth between those two values. Note also in Fig. 5.6 that the unstable fixed point of $f(x)$ located at $1 - 1/A$ is also an <u>unstable</u> fixed point of $f^{(2)}(x)$.

Before continuing our discussion, we would like to make a few remarks about fixed points and how to extend these notions to higher-order iterates. First, we point out that the nth iterate of $f(x)$ is defined as the function that results from applying f n–times:

$$f^{(n)}(x) = f(f(f(\ldots f(x))) \qquad (5.4\text{-}7)$$

with n fs on the right-hand-side. A moment's reflection should convince you that a fixed point of $f(x)$ is also a fixed point of $f^{(n)}$, but that the converse is <u>not</u> always true. Furthermore, if x^* is a stable fixed point of $f(x)$, it is also a stable fixed point of $f^{(n)}$. Again, the converse is <u>not</u> always true.

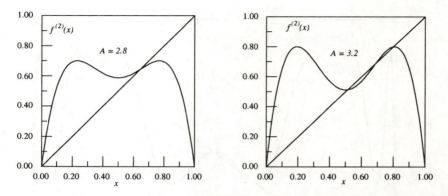

Fig. 5.6. A graph of the second-iterate of the logistic map. On the left, the value of A is 2.8. On the right, just above the first period-doubling bifurcation, the value of A is 3.2. Note that the slopes of the second iterate of f at the two fixed points near $1 - 1/A$ are the same and for $A = 3.2$ have magnitude less than 1.

The smallest n ($n > 0$) for which x^* is a fixed point defines the ***prime period*** of that fixed point. For example, the two distinct values of x that satisfy Eq. (5.4-4) have a prime period of 2. The point $x = 1 - 1/A$ is also a fixed point of $f^{(2)}$, but its prime period is 1.

> **Exercise 5.4-3.** Show that the iterates of f obey the so-called ***composition rule***: $f^{(n+m)}(x) = f^{(m)}(f^{(n)}(x))$.

> **Exercise 5.4-4.** Use the chain rule to prove that the derivative of $f^{(n)}$ evaluated at any of the fixed points of an n-cycle has the same value.

Now we are ready to continue our discussion of the behavior of the logistic map as A increases. The two 2-cycle fixed points of $f^{(2)}$ continue to be stable fixed points until $A = 1 + \sqrt{6}$. At this value of A, which we shall call A_2, the derivative of $f^{(2)}$ evaluated at the 2-cycle fixed points is equal to -1 and for values of A larger than A_2, the derivative is more negative than -1. Hence, for A values greater than A_2, these 2-cycle points are unstable fixed points. What happens to the trajectories? We find that for A values just greater than A_2, the trajectories settle into a 4-cycle, that is the trajectory cycles among 4 values, which we can label x_1^*, x_2^*, x_3^*, and x_4^*. These x values are fixed points of the fourth-iterate function $f^{(4)}$. We have another period-doubling bifurcation: The system has changed from 2-cycle behavior to 4-cycle behavior. The mathematical details of this bifurcation are exactly the same as the discussion for the birth of the 2-cycle behavior since we can view $f^{(4)}$ as the second-iterate of $f^{(2)}$. Thus, we see that the 4-cycle is born when the derivative of $f^{(2)}$ evaluated at its 2-cycle fixed points passes through the value -1 and becomes more negative while the derivative of $f^{(4)}$ evaluated at its 4-cycle

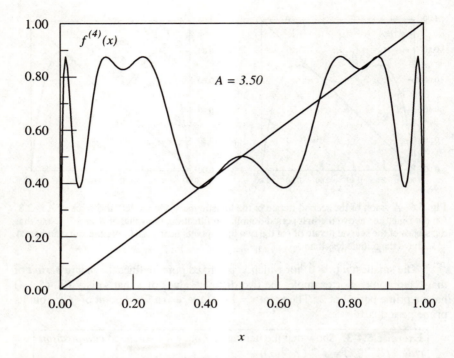

Fig. 5.7. A graph of the fourth-iterate of the logistic map function for $A = 3.50$ just greater than the value A_4 at which the period-4 cycle becomes stable.

fixed points becomes less than 1. Figure 5.7 shows the function $f^{(4)}(x)$ for the logistic map for values of A just greater than A_2.

Exercise 5.4-5. Show that the bifurcation giving rise to stable period-4 behavior occurs at $A = 1 + \sqrt{6}$. Hint: When period-4 is born, $df^{(2)}/dx = -1$.

As we saw in Chapter 1, this sequence of bifurcations continues with ever longer periods, until we reach a parameter value at which the period of the trajectory becomes infinite. That is equivalent to saying that the trajectory never repeats itself. We are at the onset of chaos. We also saw in Chapter 1 that there are periodic windows beyond the parameter value at which chaos begins. Below we will have more to say about those periodic windows.

It is important to note that as a result of the sequence of period-doubling bifurcations, the logistic map has an infinite number of unstable periodic orbits, those that correspond to period 1, 2, 4, and so on, orbits that became unstable when they gave birth to the next generation of orbits. These unstable periodic points turn out to play an important role in the behavior of other trajectories.

> **Exercise 5.4-6.** Use the methods developed in this section to study the behavior of the so-called **quadratic map**, which is given by the expression
>
> $$x_{n+1} = 1 - Cx_n^2$$
>
> In particular, find its fixed points, and study the period-doubling sequence, which occurs as the parameter C is increased.

Chaos and Lyapunov Exponents

We are now ready to explore what chaotic behavior means quantitatively. One of the signatures of chaos is the divergence of nearby trajectories. In fact, as introduced in Section 4.2, for a chaotic system, this divergence is exponential in time (or for an iterated map, exponential as a function of the iteration number). Thus stated, the property of being chaotic is a characteristic of a group of trajectories. However, a trajectory on a chaotic attractor for a bounded system also returns infinitely often, infinitely closely to any previous point on the trajectory. Thus, we could also examine the divergence of these nearby points (corresponding to quite different times) on a single trajectory.

How do we state this more formally? We begin by considering an attractor point x_0 and a neighboring attractor point $x_0 + \varepsilon$. We then apply the iterated map function n times to each value and consider the absolute value of the difference between those results:

$$d_n \equiv \left| f^{(n)}(x_0 + \varepsilon) - f^{(n)}(x_0) \right| \tag{5.4-8}$$

If the behavior is chaotic, we expect this distance to grow exponentially with n, so we write

$$\frac{d_n}{\varepsilon} = \frac{\left| f^{(n)}(x_0 + \varepsilon) - f^{(n)}(x_0) \right|}{\varepsilon} \equiv e^{\lambda n} \tag{5.4-9}$$

or

$$\lambda = \frac{1}{n} \ln \left(\frac{\left| f^{(n)}(x_0 + \varepsilon) - f^{(n)}(x_0) \right|}{\varepsilon} \right) \tag{5.4-10}$$

This last pair of equations defines what we mean by the *Lyapunov exponent* λ for the trajectory. (We will tighten up the definition shortly.)

If we now let $\varepsilon \to 0$ we recognize from elementary calculus that the ratio on the right-hand side of Eq. (5.4-10) is just the definition of the absolute value of the derivative of $f^{(n)}$ with respect to x. We have also seen that by applying the chain rule for differentiation, the derivative of $f^{(n)}$ can be written as a product of n

derivatives of $f(x)$ evaluated at the successive trajectory points x_0, x_1, x_2, and so on. Thus, we can put the definition of the Lyapunov exponent in a more intuitive form

$$\lambda = \frac{1}{n} \ln \left(\left| f'(x_0) \right| \left| f'(x_1) \right| \ldots \left| f'(x_{n-1}) \right| \right) \qquad (5.4\text{-}11)$$

where $f'(x) = df/dx$. We may rewrite Eq. (5.4-11) as

$$\lambda = \frac{1}{n} \left(\ln \left| f'(x_0) \right| + \ln \left| f'(x_1) \right| + \ldots + \ln \left| f'(x_{n-1}) \right| \right) \qquad (5.4\text{-}12$$

Eq. (5.4-12) tells us that the Lyapunov exponent (the rate of divergence of the two trajectories) is just the <u>average</u> of the natural logarithm of the absolute value of the derivatives of the map function evaluated at the trajectory points. Let us unravel this rather complicated statement. If the application of the map function to two nearby points leads to two points further apart, then the absolute value of the derivative of the map function is greater than 1 when evaluated at those trajectory points. (We are assuming that ε is small enough that the derivative is the same at both starting points.) If the absolute value is greater than 1, then its logarithm is *positive*. If the trajectory points continue to diverge, *on the average*, then the average of the logarithms of the (absolute values of the) derivatives is positive.

A practical computational question: How large must n be to give a precise value for n? n must be large enough to give us a reasonable average, but if n gets too large the rate of divergence will no longer be exponential because the range of x values is bounded. Obviously, the smaller ε is, the larger n can be and still give us exponential divergence. A good rule of thumb is the following: Choose n to be the value that makes d_n about half the size of the overall range of x values. These questions will be addressed in more detail in Chapter 9 on Quantifying Chaos.

So far, our Lyapunov exponent has been calculated for a single starting point x_0. If we compute the Lyapunov exponent for a sample of starting points and then average those results, we define the ***average Lyapunov exponent*** for the system. (Most authors call this simply <u>the</u> Lyapunov exponent.) Our quantitative definition of chaotic behavior is the following:

A one-dimensional iterated map function has chaotic trajectories for a particular parameter value if the average Lyapunov exponent is positive for that parameter value.

In most cases this exponent must be computed numerically. In Chapter 9, we will return to the consideration of the computation of Lyapunov exponents.

> **Exercise 5.4-7.** In Chapter 3, we defined average Lyapunov exponents
> for systems described by ordinary differential equations and found that the
> Lyapunov exponents were related to the derivatives of the time evolution
> functions. In this section, we saw that for iterated map functions, the
> average Lyapunov exponent is given by the <u>logarithm</u> of the absolute
> value of the derivative of the map function.   (a)   Explain why the
> definitions are different in the two cases. (b) Use Eq. (I-26) of Appendix
> I, the linearized form of the van der Pol oscillator model, and the
> approximate Poincaré map function in Eq. (I-28) to compare the two
> definitions.

5.5 Qualitative Universal Behavior: The *U*-Sequence

In this section we will discuss the generality of the types of bifurcations and period-
doublings that we found in the last section for the specific case of the logistic map.
Obviously, it is indeed the generality of these bifurcations that makes the study of
one-dimensional maps more than just a mathematical curiosity. We will focus here
on qualitative universality; that is, certain kinds of behavior occur in certain
universal sequences, but we are not concerned with the numerical values of
parameters at which such bifurcations occur. In the next section we will take up the
question of quantitative universal features.

Let us begin with the most general kinds of universality and then proceed to
others that put more restraints on the iterated map functions. First, we restrict
ourselves to map functions that take an interval of the x axis (usually the interval
between $x = 0$ and $x = 1$) and map points in that interval back into the interval. (As
long as the map function takes some finite segment of the x axis and maps it into
the same range of x values, we can always make that interval be the **unit interval** 0
$< x < 1$ by suitably shifting and rescaling coordinates.)   Second, we restrict
ourselves to so-called **unimodal** map functions.

> A *unimodal map function* (for the unit interval) is a smooth (that is,
> continuous and differentiable) function with a single maximum on that
> interval. By convention we take $f(0) = f(1) = 0$. The value of x (labeled
> x_c) at which the maximum occurs is called the *critical point*.

The existence of a maximum in $f(x)$ inside the interval [0,1] is important
because it creates the possibility that two initial points, say, x_0 and x_0' in Fig. 5.8,
lead to the same point x_1 after one iteration of the function. As we shall see this
lack of uniqueness in the "prehistory" of x_1 means that clusters of initial points will
be stretched and folded by the mapping procedure, and that this stretching and
folding, under the appropriate conditions, leads to chaotic behavior.

In 1973, Metropolis, Stein, and Stein (MSS73), pointed out that such
unimodal maps have a well-defined and apparently universal sequence of periodic
trajectories as a control parameter is changed. MSS concentrated on periodic

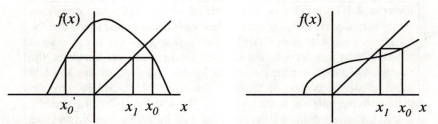

Fig. 5.8. On the left is the graph of a unimodal map function. Two different initial points, x_0 and x_0', lead to the same trajectory point x_1. On the right is a map function that is not unimodal. By rescaling and shifting the x axis, the relevant x axis interval can be made to be the unit interval [0,1].

trajectories, one member of which is the value at which the maximum (or critical value) of the map function occurs. The MSS procedure is: Find the parameter value for which x_c is a member of the periodic trajectory of length, say, n, and write down the sequence of x values:

$$x_1 = f(x_c) \qquad x_2 = f(x_1)$$
$$x_3 = f(x_2) \qquad \ldots \qquad\qquad (5.5\text{-}1)$$
$$\ldots \qquad x_c = f(x_{n-1})$$

The last equality follows because we have assumed that we have a periodic trajectory of period n; so, after n applications of f, we return to our starting value of x. Next, in place of the x values, we write a symbol R or L according to the following rule: If $x_i > x_c$, then we write R (since in this case x_i falls to the right of x_c). If $x_i < x_c$, then we write an L. By this procedure, the trajectory of period n is represented by a set of $n-1$ symbols RLLRRL and so on. (x_c is the nth member of the sequence.) Such a representation, sometimes called a **kneading sequence** (because the iteration process generates a kind of folding as in the kneading of dough), is an example of **symbolic dynamics**, to which we shall return later.

The cycles that involve x_c are sometimes called **supercycles** because the derivative of the map function is 0 for these cycles. That means that the stability is "highest" for these cycles since they are "midway" between their birth (with derivative = +1) and their death (with derivative = −1). The parameter values for these supercycles are fairly easy to compute by using $f^{(n)}(x_c) = x_c$. Given an A value close to the supercycle value, it is straightforward to have a computer program find the supercycle value to high precision by an iterative process. (Initial estimates of the values can be calculated from Eq. (2.4-1) and its obvious extension once the supercycle values for $n = 1$ and $n = 2$ have been found.)

Exercise 5.5-1. Use the graphic representation of the iteration process to show that the first symbol in the MSS sequence must be an R if $f(x)$ has a maximum at x_c.

MSS found that these RL sequences always appear in the same order for any unimodal map function as the appropriate control parameter is varied. Hence, they named this sequence the U (for universal) sequence. These results, found numerically by MSS, were put on a more rigorous basis by GUC81. Table 5.1 lists the U-sequences for periods up to period 6. Also listed are the parameter values for which these sequences are supercycles for the logistic map function.

Note that we have listed only periods up through 6 in the table. Thus, period-8, -16, and so on, which follow period-2 and period-4 in the period-doubling sequence leading to chaos at about $A = 3.5699 \ldots$ are not listed. Also note that for any period greater than 3, there are several possible RL sequences. For example, there are three different period-5 U-sequences, and these occur for three different parameter values. We should also note that these are stable periodic orbits, since they involve x_c as a member. Hence, for all iterates $f^{(n)}$, the derivative of the nth iterate function evaluated at x_c is equal to 0. Thus, we conclude that once we are beyond the period-doubling accumulation point (at $A = 3.5699 \ldots$ for the logistic map), these periodic orbits are part of the periodic windows that occur within the chaotic regime. The period-6 and period-5 windows can be seen in Fig. 5.9.

There are two important (and related) points to note for the results listed in the table. First, we can conclude that if the dynamics of the system is described by a one-dimensional iterated map function and if the system shows, for example, a period-4 RLR U-sequence for some parameter value, then for a smaller parameter value it must have had a period-2 R cycle. Second, the sequence in which various periods and the corresponding U-sequences occur is independent of the details of the map function. The quantitative values of the parameters for which the periods occur do depend on the map function, but the qualitative behavior is the same for all unimodal map functions.

Table 5.1
U-Sequences through Period 6 (from MSS 73)

Period	U-Sequence	Parameter for Logistic Map
2	R	3.2361
4	RLR	3.4986
6	RLRRR	3.6275
5	RLRR	3.7389
3	RL	3.8319
6	RLLRL	3.8446
5	RLLR	3.9057
6	RLLRR	3.9375
4	RLL	3.9603
6	RLLLR	3.9778
5	RLLL	3.9903
6	RLLLL	3.9976

What can we say about the behavior of real physical systems whose dynamics might be modeled by such a unimodal map function? Do they exhibit the same set of *U*-sequences? The answer is a qualified yes. The answer must be qualified because only a few experiments have had sufficient parameter resolution to see many of these sequences (most seem to occur over very small parameter ranges). The experiment might see some of the period-2, -4, and -5 sequences, but have difficulty resolving the intervening period-7 sequences. We can say, however, that the sequences that have been resolved do seem to occur in the MSS *U*-sequence order. For example, the first period-4 behavior that occurs in the diode circuit and illustrated in Fig. 1.5, shows the expected RLR pattern (see Exercise 5.5-2). More detailed examinations of *U*-sequences have been given in studies of oscillating chemical reactions (SWS82) and a varactor diode circuit (TPJ82). However, deviations from the *U*-sequence have been seen in oscillating chemical reactions (CMS86) when the effective one-dimensional map describing the system is not always concave downward. (See Section 5.9, The Gaussian Map, for an example.)

> **Exercise 5.5-2.** If an experimental system shows, say, period-4 behavior as illustrated in Fig. 1.5, how can we assign an RL pattern to the observations since we do not (usually) know the variable value corresponding to x_c? Hints: First we identify the peaks (or alternatively, the valleys) of the variable as the quantities corresponding to x_n. Next we note that $f(x_c)$ gives us the largest value of x in the sequences of xs for that parameter value. Even if the trajectory does not involve x_c directly, we can still assign an RL sequence if we use for the variable value separating R from L the peak value that <u>precedes</u> the largest peak in the sequence for that parameter value. Show that this procedure leads to an RLR sequence for Fig. 1.5 in agreement with the MSS pattern.

In the following subsections, we will present some more general mathematical results concerning one-dimensional iterated maps. The main point to take away from reading those sections is that a great deal of the structure of periodic windows and the existence of orbits of various periods is quite general. Many of these features are independent of the details of the iterated map function. For further details and proofs, the reader is referred to the references given at the end of the chapter.

Some Mathematical Comments

We would like to point out that some of the restrictions we have imposed on the map functions are chosen mainly for mathematical convenience; they are by no means necessary for most of the results. In this section we will discuss how some of the conditions can be relaxed.

In our discussion, we have assumed that the map function has a single maximum in the interval under discussion. Most of the results are the same if the function has a single minimum on that interval. For example, consider the map function $f(x) = 1 - Ax(1-x)$ This function is an inverted version of the logistic map function. It has a *minimum* at $x = 1/2$. You should convince yourself that this function maps the interval [0,1] onto itself, and that as A, the control parameter increases, the iterates of Eq. (5.5-2) pass through a period-doubling sequence to chaos just like that for the logistic map. However, the iterate values are different. For example, for $A < 2$, the point $x = 1$ is the stable fixed point attractor for the system, and for $2 < A < 3$, the stable fixed point attractor occurs at $x = 1/A$.

Exercise 5.5-3. Prove the results stated in the previous paragraph.

The restriction to the interval [0,1] is not necessary either. For example, consider the map function

$$f(x) = Bx(b - x) \tag{5.5-3}$$

This function has a maximum at $x = b/2$ and it goes to 0 at $x = 0$ and $x = b$. By using a new variable, say, $u = x/b$, the map function can be converted into the logistic map function. Thus, by rescaling the independent variable, we can make the interval be the interval [0,1].

Exercise 5.5-4. In most experiments, various "transducers" convert the physical signal of interest to an electrical form (usually a voltage). The electronic equipment often will amplify the signal by some amount. The signal may be inverted in sign, and shifted relative to a 0 value. If the dynamics of the system under study can be modeled with a one-dimensional map function, show that these transformations do not affect in an essential way the dynamics of the iterated map model.

Uniqueness of Stable Periods and the Schwarzian Derivative

The MSS U-sequence presented earlier says nothing about the uniqueness of the listed periodic cycles. In fact, in general there might exist several periodic cycles for a given parameter value. (Of course, only one of these can involve x_c.) In 1978, D. Singer (SIN78) proved that unimodal map functions with a negative value of the so-called *Schwarzian derivative* can have at most three stable periodic cycles for a given parameter value and in many cases there is at most one stable periodic cycle [Devaney, 1986, page 70]. Thus, if a stable periodic cycle exists for some parameter value, that periodic cycle is unique. Furthermore, if it exists, a trajectory starting at x_c is attracted to this periodic cycle. (x_c is part of the cycle only for supercycle parameter values.)

What is a Schwarzian derivative and why must it be negative for this uniqueness to hold? The notion of a Schwarzian derivative is due to H. A. Schwarz (1869) and has applications in various areas of analysis (see for example, [Hille, 1969]). Its relevance to one-dimensional iterated maps apparently went unnoticed until Singer's work in 1978. The Schwarzian derivative, denoted here as $SD(f(x))$, of a function $f(x)$ is defined as:

$$SD(f(x)) \equiv \frac{f'''}{f'} - \frac{3}{2}\left(\frac{f''}{f'}\right)^2 \tag{5.5-4}$$

where f' means the first derivative of f with respect to x; f'' is the second derivative, and f''' is the third derivative. Singer's theorem holds if $SD(f(x))$ is negative over the entire interval $[0,1]$.

Exercise 5.5-5. Prove that the logistic map function has a negative Schwarzian derivative over the entire interval $[0,1]$. Are there any restrictions on the parameter A? Prove that the sine map [see Eq. (2.2-2)] has a negative Schwarzian derivative over the interval $[0,1]$.

A proof of Singer's theorem would take us too far afield. The interested reader is referred to [Devaney, 1986], [Gulick, 1992], and [Davies, 1999]. Some intuition, however, about why $SD(f(x))$ is important can be gathered from noting the following lemma [Devaney, 1986, page 70]: If $SD(f(x)) < 0$, then $f'(x)$ cannot have a local minimum where it is positive or a local maximum where it is negative. Remarks: you may recall that a point at which $f'(x)$ has a maximum or a minimum is called an *inflection point*. This lemma states that certain types of inflection points cannot occur for $f(x)$ on the interval for which $SD(f(x)) < 0$.

Exercise 5.5-6. Sketch a graph of a function that has an inflection point with a positive Schwarzian derivative of the function at the inflection point. Sketch a graph of a function that has a negative Schwarzian derivative at an inflection point. Use the graphic iteration method to show why the first case might lead to more than one stable fixed point for the function.

Sarkovskii's Theorem

In our discussion of the *U*-sequence we were concerned with the appearance of different periodic cycles as a control parameter was changed. We can also ask about the existence of periodic cycles (some stable and some unstable) for a fixed control parameter value. As we have hinted earlier, the existence of unstable periodic cycles can have a significant influence on the behavior of trajectories that are not part of the cycles. In 1964, the Russian mathematician A. N. Sarkovskii proved the following remarkable theorem:

Consider a continuous one-dimensional map function $f(x)$. If for some parameter value, f has a periodic point with prime period m, then f also has (for the same parameter value) a periodic point of period n, where n occurs to the right of m in the following ordered set:

$3 \rightarrow 5 \rightarrow 7 \rightarrow \ldots 2\times3 \rightarrow 2\times5 \rightarrow \ldots 2^2\times3 \rightarrow 2^2\times5 \rightarrow \ldots \rightarrow 8 \rightarrow 4 \rightarrow 2 \rightarrow 1.$

What is the organization of this set? First we list all of the positive odd integers starting with 3 (3, 5, 7, . . .) in increasing order. Then we list the positive integers that are two times the odd integers (again starting with 3) in increasing order, then the positive integers that are four times the odd integers, and so on. Finally, we list in <u>decreasing</u> order the positive integers that we have not yet listed. These are the integer powers of 2: . . . 2^4, 2^3, 2^2, 2^1, 2^0. (You should convince yourself that there are no duplications and no omissions in this listing of the positive integers.)

A proof of this theorem can be found in [Devaney, 1986]. (An alternative, elementary proof based on geometrical ideas can be found in KAP87). Here we will be concerned only with the significance of the result. First (and most important for chaos) is the observation that if f has a period-3 point for some parameter value, then for that same parameter value it has periodic trajectories of all lengths, including infinitely many of infinite length! As mentioned earlier, an infinite period is equivalent to an aperiodic trajectory, and this is one signature of chaos. However, it is not the kind of chaotic trajectory we can observe. For example, in the logistic map, when $A = 3.8319. . .$, there is a stable period-3 trajectory. By Sarkovskii's theorem, the logistic map also has infinitely many periodic cycles for that parameter value, including ones of infinite length. However, they are all <u>unstable</u> by Singer's theorem. Thus, the only one we see in a computer calculation of trajectories is the stable period-3 trajectory.

Sarkovskii's theorem is essentially the result presented (apparently independently) in a famous paper by Li and Yorke in which the word chaos first appeared in its contemporary scientific meaning (LIY75). It is important to note, and this point seems to be often missed, that Li and Yorke's result, like the Sarkovskii theorem, applies specifically to a fixed parameter value. Neither result directly implies that chaotic behavior will occur for other parameter values, though the existence of trajectories with infinite periods is quite suggestive.

Although Sarkovskii's theorem applies only to a single parameter value, in some cases at least, the ordering stated by the theorem is also the order in which the periodic orbits occur in their <u>stable</u> form as a function of parameter value. For example, for the logistic map, starting with small values of the parameter A, we first have period-1, then period-2, then period-4, and so on, as A increases. Then within the chaotic bands beyond the period-doubling accumulation point are periodic windows, first with periods of very large (odd) values, gradually decreasing in length according to the Sarkovskii ordering and finally between $A = 3.7$ and $A = 3.83$, we have period-7, then period-5, and finally period-3 (see Fig. 5.9). A rationale for this behavior is given shortly.

You should note that Sarkovskii's ordering does not describe the multiple appearance of trajectories, with different U-sequences, of a particular periodicity. It describes only the order in which the periodic trajectories first appear.

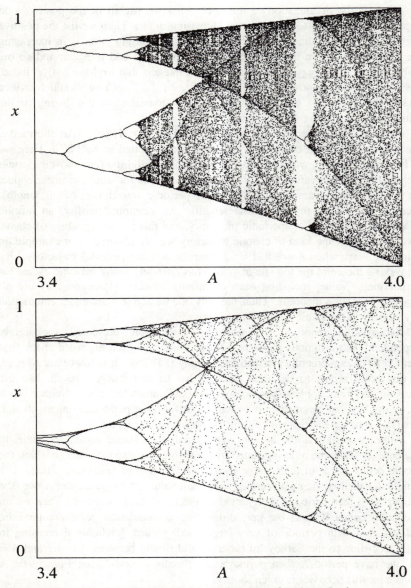

Fig. 5.9. The bifurcation diagram for the logistic map function. In the upper part 100 trajectory points have been plotted (after transients have been allowed to die away) for each value of A. In the lower part, superposed on the bifurcation diagram (with only 10 points plotted for each value of A for clarity's sake) are the first eight trajectory points that follow from $x_0 = 0.5$. Note that these points coincide with the attractor points only for the supercycle values of A. In all cases they give the upper and lower ranges for the attractor and its subbands.

The Organization of Chaotic Behavior

We now discuss some details of the bifurcation diagrams for one-dimensional iterated maps. The upper part of Fig. 5.9 shows the bifurcation diagram for the logistic map for $3.4 < A < 4$. The eye immediately picks up a great deal of order that exists even within the chaotic regimes of the diagram.

First we want to concentrate on the heavy "curves" of points that run through the chaotic regions. These heavy concentrations of points are due to trajectories that pass near the critical point ($x = 0.5$ for the logistic map function) of the iterated map function. All trajectories that pass near the critical point track each other for several subsequent iterations because the slope of the map function (and all of its higher iterates) is 0 at the critical point. Thus, those trajectories diverge rather slowly leading to concentrations of points in the bifurcation diagram.

The trajectory points that follow exactly from the critical point value x_c determine many of the properties of the bifurcation diagram. (These points are called the *images* of the critical point.) For example, $x_{max} = f(x_c)$ is the maximum x value visited by trajectories on the attractor for that particular parameter value. $x_{min} = f(x_{max}) = f^{(2)}(x_c)$ is the minimum value of x visited for that parameter value. That is, the first two iterates of f, starting from x_c, give the upper and lower limits for the attracting region. (Here, we assume that the critical point falls within the attracting region. Transient trajectories may start outside this region, and periodic attractor trajectories may exist entirely inside the boundaries, without touching the boundaries.) Further iterates, leading to higher-order images of the critical point, map out the interior boundaries of the regions visited by the trajectories for a given parameter value.

The lower part of Fig. 5.9 shows the first eight images of the critical point superposed on the bifurcation diagram. Note that these trajectory points do not coincide with the periodic attractor points except at the supercycle values of the parameter A. However, for $A > 3.569...$ these images of the critical point do delimit a set of chaotic "bands" to which the trajectories are confined. As A increases, those chaotic bands merge. Beyond $A = 3.68$ (approximately) the trajectories occupy one large chaotic band.

> **Exercise 5.5.-7.** Explain what happens near $A = 3.68$ where two chaotic bands merge into one and there seems to be a convergence of the curves of the image points. This special point is called the Misiurewicz point (MIS81). Notice that the same type of crossing (convergence) occurs where four chaotic bands merge to give two bands near $A = 3.6$. Hint: consider $f^{(3)}(x_c)$.

The lower part of Fig. 5.9 also shows that periodic windows occur when some of the images of the critical point merge. For example, near $A = 3.83$ some of the image lines merge in the region of the period-3 window. The lines actually touch at the supercycle value for period-3. Similarly, near $A = 3.74$ some of the images merge in the region of a period-5 window. The same effect is seen in a narrow period-7 window near $A = 3.7$. In general, this merging comes about because, for each periodic window, there is some parameter value, the supercycle value, for which the critical point is part of the periodic trajectory. Since the trajectory is periodic, some image of the critical point must return to the critical point value.

We can now understand the order in which the periodic windows occur. The windows are ordered as they are because the higher-order images form curves (as a function of A) rising more steeply away from the convergence point near $A = 3.68$. Thus, the higher-order images are the first to merge with the first image; this merging is a sign of a periodic window. Furthermore, the higher-order images are more sharply peaked near the A values at which they touch the first image curve. The sharpness of the peak is clearly correlated with the length of the periodic window along the A axis: Sharp peaks correlate with narrow windows. We see that the period-7 window is narrower than the period-5 window, which, in turn, is narrower than the period-3 window.

This same association of image merging and periodic windows occurs in the region between $A = 3.569...$ and $A = 3.68$, where there are multiple chaotic bands. Within each band are periodic windows, with period-3 occurring to the right (larger A value) of period-5, which occurs to the right of period-7, and so on. Because there are multiple chaotic bands, however, these windows actually occur as windows of higher periodicity. For example, near $A = 3.62$, where there are two chaotic bands, each subband shows a period-3 window, which together form a period-6 window.

If you compare the Sarkovskii order of periods as given in the previous section, you see that this order corresponds exactly to the ordering of the periodic windows established by the mergings of the critical point image curves. These curves are graphs of the functions, polynomials in the parameter A in the case of the logistic map, determined by iterating the map function starting from the critical point. The geometric properties of these functions control the behavior of the trajectories. For example $f(x_c) = A/4$ for the logistic map; so the upper boundary of the chaotic bands (for $A > 3.59$) is a straight line that hits $x = 1$ at $A = 4$. $f^{(2)}(x_c)$ is given by $(A^2/4)(1 - A/4)$. Thus, the lower boundary of the chaotic bands is a cubic curve that hits $x = 0$ at $A = 4$. This analysis can be continued to understand the behavior of each of the image curves.

As A increases, a particular periodic window, which begins with period-n, disappears through a sequence of period-doublings. Then a set of n chaotic bands is formed. Finally, the chaotic bands suddenly merge in an event called a *crisis*, which we shall explore in Chapter 7. The periodic orbits still exist, but they are unstable. The Sarkovskii ordering tells us the order in which the periodic windows are created. Thus, if we see a period-3 window for some parameter value, we know that all of the windows corresponding to the numbers that precede 3 in the Sarkovskii ordering have been created and their remnant unstable periodic orbits coexist with the period-3 orbit.

> **Exercise 5.5-8.** Work through the details of the previous argument to verify the Sarkovskii ordering.

> **Exercise 5.5-9.** Each chaotic band has its own Misiuriewicz point. Describe the condition that determines where that point will occur as a function of the parameter A.

5.6 Feigenbaum Universality

We discussed in Chapter 2 several quantitative features that are shared by many one-dimensional iterated maps. The most famous of these are the two Feigenbaum numbers α and δ. In this section we present a brief "derivation" of the Feigenbaum α based on COP99. A more detailed derivation for both α and δ is given in Appendix F. Other quantitative universal features are discussed in Appendix H.

We recall from Chapter 2 that the Feigenbaum α is defined as a size scaling factor relating "vertical" distances on a bifurcation diagram of the logistic map function for successive bifurcations as a parameter, say A for the logistic map, is varied. (See Fig. 2.3 and Eq. (2.5-1).) To be specific, we focus on supercycles in the period-doubling cascade leading to chaotic behavior. Those supercycles are the cycles that include the critical value $x_c = \frac{1}{2}$ in the trajectory. The relevant distance is the distance between $\frac{1}{2}$ and the value of x that occurs <u>halfway</u> through the cycle. This value of x is the trajectory value closest to the trajectory point $x = \frac{1}{2}$. For a cycle of period 2^n, the relevant distance is

$$d_n = \tfrac{1}{2} - f_{A_n^s}^{(2^{n-1})}(x = \tfrac{1}{2}) \tag{5.6-1}$$

That is, we start the trajectory at $x = \frac{1}{2}$, iterate 2^{n-1} times, halfway through a cycle of period 2^n. The subscript on A_n^s tells us that we are looking at a supercycle of period 2^n.

The Feigenbaum α is defined as the number that satisfies

$$-\alpha = \lim_{n\to\infty} \frac{d_n}{d_{n+1}} = \lim_{n\to\infty} \frac{\tfrac{1}{2} - f_{A_n^s}^{(2^{n-1})}(x = \tfrac{1}{2})}{\tfrac{1}{2} - f_{A_{n+1}^s}^{(2^n)}(x = \tfrac{1}{2})} \tag{5.6-2}$$

The minus sign reminds us that the "nearest neighbor" is alternately above and below $x = \frac{1}{2}$.

To find the numerical value of α, we need to establish the existence of a universal function $g(y)$ that satisfies

$$g(y) = -\alpha g(g(-y/\alpha)) \tag{5.6-3}$$

This equation tells us that there is a special function g and a special number α such that iterating the function once gives the same result as rescaling the independent variable axis by the value $-\alpha$, iterating twice and then multiplying the result by $-\alpha$. We now show the connection between the definition given in Eq. (5.6-2) and the function defined in Eq. (5.6-3).

We start with Eq. (5.6-2) re-expressed in terms of a new variable $y = \frac{1}{2} - x$. (We are just shifting the independent variable axis so that the maximum of the iterated map function occurs at $y = 0$. Then, we introduce the notation

$$y_{[n]} = d_n = \tfrac{1}{2} - f_{A_n^s}^{(2^{n-1})}(x = \tfrac{1}{2}) \tag{5.6-4}$$

Dropping the limit in Eq. (5.6-2) and assuming that the ratio still holds for values *near y = 0* we may write

$$-\alpha y_{[2n]} = y_{[n]} \tag{5.6-5}$$

COP99 provides some numerical and graphical evidence for this assumption. Since we are claiming that Eq. (5.6-5) holds for all values of n, we may also write

$$-\alpha y_{[2(n+1)]} = y_{[n+1]} \tag{5.6-6}$$

We are seeking a map function g that generates this sequence of y values:

$$y_{[n+1]} = g(y_{[n]}) \tag{5.6-7}$$

Using this definition, we may rewrite Eq. (5.6-6) as

$$-\alpha g(g(y_{[2n]}) = g(y_{[n]}) \tag{5.6-8}$$

Finally, we use Eq. (5.6-5) and assume that the result holds for all n. We may then drop the subscripts on y to get our desired result:

$$-\alpha g(g(-y/\alpha) = g(y) \tag{5.6-9}$$

If we assume that the map function has quadratic behavior near its maximum value, we write

$$g(y) = 1 - by^2 \tag{5.6-10}$$

where we have assumed without any lose of generality that the maximum value of the function is 1. We then insert Eq. (5.6-10) into Eq. (5.6-9) to find

$$1 - cy^2 = -\alpha(1-c) - 2c^2 y^2 / \alpha + c^3 y^4 / \alpha^3 \tag{5.6-11}$$

Since we are interested in small values of y, we drop the y^4 term. For the resulting equation to hold for a variety of values of y, we must have

$$1 = -\alpha(1-c) \text{ and } c = \alpha/2 \tag{5.6-12}$$

Equations (5.6-12) then lead to a quadratic equation for α:

$$\alpha^2 - 2\alpha - 2 = 0 \qquad (5.6\text{-}13)$$

whose solution is $\alpha = 2.73$, a value within 10% of Feigenbaum's value 2.502....

Appendix F shows how to improve this argument to find a more precise value for α and the function $g(y)$ itself. Similar arguments, also presented in Appendix F, show how to obtain Feigenbaum's δ. The crucial point is that the values of α and δ depend only on very general features of the iterated map function, specifically its mathematical behavior near the critical point. Hence, we expect to find the same values of α and δ for a wide variety of iterated map functions and for physical systems whose dynamics are well modeled by such map functions.

5.7 Tent Map

In this section we examine the behavior of iterates of another one-dimensional map function called the ***tent map***, because its graph reminds one of the front view of a tent. (Because of its shape, it is also called a ***triangle map***.) The tent map is an example of a class of map functions that are called ***piece-wise linear***. This terminology means that the map function graph is made up of sections of straight line segments. The tent map function is continuous, but it does not have a derivative at the point where the straight-line segments of different slopes meet. The derivative has one value to the left of the meeting point and changes discontinuously to another value to the right of the meeting point. As we shall see, this lack of continuity in the derivative of the map function makes the behavior of the iterates quite different from the behavior of the iterates of the smooth, unimodal map functions discussed thus far.

The tent map function is given by the equation

$$x_{n+1} = f(x_n) = r\left(1 - 2\left|x_n - \tfrac{1}{2}\right|\right) \qquad (5.7\text{-}1)$$

where r is the control parameter and as usual, we restrict ourselves to x values in the interval $[0,1]$. Figure 5.14 shows a graph of the tent map function for two different values of r: one with $r < 1/2$; the other with $r > 1/2$.

It should be clear from Fig. 5.10 that there is only one fixed point for the tent map function for $r < 1/2$. That fixed point is located at the origin. Since the slope of the map function is less than one at the origin (for $r < 1/2$), that fixed point is stable; that is, trajectories starting anywhere on the interval $[0,1]$ are attracted toward $x = 0$ for $r < 1/2$.

Exercise 5.7-1. Use the graphic iteration method to show that tent map trajectories starting between 0 and 1 are attracted to $x = 0$ for $r < 1/2$. What happens if $x_0 = 1$? What happens if x_0 is outside the interval $[0,1]$?

For $r > 1/2$, there are now two fixed points; one at $x = 0$; the other at

$$x = \frac{2r}{1 + 2r} \qquad (5.7\text{-}2)$$

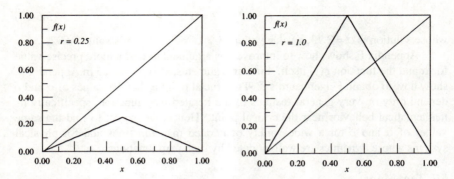

Fig. 5.10. The tent map function is plotted as a function of x. On the left the control parameter $r = 1/4$. On the right $r = 1$. For reference the line $y = x$ is also plotted on the graphs.

From Fig. 5.10, we see that the magnitude of the slope of the map function is greater than 1 at each fixed point. Thus, we conclude that both fixed points are unstable; trajectories are repelled by <u>both</u> fixed points.

Exercise 5.7-2. (a) Show that Eq. (5.7-2) gives the location of the second fixed point for the tent map for $r > 1/2$. (b) Show that the magnitude of the slope of the tent map function is greater than 1 for $r > 1/2$. (The magnitude of the slope for the tent map is the same everywhere, except at $x = 1/2$, where it is not defined.) (c) What happens to a trajectory that reaches $x = 1/2$ for $r = 1$? Do you expect this behavior to cause a problem in a graphical or numerical calculation of trajectories?

What happens to trajectories for $r > 1/2$? You should convince yourself, by using either the graphic iteration method or a computer program, that trajectories starting between 0 and 1 remain bounded (for $r \le 1$) in the interval [0,1] and, more importantly, that two trajectories starting close to each other, say, to the left of $x = 1/2$, diverge until they are "folded back" by mapping from the right-hand side of the map function. In fact, these initially close trajectories diverge exponentially as a function of the iteration number. These trajectories are chaotic. The constancy of the magnitude of the slope of the tent map allows us to find the rate of divergence quite easily (Exercise 5.7-3). The crucial point to be noted here, however, is that in the case of the tent map the behavior of the iterates changes dramatically from stable fixed point behavior for $r < 1/2$ directly to chaotic behavior of $r > 1/2$. The iterates do not pass through a sequence of period-doublings to reach chaos.

Exercise 5.7-3. Use Eq. (5.4-12) to compute the average Lyapunov exponent for the tent map. Show that it is positive for $r > 1/2$. Hint: The absolute value of the slope of the tent map function is the same at every trajectory point.

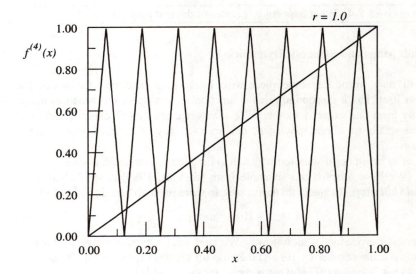

Fig. 5.11. A graph of the fourth-iterate of the tent map function with $r = 1.0$. The line $y = x$ is included for reference.

We can get some understanding of why the period-doublings are missing by plotting some higher iterate of the tent map function, say $f^{(4)}(x)$ as shown in Fig. 5.11. We see that the magnitude of the slope of $f^{(n)}(x)$ is greater than the slope of $f(x)$ (except at the peak and valley points where the slope is not defined). In fact the magnitude of the slope of the nth iterate of f is the magnitude of the slope of f raised to the nth power.

$$\left| \frac{df^{(n)}(x)}{dx} \right| = \left| \frac{df(x)}{dx} \right|^{n} \tag{5.7-3}$$

This result is easy to demonstrate: Look at the first segment of the original tent map function; that is, the segment from 0 to 1/2. For the nth iterate, the first segment goes from $x = 0$ to $x = (1/2)^{n}$, arriving at the same ordinate value, but in a shorter horizontal distance. Hence, its slope must satisfy Eq. (5.7-3).

We conclude from the preceding analysis that all the fixed points of the higher-order iterates of the tent map are unstable if the fixed points of $f(x)$ are unstable. By way of contrast, we should recall that what made period-doubling possible for the logistic map was the stability of some of the fixed points of $f^{(2)}(x)$ when the fixed points of $f(x)$ became unstable. Thus we see that the smoothness (continuous differentiability) of the map function is crucial in having a period-doubling sequence precede the onset of chaos. In more practical terms, if we use a piece-wise linear map function to model the dynamics of some system, we should not expect to see a period-doubling sequence for that system.

Exercise 5.7-4. What happens to iterates of the tent map for $r > 1$?

5.8 Shift Maps and Symbolic Dynamics

We will now introduce some piece-wise linear map functions for which the function itself has a discontinuity as a function of x; that is, the function jumps suddenly from one value to another. We shall see that a certain class of such map functions leads to a very powerful, but abstract way of characterizing chaotic behavior.

Let us begin our discussion with a map function whose graph is shown in Fig. 5.12. (We have adapted this example from [Stewart, 1989].) We call this the *decimal shift map*. In algebraic terms, we can write the map function as follows:

$$x_{n+1} = 10x_n \quad \mod[1] \tag{5.8-1}$$

The iteration procedure is as follows: We start with a number between 0 and 1. We multiply that number by 10, and then lop off the digit to the left of the decimal point. (That "lopping off" is what is meant by the "mod [1]." "mod" is short-hand for *modulus*.) For example, if we start with $x_0 = 0.89763428$, we multiply by 10 to get 8.97634280, and then drop the 8 to the left of the decimal point to arrive at $x_1 = 0.97634280$. Following the procedure again leads to $x_2 = 0.76342800$. As you can see, the procedure shifts the decimal digits left by one location each iteration and chops off the digit that goes to the left of the decimal point. (We have added zeroes to fill in the places that have been vacated by the shift; that is, we have assumed that the original number was actually 0.897634280000000...)

We can now ask the important question: What happens to the trajectories that begin with various values of x_0? It should be obvious from the example in the previous paragraph that any x_0 represented by n digits followed by an infinite string of zeroes will give rise to a trajectory that ends up at $x = 0$ after n iterations. Can we conclude that $x = 0$ is an attractor for the system? Let us answer this by looking at the trajectory that begins with $x_0 = 1/7 = 0.142857142857142$ .... It should be obvious that this trajectory will cycle among seven values of x forever since the same seven digits appear over and over again in the decimal representation of 1/7. In fact any *rational* number, that is, any number that can be presented by a *ratio* of two integers, say m and n, results in a decimal representation that ends in repeating digits (which may be zeroes as in the cases of 3/8 or 1/10, for example). Thus we can conclude that any initial x_0 that is a rational number will lead to a periodic trajectory under the action of the decimal shift map. (Settling to $x = 0$ is obviously just a special case of a periodic trajectory.)

What happens, however, to trajectories that start on x values that are not rational numbers? (These numbers are called irrational. These numbers cannot be represented exactly as ratios of integers.) For example, what happens if $x_0 = \pi/10$? Since π is an irrational number, it consists of a sequence of digits that never repeats. Hence, the trajectory starting from this x_0 will never repeat. In fact,

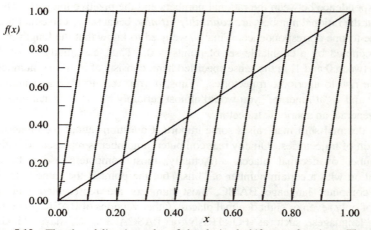

Fig. 5.12. The dotted line is a plot of the decimal shift map function. The function is discontinuous at $x = 0.1$, 0.2, and so on. The solid line $y = x$ is included for reference.

as we shall see later, it wanders, apparently randomly, over the entire interval between 0 and 1.

What happens to nearby trajectories? Suppose we start one trajectory on $x_0 = \pi/10$ and a second trajectory on $x_0 = 22/70$, a rational number close to $\pi/10$, an irrational number. By the arguments presented earlier, we know that the second trajectory will eventually settle down to a periodic cycle, while the first will wander randomly throughout the interval. Thus, we can conclude that this system displays sensitive dependence on initial conditions. Nearby starting points can lead to completely different long term behaviors.

Are the trajectories chaotic? This would appear to be a subtle question because some trajectories—those that begin on rational values of x—lead to periodic trajectories, while those beginning on irrational values of x wander over the interval never repeating. However, we have adopted the definition of chaos given in Section 5.5; namely, the average Lyapunov exponent must be positive. Thus, we see that the decimal shift map has chaotic trajectories because the slope of the map function is always greater than 1, except at those isolated points of discontinuity at which the derivative is not defined.

Another important question is: Do most of the starting values x_0 lead to periodic or aperiodic trajectories? The answer is that most values of x_0 lead to aperiodic trajectories. What do we mean by "most"? To understand this issue, we will rely on some results from number theory in mathematics. (The interested reader is referred to [Wright and Hardy, 1980] and [Zuckerman, Montgomery, Niven, and Niven, 1991] for more details.) First, let us note that between 0 and 1 there are an infinite number of rational values of x. Thus, it might seem that there is no "room" left for the irrational numbers. However, we can measure how much of the interval is taken up by rational numbers by noting that we can form a one-to-one correspondence between the rational numbers and the positive integers. Thus, we say that the rational numbers are ***countably infinite***, because we can count them by this one-to-one correspondence. Finally, we need to know that the length of an interval occupied by a countable set of points is 0. Thus, essentially all of the interval between 0 and 1, in the sense specified here, consists of irrational numbers, which give rise to aperiodic trajectories. Thus, if you were to choose a number between 0 and 1 "at random," you would almost certainly hit an irrational number and thus generate an aperiodic trajectory.

This decimal shift map raises some important questions about the numerical computation of trajectories. On any real computer, a number is represented with a <u>finite</u> number of decimal places. (Actually, most computers use a binary representation with a certain number of "bits," but the point is the same.) In the common computer language BASIC, most numbers are represented with an accuracy of only seven or eight decimal places. Other versions of BASIC and other computer languages, such as FORTRAN or PASCAL or C, have "double precision" representations that use 15 or 16 decimal places. The main point is that no matter what precision you use, you represent any number, rational or irrational, by a finite string of digits. Hence, when you apply the decimal shift map to a number on the computer, after seven or eight shifts in BASIC, for example, you will have shifted away all the numbers you originally specified. What happens after that depends on the computer language you are using. Most languages (but not all) will fill in zeroes and hence all trajectories, even those starting from what you think might be irrational numbers, will eventually end up at $x = 0$.

We now introduce a shift map that is based on a binary number representation. In algebraic terms, the map function can be written

$$x_{n+1} = 2x_n \quad \mathrm{mod}[1] \tag{5.8-2}$$

This map function is called the ***Bernoulli shift map*** (after the brothers Jakob and Johann Bernoulli, Swiss mathematicians active in the late 1600s). The graph of the Bernoulli shift map function is shown in Fig. 5.13.

The Bernoulli shift map tells us to multiply our starting number by 2 and again lop off any part that ends up to the left of the decimal point. If we use a decimal representation of x_0, it may not be so obvious that we are performing a shift operation. However, if we use a binary number representation of the xs, where the possible symbols are 0 and 1 and the place values to the right of the "binary point"

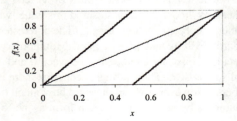

Fig. 5.13. A graph of the Bernoulli shift map. The line $y = x$ is included for reference.

are $1/2$, $(1/2)^2$, $(1/2)^3$, and so on, then multiplying the original value by 2 is equivalent to shifting the string of 1s and 0s left one place, and mod [1] is equivalent to dropping anything to the left of the binary point. For example, if we start with $x_0 = 0.111000$ (this is equivalent to the decimal $0.875000 = 7/8$), then multiplying by 2 yields 1.110000, and the mod [1] gives us 0.110000. We see that the overall process is equivalent to a shift of the 1s and 0s one place to the left.

Which values of x_0 lead to which kinds of trajectories? The reasoning is exactly that used for the decimal shift map: Any rational number is represented by a repeating sequence of 1s and 0s and hence leads to a periodic trajectory. Any irrational number is represented by a nonrepeating sequence of 1s and 0s and hence leads to an aperiodic trajectory. Since there is an irrational number close to every rational number (and vice versa), the map exhibits sensitive dependence on initial conditions. Again, we have an infinite number of initial points leading to periodic trajectories and an infinite number of initial points leading to aperiodic trajectories. If you choose an x_0 at random though, you are likely to choose one leading to an aperiodic trajectory.

Exercise 5.8-1. Show that the Lyapunov exponent for the Bernoulli shift map is equal to $\ln 2$.

The Bernoulli shift map is easily tied to the notion of randomness. Suppose we flip a coin many times and record the sequence of heads and tails that we obtain by the sequence of symbols HTTHHHTHTHTTHH, and so on. Let us now replace H by 1 and T by 0. This leads to a sequence of 1s and 0s that look just like the sequence of 1s and 0s from the binary representation of an irrational number, say $\sqrt{2} - 1$. Another way of saying this is: If we give you a "random" sequence of 1s and 0s, you cannot tell whether the sequence comes from the procedure of tossing a coin, which we take as a paradigm for randomness, or from reading the 1s and 0s that appear next to the binary point when the Bernoulli shift operation is applied to the binary representation of an irrational number. The sequence of 1s and 0s generated by the Bernoulli shift applied to an irrational number is as random as a coin toss.

The Bernoulli shift map is an example of *symbolic dynamics*, a formalism in which our attention is focused solely on a sequence of a finite number of symbols (usually just two). Obviously we are far from any direct connection to the physical world, but the dynamics of the symbols does allow us to say something about the periodicity or chaoticity of the dynamics. The RLRR-type of sequence used in the description of the U-sequence for one-dimensional maps is another example of symbolic dynamics. As we have seen, the nature of the symbolic sequence can be analyzed in terms of the sequence of symbols used to represent rational or irrational numbers. Hence, all the results of number theory can be used to help characterize the kinds of possible sequences. For more information on symbolic dynamics and its use in the theory of chaos, the reader is referred to [Devaney, 1986].

A Special Case of the Logistic Map

A rather special situation occurs for the logistic map function when $A = 4$: By a clever change of variables, we can show that the logistic map function for this value of A is equivalent to the Bernoulli shift map! To see this connection, we introduce a new variable θ, which is related to the logistic map variable by

$$x = \frac{1 - \cos(\pi\theta)}{2} \tag{5.8-3}$$

where the variable θ lies in the range 0 to 1. If Eq. (5.8-3) is used in the logistic map Eq. (5.3-2) with $A = 4$, then we find after some algebraic manipulation that

$$\cos(\pi\theta_{n+1}) = \cos(2\pi\theta_n) \tag{5.8-4}$$

This equation is satisfied if

$$\theta_{n+1} = 2\theta_n \quad \text{mod} \, [1] \tag{5.8-5}$$

which we recognize as a Bernoulli shift map function. We conclude that for $A = 4$, the logistic map function is equivalent to the Bernoulli shift map function.

From a sequence of the θs generated from Eq. (5.8-5), we could find the corresponding sequence of xs for the logistic map by using Eq. (5.8-3). We also see that the Lyapunov exponent for the logistic map for $A = 4$ must be equal to the Lyapunov exponent for the Bernoulli shift map—namely, ln 2. Furthermore, all the statements made earlier about the randomness of the sequences of values generated by the Bernoulli shift map apply immediately to the sequences of x values generated by the logistic map for $A = 4$. In fact, the logistic map function iteration can be used as a random number generator for computers [CFH92].

5.9 The Gaussian Map

Do all smooth one-dimensional map functions with a single maximum lead to period-doubling, chaos, chaotic bands, and periodic windows? The answer is *no*. In this section, we examine a relatively simple mapping function whose iterates behave, for some range of parameter values, quite differently from the iterates of the logistic or sine maps. We shall see that there is a simple geometric reason to explain these differences.

The map function we will discuss is one we call a "Gaussian map" since it is based on an exponential function, which is often called a Gaussian function. This map function is characterized by two control parameters, which we shall label b and c:

$$x_{n+1} = e^{-bx_n^2} + c \tag{5.9-1}$$

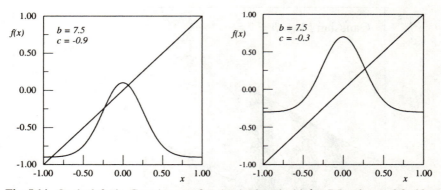

Fig. 5.14. On the left, the Gaussian map function is plotted with $b = 7.5$ and $c = -0.9$. Note that there are three fixed points. On the right, $b = 7.5$ and $c = -0.3$. Here there is only one fixed point.

Based on our discussion of Feigenbaum scaling, we might expect the behavior of this map function to be similar to the behavior of the logistic map function since the map functions are very similar close to their maximum values. (They are both quadratic.) As we shall see, however, the global structure of the map function is also important. Since there are two control parameters, the behavior of the iterates of this map function is much more complicated than is the behavior for the logistic map.

First, let us explore the properties of this function. Figure 5.14 shows the Gaussian map function for $b = 7.5$ and two different values of c. The parameter b controls the "width" of the Gaussian map function: The maximum value occurs at $x = 0$; that maximum value is $c + 1$. For very large values of $|x|$, the function drops to the value given by the parameter c. When $x = 1/\sqrt{b}$ the function has fallen off to e^{-1} of its maximum value above the "baseline" value set by the parameter c. Thus, a rough measure of the width of the function is $2/\sqrt{b}$. It is important to note that this Gaussian map function has two *inflection points*; that is, points at which the slope of the map function has maximum positive and negative values. For large values of x, the magnitude of the slope is small. As x approaches 0, either from the left or from the right, the slope first increases in magnitude, reaches a maximum value at the inflection points and then decreases as x gets close to 0.

Exercise 5.9-1. Show that the inflection points for the Gaussian map occur at $x = \pm 1/\sqrt{2b}$. *Hint*: at an inflection point, the second derivative of the function is equal to 0. Evaluate the magnitude of the slope of the map function at the inflection point in terms of the parameter b.

To display some of the behavior of the iterates of the Gaussian map, we plot some bifurcation diagrams in which the long-term values of the iterates of Eq. (5.9-1) are displayed as a function of the parameter c. We will then examine what happens to these bifurcation diagrams for various values of the parameter b. First,

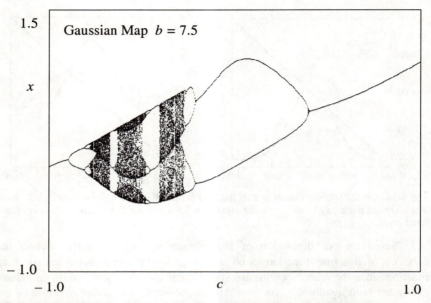

Fig. 5.15. The bifurcation diagram for the Gaussian map with $b = 7.5$ and $x_0 = 0$.

let us begin with a set of parameter values that leads to a bifurcation diagram somewhat reminiscent of the diagram for the logistic map function.

Figure 5.15 shows the bifurcation diagram for this value of b. You should compare this figure with the bifurcation diagram for the logistic map in Fig. 1.14. You will note, starting on the left, that the Gaussian map diagram shows a period-doubling sequence leading to chaos, chaotic bands, and periodic windows (period-3 is particularly clear), just like the logistic map. For larger values of c, however, the two diagrams look quite different. The diagram for the logistic map ends abruptly at $A = 4$. (Recall that for $A > 4$, most trajectories of the logistic map lead to $x \to -\infty$.) For the Gaussian map, the diagram goes through chaotic band mergings and period-doublings—should we call them "period-undoublings"?—and finally ends up with period-one behavior again.

We can understand why the diagram recollapses back to period-one by looking at the location of the fixed points of the map function. It should be clear from Fig. 5.14, that for the extreme values of c, the fixed points occur on the "wings" of the Gaussian function, where the slope is very small. This means that the fixed points are stable and correspond to period-one behavior. By way of contrast, the magnitude of the slope of the logistic map function continues to increase at the fixed point as the parameter A is increased; therefore, period-one behavior occurs for only one range of parameter values. Thus, we see that the existence of the inflection point for the Gaussian map function is important in allowing period-one behavior to occur for two ranges of control parameter values.

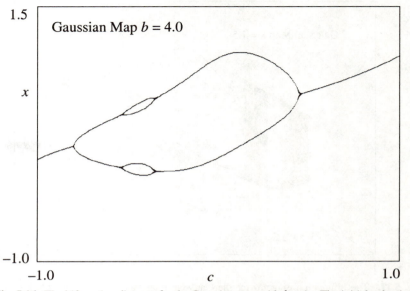

1.5

x

−1.0

−1.0 c 1.0

Fig. 5.16. The bifurcation diagram for the Gaussian map with $b = 4$. The initial value in the iteration is $x_0 = 0$.

Let us now look at the bifurcation diagram for a smaller value of b, namely, $b = 4$. Recall that smaller b values mean that the Gaussian function is wider, as a function of x, and hence the maximum value of the slope of the function is smaller. The resulting bifurcation diagram is shown in Fig. 5.16. As a function of c, the behavior undergoes a period-doubling bifurcation from period-one to period-two and then another from period-two to period-four. Instead of continuing on to chaos, however, the system period-undoubles to period-two and then finally back to period-one. In a whimsical mood, you might call this "period-bubbling" (BIB84). In a more biological frame-of-mind, you might call it the "mouse map."

> **Exercise 5.9-2.** Describe qualitatively what happens to the fixed point of the Gaussian map function and the slope of the function at that fixed point for the range of c values shown in Fig. 5.16.

The Gaussian map, however, has more surprises in store for us. In Fig. 5.17, we show the bifurcation diagram with $b = 7.5$, the same value used in Fig. 5.15. The diagrams, however, are different! For values of x near − 1, the behavior is period-one but with a value different from that shown in Fig. 5.15. There is, then, a sudden change to a diagram that looks just like that in Fig. 5.15. What is the difference between the two figures? In Fig. 5.17, the initial value for the iterations was $x_0 = 0.7$. In Fig. 5.15, we used $x_0 = 0$.

If we look at Fig. 5.14, we can see that for c near −1 and $b = 7.5$, the Gaussian map has three fixed points, two of which are stable. Some initial values of x lead to trajectories that end up on one of the stable fixed points; others lead to trajectories

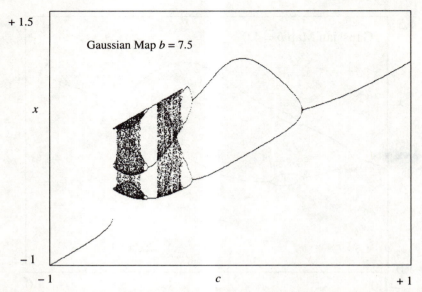

Fig. 5.17. The bifurcation diagram for the Gaussian map function with $b = 7.5$, but with trajectories starting at $x_0 = 0.7$.

heading to the other stable fixed point. Figure 5.18 shows the graphic iteration procedure applied to a starting value of $x_0 = 0$. The trajectory obviously approaches the fixed point that occurs at positive x values. Figure 5.19 shows the graphic procedure applied to a trajectory starting at $x_0 = 0.5$. That trajectory approaches the other stable fixed point. Thus, for these values of b and c, the Gaussian map has two stable fixed points, each of which has its own basin of attraction. For larger values of c, say $c = -0.3$, we can see from the right-hand side of Fig. 5.14 that there is only one fixed point. Hence, all initial points lead to the same attractor.

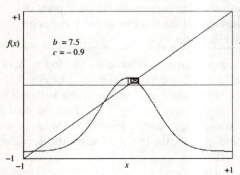

Fig. 5.18. A graphic iteration of a trajectory of the Gaussian map starting at $x_0 = 0$. The trajectory approaches the fixed point just to the right of $x = 0$.

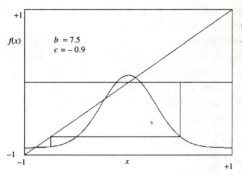

$f(x)$ $b = 7.5$
         $c = -0.9$

Fig. 5.19. A graphic iteration of a trajectory of the Gaussian map starting at $x_0 = 0.5$. The trajectory approaches the fixed point at the left of the diagram.

Exercise 5.9-3. Write a computer program to find the basin of attraction for each of the fixed points shown on the left side of Fig. 5.14. What kind of bifurcation event marks the disappearance of the fixed point located at negative x values?

Should we have anticipated this unusual behavior for the Gaussian map? The crucial feature here is the decrease in the magnitude of the slope of the function for large values of x. By way of contrast, the magnitude of the slope of the logistic map functions increases monotonically as we go away from the maximum at $x = 1/2$. Thus, the logistic map does not display a set of period-undoublings. We would expect that a physical system such as the diode circuit of Chapter 1 that displays these period-undoublings (see Figs. 1.8 and 1.9) should be modeled by an iterated map function that has this inflection point property. The return map (Fig. 5.3) for the diode circuit does show such an inflection point.

The Gaussian map and other maps that share its characteristic geometric features violate the MSS U-sequence. Some of the U-sequences occur for more than one range of parameter values. A set of reverse bifurcations (period-undoublings) seems to herald this type of behavior. In addition to the diode circuit, this type of behavior has been seen in oscillating chemical reactions (CMS86).

A similar type of behavior has been called ***antimonotonicity*** because the sequence of bifurcations does not necessarily follow the monotonic sequence found in the logistic map model. Such effects can occur when there is more than one critical point within the attracting region of state space. The signature of antimonotonicity is the formation of a "dimple" in the return map near the critical point. NKG96 discusses these effects for the diode-inductor circuit.

5.10 Two-Dimensional Iterated Maps

In this section we extend our discussion of iterated maps to map functions of two variables, say, x and y. Just as we saw for systems described by differential equations, increasing the number of dimensions (here, variables) for iterated maps increases greatly the range of possible behaviors. For the most part, the systematic study of two- (or higher-) dimensional iterated map functions is unexplored territory. For more details, refer to [Devaney, 1986, Chapter 2].

To explore the behavior of two-dimensional maps, let us consider a map function, which has become something of a classic in the literature of nonlinear dynamics, the Hénon map. This map function was first discussed by Hénon (HEN76), who introduced it as a simplified model of the Poincaré map for the Lorenz model. The Hénon map function is essentially a two-dimensional extension of the one-dimensional quadratic map (Exercise 5.4-6):

$$x_{n+1} = 1 + y_n - Cx_n^2$$
$$y_{n+1} = Bx_n \tag{5.10-1}$$

C is a positive parameter. We see that when $B = 0$, the Hénon map function reduces to the quadratic map function. The map function applies to the entire xy plane. Depending on the values of the parameters B and C, some regions of the state space give rise to bounded trajectories; other initial points may lead to trajectories that escape to infinity. See Exercise 5.10-2.

The Hénon map is an invertible map function. That is, if we are given x_n and y_n, we can find the unique pair of values x_{n-1} and y_{n-1}, which gave rise to these values. Thus, we can follow both the forward and backward iterations of the Hénon map. By way of contrast, for the logistic, quadratic, and Gaussian maps, there are two possible x_{n-1} values for each x_n.

> **Exercise 5.10-1.** Derive the general expression that gives x_{n-1} and y_{n-1} in terms of x_n and y_n for the Hénon map.

We will restrict ourselves to values of B such that $0 < |B| < 1$; so that the iterations of the Hénon map, when applied to a cluster of initial conditions in the xy plane, collapse that cluster in the y direction. The value $B = 0.3$ has been well studied in the literature, and we shall use that value. Figure 5.20 shows the action of the Hénon map on a rectangular cluster of initial points. Note that the Hénon map induces an effective stretching and folding of the cluster of initial points.

> **Exercise 5.10-2.** (After [Strogatz, 1995]). For fixed values of the parameters B and C, the Hénon map traps some regions of state space in the sense that trajectories starting within those regions stay within that region. Let $B = 0.3$ and $C = 1.4$. Show that the quadrilateral with vertices $(-1.33, 0.42)$, $(1.32, 0.133)$, $(1.245, -0.14)$, $(-1.06, -0.5)$ is mapped to a new region that is entirely inside the original quadrilateral. Hint: write the equation for the straight line segments that form the boundary of the quadrilateral and apply the Hénon map to those lines. Other trajectories may escape to infinity. Find an initial point of the Hénon map that leads to a trajectory that escapes to infinity.

We can construct a bifurcation diagram for the Hénon map by plotting the values of x_n (after allowing transients to die away) as a function of the control

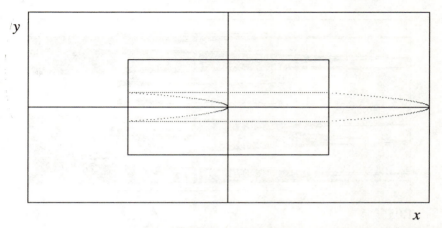

Fig. 5.20. The result of applying the Hénon map function once to the borders of the inner rectangular region in the *xy* plane. The origin is at the center of the diagram. Here $B = 0.3$ and $C = 1$. The area contained inside the initial rectangle is stretched and folded by the action of the map function. The new boundary is indicated by the dotted curve.

parameter C for a fixed value of B. Figure 5.21 shows such a bifurcation diagram. The important feature to note is the presence of two chaotic attractors for C values near 1.08. One of the attractors consists of four chaotic bands, the other of six chaotic bands. Some initial conditions will lead to one attractor; other initial conditions lead to the other. When several attractors coexist for a given parameter value, the system will show *hysteresis*; that is, its behavior depends on its past history.

You may also notice that near $C = 1.08$, one of the chaotic attractors abruptly disappears. This type of event has been called a *crisis* and will be discussed in detail in Chapter 7.

Figure 5.21 shows several periodic windows. We might ask for what range of the two parameters B and C do those periodic windows occur. This question has been addressed in rather general terms in BHG97.

5.11 The Smale Horseshoe Map

In Chapter 4, we described qualitatively how homoclinic and heteroclinic orbits lead to "horseshoe dynamics" in the corresponding Poincaré sections. We will now describe a two-dimensional map, which captures the essence of that stretching, compression, and folding. This map, called the *Smale horseshoe map*, is defined in such a way that the methods of symbolic dynamics can be applied to describe the trajectories that iterates of the map follow. However, in constructing such a map, we give up the notion of an attractor; that is, under the Smale horseshoe map, an area of initial conditions in the *xy* plane does not collapse to some attracting region. We say that the Smale horseshoe map is "area-preserving." Thus, this map cannot be a good model for a dissipative system. On the other hand, the resulting

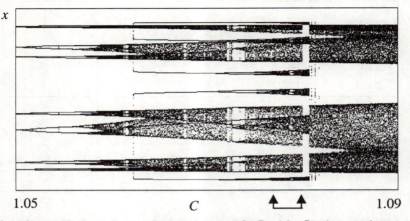

x

1.05 C ↟ ↟   1.09

Fig. 5.21. A bifurcation diagram for the Hénon map for $B = 0.3$. C is the control parameter. Note the coexistence of two attractors for a certain range of C values indicated by the arrows. Different initial conditions give rise to trajectories that lead to the two attractors.

characterization of trajectories is so important and powerful (and it can be applied as a model for nondissipative systems) that we need to understand how the horseshoe map works.

First, we should point out that the horseshoe map is defined geometrically; we do not write down any formulas for the map functions. We begin by considering a square S of initial conditions in the xy plane. The horseshoe map operation can be broken down into two stages: First, the initial square is stretched in the y direction by a factor s and simultaneously compressed in the x direction by a factor $1/s$. (We could use different stretching and compression factors without changing the essence of the argument.) The new long, thin rectangle, whose area is the same as the area of the original square, is then folded over and shifted back so that the rectangular parts of the "legs" overlap the original area. This sequence is shown in Fig. 5.22. (To keep the curved part of the stretched rectangle out of the original rectangle, we need to use $s > 2$.) We want to concentrate on those sections of the horseshoe that overlap the original square. We label these "vertical" rectangles V_0 and V_1. Where did these regions originate? Figure 5.22 shows that the points in V_0 and V_1 came from two horizontal strips, labeled H_0 and H_1. Points in H_0 are mapped into V_0 and points in H_1 are mapped into V_1.

We will also be concerned with the inverse of the horseshoe map, which we shall label as $h^{(-1)}$. In particular, we want to know what happens when we apply $h^{(-1)}$ to the original square S. We can see what this action is by thinking of applying $h^{(-1)}$ to the two vertical strips that we obtained by applying h to the original square. If we use the set theory symbol $\cap$ to indicate the intersection or overlap of two sets of points, we may write

$$h^{(-1)}(S \cap h(S)) = h^{(-1)}(S) \cap S \qquad (5.11\text{-}1)$$

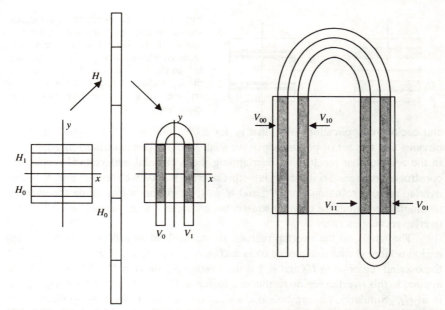

Fig. 5.22. On the left, a picture of the compression-stretching and folding-shifting stages of the horseshoe map. On the right is shown the results of a second iteration of the map, enlarged for the sake of clarity. The subscripts, read from left to right indicate the "history" of that strip, as explained in the text. We focus our attention on those regions of the horseshoe that overlap the original square. These are indicated by the filled-in rectangles.

The previous equation says that applying $h^{(-1)}$ to the two vertical rectangles [which are the overlap of S and $h(S)$] gives us the overlap of S with what we get by applying $h^{(-1)}$ to S. The result of this operation is just the two horizontal rectangles, H_0 and H_1 shown on the left in Fig. 5.23. If the inverse map is applied to these two horizontal rectangles, we arrive at the four thin horizontal rectangles shown in Fig. 5.23. The subscript labels are read from the right. For example, H_{01} is that set of points that were in H_1 after the first application of $h^{(-1)}$, but end up in H_0 after the next application of $h^{(-1)}$.

A second <u>forward</u> iteration of the horseshoe map leads to results shown in the right-hand side of Fig. 5.22. The meaning of the subscripts is: V_{01}, for example, labels the strip that results from applying $h^{(2)}$ to H_{01}. More formally, we write $V_{ij} = h^{(2)}(H_{ij})$. Thus, we see that forward iteration of the horseshoe map leads to an intricate sequence of narrow, vertically oriented rectangular regions. The width of each of the rectangles after n iterations is $(1/s)n$. The horseshoe map is constructed so that in the overlap regions, the map function is linear. Thus, we get the same stretching and the same compression factors for each iteration of the map. Likewise, repeated application of $h^{(-1)}$ leads to an intricate sequence of thin horizontally oriented rectangles.

Now we are ready to be more formal about the horseshoe map. We want to concern ourselves with those points that remain in the square S under all forward

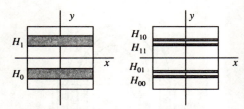

Fig. 5.23. The horizontal rectangles shown on the left transform under the action of the horseshoe map into the rectangle shown in Fig. 5.22. Applying $h^{(-1)}$ to the two horizontal rectangles on the left yields the four rectangles shown on the right. The subscripts are explained in the text.

and backward applications of h, that is, for all $h^{(n)}$ and all $h^{(-n)}$. It should be fairly obvious that this set of points, which we shall label **IN** (for invariant), is contained in the regions that result from overlapping the horizontal and vertical rectangles constructed earlier. To illustrate how this overlap works, we show in Fig. 5.24 the overlap regions resulting from $h^{(2)}$ and $h^{(-2)}$. (Not all the points inside the overlap rectangles remain in the original square, but all those that do remain, are inside the overlap rectangles.)

The labels on the overlap regions are determined as follows. Each overlap region is labeled with a sequence of 1s and 0s: $a_{-2}, a_{-1} . a_0, a_1$. The symbol a_0 is 0 if the overlap region is in H_0 and is 1 if the overlap region is in H_1. Then $a_1 = 0$ if h applied to this overlap region results in a region in H_0 and is 1 if the resulting region is in H_1. Similarly, the symbols a_{-1}, and so on, are determined by application of $h^{(-1)}$. Thus, all overlap blocks across the top of Fig. 5.24, should have labels $xx.10$, while the second row has labels $xx.11$.

Exercise 5.11-1. Verify that the labels of the overlap regions correctly describe the actions of h and $h^{(-1)}$ on those regions. Verify that the labels are also given by combining the subscripts of the H and V regions which overlap. For example, the overlap region 01.00 occurs at the intersection of V_{01} (which gives the subscripts to the left of the period) and H_{00}, which gives the labels to the right of the period.

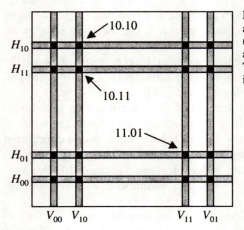

Fig. 5.24. The overlap regions for $h^{(2)}$ and $h^{(-2)}$. The points that remain in S under the action of $h^{(n)}$ and $h^{(-n)}$ for all n are contained in these overlap regions. The labeling of the regions is explained in the text.

If we extend the construction of the set IN by letting the number of forward and backward iterations become very large, then the labeling of the overlap regions, which contain the points of IN, proceeds exactly as described earlier. Thus, we see that each point of IN can be labeled with a doubly infinite sequence of 1s and 0s. This labeling leads to an important connection between the dynamics of points under the action of the horseshoe map and symbolic dynamics. The trajectory of one of the points that remains in S under the action of the horseshoe map can be given by the sequence of labels $\dots a_{-2}, a_{-1}. a_0, a_1, a_2, \dots$ meaning that the point (x,y) (assumed to be a member of IN) is in region H_i for $h^{(i)}(x,y)$.

Note that by our overlap construction, all possible sequences of the symbols 0 and 1 occur. [For the regions shown in Fig. 5.24, there are 16 possible combinations. Also note that symbol sequences that are identical in their values around the period (.) are close to each other on the diagram. For example, the four intersection regions at the top left of Fig. 5.24 all have the labels _0.1_ .] Finally, note that the application of the horseshoe map is equivalent to shifting the symbol sequence one unit to the left. The horseshoe map is, therefore, equivalent to a symbol shift.

> **Exercise 5.11-2.** Verify that the horseshoe map is equivalent to shifting the symbol sequence, defined earlier, left one position.

Since we may view the sequences of 0s and 1s as binary representations of numbers, we can invoke number theory to make the following powerful statements:

1. There is an infinite number of periodic points in the set IN. A point is periodic if repeated application of h bring us back to the same point. This occurs if the sequence symbol contains a repeating pattern of 0s and 1s. (This is expressed formally by saying that the point has period n if $a_i = a_{i+n}$ for all i. The period is said to be a primary period n if the previous equation does not hold for any smaller value of n.) From number theory, we know that these repeating sequences occur when the number is a rational number. Since there is an infinity of rational numbers, there is an infinity of periodic points in the set IN. Note, however, that these are unstable periodic points in the sense that nearby points either leave the square under repeated iterations of h or go off on quite distinct (and in general nonperiodic) trajectories.
2. There is an infinite number of aperiodic points in the set IN. A point is aperiodic if repeated applications of h never bring us back exactly to that point. The corresponding symbol sequence is that of an irrational number. Since there is an infinite number of irrational numbers, there is an infinite number of aperiodic points in the set IN.
3. There is at least one point in the set IN whose trajectory comes arbitrarily close to every point in the set IN. (This trajectory is then called a dense trajectory.) We find this trajectory by construction. That is, we specify the "closeness" by specifying the number (say, n)

of binary symbols around the full stop that must agree when the two points are close to each other. We then construct all possible sequences of n 0s and 1s and string them together. This string describes a point whose trajectory comes "close" to all the points of *IN* as we shift the symbols to the left by the application of h.

What is the dynamical significance of the horseshoe map? The crucial point is that a trajectory entering a region in which the dynamics are (at least approximately) described by a horseshoe map will undergo very complicated behavior under the influence of the unstable periodic and aperiodic points. Since nearby trajectories will behave very differently, this complex behavior is equivalent to what we have been calling "chaotic behavior."

5.12 Summary

We introduced several types of iterated map functions and explored them as dynamical systems in this chapter. The theory of one-dimensional iterated map functions is well-established. A combination of analysis, geometric constructions, and number theory allows us to make many general statements about the attractors and bifurcations associated with these map functions. In two or more dimensions, however, much remains to be learned. In later chapters, we will introduce yet more iterated map functions as models for dynamical systems, and we will often return to the functions introduced in this chapter as familiar friends on which to try out new ideas for characterizing nonlinear dynamics.

5.13 Further Reading

General Treatments of Iterated Map Functions

Most of the texts listed at the end of Chapter 1 have significant sections devoted to iterated maps as dynamical systems. We particularly recommend the books [Devaney, 1986] and [Gulick, 1992].

P. Collet and J. P. Eckmann, *Iterated Maps on the Interval as Dynamical Systems* (Birkhauser, Cambridge, MA, 1980). A thorough introduction to the mathematics of iterated maps.

[Schuster, 1995] gives a detailed treatment of piece-wise linear maps and quadratic maps at a level just slightly more sophisticated than the treatment here.

[Jackson, 1989, Chapter 4] covers one-dimensional iterated maps ("difference equations"). Appendix E shows how a "digraph" method can be used to prove Sarkovskii's Theorem.

R. M. May, "Simple mathematical models with very complicated dynamics," *Nature* **261**, 459–67 (1976) (reprinted in [Cvitanovic, 1984] and [Hao, 1984]). A stimulating introduction to iterated maps *ante* Feigenbaum.

J. Guckenheimer, "One-dimensional Dynamics," *Ann. N.Y. Acad. Sci.* **357**, 343–347 (1981). Provides proofs of many properties of iterated map functions and their trajectories.

Reduction of Dynamics to Iterated Maps

LOR63 shows numerically that the dynamics of the Lorenz model can be reduced to a one-dimensional iterated map function.

The driven damped pendulum mentioned in Section 5.2 is discussed in detail in [Baker and Gollub, 1996]. See also E. G. Gwinn and R. M. Westervelt, "Horseshoes in the driven, damped pendulum," *Physica D* **23**, 396–401 (1986).

Universality in Iterated Maps

N. Metropolis, M. L. Stein, and P. R. Stein, "On Finite Limit Sets for Transformations of the Unit Interval," *J. Combinatorial Theory* (A) **15**, 25–44 (1973) (reprinted in [Cvitanovic, 1984]). Introduced the U-sequence.

T.-Y. Li and J. A. Yorke, "Period Three Implies Chaos," *Amer. Math. Monthly* **82**, 985–992 (1975)

M. Feigenbaum, "The Universal Metric Properties of Nonlinear Transformations," *J. Stat. Phys.* **21**, 669–706 (1979) (reprinted in [Hao, 1984]). Provides a proof of the universality of α and δ.

M. J. Feigenbaum, "Universal Behavior in Nonlinear Systems," *Los Alamos Science* **1**, 4–27 (1980) (reprinted in [Cvitanovic, 1984]). Gives a quite readable introduction to the universal features of one-dimensional iterated maps.

O. E. Lanford III, "A Computer-Assisted Proof of the Feigenbaum Conjectures," *Bull. Am. Math. Soc.* **6**, 427–34 (1982) (reprinted in [Cvitanovic, 1984]). Generates a power series representation of the universal $g(y)$ function.

H. Kaplan, "A Cartoon-Assisted Proof of Sarkovskii's Theorem," *Am. J. Phys.* **55**, 1023–32 (1987). This paper gives a nice introduction to the behavior of the logistic map function as well as an introductory-level proof of Sarkovskii's Theorem.

K. T. Alligood, E. D. Yorke, and J. A. Yorke, "Why Period-Doubling Cascades Occur: Periodic Orbit Creation Followed by Stability Shedding," *Physica D* **28**, 197–203 (1987).

E. Hille, *Lectures on Ordinary Differential Equations* (Addison–Wesley, Reading, Mass. 1969). Shows how the Schwartzian derivative is used in a wide variety of contexts in classical analysis.

D. Singer, "Stable Orbits and Bifurcations of Maps of the Interval," *SIAM J. Appl. Math.* **35**, 260–7 (1978). Proves "Singer's Theorem" on the number of stable orbits possible for iterated maps.

R. Delbourgo, W. Hart, and B. G. Kenny, "Dependence of Universal Constants upon Multiplication Period in Nonlinear Maps," *Phys. Rev. A* **31**, 514–6 (1985).

Experiments Showing Universal Features of Iterated Maps

M. Giglio, S. Musazzi, and U. Perini, "Transition to Chaotic Behavior Via a Reproducible Sequence of Period-Doubling Bifurcations," *Phys. Rev. Lett.* **47**, 243–46 (1981).

R. H. Simoyi, A. Wolf, and H. L. Swinney, "One-dimensional Dynamics in a Multicomponent Chemical Reaction," *Phys. Rev. Lett.* **49**, 245–48 (1982).

J. Testa, J. Perez, and C. Jeffries, "Evidence for Universal Chaotic Behavior of a Driven Nonlinear Oscillator," *Phys. Rev. Lett.* **48**, 714–17 (1982).

W. J. Yeh and Y. H. Kao, "Universal Scaling and Chaotic Behavior of a Josephson-Junction Analog," *Phys. Rev. Lett.* **49**, 1888–91 (1982).

K. Coffman, W. D. McCormick, and H. L. Swinney, "Multiplicity in a Chemical Reaction with One-dimensional Dynamics," *Phys. Rev. Lett.* **56**, 999–1002 (1986). Gives a quite readable description of the observations of violations of the MSS U-sequence for periodic windows in an experimental study of oscillating chemical reactions.

Images of the Critical Point and Related Issues

M. Misiurewicz, *Publ. Math. I.H.E.S.* **53**, 17 (1981). Misiurewicz was the first to point out the importance of the chaotic band merging points in the bifurcation diagrams of one-dimensional iterated maps.

R. V. Jensen and C. R. Myers, "Images of the Critical Points of Nonlinear Maps," *Phys. Rev. A* **32**, 1222–24 (1985). Gives further information on the importance of the images of the critical point.

J. A. Yorke, C. Grebogi, E. Ott, and L. Tedeschini-Lalli, "Scaling Behavior of Windows in Dissipative Dynamical Systems," *Phys. Rev. Lett.* **54**, 1095–98 (1985). A scaling law for the sizes (as a function of parameter) of the various periodic windows of various mappings.

Feigenbaum Universality

S. N. Coppersmith, "A simpler derivation of Feigenbaum's renormalization group equation for the period-doubling bifurcation sequence," *Am. J. Phys.* **67**, 52–54 (1999).

See the references at the end of Appendix F.

Other Map Functions

J. Heidel, "The Existence of Periodic Orbits of the Tent Map," *Phys. Lett. A* **43**, 195–201 (1990). Gives a detailed treatment of periodic points in the tent map, including a demonstration of Sarkovskii's ordering.

The Hénon map was introduced by M. Hénon "A Two-Dimensional Mapping with a Strange Attractor," *Comm. Math. Phys.* **50**, 69–77 (1976).

M. Bier and T. C. Bountis, "Remerging Feigenbaum Trees in Dynamical Systems," *Phys. Lett. A* **104**, 239–44 (1984). This paper discusses "period-bubbling" and the remerging of the period-doubling cascades observed in iterated maps with more than one parameter.

The Smale horseshoe map is discussed in some detail in [Guckenheimer and Holmes, 1990] and [Ott, 1993].

I. Proccacia, S. Thomae, and C. Tresser, "First-Return Maps as a Unified Renormalization Scheme for Dynamical Systems," *Phys. Rev. A* **35**, 1884–90 (1987). Gives a unified, but fairly abstract, treatment of very general one-dimensional iterated maps that include the unimodal maps and circle maps as special cases. The dynamical behavior of these general maps is far richer than that of unimodal and circle maps (to be introduced in Chapter 6) and includes such oddities as *period-tripling*.

If a two-dimensional mapping operates on the real and imaginary parts of a complex number, the results lead to surprisingly complex and beautiful patterns. Variations on this kind of two-dimensional mapping scheme lead to the so-called Julia sets and the now infamous Mandlebrot set. For an introduction to these kinds of mappings, see [Devaney, 1986], [Devaney, 1990], and the references to fractals at the end of Chapter 9.

T. C. Newell, V. Kovanis, and A. Gavrielides, "Experimental Demonstration of Antimonotonicity: The Concurrent Creation and Destruction of Periodic Orbits in a Driven Nonlinear Electronic Resonator," *Phys. Rev. Lett.* **77**, 1747–50 (1996).

E. Barreto, B. R. Hunt, C. Grebogi, and J. A. Yorke, "From High Dimensional Chaos to Stable Periodic Orbits: The Structure of Parameter Space," *Phys. Rev. Lett.* **78**, 4561–64 (1997). Under what conditions do periodic windows occur in higher dimensional chaotic systems with more than one parameter?

Number Theory

E. M. Wright and G. H. Hardy, *An Introduction to the Theory of Numbers* (Oxford University Press, Oxford, New York, 1980). A readable and intriguing introduction to some of the results of number theory used in this chapter.

H. S. Zuckerman, H. L. Montgomery, I. M. Niven, and A. Niven, *An Introduction to the Theory of Numbers*, 5th ed. (John Wiley, New York, 1991).

Symbolic Dynamics

C. Robinson, *Dynamical Systems: Stability, Symbolic Dynamics, and Chaos* (CRC Press, Boca Raton, 1995).

5.14 Computer Exercises

CE5-1. Write a computer program to find the average Lyapunov exponent for the logistic map function. Show that for $A = 4$, that exponent is equal to $\ln 2$ as stated at the end of Section 5.8.

CE5-2. Write a computer program to plot the first eight images of the critical point for the logistic map as a function of the control parameter A. If possible, use different colors to keep track of the successive images. Verify that the images form the boundaries of the attracting regions and that images merge in the periodic windows. Explain what happens at the Misiuriewicz points.

CE5-3. Write a computer program to generate bifurcation diagrams for the Gaussian map introduced in Section 5.9. Explore behavior over a range of b and c values. Explain what you observe.

CE5-4. Use the program Bifur in Appendix E or *Chaos for Java* to zoom in on the chaotic bands for the logistic map. Verify that the parameter values for band mergings are at least approximately described by the Feigenbaum δ. Use the program to examine the bifurcations that take the large period-3 window into chaotic behavior. Are the values of δ and α the same as they are for the "main" period-doubling sequence?

CE5-5. Modify the program Bifur to add some "noise" to the trajectory values for the logistic map function. You can do this by means of the BASIC function RDN, the random number generator. Then produce a bifurcation diagram and observe the effects of different average sizes of the noise. See Appendix H for some ideas of what to expect.

CE5-6. Write a computer program to generate a bifurcation diagram for the tent map of Section 5.7. Explain how and why it differs from the bifurcation diagram of the logistic map.

CE5-7. Use *Chaos Demonstrations* Hénon map section to study the chaotic attractor for various values of the two control parameters. Zoom in on sections of the map to see some of the fine structure.

CE5-8. Write a computer program to implement the so-called *baker's transformation*, which is a two-dimension map function defined as follows

$$y_{n+1} = 2y_n \pmod 1$$
$$x_{n+1} = bx_n \quad \text{for} \quad 0 \le y_n \le 1/2$$
$$= 1/2 + bx_n \quad \text{for} \quad 1/2 < y_n \le 1$$

where the parameter $b \le 1/2$. First, sketch what one iteration of this map does to a square of unit length whose lower left corner is at the origin of the xy axes for $b = 1/2$. Then show that this iteration procedure is similar to that of the Smale horseshoe map. Finally, show that the map has two Lyapunov exponents, one for the x direction and one for the y direction with $\lambda_y = \ln 2$ and $\lambda_x = \ln b$. [Moon, 1992, pp. 317–19] discusses a slightly fancier version of the baker's transformation.

FOY83 also use this transformation as an example for calculations of fractal dimensions.

CE5-9. Use *Chaotic Dynamics Workbench* to study Poincaré sections for the driven damped pendulum. Show that near torque $F = 2.048$, the dynamics of the Poincaré map are equivalent to those of a one-dimensional unimodal map when the frequency ratio is 2/3 and the damping is 1.0. Explore other values of the torque to find more complicated dynamics.

CE5-10. Using the supercycle parameter values given in Table 2.1, write a computer program to find d_n as defined in Eq. (5.6-1) for the logistic map function. Then use the d_n values to estimate the value of the Feigenbaum α as defined in Eq. (5.6-2).

CE5-11. Write a computer program to create a bifurcation diagram for the logistic map model with its parameter $A < 0$. Explore the range $0 > A > -2$. Compare that diagram with the bifurcation diagram for the logistic map model with $A > 0$. Explore similarities and differences between the two diagrams.

6

Quasi-Periodicity and Chaos

Chaos is the score upon which reality is written. Henry Miller, *Tropic of Cancer*

6.1 Introduction

In this chapter, we discuss another important scenario of bifurcations that leads from regular (periodic) behavior to chaotic behavior. This scenario, which involves motion described by several fundamental frequencies, is called the quasi-periodic route to chaos. It has been observed in experiments on several different kinds of systems, some of which we will describe briefly as we proceed. A considerable body of theory has been developed for this route to chaos. As in the case of the period-doubling route to chaos, the theory tells us that there should be some universal quantitative features associated with this scenario. Unfortunately, as is also the case with period-doubling, the theory fails to tell us when we might expect such a scenario to occur for a particular system. As we shall see, there are some "weak" theoretical statements that tell us when this scenario is "likely to occur." However, the types of systems that exhibit quasi-periodicity also show several kinds of complex behavior, some of which are chaotic and some are not. Our understanding of this complex behavior is incomplete.

The quasi-periodic scenario involves competition, in a rough sense, between two or more independent frequencies characterizing the dynamics of the system. This scenario occurs in (at least) two kinds of systems:

1. A nonlinear system with a "natural" oscillation frequency, driven by an external periodic "force." Because the system is nonlinear, the natural oscillation frequency, in general, depends on the amplitude of the oscillations. Here the competition is between the externally applied frequency and the natural oscillation frequency.
2. Nonlinear systems that "spontaneously" develop oscillations at two (or more) frequencies as some parameter of the system is varied. In this case we have competition among the different modes or frequencies of the system itself.

In both cases, there are two (or sometimes more) frequencies that characterize the behavior of the system, and as these frequencies compete with each other, the result may be chaos.

In our discussion, the "independence" of the two (or more) frequencies will be important. The following terminology, first introduced in Chapter 4, is used to describe the ratio of the two frequencies, say, f_1 and f_2. If there are two (positive) integers, p and q, that satisfy

$$\frac{f_2}{f_1} = \frac{p}{q} \tag{6.1-1}$$

then we say that the frequencies are **commensurate** or, equivalently, that the frequency ratio is **rational**. If there are no integers that satisfy Eq. (6.1-1), then we say that the frequencies are **incommensurate** or, equivalently, that the frequency ratio is **irrational**. As we saw in Chapter 4, if the ratio is rational, then we say that the system's behavior is **periodic**. If the ratio is irrational, then we say that the behavior is **quasi-periodic**. (The terms **conditionally periodic** or **almost periodic** are sometimes used in place of quasi-periodic.)

> A note on conventions: We usually assume that the ratio p/q has been reduced to its simplest form; that is, any common factors in the ratio have been removed. Hence, if $f_2/f_1 = 4/6$, we would remove the common factor of 2 and write $f_2/f_1 = 2/3$. We shall also refer to the p/q frequency ratio as the $p{:}q$ frequency ratio.

The term *quasi-periodic* is used to describe the behavior when the two frequencies are incommensurate because, in fact, the system's behavior never exactly repeats itself in that case. Indeed, the time-behavior of a quasi-periodic system can look quite irregular. Figure 6.1 shows the time evolution of a system described by two frequencies. On the left, the frequencies are commensurate, and the behavior is obviously periodic. On the right, the frequencies are incommensurate, and the behavior looks quite irregular. However, a power spectrum measurement of the quasi-periodic behavior shows clearly that only two frequencies are present. If the behavior on the right were chaotic, then the power spectrum would involve a continuum of frequencies. See Appendix H, Figure H.2.

As we shall see in our subsequent discussion, whether the frequencies are commensurate or incommensurate plays an important role in the behavior of the system. We must ask, therefore, how do we know in practice whether two frequencies f_1 and f_2 are commensurate or incommensurate? The problem is that any actual measurement of the frequencies has some finite precision. Similarly, any numerical calculation, say on a computer, has only finite arithmetical precision: any number used by the computer is effectively a rational number. All we can say is that to within the precision of our measurements or within the precision of our numerical calculations, a given frequency ratio is equal to a particular irrational number or a particular rational number, which is close to that irrational number. Beyond that we cannot say whether the ratio is "really" rational or irrational.

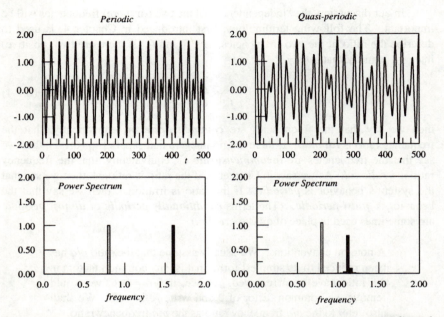

Fig. 6.1. On the left is the time evolution of a system with two frequencies. Here $f_1 = 2\,f_2$. On the right is the time evolution when the two frequencies are incommensurate $f_2 = \sqrt{2}\,f_1$. The behavior on the right looks quite irregular, but the power spectrum shown in the lower part of the figure indicates that only two frequencies (with different amplitudes) are contributing to the behavior. The crucial point is that in the case on the right has two incommensurate frequencies. The widths of the power spectrum "peaks" are due to the relatively short time interval of data used in the analysis.

The question just raised is similar to the one raised in earlier chapters: Is the behavior of a system really chaotic or does it merely have a very long periodicity, that is, long compared to the time duration of our experiment? All we can say is that if the behavior is nonperiodic at a certain level of measurement or computational precision and displays, within the precision of the measurements, divergence of nearby trajectories, then the system is behaving chaotically at that level of precision. Similarly, if two frequencies appear to be incommensurate at a certain level of measurement or computational precision, then we expect the corresponding behavior to occur roughly at that same level of precision.

6.2 Quasi-Periodicity and Poincaré Sections

As we have seen before, the use of Poincaré sections will reduce considerably the geometric complexity of the description of the state space behavior of the system. Let us review how commensurate and incommensurate frequencies lead to rather distinct Poincaré sections.

Poincaré Sections

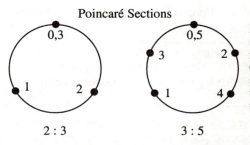

Fig. 6.2. On the left is the Poincaré section for a trajectory with frequency ratio = 2:3. The trajectory starts at point 0, then hits at point 1, then 2, and finally arrives back at its starting point at 3. On the right is the Poincaré section for a trajectory with a frequency ratio of 3:5.

As we learned in Chapter 4, behavior with two frequencies can be described by trajectories confined to the surface of a torus. One frequency, say f_1, is associated with the motion of the trajectories around the large circumference of the torus; the other frequency, f_2, is associated with the motion around the small cross section of the torus. In the cases for which one of the frequencies is the frequency of some modulation or disturbance that we apply to the system, then that frequency is under our control. For those conditions, we will associate that frequency with f_1 because Poincaré sections are most easily constructed by sampling the behavior of the system at a fixed phase of the externally controllable frequency.

If we choose a Poincaré section that slices across the torus as shown in Fig. 4.9, we then arrive at the following description of the Poincaré intersection points for different values of the frequency ratio p/q: Let $T_1 = 1/f_1$ be the time period for one trip around the outer circumference of the torus. $T_2 = 1/f_2$ is the time for a trip around the small cross section. Since the frequency ratio is p/q, the ratio of the periods is

$$\frac{T_1}{T_2} = \frac{p}{q} \qquad (6.2\text{-}1)$$

For example, if $p/q = 2/3$, then the time to go around the small cross section is 3/2 the time period to go around the large circumference.

How does this ratio show up in the Poincaré section? Figure 6.2 shows the Poincaré intersection points for two different frequency ratios. The trajectories start at the point labeled 0. After one time around the large circumference of the torus, the trajectory point is back to the Poincaré plane. For the graph on the left of Fig. 6.2, the trajectory has traveled 2/3 of the way around the small cross section during that time. Hence, it arrives at the point marked 1 in the diagram. After one more trip around the outside of the torus, the trajectory point is 4/3 of the way around and arrives at point 2. Finally, after three times around the outer circumference, the trajectory point arrives back at its starting point. The sequence of Poincaré intersection points for a frequency ratio of 3/5 is shown on the right side of Fig. 6.2. Since we associate f_1 with the motion around the outer circumference of the torus and f_2 with the motion around the cross section of the torus, and since we have chosen a Poincaré plane that cuts the torus perpendicular to the large circumference, the Poincaré section will consist of q points if the frequency ratio

$f_2/f_1 = p/q$. The trajectory "skips over" $p - 1$ points from one intersection with the plane to the next.

> **Exercise 6.2-1.** Draw the Poincaré intersection points for motion on a torus with frequency ratio $f_2/f_1 = 2/7$. How could you distinguish that Poincaré section from one with frequency ratio $= 3/7$? Do the same for $f_2/f_1 = 9/7$.

If the two frequencies are <u>incommensurate</u>, the Poincaré points will never (in principle) repeat. Eventually, the Poincaré points fill in a curve in the Poincaré plane as shown in Fig. 4.9. The intersection points drift around the curve forming what is called a ***drift ring***.

Let us consider once more the problem of distinguishing periodic from quasi-periodic motion. Suppose our experimental resolution is such that we can distinguish 79 points around the curve formed by the intersection of the state space torus with the Poincaré plane. Then, if the motion consists of two commensurate frequencies with frequency ratio of 67/79, we can identify that motion clearly as one with a rational ratio of frequencies. However, if the frequency ratio is 82/93, then we cannot distinguish this motion from quasi-periodic motion with an irrational frequency ratio simply by looking at the Poincaré section; both motions seem, at this level of resolution, to fill in a drift ring.

6.3 Quasi-Periodic Route to Chaos

We are now ready to describe the quasi-periodic route to chaos. This scenario is shown schematically in Fig. 6.3 in terms of the attractors of the system in state space.

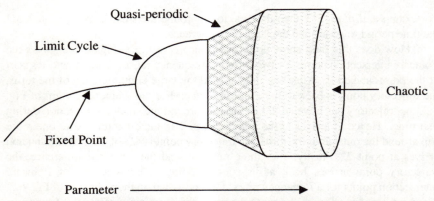

Fig. 6.3. A schematic representation of the evolution of attractors in state space for the quasi-periodic route to chaos. A periodically driven system does not have a fixed point, but begins its evolution with a limit cycle. As a parameter of the system is changed, a second frequency may emerge. If that frequency is incommensurate with the first, then quasi-periodic behavior results. As the parameter is changed further, the behavior may become chaotic.

The system, if it is not externally driven by a periodic force, may start with a fixed-point attractor (i.e., a time-independent state). As a control parameter is changed, the system may undergo a Hopf bifurcation to develop periodic behavior characterized by a limit cycle in state space. A second frequency may appear with a further change in the control parameter. The state space trajectories then reside on the surface of a torus. If the second frequency is incommensurate with the first, then the trajectory eventually covers the surface of the torus. Further changes in the control parameter then lead, for some systems, to the introduction of a third frequency. In state space the trajectories live on a three-dimensional torus (which is not easy to visualize). With further parameter changes, the system's behavior may become chaotic. (Some systems may apparently go directly from two-frequency behavior to chaotic behavior. We shall discuss this issue in the next section.)

This scenario displays the same geometric requirements for chaotic behavior that we have seen before: We need at least three state space dimensions for chaos. If the state space trajectories are confined to the two-dimensional surface of a torus, then the system's behavior cannot be chaotic. We need a third state space dimension, signified either by the third frequency or by the "destruction" of the two-dimensional torus surface as the trajectories move off the surface and show three-dimensional behavior.

To see the significance of this scenario, we must learn something about an older scheme proposed to explain the appearance of chaotic (turbulent) behavior in such systems. This scheme was originally proposed by the eminent Soviet physicist L. Landau in 1944 (LAN44) [Landau and Lifshitz, 1959] to explain the production of turbulence in fluid systems as they are driven away from equilibrium. In Landau's scheme, the system's behavior would be characterized by an *infinite* sequence of Hopf bifurcations, each generating a new frequency incommensurate with the others, as a control parameter is changed. With an infinite number of incommensurate frequencies, Landau hoped to describe the complex behavior of turbulent fluids. In contrast, the quasi-periodic scenario leads to chaos (which is not the same as fully developed turbulence) with the generation of only two (or perhaps three) incommensurate frequencies. Some experiments, described later, indicate that nature seems to take the quasi-periodic route. Other experiments (WKP84), however, show that in spatially extended systems, we may have nonchaotic behavior with more than three incommensurate frequencies.

6.4 Universality in the Quasi-Periodic Route to Chaos

As we mentioned earlier, the theory underlying the quasi-periodic route to chaos tells us only that this scenario is likely to lead to chaotic behavior, not that it must. In 1971, Ruelle and Takens (RUT71) first proposed the quasi-periodic scenario. In 1978, Newhouse, Ruelle, and Takens (NRT78) proved more rigorously that if the state space trajectories of a system are confined to a three-dimensional torus (corresponding to three-frequency quasi-periodic motion in a state space with four or more dimensions), then even a small perturbation of the motion (due to external noise, for example) will "destroy" the motion on the torus and lead to chaos and a strange attractor.

We will not go into the details of what constitutes a suitable small perturbation, but only say that in practice the transition to chaos does seem to occur as described. In the cases when only two frequencies are observed before chaos sets in, we might say that the three-dimensional torus is destroyed by extremely small perturbations. In an experiment using two coupled nonlinear electrical oscillators driven by a sinusoidal signal source (so, there are generally three independent frequencies), Cumming and Linsay (CUL88 and LIC89) showed that for weak coupling between the oscillators, two-frequency and three-frequency quasi-periodicity were easily observable, but that chaos also occurred for certain ranges of parameter values.

Other experiments that display a transition from two-frequency quasi-periodicity directly to chaos seem to involve stronger couplings between the oscillators and happen to have a set of parameter values that place them in a chaotic regime of the parameter space. In these regions, bands of chaotic behavior generally alternate with bands of two-frequency quasi-periodicity. Hence, we can say that the smallest noise perturbation destroys the three-frequency quasi-periodicity for these systems, and chaotic behavior occurs. The system is bumped immediately to a chaotic attractor. This scenario seems to dominate when the coupling between the oscillators is moderately strong. When the coupling is weak, the three-frequency quasi-periodicity is more likely. In general there seems to be a complex interplay between the strength of the coupling and the amplitude of the individual oscillators in determining the details of the scenarios that are observed.

In 1982, on the basis of numerical calculations, Shenker (SHE82) suggested that the quasi-periodic route to chaos displays universal quantitative features much like the Feigenbaum universality for one-dimensional iterated maps. Shortly thereafter, two groups (ROS82 and FKS82) published papers in which they established, using renormalization theory techniques similar to those outlined in Appendix F for the period-doubling route, that one would expect universal quantitative features in the quasi-periodic route to chaos if the transition to chaos were made with the frequency ratio approaching a specified irrational number. (In Section 6.10, we will discuss this matter in more detail.) Here *universal* means that the same numbers should occur for different systems as long as the frequency ratios are held fixed to the same values. Later we shall see that systems can also become chaotic via a ***frequency-locking*** route and that there are different features for that route. Before we can understand the significance of these universal features, however, we need to discuss the notion of frequency-locking to which we turn in the next section.

6.5 Frequency-Locking

Frequency-locking is a common phenomenon whenever two or more "oscillators" interact nonlinearly. Each oscillator is characterized by some frequency f_1 for oscillator number 1 and f_2 for oscillator number 2. (The oscillators may be physically distinct oscillators—two different pendulum clocks, for example—or they may be different "modes of motion" within the same physical system.) If the two frequencies are commensurate over some range of control parameter values, that is, if $f_2/f_1 = p/q$ (p,q integers) over this parameter range, then we say that the two oscillators are *frequency-locked* (or equivalently, *mode-locked* or *phase-locked*). One frequency is locked into being a rational number times the other frequency. For example, frequency-locking explains the fixed relationship between rotation frequency and orbital frequency for the Moon's motion around the Earth (we, on the Earth, get to see only one side of the Moon) and the motion of Mercury about the Sun. In both cases, tidal forces cause an interaction between the axial motion of rotation and the orbital motion. The two motions become locked together.

To understand the physical significance of frequency-locking, we need to keep in mind two facts about nonlinear systems:

First, for a nonlinear oscillator, in general, the actual oscillation frequency depends on the oscillator's amplitude of motion. Thus, under normal conditions, we would expect the frequency of a nonlinear oscillator to change if some parameter of the system changes, because this change in parameter will cause the amplitude of the oscillation to change. On this account, we might expect that we need a very precise setting of parameter values to get a particular frequency ratio, say p/q. However, in frequency-locking the same ratio p/q holds over some range of parameter values.

Second, for a fixed set of parameter values, the time behavior of an oscillator can be characterized by a Fourier series of sinusoidal oscillations (see Appendix A). Thus, for oscillator number 1, we can write

$$x_1(t) = \sum_{k=1}^{\infty} B_k \sin(2\pi k f_1 t + \phi_k) \qquad (6.5\text{-}1)$$

where k is a positive integer. B_k is the amplitude associated with the harmonic kf_1, and ϕ_k is the phase for that frequency. The crucial point here is that the motion can be thought of as being made up of periodic motion with frequencies $f_1, 2f_1, 3f_1, \ldots$, and amplitudes $B_1, B_2, B_3, \ldots$. Similarly, oscillator number 2 can be described by

$$x_2(t) = \sum_{j=1}^{\infty} C_j \sin(2\pi j f_2 t + \Psi_j) \qquad (6.5\text{-}2)$$

Now, if the two frequencies f_1 and f_2 are commensurate with the frequency ratio $f_2/f_1 = p/q$, then the pth harmonic of f_1 (that is, $p f_1$) is the same as the qth of f_2 ($p f_1 = q f_2$). Since the two oscillators are interacting, this equality means that the

qth harmonic of f_2 can "tickle" the motion associated with the pth harmonic of f_1 to generate a kind of resonance effect. (Note that if $q f_2 = p f_1$, then there will be an infinite number of overlapping frequencies with $nqf_2 = np f_1$ and $n = 1, 2, 3, \ldots$ If the frequencies were incommensurate, then none of the harmonics would coincide, and the mutual resonance would not occur.)

These notions allow us to say that the frequency-locking occurs whenever the resonance interaction of harmonics (due to nonlinearities) wins out over the tendency of the oscillators' frequencies to change (also due to nonlinearities). Based on this picture, we would expect that as one of the frequencies is changed (say, f_1, which might be the frequency of the oscillator driving the system) with the strength of the nonlinearities held fixed, the frequencies would lock for some range of f_1 values, but then would either unlock (become incommensurate) or jump to another integer ratio for the new value of f_1. We might also guess that if the ratio p/q has small p and small q, for example 1/2, 2/3, but not 17/19, then the range of f_1 over which frequency-locking occurs might be larger than if p and q (again assuming we have removed all common divisors) were large. The reasoning here is that the amplitudes of the motion associated with the low harmonics (the low values of k and j in Eqs. (6.5-1) and (6.5-2) are usually (but not always) larger than those for the higher frequency harmonics. Thus, if the fundamental of f_1 is the same as the second harmonic of f_2, that is, $f_1 = 2 f_2$, then we would expect greater interaction between the oscillators than if $17f_1 = 19f_2$, where only the seventeenth and nineteenth (and higher) harmonics interact. Although this is just a hand-waving argument, we shall see that it does describe what happens in many systems.

6.6 Winding Numbers

As we indicated in the previous section, the ratio of frequencies involved in periodic or quasi-periodic behavior is fundamental to the description of that behavior. In fact, we shall need (and make use of) two different frequency ratios. The first is traditionally labeled Ω. We shall call Ω the *frequency-ratio parameter* because it specifies the ratio of the two frequencies f_1 and f_2 that would occur in the limit of vanishingly small nonlinearities and couplings between the two oscillators. A concrete example might be helpful: Consider a sinusoidally driven, damped pendulum. The pendulum is driven by an externally applied torque whose time variation is sinusoidal with a frequency f_1. Let f_2 be the small-amplitude oscillation frequency of the pendulum $f_2 = (1/2\pi)\sqrt{g/L}$, where, as usual, L is the length of the pendulum and g is local gravitational field strength. ($g = 9.8$ N/kg near the surface of the Earth.) The expression

$$\Omega = \frac{f_2}{f_1} \tag{6.6-1}$$

gives us the ratio of these two frequencies.

As the amplitude of the pendulum's oscillations grows, the actual frequency of its oscillatory motion decreases, a typical nonlinear oscillator effect. Let us denote the actual frequency by f_2'. The ratio of f_2' to f_1 (the oscillator frequency, which we shall consider to be under our control) is called the **winding number** (or **rotation number**):

$$w = \frac{f_2'}{f_1} \tag{6.6-2}$$

The term *winding number* comes from the state space picture of trajectories winding around a torus. The winding number tells us how many times the trajectory winds around the small cross section of the torus for each time around the large circumference of the torus. If w is a rational number, then the resulting motion is periodic: The trajectory closes on itself. If w is irrational, the motion is quasi-periodic, and the trajectory wanders over the entire surface of the torus, never exactly repeating itself.

If we focus our attention on Poincaré sections, winding numbers and frequency ratios that differ by an integer value will give the same Poincaré section. Hence, it is sometimes useful to remove any integer part of Ω and w. That is, we give just the fractional part. In mathematical terminology, we are expressing Ω and w **modulo 1**. For example, 2.3 mod [1] = 0.3 and 16.77 mod [1] = 0.77. In general, a mod $[b] = N - a/b$, where N is the smallest integer satisfying $Nb \geq a$. We implement this truncation because the resulting Poincaré section looks exactly the same if the winding number is 0.3, 4.3 or 1731.3. Since we will be focusing our attention on the Poincaré section, it is appropriate to give the winding number modulo 1. (We can, of course, distinguish among these cases by seeing how the Poincaré section points change if we move the location of the Poincaré plane around the torus.)

6.7 Circle Map

The discussion in the previous section can be made more concrete by looking at a mathematical model that exhibits the phenomenon of frequency-locking in a relatively straightforward way. In fact, this particular model has become the standard tool for investigating the quasi-periodic route to chaos.

The model is an iterated map scheme similar to the one-dimensional iterated maps we studied in Chapter 5. Now, however, the iteration variable is interpreted as the measure of an angle that specifies where the trajectory is on a circle. This notion is motivated by consideration of Poincaré sections of state space motion on a torus. If the state space variables are properly scaled, then the intersection points will lie on a circle in the Poincaré plane. Of course, for models of many physical systems and for many models based on differential equations, the torus cross section will <u>not</u> be circular. Even in those cases, we can use an angle to specify where the trajectory lies on the cross section (see Fig. 6.4). For almost all of our

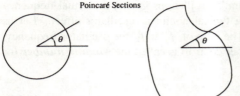

Fig. 6.4. Two Poincaré sections for state space motion on a torus. On the left the torus cross section is (almost) circular. On the right, the cross section is not circular. In both cases, we can use an angle variable to tell us where the trajectory point is on the cross section.

discussion, the details of the shape of the torus cross section are not important, and we gain a great deal of simplicity by using a circular cross section.

A general iterated map of this type can be written as

$$\theta_{n+1} = f(\theta_n) \tag{6.7-1}$$

where the function $f(\theta)$ is periodic in the angle θ. We define a trajectory (or orbit), just as we did for one-dimensional iterated maps, as the sequence of angles θ_0, θ_1, θ_2, ...

It is conventional to measure the angle θ in units such that $\theta = 1$ corresponds to one complete revolution around the circle. (More familiar units would have $\theta = 360°$ or $\theta = 2\pi$ radians for one complete revolution.) The map variable θ in Eq. (6.7-1) is defined modulo 1. With this convention, the angles $\theta = 0.7$ and $\theta = 1.7$ refer to the same point on the circle.

Before we introduce the map function used to study the quasi-periodic route to chaos, let us look at a simpler case, a case in which the map function $f(\theta)$ is linear:

$$\theta_{n+1} = f(\theta_n) = \theta_n + \Omega \quad \mathrm{mod}\,[1] \tag{6.7-2}$$

Here the mapping operation consists simply of adding the number Ω to the previous angle value. Thus, the mapping operation moves the trajectory around the circle in steps of size Ω. Figure 6.5 shows two examples of such linear map functions.

To define the winding number for such maps, we first compute

$$f^{(n)}(\theta_0) - \theta_0 \tag{6.7-3}$$

<u>without</u> taking θ mod [1], thereby computing the angular distance traveled after n iterations of the map function. The winding number for the map is defined to be the limit of the ratio

$$w = \lim_{n \to \infty} \frac{f^{(n)}(\theta_0) - \theta_0}{n} \tag{6.7-4}$$

The limit is included to allow any transient behavior, if it is present, to die away. There are no transients for the linear map given in Eq. (6.7-2).

It is easy to see that w and Ω are the same for the linear map of Eq. (6.7-2):

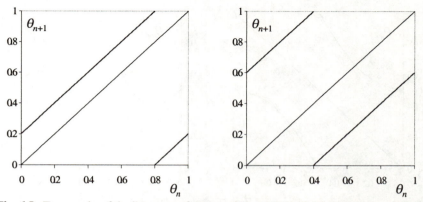

Fig. 6.5. Two graphs of the linear map function of Eq. (6.7-2). θ_{n+1} is plotted as a function of θ_n. The $\theta_{n+1} = \theta_n$ line is shown for reference here and in the following three figures. On the left $\Omega = 0.2$; on the right $\Omega = 0.6$. The sudden jump from $f(\theta) = 1$ down to 0 is a result of $\theta \bmod [1]$. There is no physical jump in the trajectory point's location.

$$w = \lim_{n \to \infty} \frac{n\Omega + \theta_0 - \theta_0}{n} = \Omega \tag{6.7-5}$$

As we shall see, for nonlinear map functions, w and Ω will generally be different. For the reasons discussed in Section 6.6, we shall restrict ourselves to values of Ω between 0 and 1. If Ω is a rational number, then w is rational and the trajectory is periodic. If Ω is irrational, then the trajectory is quasi-periodic.

With these preliminaries out of the way, we are now ready to introduce the famous sine-circle map, sometimes called simply the circle map. The sine-circle map is a nonlinear map function with the nonlinearity taking the specific form of a sine function:

$$\theta_{n+1} = \theta_n + \Omega - \frac{K}{2\pi}\sin(2\pi\theta_n) \quad \bmod [1] \tag{6.7-6}$$

The parameter K (with $K > 0$) is a measure of the strength of the nonlinearity. When $K = 0$, the circle map reduces to the simple linear map of Eq. (6.7-2). The 2π in the denominator is just a convention tied in with our using [0,1] as the range of angle values.

Note that the sine-circle map in Eq. (6.7-6) differs from most of the one-dimensional iterated maps discussed in Chapter 5 in having two control parameters: Ω for the frequency ratio and K for the strength of the nonlinearity. Since Ω is the winding number when $K = 0$, we call Ω the **bare winding number**, or, preferably, the *frequency-ratio parameter*.

$K = 1$ corresponds to an interesting change in behavior of the map function and its iterates, as we shall see shortly. For $K > 1$, some values of θ_{n+1} have more than one possible precursor θ_n. (We say that for $K > 1$, the sine-circle map is not

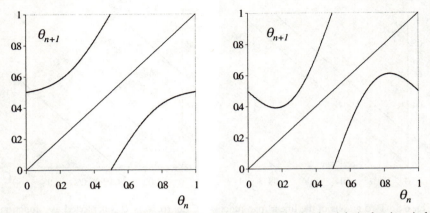

Fig. 6.6. On the left is the sine-circle map function for $K = 0.8$. On the right is the sine-circle map function for $K = 2.0$. θ_{n+1} is plotted as a function of θ_n with $\Omega = 0.5$. Note that for $K > 1$, there are some $f(\theta)$ values that have two possible θs as precursors. Hence, we say that for $K > 1$, the sine-circle map is not invertible.

invertible.) Hence, as we know from the Chapter 5 discussion of one-dimensional iterated maps, the iterated map function will introduce a "folding" of the trajectories, which folding may lead to chaotic behavior.

Again, the discontinuity in the map function as seen in Fig. 6.6 is only apparent. When the angle value exceeds 1, we subtract 1 from that value and return to the lower branch of the function as plotted in the figure.

Given our discussion of iterated maps in Chapter 5, the first question to ask is obviously: What are the fixed points of the sine-circle map? The fixed point values satisfy the equation

$$\theta = \theta + \Omega - \frac{K}{2\pi} \sin(2\pi\theta) \tag{6.7-7}$$

or equivalently

$$\frac{2\pi\Omega}{K} = \sin 2\pi\theta \tag{6.7-8}$$

Thus, we see that if

$$K \geq 2\pi\Omega \tag{6.7-9}$$

then there will be at least one fixed point for the circle map. The important result is that fixed points can occur only for certain combinations of K and Ω. If there are fixed points, then we can examine their stability (that is, the behavior of trajectories in their neighborhood) by examining the derivative of the map function evaluated at those fixed points:

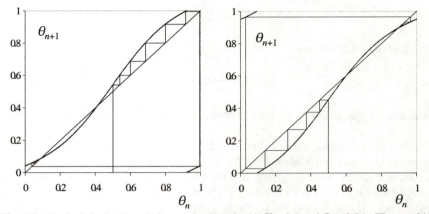

Fig. 6.7. On the left, the sine-circle map is plotted with $K = 0.5$ and $\Omega = 0.04$. The graphic iteration method shows that trajectories approach the stable fixed point near $\theta = 0.1$. The fixed point near $\theta = 0.4$ is unstable. On the right, $K = 0.5$ and $\Omega = 0.95$. The map now has a stable fixed point at $\theta = 0.9$ and an unstable fixed point near $\theta = 0.6$.

$$\frac{df}{d\theta} = 1 - K \cos 2\pi\theta \qquad (6.7\text{-}10)$$

Thus, we see that if K gets too large, so that the magnitude of the derivative exceeds 1, then the fixed point, if it exists, may be unstable. A fixed point θ^* of the sine-circle map will be stable if $0 < K \cos 2\pi\theta^* < 2$. Otherwise, the fixed point will be unstable.

To get some feel for the behavior of the sine-circle map, let us look at the fixed points of the map for $K < 1$. As we shall see, fixed points of the sine-circle map give rise to frequency-locking ratios 0:1, 1:1, 2:1, and so on, which will occur over some range of Ω. For example, according to Eq. (6.7-9), for a specific value of K, there will be a stable fixed point of the map function for a range of small values of Ω, that is, for $0 \le \Omega \le K/2\pi$. Within this range, the winding number [as defined by Eq. (6.7-3)] is equal to 0 and is independent of the starting value of θ for the trajectory.

It is instructive to use the graphical iteration technique to get a better sense for what is occurring. Figure 6.7 shows on the left a graph of the sine-circle map with $K = 0.5$ and $\Omega = 0.04$. A graphic iteration of the map is started from $\theta = 0.55$. We see that the trajectory converges to the fixed point near $\theta = 0.1$. We also see that there is another fixed point at $\theta = 0.4$, which we note is just 0.5 minus the value of the first fixed point; however, that second fixed point is unstable.

For a different range of Ω values, we get 1:1 frequency-locking. For Ω values near 1, we need to take into account that we are measuring the angles modulo 1. Thus, a fixed point can occur that satisfies

$$\theta + 1 = \theta + \Omega - \frac{K}{2\pi}\sin(2\pi\theta) \qquad (6.7\text{-}11)$$

This leads to the condition

$$1 - \frac{K}{2\pi} \le \Omega \qquad (6.7\text{-}12)$$

Note that for this situation $\sin 2\pi\theta$ is negative because θ is close to (but less than) 1.

Let us summarize what we have found: For a fixed value of K, there are two ranges of Ω values that lead to stable fixed points of the sine-circle map. One range is $0 \le \Omega \le K/2\pi$ corresponding to 0:1 frequency-locking; the other range is $1 - K/2\pi \le \Omega \le 1$ and is the 1:1 frequency-locking region. Thus, we see that as K approaches 1, the range of Ω values over which frequency-locking occurs increases. In the lower range of Ω values, θ remains fixed once transients have died away. In the higher Ω range, θ increases by 1 for each iteration. Both ranges lead to a single fixed point in the Poincaré section. If we imagine the circle map as representing the Poincaré section of the trajectory motion on the surface of a torus in three-dimensional state space, the lower range corresponds to trajectories that do not rotate at all about the small cross section of the torus. The upper range corresponds to trajectories that go once around the small cross section for each trip around the large circumference of the torus.

Exercise 6.7-1. Show that the two fixed points identified earlier as stable fixed points are indeed stable. Show that the two other fixed points are unstable.

Exercise 6.7-2. If we allow Ω to be greater than 1, find the ranges of K for which 2:1 and 2:2 frequency-locking occur.

Now let us turn our attention to 1:2 frequency-locking. On the left Fig. 6.8 shows the iterates of the sine-circle map with $K = 0.8$ and $\Omega = 0.5$. Note that the trajectory settles into a cycle of two values corresponding to 1:2 frequency locking. The right-hand side of Fig. 6.8 shows that these two cycle values correspond to the (stable) fixed points of the second-iterate of the sine-circle map function. It is easy to see that as K increases, the range of Ω values for which 1:2 frequency-locking occurs also increases. Although we cannot find the range of 1:2 frequency-locking analytically, it is reasonably straightforward to find it numerically for a given value of K. We simply find the range of Ω that gives fixed-points for $f^{(2)}$. For example, for $K = 0.5$, the range is approximately from 0.491 to 0.509; for $K = 1.0$, the range is approximately from 0.464 to 0.562 (with an uncertainty of ± 0.0005).

It is traditional to illustrate the range over which the various ratio frequency-lockings occur by drawing a diagram in the K-Ω plane and indicating those areas

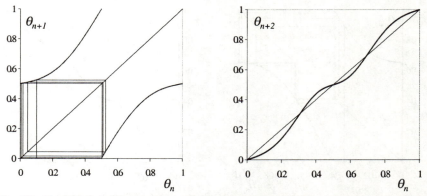

Fig. 6.8. On the left the sine-circle map function is shown with $K = 0.8$ and $\Omega = 0.5$. The graphic iteration technique shows that this value yields a trajectory corresponding to 1:2 frequency-locking. The initial θ value is 0.1. On the right, the second-iterate of the sine-circle map function is shown, indicating that the two cycle values correspond to the second-iterate function's two stable fixed points located near $\theta = 0$ and $\theta = 0.5$.

corresponding to frequency-locking. Figure 6.9 shows the 0:1, 1:2, and 1:1 frequency-locking regions. As K increases, the frequency-locking regions expand to fill finite intervals along the Ω axis. These regions are called, rather imaginatively, **Arnold tongues** after the Russian mathematician who pioneered the study of frequency-locking via circle maps [Arnold, 1983].

As a further example, let us look at the iterations with $K = 0.9$ and $\Omega = 0.65$. In Fig. 6.10, we see that the iterations lead to a three point cycle, which in this case represents 2:3 frequency-locking. On the right side of that figure, we see that the cycle points correspond to the stable fixed points of $f^{(3)}$.

By comparing Fig. 6.10 with Fig. 6.8, we can see why the range of Ω for which $p:q$ frequency-locking occurs gets smaller as q increases. Changing Ω corresponds to shifting the graph vertically. We get $p:q$ frequency locking when the qth iterate of the map function overlaps the 45° line. For larger values of q, the amplitude of the oscillatory part of the function decreases; hence, the range of Ω

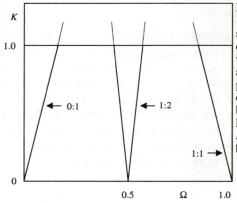

Fig. 6.9. A sketch of the 0:1, 1:2, and 1:1 frequency-locking regions for the sine-circle map. The shaded regions, called *Arnold tongues*, are the regions in which frequency-locking occurs. There are frequency-locking regions for all positive integer ratios $p:q$. At $K = 0$, most of the Ω axis consists of quasi-periodic behavior. As K increases the frequency-locked regions broaden. Above the line $K = 1$, the frequency-locking regions begin to overlap and chaos may occur.

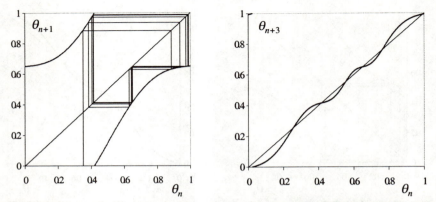

Fig. 6.10. On the left the sine-circle map with the parameter values $K = 0.9$ and $\Omega = 0.65$. The initial θ value is 0.35. The trajectory corresponds to 2:3 frequency-locking. On the right, the third-iterate of the map function shows that the cycle values correspond to the three stable fixed points of that iterate, located near 0.4, 0.65, and 0.98.

over which the fixed points for that iterate exist is reduced.

We can also understand why the frequency-locking tongues broaden as K increases. As K approaches 1, the graph of the appropriate iterate of the mapping function develops larger oscillations around the monotonic increase due to the $\theta +$ Ω part of the map function. This larger set of oscillations means that there is a larger range of Ω values for which the oscillations overlap the 45° line in the graph of the map function, hence, a larger range of Ω values over which the fixed points exist. On a more physical level, as we discussed earlier, the larger nonlinearity implied by the larger value of K means that there is stronger coupling between the modes of oscillation for the physical system being described and that leads to larger ranges of parameter values over which frequency-locking occurs.

Exercise 6.7-3. Review the arguments in Section 6.2 about frequency-locking to show that the cycle shown in Fig. 6.10 does indeed correspond to 2:3 frequency-locking, not 1:3. Carry out graphical iteration of the sine-circle map for $K = 0.9$ and $\Omega = 0.34$, and verify that you get 1:3 frequency-locking for those parameter values.

To get more of an overview of the kind of behavior that occurs with the sine-circle map, we have plotted in Fig. 6.11 the winding number, computed from Eq. (6.7-3), as a function of the frequency parameter Ω for a fixed value of $K = 1.0$. Quite noticeable in Fig. 6.11 are the "plateaus" (e.g., near $\Omega = 1/2$) where w does not change over a substantial range of Ω. In fact, we find that as K increases (at least up to $K = 1$), these plateaus broaden out until they apparently fill the entire interval between $\Omega = 0$ and $\Omega = 1$. These plateaus are called frequency-locking steps because here the winding number is a rational fraction (for example, 1/3 or 2/7). We say that the "response frequency" (f_2 in our previous notation) is rationally related to the "drive frequency" (f_1).

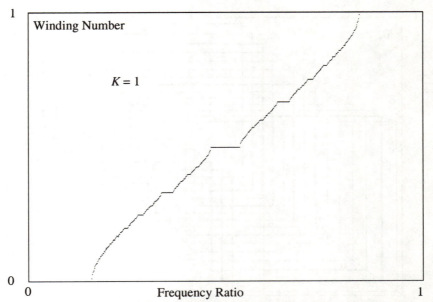

Fig. 6.11. The winding number, as defined by Eq. (6.7-3), plotted as a function of the frequency parameter Ω for the sine-circle map with $K = 1.0$. Note the frequency-locking plateaus over which the winding number does not change for a substantial range of Ω. These plateaus form what is called the *Devil's Staircase*.

One question we might readily ask is: How many frequency-locking plateaus are there between $\Omega = 0$ and $\Omega = 1$? It turns out there is an infinite number of these plateaus! There is one plateau for each rational fraction p/q between 0 and 1. Since there is an infinite number of such rational fractions, there is an infinite number of plateaus or steps. If we imagine a small creature trying to climb this set of stairs to get from $w = 0$ to $w = 1$, we see that the creature would need to make an infinite number of steps. Hence, this set of steps is often called the *Devil's Staircase*.

How do we know that w is precisely equal to a rational fraction (e.g., 1/2) in the center frequency-locked plateau? We can verify this rationality by plotting the resulting trajectory points (after allowing transients to die away). For example, for any Ω value under that center step, we should have just two trajectory points, since there $2f_2 = f_1$. For Ω values that do not correspond to a locking plateau, the trajectory values should eventually fill in a complete curve (drift ring) on a θ_{n+1} versus θ_n plot. Figure 6.12 shows the results of graphically iterating the sine-circle map for parameters that give quasi-periodic behavior.

6.8 The Devil's Staircase and the Farey Tree

In this section we will explore some intriguing connections between the sine-circle map and simple numerology. It may at first seem surprising that number theory

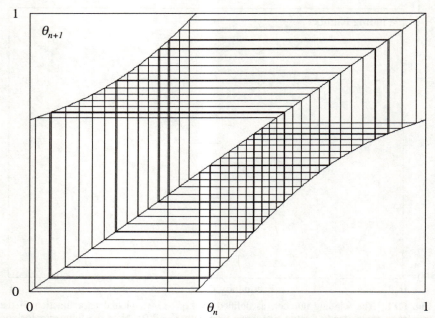

Fig. 6.12. The graphic iteration of the sine-circle map with $K = 0.5$ and $\Omega = 0.6180339$. $\theta_0 = 0.35$. The behavior is quasi-periodic (not frequency-locked). The trajectory would eventually fill in a complete range of θ values.

can tell us anything about dynamics. We have had a hint of this connection in our earlier discussions of symbolic dynamics; here, we will see more results of this nature.

In a previous section we gave a rough argument that indicated we might expect to see broader frequency-locking plateaus when p and q are relatively small integers. A glance at Fig. 6.11 seems to indicate the result is indeed true. We would now like to describe how results from number theory account for this size ordering. Please keep in mind that these results do not explain (physically) why the frequency-locking occurs (that explanation lies in the competition between resonance and nonlinear frequency shifts), but these results do allow us to understand the ordering, which is observed in almost all systems that display frequency-locking.

The basic observation we want to take into account is that the frequency-locking plateaus <u>decrease</u> in length (along the Ω axis) when the denominator in the fraction p/q increases. For example, the plateau for $w = 1/4$ is shorter than the plateau for $w = 1/3$. (For the sine-circle map the plateaus for a given q but different ps, all have the same length.)

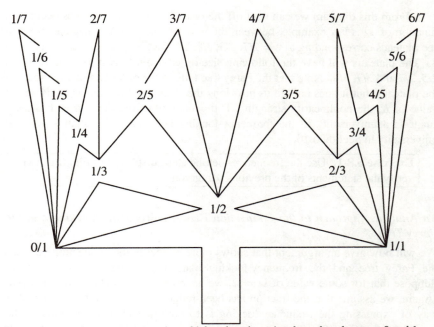

1/7   2/7   3/7   4/7   5/7   6/7

1/6   5/6

1/5   2/5   3/5   4/5

1/4   3/4

1/3   2/3

1/2

0/1   1/1

Fig. 6.13. The Farey tree construction which orders the rational numbers between 0 and 1.

From the theory of numbers (see, for example, [Wright and Hardy, 1980]) we know that if we have two rational fractions p/q and p'/q', the rational fraction that lies between them and has the smallest denominator is the rational fraction

$$\frac{p+p'}{q+q'}$$

Since the denominator controls the frequency-locking plateau size, we expect that the plateau corresponding to this fraction will be the largest plateau to be found between the p/q and the p'/q' plateaus.

We also know that between any two rational fractions lie an infinite number of other rational fractions; hence, between any two frequency-locking plateaus lie an infinite number of other frequency-locking plateaus. However, for the larger denominators, the frequency-locking plateaus are so short along the Ω axis that they are essentially unobservable.

We can order the rational fractions that lie between 0 and 1 according to their increasing denominators by constructing the so-called Farey tree, shown in Fig. 6.13. As we go up the Farey tree, each new fraction is formed by adding the numerators and denominators of the rational fractions to which the corresponding "branches" are attached. At each horizontal layer, all the fractions have the same denominator and correspond to frequency-locking plateaus of the same size for the sine-circle map. The plateau sizes decrease as we move up the tree.

From this diagram we can read off the order in which the plateaus occur as a function of Ω. For example, between the $w = 1/4$ and $w = 1/2$ plateaus, we will see plateaus corresponding to $w = 2/7$, $1/3$, $3/8$, $2/5$, and $3/7$ for denominators up to 8. The plateaus will have the following size order (from largest to smallest): $1/3$, $2/5$, $2/7$ and $3/7$, $3/8$. Note that the Farey tree tells us the order and relative sizes of the plateaus, but it does not tell us how long the steps are; that size depends on the value of K, the nonlinearity strength. The generality of the Farey tree, however, suggests a universality in the frequency-locking behavior. We shall study that universality in Section 6.10.

Exercise 6.8-1. Use the frequency-locking steps displayed in Fig. 6.11 to verify the statements of the previous paragraph.

An Analytic Approach to the Connection Between Frequency-Locking and the Farey Tree

We will now give an argument that shows more explicitly the connection between the Farey tree and the frequency-locking plateaus. (After [Schuster, 1995].) Suppose that for some value of $\Omega = \Omega_1$ we have frequency-locking with $w = p/q$. (Again, we assume that the fraction has been reduced to simplest form.) Another way of expressing the frequency-locking is to note that after q iterations of the circle map function f_{Ω_1}, we have

$$f_{\Omega_1}^{(q)}(\theta) = p + \theta \tag{6.8-1}$$

That is, after q iterations of the map function, the angle value we arrive at is the original angle value plus p. We have added a subscript to the function to indicate explicitly the dependence on Ω. Suppose for a somewhat larger value of Ω we get locking with $w = p'/q'$, that is

$$f_{\Omega_2}^{(q')}(\theta) = p' + \theta \tag{6.8-2}$$

Now let us formally combine the two iterations to write

$$f_{\Omega_1}^{(q)}\left(f_{\Omega_2}^{(q')}(\theta)\right) = p + p' + \theta \tag{6.8-3}$$

(Note that the two parts of the iteration use <u>different</u> values of Ω.) We want to be able to write this result as a single $q + q'$ iteration with a single value of $\Omega = \Omega_3$. To find this value, we note that if we increase Ω_1 we would get a result larger than $p + p' + \theta$ because $f(\theta)$ increases monotonically with Ω. We can compensate for the overshoot, however, by decreasing the value of Ω_2. Thus, by increasing Ω_1 while simultaneously <u>decreasing</u> Ω_2, we can finally end up with a single value Ω_3 for which we can write

$$f_{\Omega_3}^{(q+q')}(\theta) = p + p' + \theta \tag{6.8-4}$$

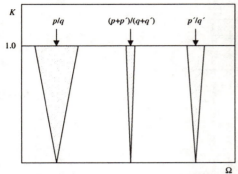

Fig. 6.14. The ordering of the Arnold tongues (frequency-locking regions) based on the ordering of rational fractions and the Farey tree. The intermediate region covers a smaller range of Ω because its denominator is larger.

This last equation tells us that there is an Ω_3 value located between Ω_1 (where p/q locking occurs) and Ω_2 (where p'/q' locking occurs) for which we get locking with $(p + p')/(q + q')$. We note, however, that this is just the ordering of rational numbers produced by the Farey tree. Figure 6.14 illustrates this ordering of the Arnold tongues.

6.9 Continued Fractions and Fibonacci Numbers

In our later discussion, we will be concerned particularly with situations in which the winding number is an irrational number (i.e., the system is not frequency-locked). It will be helpful in that discussion to have in hand some ways of approximating a given irrational number as a sequence of rational numbers. One way to do this is the method of continued fractions. Let us illustrate what this means by use of a specific example. The irrational number

$$G = \frac{\sqrt{5} - 1}{2} = 0.6180339...\qquad(6.9\text{-}1)$$

(often called the "Golden Mean," for reasons to be discussed later) can be written as

$$G = \cfrac{1}{1 + \cfrac{1}{1 + \cfrac{1}{\cdots}}}\qquad(6.9\text{-}2)$$

The ellipsis in the previous equation means that we continue the process of putting more fractions in the denominator ad infinitum. If we stop with n denominators (i.e., we have n "fraction lines"), we have the nth order approximation to G. The first few approximations are:

$$G_1 = \frac{1}{1} = 1$$

$$G_2 = \cfrac{1}{1+\cfrac{1}{1}} = \frac{1}{2}$$

$$G_3 = \cfrac{1}{1+\cfrac{1}{1+\cfrac{1}{1}}} = \frac{2}{3}$$

(6.9-3)

$$G_4 = \frac{3}{5}$$

$$G_5 = \frac{5}{8}$$

We say that the sequence of rational numbers G_n converges to the irrational number G as $n \to \infty$. G_n is called the *n*th **convergent** for G.

Two somewhat more compact notations for the continued fraction are

$$G = a_0 + \frac{1}{a_1 +}\frac{1}{a_2 +}\frac{1}{a_3 +}$$

$$= a_0 + (a_1, a_2, a_3 ...)$$

(6.9-4)

For G, we have $a_0 = 0$, $a_n = 1$ for all $n > 0$.

We see from the previous construction that in general

$$G_n = \frac{1}{1 + G_{n-1}}$$

(6.9-5)

As $n \to \infty$, the sequence of G_n values approaches the limiting value G, which must then satisfy

$$G = \frac{1}{1 + G}$$

(6.9-6)

Eq. (6.9-6) yields a quadratic equation for G:

$$G^2 + G - 1 = 0$$

(6.9-7)

Thus, we see that indeed G is the value given in Eq. (6.9-1).

Exercise 6.9-1. The number G is often called the Golden Mean because it can be defined by the following geometric argument: Divide a line segment of length L into two parts, one of length l_1, the other of length l_2 (with $l_1 + l_2 = L$), such that the ratio l_1/L is equal to the ratio l_2/l_1. Show that this ratio is equal to G as defined in Eq. (6.9-1). A rectangle whose sides have lengths whose ratio is the Golden Mean seems to be particularly appealing visually. *N.B.* Sometimes the reciprocal of G ($1/G$ = 1.61803...) is called the Golden Mean.

You may have observed that the sequence G_3, G_4, G_5...could also be generated by combining the numerators and denominators of the previous two numbers in the sequence, once we have the numbers G_1 and G_2:

$$G_1 = \frac{1}{1}$$

$$G_2 = \frac{1}{2}$$

$$G_3 = \frac{1+1}{1+2} = \frac{2}{3}$$

$$G_4 = \frac{1+2}{2+3} = \frac{3}{5}$$ (6.9-8)

$$G_5 = \frac{2+3}{3+5} = \frac{5}{8}$$

Note that this construction is reminiscent of the Farey tree construction.

Another related construction uses the so-called **Fibonacci sequence**. (Fibonacci, also known as Leonardo of Pisa, learned arithmetic from Arab scholars in North Africa around 1200. His book *Liber Abaci* was very influential in the development of mathematics in Europe.) The Fibonacci sequence is defined using two base numbers $F_0 = 0$ and $F_1 = 1$ and the recursion relation

$$F_{n+1} = F_n + F_{n-1}$$ (6.9-9)

that is, the nth number in the sequence is equal to the sum of the two previous numbers in the sequence. Thus, the Fibonacci sequence based on 0 and 1 is 0, 1, 1, 2, 3, 5, 8, 13,...It is easy to see that the ith approximation to the Golden Mean is given by

$$G_i = \frac{F_i}{F_{i+1}}$$ (6.9-10)

We see that both the numerators and denominators of the Gs are part of the Fibonacci sequence.

Continued fractions that have all the as in Eq. (6.9-4) equal to integers are particularly simple fractions. A further simplification occurs if there are only a finite number of different a values and if these values occur in a periodic fashion. A continued fraction is said to be periodic if

$$a_m = a_{m+k}$$ (6.9-11)

for some fixed k (called the *period of the fraction*) and for all m greater than some number M. A periodic continued fraction represents an irrational number that can be expressed as the solution of a quadratic equation with integer coefficients [e.g., Eq. (6.9-7) for G]. The converse is also true: The solution of any quadratic equation with integer coefficients can be expressed as a periodic continued fraction.

Exercise 6.9-2. Show that the following periodic continued fractions are equal to the irrational numbers given:

(a) $\dfrac{1}{2+}\dfrac{1}{2+}\dfrac{1}{2+}\ldots = \sqrt{2}-1$

(b) $\dfrac{1}{1+}\dfrac{1}{2+}\dfrac{1}{1+}\dfrac{1}{2+}\ldots = \sqrt{3}-1$

N. B.: (a) is sometimes called the ***Silver Mean***.

Exercise 6.9-3. The continued fraction expansion for any number x can be constructed as follows: First, define the "largest integer function" $y = \text{INT}(x)$, where y is the largest integer less than x. The continued fraction expansion Eq. (6.9-4) is then constructed by setting $a_0 = \text{INT}(x)$. We define $b_0 = x - a_0$. Then $a_1 = \text{INT}(1/b_0)$. Next we set $b_1 = (1/b_0) - a_1$ and $a_2 = \text{INT}(1/b_1)$. In general, we have $a_n = \text{INT}(1/b_{n-1})$ and $b_n = ((1/b_{n-1}) - a_n)$. The algorithm continues as long as b_n is not equal to 0. Use this method to find the continued fraction expansion for the Golden Mean, the Silver Mean, and for π.

The results stated in this section will prove to be important in quasi-periodic dynamics because we will want to approximate the irrational ratio of frequencies associated with quasi-periodic behavior by a series of rational numbers. The continued fraction scheme allows a systematic way of carrying out that approximation. The reader intrigued by the mathematics of continued fractions is referred to [Wright and Hardy, 1980] for an elegant and rigorous discussion.

6.10 On to Chaos and Universality

After our detour through the land of frequency-locking, we are ready to resume our journey to chaos via quasi-periodicity. Let us focus on the sine-circle map for our initial tour. For the sine-circle map, we reach a critical value of nonlinearity when the parameter $K = 1$. What is special about this value? For $K < 1$, there is a unique value of θ for each value of $f(\theta)$. We say the map is invertible. That is, given a value of θ_{n+1}, we can find the unique value of θ_n that led to it. However, for $K > 1$, this invertibility no longer holds. For $K > 1$, there may be two or more values of θ_n that lead to the same value of θ_{n+1}. This lack of invertibility is tied up with the possibility of chaotic behavior because it induces the trajectory folding, which is necessary for chaos.

For $K < 1$, the sine-circle map leads to either periodic (frequency-locked) behavior, characterized by a rational winding number, or quasi-periodic behavior, characterized by an irrational winding number. In fact as K approaches 0, the fraction of the Ω axis occupied by frequency-locked steps goes to 0; that is, almost all Ω values lead to quasi-periodic behavior. As K approaches 1, the length of the frequency-locking steps increases, and at $K = 1$, where chaotic behavior becomes possible, the frequency-locked steps occupy essentially the entire Ω axis. The quasi-periodic trajectories still occur at $K = 1$, but they occur for a vanishingly small fraction of the axis. (To characterize how much of the Ω axis is occupied by quasi-periodic behavior requires the notion of fractal dimension, to be taken up in Chapter 9.) This fraction turns out to be independent of the specific circle map and is an example of the universal behavior of circle maps, much like the Feigenbaum universality of one-dimensional unimodal iterated map functions discussed in Chapter 5. This particular fraction is called a <u>global</u> universal characteristic because it involves the entire Ω axis. Similarly, the fraction of the Ω axis occupied by quasi-periodic behavior approaches 1 as K approaches 0 with a (global) universal behavior given by the expression (JBB84)

$$f_q = c(1 - K)^{\beta} \tag{6.10-1}$$

where c is a constant and $\beta \approx .034$ is a universal exponent for a wide class of circle maps. (As we shall explain later, the map classes are determined by a particular geometric property of the map functions.)

For $K > 1$, some values of Ω lead to chaotic behavior. In those cases the winding number defined in Eq. (6.7-3) does not exist because the limit used in the definition does not exist. Other values of Ω lead to frequency-locked behavior. The regions in which these two types of behavior occur are intertwined in the K–Ω diagram in a complicated fashion. If K becomes larger than 1 with Ω set so that quasi-periodic behavior occurs, then chaotic behavior sets in for $K > 1$. If, on the other hand, Ω is set so that the behavior is frequency-locked at $K = 1$, then the behavior generally remains periodic for $K > 1$ with a sequence of period-doublings leading to chaotic behavior as K continues to increase.

In addition to these global features, there are also some local universal features (i.e., features that occur for a small range of Ω values). The most interesting of these local features involves the Ω values that give rise to a particular sequence of winding numbers (for $K \leq 1$). The most studied of these sequences is the sequence of winding numbers that approaches the Golden Mean winding number by following the sequence of ratios of Fibonacci numbers given in Eq. (6.9-10). To show how this universality is defined, we introduce a particular value of $\Omega_n(K)$, which yields a winding number equal to the nth approximation to the golden mean and has $\theta = 0$ as a point on the trajectory

$$\Omega_n(K) = \frac{F_n}{F_{n+1}} \tag{6.10-2}$$

[We include $\theta = 0$ because the slope of $f(\theta)$ is 0 there for $K = 1$, which makes the cycle a supercycle, in the language of Chapter 5.] In Eq. (6.10-2), we have used the notation of the previous section for Fibonacci numbers. We then introduce two scaling exponents that play the role for the circle map that the Feigenbaum δ and α play for the iterated maps introduced in Chapter 5. The first gives the ratio of successive differences in Ω values as we let the winding number approach the Golden Mean value:

$$\delta(K) = \lim_{n \to \infty} \frac{\Omega_n(K) - \Omega_{n-1}(K)}{\Omega_{n+1}(K) - \Omega_n(K)} \qquad (6.10\text{-}3)$$

The second quantity gives the ratio of distances d_n between the point $\theta = 0$ and the nearest element in the cycle:

$$\alpha(K) = \lim_{n \to \infty} \frac{d_n(K)}{d_{n+1}(K)} \qquad (6.10\text{-}4)$$

The values of $\delta(K)$ and $\alpha(K)$ can be found by the use of arguments much like those used to find the values of the Feigenbaum numbers (FKS82 and ROS82). For the sine-circle map with $K = 1$, we have

$$\delta(K) = -2.83360$$
$$\alpha(K) = -1.28857 \qquad (6.10\text{-}5)$$

For $K < 1$, $\alpha(K) = -G^{-1}$ and $\delta(K) = -G^{-2}$, where G is the Golden Mean.

For sequences of Ω values constructed from sequences of rational fractions leading to other irrational values, the values of $\delta(K)$ and $\alpha(K)$ are different (SHE82). However, they can all be expressed in a similar form when $K = 1$, the critical value:

$$\delta(K) = -N^{-a}$$
$$\alpha(K) = -N^{-b} \qquad (6.10\text{-}6)$$

where a and b are constants that depend only weakly on the irrational number. For example, for the Golden Mean, $a = 2.16443...$ and $b = 0.52687...$, while for the Silver Mean, $a = 2.1748...$ and $b = 0.5239...$ (SHE82). N is defined in terms of the ratio of differences in rational winding numbers used in the approach to the irrational number. If we let W_i represent the ith rational number in the sequence, analogous to the sequence in Eq. (6.9-3), then N is defined by

$$N^2 = \lim_{i \to \infty} \frac{W_i - W_{i+1}}{W_i - W_{i-1}} \qquad (6.10\text{-}7)$$

For the Golden Mean and the Silver Mean, N turns out to be equal to the irrational number itself.

Exercise 6.10-1. Show that for $p{:}q$ frequency-locking trajectory $f^{(p)}(0)$ gives the trajectory point closest to $\theta = 0$.

Exercise 6.10-2. Prove that for $K < 1$ and the sequence of Ω values given in Eq. (6.10-2), we get $\alpha = -G^{-1}$ and $\delta = -G^{-2}$.

Exercise 6.10-3. Show that the values of a and b cited in the text when used in Eq. (6.10-6) give the values listed in Eq. (6.10-5).

In Chapter 5 and Appendix F, we pointed out that the "standard" Feigenbaum values apply only to those one-dimensional iterated map functions that have quadratic behavior near their maximum values. For other functional dependences, the values of $\delta(K)$ and $\alpha(K)$ are different. Thus, these one-dimensional maps fall into various universality classes depending on their behavior near the maximum value. Similarly, we can define various universality classes of circle maps. For circle maps, it is the functional dependence of the map function near the inflection point at $\theta = 0$, which occurs for $K = 1$. (Recall that at an inflection point, the second derivative is equal to 0.) Hu, Valinia, and Piro (HVP90) have studied these universality classes by calculating $\delta(K)$ and $\alpha(K)$, as defined earlier, for the map functions

$$f(\theta) = 2\pi\theta - 2^{z-1}\theta \left| \theta \right|^{z-1} \qquad (6.10\text{-}8)$$

Here θ is defined to be in the interval [-1/2, 1/2]. The parameter z controls the "degree" of the inflection. For $z = 3$, we have a so-called cubic inflection and the

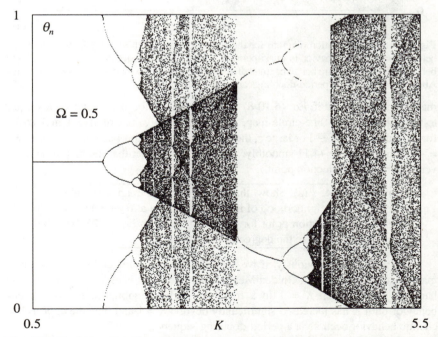

Fig. 6.15. The bifurcation diagram for the sine-circle map with $\Omega = 0.5$. The value of K is plotted along the horizontal axis. The initial value of θ is 0. Note that no chaotic behavior occurs until well above the critical value of $K = 1$. Compare the period-doubling sequences and chaotic band-mergings to those of the one-dimensional iterated maps of Chapter 5.

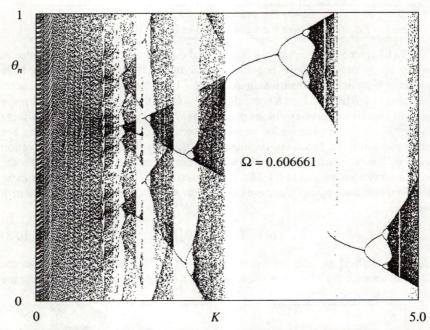

Fig. 6.16. The bifurcation diagram for the sine-circle map with $\Omega = 0.606661$. Note that below $K = 1.0$ the behavior is quasi-periodic. Immediately above $K = 1$ there is chaotic behavior, but this chaotic behavior immediately gives way to a window of periodic behavior. Above $K = 1$, there are period-doublings and more chaos.

map function defined in Eq. (6.10-8) yields the values for $\delta(K)$ and $\alpha(K)$ equal to the values for the sine-circle map. HVP90 list the values of $\delta(K)$ and $\alpha(K)$ for various values of z. For large z, these values seem to approach the values $\alpha(K) = -1.0$ and $\delta(K) = -4.11$ smoothly. (Large z means that that the map function is very flat near the inflection point.)

Exercise 6.10-4. (a) Show that the sine-circle map has cubic type behavior in the neighborhood of its inflection point at $\theta = 0$ and $K = 1$. (b) Where is the inflection point for the $f(\theta)$ given in Eq. (6.10-8)? Show that z does indeed give the degree of the inflection.

As mentioned previously, if we keep the winding number for the sine-circle map tuned to the irrational Golden Mean value, chaos sets in at $K = 1$. On the other hand, if we approach $K = 1$ in a frequency-locked region, we see no chaotic behavior until K has increased significantly beyond $K = 1$ and then the approach to chaotic behavior occurs via a period-doubling sequence.

Above $K = 1$, the bifurcation diagram for the sine-circle map becomes quite complicated. Figures 6.15 and 6.16 show the bifurcation diagrams for a range of K values and two values of Ω, 0.5 and 0.606661, the latter of which gives a winding number close to the Golden Mean.

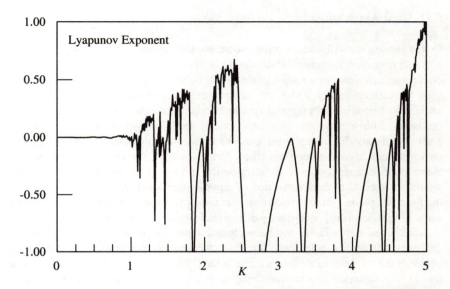

Fig. 6.17. The (average) Lyapunov exponent is plotted as function of K for $\Omega = 0.606661$. By comparing this figure with Fig. 6.16, we see that the Lyapunov exponent is 0 for quasi-periodic behavior and positive for chaotic behavior.

Three important points emerge from those diagrams. First, we must be careful to distinguish the smear of points that occurs for $K < 1$ due to quasi-periodicity from the smear of points for $K > 1$ in the chaotic regions. One should notice the banded structure of the quasi-periodic behavior and compare that to the "random scatter" of the chaotic areas. Second, we note that for $\Omega = 0.5$, we pass $K = 1$ in a frequency-locking tongue; chaotic behavior does not emerge directly at $K = 1$ but occurs only above $K = 1$. Third, we note that the diagrams depend significantly on the Ω value. To some extent, the behavior also depends on the trajectory starting point in the frequency-locking tongues. Above $K = 1$, the Arnold tongues overlap, and different starting points may lead to different frequency-locking ratios.

To verify that the regions above $K = 1$ with a scattering of trajectory points are actually chaotic, we should compute the (average) Lyapunov exponent for the trajectory. Figure 6.17 shows the Lyapunov exponent, calculated according to the method described in Chapter 5, as a function of K corresponding to the bifurcation diagram of Fig. 6.16. We see that the Lyapunov exponent is positive in those regions above $K = 1$ that appear chaotic. The quasi-periodic regions give a Lyapunov exponent of 0. Hence, we can use the Lyapunov exponent to distinguish between quasi-periodic and chaotic behavior, which to the eye appear to be very similar.

6.11 Some Applications

In this section we will discuss briefly some physical systems that display the quasi-periodic transition from regular to chaotic behavior. We should note that the sine-circle map can serve as a useful guide to this behavior for parameter values below what is called **criticality**, which for the sine-circle map means below $K = 1$. Below criticality, frequency-locking and quasi-periodicity are the only possible types of behavior. Above criticality, chaos may occur. For a physical system described by a set of ordinary differential equations, we know that chaos requires trajectories to move through at least three state space dimensions. However, even above $K = 1$, the sine-circle map continues to describe the motion of points on a circle, which would correspond to the intersection of a two-dimensional state space surface with the Poincaré plane. If the Poincaré section were just a circle (or other closed curve) for a system described by ordinary differential equations, then the motion could be at worst quasi-periodic. If we have chaotic behavior for a system described by ordinary differential equations, the Poincaré section corresponding to chaotic behavior must fill out more than a curve in the Poincaré plane. Thus, we cannot use the $K > 1$ behavior of the sine-circle map as a guide to what happens above "criticality" for those kinds of systems. In fact, the region above the critical line is largely unexplored territory, both theoretically and experimentally. Most of the applications described later have focused on the frequency-locking regions below

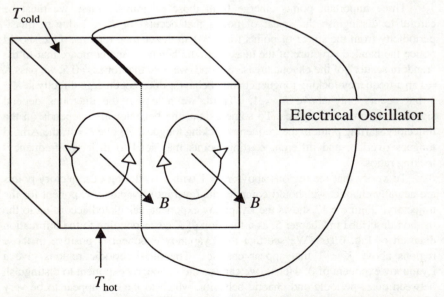

Fig. 6.18. A schematic diagram of the periodically-modulated Rayleigh–Bénard experiment. The fluid is mercury. A magnetic field (B), indicated by the arrows, is applied parallel to the convection cell axes. An alternating electrical current is applied in a sheet dividing the two convection rolls. The dotted line indicates the main flow of electrical current.

the critical line and on the transition to chaos via quasi-periodic behavior with an irrational winding number. In those experiments, at least two parameters must be adjusted to maintain the required winding number.

Forced Rayleigh–Bénard Convection

We introduced Rayleigh–Bénard fluid convection in Chapter 1 in the discussion of the Lorenz model. As we mentioned there, if the temperature difference between the bottom and top of the cell is made large enough, the convective rolls begin to oscillate. This type of behavior would be described by a limit cycle in state space. Stavans, Heslot, and Libchaber (SHL85) have carried out a Rayleigh–Bénard experiment in a small cell (so there are only two convection rolls) filled with mercury. They applied a steady magnetic field of about 200 gauss to the system with the magnetic field direction parallel to the axis of the convective cells as illustrated in Fig. 6.18. By sending an alternating electrical current through the vertical plane separating the two convection rolls, they induced a periodic modulation of the fluid flow. (Recall that charged particles moving through a magnetic field feel a magnetic force that is perpendicular to both the velocity of the charged particle and the magnetic field.) The frequency and amplitude of the alternating current could be varied. The ratio of the natural oscillatory frequency of the fluid to the frequency of the alternating current corresponds to Ω for the circle map. The amplitude of the alternating electrical current corresponds to K in the sine-circle map.

For small values of the alternating current amplitudes (say, less than 10 milliamp), the system exhibited frequency-locking and quasi-periodic behavior organized by Arnold tongues, much like those for the sine-circle map. With the frequency-ratio set to approach the Golden Mean value via the sequence described in Section 6.10, SHL determined the value of $|\delta(K)|$ to be 2.8±0.3, a value that agrees well with the universality prediction. They also measured $|\delta(K)|$ for the sequence of frequency-ratio values approaching the silver mean value and found $|\delta(K)| = 7.0±0.7$, also in good agreement with the theoretically expected value. In addition, they investigated the so-called fractal dimension of the quasi-periodic regions just at the onset of chaos. We shall discuss those measurements in Chapter 9.

Periodically Perturbed Cardiac Cells

In some sense, most living systems owe their continued existence to oscillatory behavior. For humans, the most obvious oscillations are the repetitive beating of the heart and the (more or less) regular respiratory behavior of the lungs. There are many others, however, including regular oscillations of electrical signals in the brain and many chemical oscillations with a period of about 24 hours. (The latter lead to what are called *circadian rhythms*.) It is not surprising that over the last fifteen years or so, many scientists have begun to apply the methodology of nonlinear dynamics to understand the behavior of these oscillating systems. In a

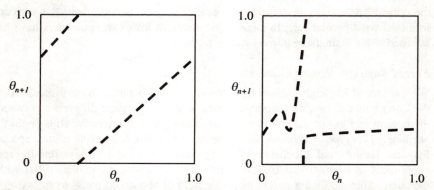

Fig. 6.19. Graphs of the circle map function used to model the periodically stimulated heart cell dynamics. On the left, the conditions yielded quasi-periodic behavior. On the right, the conditions led to chaotic behavior. Note that on the right the map function is not invertible and leads to the folding of trajectories. The thick dashed lines give a rough idea of the scatter of experimental points. [Based on (GSB86, pp. 248-249).)]

living organism, these oscillating systems interact with each other and with external perturbations, so we might expect the phenomena of frequency-locking, quasi-periodicity, and perhaps chaos to apply to them.

As an example of how nonlinear dynamics is applied to the study of these systems, we will describe a series of elegant experiments carried out at McGill University [Glass and Mackey, 1988] (GGB84) (GSB86). There, L. Glass and co-workers applied an external, periodic electrical stimulation to a culture of chick embryo heart cells. The culture consisted of small aggregates, each about 100 μm in diameter, of cells taken from 7-day-old embryonic chicks. Each aggregate beats spontaneously with a period of about 0.5 sec, much like a mature heart in a living chicken. In the experiment, the amplitude and frequency of the external stimulation, which was an electrical pulse applied through an intracellular electrode, were varied.

The qualitative results of the experiment were in agreement with expectations based on the behavior of the sine-circle map. For moderate amplitude stimulation, the cardiac cells exhibited frequency-locking and quasi-periodic behavior organized in Arnold tongues. For large amplitude stimulation, regions of chaotic behavior introduced by period-doubling were observed. In these experiments, the variable recorded is the time between successive beats of the heart cell culture. Chaotic behavior means that these time intervals seem to have no recognizable periodicity. This behavior is again qualitatively like the large K behavior of the sine-circle map (see Figs. 6.15 and 6.16).

Glass and co-workers modeled the dynamical behavior of the cardiac cell system by making a connection with a circle map: The external perturbation can be said to occur at a particular phase of the spontaneous oscillation cycle, given by the angle variable θ, where $0 < \theta < 1$, as in our discussion of circle maps. The effect of the pulse perturbation is to reset the oscillator to a new phase value θ' :

$$\theta' = g(\theta) \qquad (6.11\text{-}1)$$

where the function g is given by the so-called phase transition curve. The latter can be measured by applying a single pulse to the system at various values of θ.

When the system is subject to a periodic sequence of stimulus pulses, the behavior of the phase of the spontaneous oscillations just before the nth pulse is given by

$$\theta_n = g(\theta_{n-1}) + \Omega \qquad (6.11\text{-}2)$$

where Ω is the ratio of the spontaneous oscillation frequency to the stimulus frequency. This is exactly the form of a circle map. Figure 6.19 shows on the left a graph of the circle map determined from the heart cell data in the range of parameters in which quasi-periodic behavior occurs. On the right is the corresponding graph for parameter range in which chaotic behavior occurs. The qualitative similarities to the sine-circle map are obvious.

Comments on Biological Models

This is a good point to inject a few words of commentary. The heart cell example raises a number of important methodological (and philosophical) questions. What does it mean to say that the chick heart cell dynamics is modeled by a circle map? Are we to infer that there is (the equivalent of) a small computer inside the heart cells rapidly iterating a circle map to determine their behavior? In more general, and perhaps less facetious terms, what is the role of a mathematical model of a biological system?

At one level, we can say that we expect on the basis of our study of nonlinear dynamics to see many universal features among perturbed oscillating systems, whether they are physical, chemical, or biological. Since these features are present in simple mathematical schemes, such as circle maps, we can use the circle map to guide our investigation. Indeed this procedure of learning from models has been extremely important in the development of our understanding of nonlinear dynamics and chaos. As we have seen, the behavior of nonlinear systems is exceedingly complex, and we need some guidance just to develop a useful taxonomy of behavior. Moreover, we have learned that certain types of transitions, for example, between frequency-locking, organized in Arnold tongues, and quasi-periodicity, or between periodic behavior and chaotic behavior via a sequence of period-doublings, are common to wide classes of models. Hence, we can use any one of the models to guide our investigations of the "real" system.

On the other hand, we would like our models to be "realistic" (i.e., to give us some information about what is "really" going on in the chick heart cells). A realistic description of the experiment would presumably tell us what goes on with the cell membrane electrical potentials that are influenced by the external stimulation, which the circle map does not provide. It might be argued that using a circle map to model the dynamics is just playing with numbers. The circle map gives us no information on what is really going on with the heart cells and that we would be better off, as scientists, focusing on a more realistic description. Moreover, the natural biological variability between systems and even within a system (sometimes called *biological noise*) means that the circle map used for one heart cell aggregate is different from (at least in detail) the circle map for another aggregate or even for the same aggregate at a later time. The systems are living, often growing, or dying, always changing.

We would not disagree with the goal of obtaining a realistic picture of the system's behavior. We would say, however, that mathematical modeling with simplified models such as the circle map provides an important intermediate step in our understanding of the biological system. The mathematical model, if it captures the correct, but gross, features of the dynamics, gives us a guide for organizing our study of the behavior. It is not a replacement for a realistic model, however, which gives us an understanding of the detailed biological mechanisms of the system.

Biological systems are complex systems, at least compared to those commonly studied by most physicists and chemists. It may be surprising that anything at all can be said quantitatively about their behavior. Perhaps the most important discovery is that biological dynamics, at least for relatively simple systems (on the biological scale), can be organized in terms of models with only a few degrees of freedom. [Glass and Mackey, 1988] and [Winfree, 1980] provide an excellent introduction to the use of nonlinear dynamics in understanding some aspects of biological oscillators.

Periodically Driven Relaxation Oscillator

As a final example of quasi-periodic behavior, we turn to a variation of the van der Pol oscillator, introduced in Appendix I. There we saw that solutions of the van der Pol equation (I-4) describe spontaneously generated limit cycles; that is, with time-independent voltage inputs, an electrical circuit described by the van der Pol equation will exhibit periodic oscillations. However, as we argued in Appendix I, the dynamical possibilities of the van der Pol model are limited. We can enrich the range of dynamical possibilities by adding another term to the equation, a term describing a periodic modulation of the circuit. This modulation can take many forms but a common one is to add a periodic "force" term to the right-hand side of Eq. (I-4):

$$\frac{d^2Q}{d\tau^2} - (R - Q^2)\frac{dQ}{d\tau} + Q = F\cos(2\pi f\tau) \qquad (6.11\text{-}3)$$

where F is the magnitude of the force and f is its frequency of oscillation.

The *forced van der Pol oscillator* has a venerable history. In 1928, van der Pol and van der Mark (VAV28) used this equation to model some aspects of heartbeats and their abnormalities known as arrhythmias. As an outgrowth of the study of electrical oscillators used in the development of radar, Mary Cartwright and J. Littlewood published a mathematical study of the forced van der Pol oscillator equations in 1945 (CAL45). It is apparent they noted the possibilities of what we would now call chaotic solutions (see also CAR48). N. Levinson published similar results in 1949 (LEV49). [S. Smale was led to his horseshoe map (SMA63) in his study of Levinson's work.] More recent work, but before the recognition of chaotic behavior in a formal way, is summarized in [Hayashi, 1964]. It turns out that steady-state (as compared to transient) chaotic behavior is more readily observable if the periodic modulation is put into the "velocity" term (the term involving $dQ/d\tau$) rather than being introduced as a periodic force. This model has been extensively studied by R. Shaw (SHA81). More recently Abraham and Simo (ABS86) have studied forced van der Pol systems with asymmetric forcing terms in which chaos appears more readily.

With this brief (and inadequate) survey of some theoretical studies of forced relaxation oscillators, we now turn to an experiment. By adding periodic modulation to an operational amplifier relaxation oscillator, Cumming and Linsay (CUL88) provided a detailed quantitative study of the quasi-periodic route to chaos. They were able to locate over 300 Arnold tongues. Somewhat surprisingly they found significant deviations from the supposedly universal sine-circle map predictions at the onset of chaos. In particular, the value of $\delta(K)$ found using the sequence of winding numbers approaching the Golden Mean did not converge, as expected from Eq. (6.10-3), but oscillated between two values, -3.3 ± 0.1 and -2.7 ± 0.2. Moreover, the fractal dimension (to be discussed in Chapter 9) did not agree with the value expected on the basis of the sine-circle map.

An explanation of these deviations was supplied by Alstrom, Christiansen, and Levinsen (ACL88), who used an "integrate and fire" model to show that in some relaxation oscillators, the transition to complete frequency-locking and the transition to chaos are distinct transitions. In the sine-circle map model, chaos first begins at $K = 1$ when the Arnold tongues expand to cover the entire Ω axis. Thus, for the sine-circle map the two transitions are identical. ACL argue that for a periodically-modulated integrate-and-fire relaxation oscillator, which is a better physical model for the operational amplifier relaxation oscillator, the two transitions can be distinct and hence the quantitative features of the sine-circle map need not apply. However, they also argue that in a real system the gap predicted by the integrate-and-fire models will be smoothed over and the numerical values for the characterization of the transition to chaos will not be too different from those predicted by the sine-circle map. (For a discussion of biological oscillators and integrate-and-fire models, see [Glass and Mackey, 1988].)

This last example provides us with a cautionary tale: We should not expect too much from simple models. The sine-circle map does provide us with a useful guide to quasi-periodic behavior, but it does not capture all of nature's possibilities. As in many other areas of nonlinear dynamics, we lack a complete categorization and understanding of these possibilities.

6.12 Further Reading

The quasi-periodic route to chaos is discussed in almost every book on nonlinear dynamics. See Chapter 1 for an extensive listing. Particularly appropriate at the level of this book are [Berge, Pomeau, and Vidal, 1984] and [Schuster, 1995].

T. W. Dixon, T. Gherghetta, and B. G. Kenny, "Universality in the quasiperiodic route to chaos," *Chaos* **6**, 32–42 (1996). A very nice overview of quasiperiodicity, the sine-circle map, and the theory of universality in the quasi-periodic route to chaos.

Early Studies of Nonlinear Oscillators and Frequency-Locking

B. Van der Pol and J. Van der Mark, "The Heartbeat Considered as a Relaxation Oscillation and an Electrical Model of the Heart," *Phil. Mag.* **6**, 763–75 (1928).

M. L. Cartwright and J. E. Littlewood, "On Nonlinear Differential Equations of the Second Order. I. The Equation $\ddot{y} - k(1 - y^2)\dot{y} + y = b\lambda k \cos(\lambda t + \alpha)$, k Large, " *J. London Math. Soc.* **20**, 180–89 (1945).

M. L. Cartwright, "Forced Oscillations in Nearly Sinusoidal Systems," *J. Inst. Electr. Eng.* **95**, 88–96 (1948).

N. Levinson, "A Second-Order Differential Equation with Singular Solutions," *Annals of Mathematics* **50**, 127-53 (1949).

C. Hayashi, *Nonlinear Oscillations in Physical Systems* (McGraw–Hill, New York, 1964; reprinted by Princeton University Press, 1985).

Frequency-locking in Biological Systems

A. T. Winfree, *The Geometry of Biological Time* (Springer-Verlag, New York, 1980).

L. Glass and M. C. Mackey, *From Clocks to Chaos* (Princeton University Press, Princeton, NJ, 1988).

Quasi-Periodic Route to Chaos

D. Ruelle and F. Takens, "On the Nature of Turbulence," *Commun. Math. Phys.* **20**, 167–92 (1971). This article marks the first appearance of the terms *chaotic* and *strange attractor* in the context of nonlinear dyanmics. (Reprinted in Hao, 1984.)

S. E. Newhouse, D. Ruelle, and R. Takens, "Occurence of Strange Axiom A Attractors near Quasi-Perioidc Flows on T_m ($m = 3$ or more)," *Commun. Math. Phys.* **64**, 35 (1978). Puts the conditions for the quasiperiodic route to chaos on a firm mathematical basis.

L. D. Landau, "On the Problem of Turbulence," *Akad. Nauk. Doklady* **44**, 339 (1944). English translation reprinted in [Hao, 1984]. A proposal for a cascade of an infinite number of frequencies to explain turbulence.

R. W. Walden, P. Kolodner, A. Passner, and C. Surko, "Nonchaotic Rayleigh–Bénard Convection with Four and Five Incommensurate Frequencies," *Phys. Rev. Lett.* **53**, 242–45 (1984).

Circle Maps

V. I. Arnold, *Geometrical Methods in the Theory of Ordinary Differential Equations* (Springer, New York, 1983). Section 11.

Universal features for circle maps at the critical point but with a focus on the Golden Mean:

S. Shenker, "Scaling Behavior in a Map of a Circle onto Itself: Empirical Results," *Physica D* **5**, 405–11 (1982). (Reprinted in [Cvitanovic, 1984].)

M. J. Feigenbaum, L. P. Kadanoff, and S. J. Shenker, "Quasiperiodicity in Dissipative Systems: A Renormalization Group Analysis," *Physica D* **5**, 370–86 (1982). (Reprinted in [Hao, 1984].)

D. Rand, S. Ostlund, J. Sethna, and E. Siggia, "Universal Transition from Quasi-Periodicity to Chaos in Dissipative Systems," *Phys. Rev. Lett.* **49**, 132–35 (1982). (Reprinted in [Hao, 1984].)

M. H. Jensen, P. Bak, and T. Bohr, "Transition to Chaos by Interaction of Resonances in Dissipative Systems I, II," *Phys. Rev. A* **30**, 1960–69 and 1970–81 (1984).

The following papers extend the universal features to other irrational numbers:

P. Cvitanovic, M. H. Jensen, L. P. Kadanoff, and I. Procaccia, "Renormalization, Unstable Manifolds, and the Fractal Structure of Mode Locking," *Phys. Rev. Lett.* **55**, 343–46 (1985).

S. Kim and S. Ostlund, "Universal Scaling in Circle Maps," *Physica D* **39**, 365–92 (1989).

B. Hu, A. Valinai, and O. Piro, "Universality and Asymptotic Limits of the Scaling Exponents in Circle Maps," *Phys. Lett. B* **144**, 7–10 (1990).

Number Theory and Continued Fractions

H. E. Huntley, *The Divine Proportion* (Dover, New York, 1970). A delightful source of information on the Golden Mean.

[Wright and Hardy, 1980]

H. S. Zuckerman, H. L. Montgomery, I. M. Niven, and A. Niven, *An Introduction to the Theory of Numbers*, 5th ed. (John Wiley, New York, 1991).

A. Ya Khinchin, *Continued Fractions* (The University of Chicago Press, Chicago, 1992).

Forced Van der Pol Equation

S. Smale, "Diffeomorphisms with many periodic points," in *Differential and Combinatorial Topology*, (S. S. Cairns, ed.) (Princeton University Press, Princeton, NJ, 1963).

R. Shaw, "Strange Attractors, Chaotic Behavior, and Information Flow," *Z Naturf.* **36a**, 80–112 (1981).

R. H. Abraham and C. Simo, "Bifurcations and Chaos in Forced van der Pol Systems," in *Dynamical Systems and Singularities* (S. Pnevmatikos, ed.) (North-Holland, Amsterdam, 1986) pp. 313–23.

Applications of the Quasi-Periodic Route to Chaos

J. Stavans, F. Heslot, and A. Libchaber, "Fixed Winding Number and the Quasiperiodic Route to Chaos in a Convective Fluid," *Phys. Rev. Lett.* **55**, 596–99 (1985).

M. H. Jensen, L. Kadanoff, A. Libchaber, I. Proccacia, and J. Stavans, "Global Universality at the Onset of Chaos: Results of a Forced Rayleigh–Bénard Experiment," *Phys. Rev. Lett.* **55**, 2798–801 (1985).

L. Glass, M. R. Guevar, J. Belair, and A. Shrier, "Global Bifucations and Chaos in a Periodically Forced Biological Oscillator," *Phys. Rev. B* **29**, 1348–57 (1984).

L. Glass, A. Shrier, and J. Belair, "Chaotic Cardiac Rhythms," in *Chaos* (A. V. Holden, ed.) (Princeton University Press, Princeton, NJ, 1986).

S. Martin and W. Martienssen, "Circle Maps and Mode Locking in the Driven Electrical Conductivity of Barium Sodium Niobate Crystals," *Phys. Rev. Lett.* **56**, 1522–25 (1986).

S. Martin and W. Martienssen, "Small-signal Amplification in the Electrical Conductivity of Barium Sodium Niobate Crystals," *Phys. Rev. A* **34**, 4523–24 (1986).

P. Bryant and C. Jeffries, "The Dynamics of Phase Locking and Points of Resonance in a Forced Magnetic Oscillator," *Physica D* **25**, 196–232 (1987).

M. Courtemancho, L. Glass, J. Belari, D. Scagliotti, and D. Gordon, "A Circle Map in a Human Heart," *Physica D* **40**, 299–310 (1989).

The following two papers presented experimental results that seem to violate the Ruelle–Takens quasi-periodic scenario to chaos:

A. Cumming and P. S. Linsay, "Quasiperiodicity and Chaos in a System with Three Competing Frequencies," *Phys. Rev. Lett.* **60**, 2719–22 (1988).

P. S. Linsay and A. W. Cumming, "Three-Frequency Quasiperiodicity, Phase Locking, and the Onset of Chaos," *Physica D* **40**, 196–217 (1989).

These results were explained in P. Alstrom, B. Christiansen, and M. T. Levinsen, "Nonchaotic Transition from Quasiperiodicity to Complete Phase Locking," *Phys. Rev. Lett.* **61**, 1679–82 (1988).

6.13 Computer Exercises

CE6-1. Use *Chaotic Mapper* to explore the sine-circle map by looking at the results of graphic iteration, its bifurcation diagram and Lyapunov exponent as a function of the nonlinearity parameter K for various values of Ω.

CE6-2. Write a computer program to plot the Devil's staircase for the sine-circle map.

7

Intermittency and Crises

Chaos, a rough and unordered mass. Ovid, *Metamorphoses*.

7.1 Introduction

For the third (and final) chapter on routes to chaos in dissipative systems, we will discuss two more scenarios. One involves a type of behavior called *intermittency* because, as we shall see, the behavior of the system switches back and forth intermittently between apparently regular behavior and chaotic behavior. The second scenario is signaled by a so-called *crisis*, during which a strange attractor in state space suddenly changes in size or suddenly disappears. The intermittency route to chaos was first described by Pomeau and Manneville (MAP79 and POM80) and is sometimes called the Pomeau–Manneville scenario. The notion of crisis—another important class of bifurcation events—was introduced by Grebogi, Ott, and Yorke (GOY82) and (GOY83).

Both scenarios have been observed in several experiments. However, like the theory of period-doublings and the quasi-periodic route to chaos, the theory of intermittency and crises has been based primarily on the study of iterated maps, viewed either as dynamical systems in their own right or as models of Poincaré map functions for systems described by differential equations. We will introduce the basic phenomenology of the behavior for both intermittency and a crisis and then discuss (briefly) some of the theoretical analysis.

7.2 What Is Intermittency?

Intermittency occurs whenever the behavior of a system seems to switch back and forth between two qualitatively different behaviors even though all the control parameters remain constant and no significant external noise is present. The switching appears to occur "randomly" even though the system is (supposedly) described by deterministic equations. We shall focus on two types of intermittency (with brief discussions of two other types as well). In the first type, the system's behavior seems to switch between periodic behavior and chaotic behavior. The behavior of the system is predominantly periodic for some control parameter value with occasional "bursts" of chaotic behavior. As the control parameter value is changed, the time spent being chaotic increases and the time spent being periodic decreases until, eventually, the behavior is chaotic all the time. As the parameter is changed in the other direction, the time spent being periodic increases until at some value, call it A_c, the behavior is periodic all the time. In the second type of intermittency, the system's behavior seems to switch between periodic and quasi-periodic behavior.

We have used the word *seem* quite consciously in the previous paragraph to describe intermittency. Of course, for fixed control parameters, the behavior of the system is whatever it is. In principle, no switching occurs. However, in an intermittent situation, the behavior appears to have a certain character (e.g., periodic) for a long time (long compared to the typical time period associated with the system) and then "abruptly" (again, compared to the typical time period for the system) switches to behavior of a qualitatively different character (e.g., chaotic).

To get a better feeling for what intermittency means, we have plotted in Fig. 7.1 a long stretch of a "signal" that was computed from the now familiar logistic map function (Eq. 1.4-5.) for two different parameter values. On the left, the behavior is periodic with period-5. On the right, for a slightly smaller value of the parameter A, the behavior is intermittent. It is apparent from the right-side of Fig. 7.1 that the trajectory appears to be periodic for a while (with period-5 behavior) and then chaotic for a while, with no apparent periodicity. The switching between the two behaviors appears to be random. If we decrease the parameter A a small amount, the chaotic regions expand and the periodic regions decrease in size.

The reason for the cautious description of the behavior should be obvious: If the signal were exactly periodic, there would be no possibility of a "switch" to chaotic behavior. Also, the "chaotic" part looks chaotic, but we have not established that it has a positive Lyapunov exponent. The observation that each of the chaotic sections appears to be quite distinctive in character and in length even though the preceding periodic parts appear to be (nearly) identical is a strong suggestion, but not a proof, that the irregular behavior is chaotic.

Intermittency also occurs in systems described by differential equations. Figure 7.2 shows the time dependence of the Lorenz model variable Z (see Eqs.

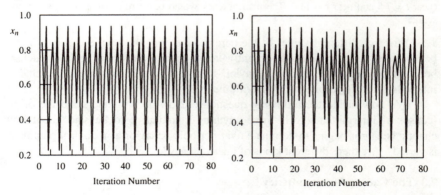

Fig. 7.1. On the left is a plot of the x values for successive iterations of the logistic map function for $A = 3.74$. Period-5 behavior is evident. On the right $A = 3.7375$ and the behavior is intermittent. Parts of the behavior appear to be periodic with period-5. Other parts appear to be chaotic. The switching between the two types of behavior appears to be random.

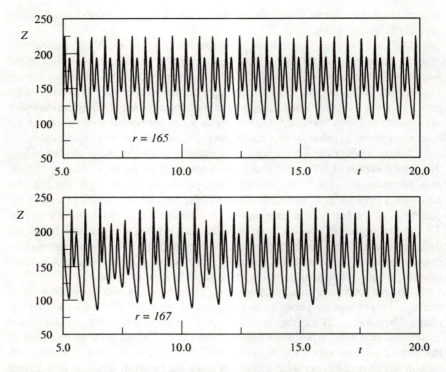

Fig. 7.2. The time dependence of the Z variable in the Lorenz model (Eqs. 1.5-1). On top, the parameters are $r = 165$, $b = 8/3$, and $p = 10$. The trajectory shows periodic behavior after a short initial transient. In the bottom half, $r = 167$, and the trajectory shows intermittent behavior, switching between periodic behavior and (very) short bursts of chaotic behavior (near $t = 7.5$ and near $t = 11$). The intermittency would be more obvious in a longer time sequence.

1.5-1) for two different values of the parameter r. For the smaller value of r, the behavior is periodic. For the slightly larger value, we see irregular switching between periodic and chaotic behavior. As r is further increased, the chaotic intervals gradually expand to fill all of time.

We now tackle the following questions: How do we understand the origin (cause) of intermittency? Is there anything quantitative to be said about how chaotic behavior begins to dominate over periodic behavior as a control parameter is varied? Are there different kinds of intermittency?

7.3 The Cause of Intermittency

The switch to chaotic behavior via intermittency should be contrasted with the two previously discussed routes to chaos: period-doubling and quasi-periodicity. In both the previous scenarios, the long-term behavior of the system was either

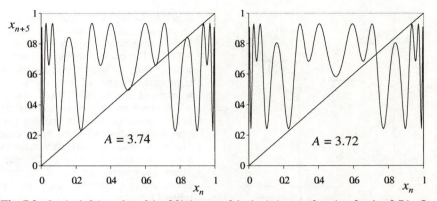

Fig. 7.3. On the left is a plot of the fifth iterate of the logistic map function for $A = 3.74$. On the right is a plot of the same function for a slightly smaller value $A = 3.72$. Iteration of the logistic map function on the left leads to period-5 behavior. For the value used on the right, we get intermittency behavior. Notice the small "gaps" on the right between the function and the $y = x$ (diagonal line) near the locations of the period-5 fixed points. At the intermittency transition, the fifth iterate function is just tangent to the diagonal line at 5 points.

completely periodic (or quasi-periodic) or completely chaotic, depending on the parameter value. In the case of intermittency, the behavior apparently switches back and forth. We emphasize "apparently" again because we shall see that the behavior in the intermittent regime, for our first type of intermittency, is completely aperiodic (and chaotic). What we need to explain is why some parts of the behavior are apparently periodic.

The general scheme is the following: For parameter values above (for example) some critical parameter value A_c, the behavior of the system is completely periodic. For an iterated map function, the behavior is determined by the fixed points of the appropriate nth iterate if we have period-n behavior. For a system, such as the Lorenz model, described by a set of differential equations, we can use the Poincaré section technique and focus our attention on the n points of the Poincaré section, which are fixed points of the (generally unknown) Poincaré map function. The general feature that gives rise to intermittent behavior is the "disappearance" of these fixed points as the relevant parameter is changed. In contrast to period-doubling, the previously stable fixed points are not replaced by new stable fixed points. Hence, the motion becomes irregular (in fact, chaotic) and the trajectories wander over a considerable region of the state space.

What then causes the episodes of apparently periodic behavior? Perhaps a concrete example would be useful. Figure 7.3 shows the fifth iterate of the logistic map for two parameter values. On the left is an A value that leads to period-5 behavior. After transients die away, trajectories cycle among the five stable fixed points. (There are also five unstable fixed points close to the stable fixed points. There are also two other unstable fixed points: one at $x = 0$, and the other at $x = 1 - 1/A$, which correspond to the unstable fixed points of the original map function.)

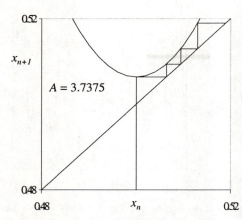

Fig. 7.4. An expanded view of the "gap" in the plot of the fifth iterate of the logistic map near $x = 1/2$. Here $A = 3.7375$. The graphic iteration technique shows that a trajectory spends a significant amount of time near the location of the period-5 fixed points, which come into existence for slightly larger values of A when the function intersects the diagonal line. In the case illustrated here, the trajectory requires about four steps to traverse the gap. These four steps correspond to four cycles of five iterations each of the original function.

On the right, for a slightly smaller value of A, there are no stable fixed points, and the behavior is chaotic.

We develop more insight concerning the apparent periodicity by examining the behavior of trajectories that go through the "gaps" between the appropriate iterate of the map function and the 45° line. Figure 7.4 shows an expanded view of one of these gaps for the logistic map. The graphic iteration technique shows that a trajectory spends a significant amount of time (many successive iterations) near the previously stable period-5 fixed point. Eventually, however, the trajectory is repelled from this region and wanders off to another region of state space.

Exercise 7.3-1. Carry out the graphic iteration technique for several size "gaps" between $f^{(5)}(x)$ and the diagonal (45°) line. Convince yourself that the smaller the gap size, the longer the time (that is, the larger the number of iterations) required to pass through the gap.

As the gap size decreases, the trajectory spends more time (on the average) in the gap region. This is qualitative evidence that more time is spent in "periodic" behavior as the parameter approaches the critical value A_c, at which point the gap vanishes and the behavior becomes exactly periodic. Since the map function (or its appropriate iterate) becomes tangent to the 45° line at A_c, the beginning of intermittency is sometimes called a **_tangent bifurcation_**. Since a stable fixed point and an unstable fixed point annihilate each other at the beginning of intermittency, this event is also a type of saddle-node bifurcation.

It is clear that for the "periodic" behavior to repeat in intermittency, the wandering trajectory must be re-injected into the vicinity of the narrow gap. We can use the graphic iteration technique "backward" to see how re-injection occurs: We ask what regions of the x axis lead upon the next iteration to trajectories that land near one of the small gaps. Figure 7.3 shows that there are two such regions for each gap for the logistic map. (Recall that if there is more than one possible antecedent for a given trajectory point, we say that the map function is not invertible.) If a trajectory lands in one of these two re-injection regions, it will find itself back in the neighborhood of the small gap on the next iteration. The next few iterations will be confined to the vicinity of the gap, and the behavior will appear to be (almost) periodic since this next set of x values will all be close in value. However, the x values will not be <u>exactly</u> the same because there is no stable fixed point and the trajectory does slowly work its way through the gap. Exactly how long it takes clearly depends on where it is re-injected; therefore, we would expect to see (and we do see) considerable irregularity in the length of the "periodic" intervals.

The weak point in our argument is the assumption that the wandering trajectory will hit one of the re-injection zones. In a rough sense, one might expect that to happen, but it seems difficult to prove in general that it should occur because we need to consider how a trajectory behaves as it wanders over an extended region of state space. Using terminology introduced before, we say that the periodic behavior is determined by the "local" behavior near the gaps while the wandering trajectories and subsequent re-injection are determined by the "global" behavior over an extended region of the state space. For a more detailed treatment of the re-injection process see [Berge, Pomeau and Vidal, 1986] and [Schuster, 1995].

These tangent bifurcations leading to intermittency are in fact quite common. For the logistic map function, the periodic windows seen in the bifurcation diagram in Figs. 1.14, 2.4, and 2.5 are "born" by tangent bifurcation events as the parameter A increases. At a tangent bifurcation birth for the period-n window, we have

$$\frac{df^{(n)}(x^*)}{dx} = 1$$

$$f^{(n)}(x^*) = x^*$$

(7.3-1)

where the second equation tells us that x^* is a fixed point of the nth iterate of the map function. As the parameter A decreases (for the logistic map function), the slope of $f^{(n)}$ near x^* increases above $+1$ as the fixed point disappears. (By way of contrast, recall that for a period-doubling, the slope of the appropriate iterate of the map function became more negative than -1.)

Let us summarize what we have found: Intermittency behavior is in fact aperiodic (chaotic) behavior characterized by irregularly occurring episodes of (almost) periodic behavior. The cause of the periodic behavior is a "trapping" of trajectories in the gaps that open up between the appropriate iterate of the map function and the diagonal line after a tangent bifurcation.

We might also point out that intermittency of a slightly different character occurs in systems that show quasi-periodicity and frequency-locking. As we saw in Chapter 6, $p{:}q$ frequency-locking in the sine-circle map occurs when the qth iterate of the map function overlaps the $x_{n+1} = x_n$ diagonal line. The frequency-locking clearly begins and ends, as the control parameter is changed, with tangent bifurcations just like the case of intermittency discussed here. However, in the case of the sine-circle map for $K < 1$, the intermittent behavior is a switching between apparently frequency-locked behavior and quasi-periodic behavior. The apparent frequency-locking occurs when the trajectory is temporarily trapped in the narrow gaps between the qth iterate of the map function and the diagonal line. Strictly speaking, we must say the behavior is quasi-periodic, but there may be long intervals of behavior that appears to be frequency-locked.

Intermittency and 1/f Noise

A wide variety of physical systems have "noisy" behavior for which the noise has a nearly identical power spectrum: The noise level increases as the power spectrum frequency decreases with an approximately $1/f$ frequency dependence. (We trust that using the symbol f for frequency will not cause confusion with its use as a function in other sections.) The ubiquity of this so-called **1/f noise** obviously demands a very general explanation. Although no completely satisfactory explanation of $1/f$ noise has been developed, recent work suggests that intermittent-type behavior may be the common feature linking all of these systems. The basic notion is that intermittency is essentially characterized by switching back and forth, apparently randomly as we have seen, between two (or perhaps, more) different types of behavior. This slow switching is then the "cause" of the $1/f$ fluctuations, which we call noise. The relationship between intermittency and $1/f$ noise is discussed in more detail in [Schuster, 1995], MAN80, PRS83, and GWW85.

7.4 Quantitative Theory of Intermittency

In this section we will present several related arguments to establish some universal features associated with the intermittency route to chaos. We begin by focusing attention on a one-dimensional iterated map model. In particular, we look at the behavior of the appropriate iterate of the map function for parameter values close to the value at which a tangent bifurcation occurs and in the region close to one of the stable fixed points—the one that is about to disappear. In that region, the nth iterate of the map function can be approximated by the expression

$$f_A^{(n)}(x) = x^* + (x - x^*) + a(x - x^*)^2 + b(A_c - A) \qquad (7.4\text{-}1)$$

where a and b are constants that depend on the particular map function, iterate, and fixed point. Their exact values will play no role in our discussion. We have chosen the form of the parameter dependence term to match the behavior of the logistic map: We have periodic behavior for $A > A_c$, and we have intermittent (chaotic) behavior for $A < A_c$.

It is traditional to put this approximate equation into a standard form by introducing new variables $y = (x - x^*)/b$, $c = ab$, and $\mu = A_c - A$. These changes shift the fixed point location to $y = 0$. The parameter μ is chosen so that when $\mu = 0$, the tangent bifurcation takes place. For $\mu < 0$, the behavior is periodic. For $\mu > 0$, the behavior is intermittent. The approximate form for the map function is then

$$y_{n+1} = h(y_n) = y_n + cy_n^2 + \mu \qquad (7.4\text{-}2)$$

A graph of Eq. (7.4-2) looks like the graph displayed in Fig. 7.4.

> **Exercise 7.4-1.** Verify the algebraic substitutions that lead from Eq. (7.4-1) to Eq. (7.4-2).

We will now use a renormalization (scaling) argument (based on [Guckenheimer and Holmes, 1990]) to determine how the average duration of the periodic bursts depends on the parameter μ for values just above 0. We expect that the length should go to 0 as μ increases and should increase to infinity as μ approaches 0 from above. The first step in the argument is to recognize that the length of these periodic bursts can be determined by finding how many iterations it takes to move a trajectory through the gap between the map function and the diagonal line. Let us call that number $n(\mu)$. Of course, $n(\mu)$ depends on where the trajectory enters the gap and on what we mean precisely by coming through the gap. Our arguments, therefore, will give us an average value. Next, we note that if we use $h^{(2)}(y)$ in place of $h(y)$ to step the trajectory through the gap, then we need only $n(\mu)/2$ steps to get through the gap because each step of $h^{(2)}(y)$ corresponds to two steps of $h(y)$.

We are now ready for the crucial part of the argument: We show that if we rescale the y axis values by a factor α and the parameter value μ by a factor δ and multiply the function by α (just as we did in the renormalization arguments for period-doubling in Appendix F), then the second iterate $h^{(2)}(y)$ looks just like $h(y)$. We will then relate the effect of a change in μ to a change in the number of steps required to get through the gap.

To see how this works, let us evaluate $h^{(2)}(y)$ directly:

$$h^{(2)}(y) = \mu + (\mu + y + cy^2) + (\mu + y + cy^2)^2 \qquad (7.4\text{-}3)$$

If we multiply out the last term and keep only terms linear in μ and terms linear and quadratic in y (in the spirit of our original approximation), we find that

$$h^{(2)}(y) \approx 2\mu + y + 2cy^2 \qquad (7.4\text{-}4)$$

Hence, we see that if we replace y with $Y = \alpha y$ and μ with $M = \delta\mu$ and then multiply the function by α with $\alpha = 2$ and $\delta = 4$, we arrive at a new function that looks just like the original function $h(y)$:

$$g_M(Y) \equiv 2[h^{(2)}_{M/4}(Y/2)] = M + Y + cY^2 \qquad (7.4\text{-}5)$$

The conclusion we draw from the previous argument is that changing to the second iterate of the function is equivalent to multiplying the parameter value by a factor of four.

Let us now go back to counting the number of steps required to get through the gap. The last sentence in the previous paragraph tells us that $n(\mu)/2 = n(4\mu)$. More generally, we can say that $n(\mu)/2^n = n(4^n\mu)$. This equation is satisfied if

$$n(\mu) = k \frac{1}{\sqrt{\mu}} \qquad (7.4\text{-}6)$$

where k is a constant. Eq. (7.4-6) tells us that as μ approaches 0 (from above) the average duration of the periodic bursts gets longer and longer. The average length of the periodic bursts increases as the reciprocal of the square root of μ. For example, if we reduce μ by a factor of four, making the "gap" smaller, the trajectories spend twice as long, on the average, in the region of the gap. This proportionality has been confirmed in experiments (JEP82 and YEK83) using semiconductor diode circuits similar to the one described in Chapter 1.

We can recast this argument in more general terms using the language of renormalization theory. Based on scaling arguments, we can assert that there should be a universal function (with the possibility of different universality classes as in Chapter 5 and Appendix F) that satisfies

$$g(x) = \alpha g\left(g\left(x/\alpha\right)\right) \qquad (7.4\text{-}7)$$

Eq. (7.4-7) is identical to Eq. (5.6-3) but the minus signs are missing. The minus signs were included in Eq. (5.6-3) to account for the alternation of trajectory points from one side of the unstable fixed point to the other after a period-doubling occurs. (Recall that this alternation is linked to the negative value of the derivative of the map function evaluated at the fixed point with parameter values near the bifurcation value.) For the tangent bifurcation described here, the derivative of the map function is near +1; so, no minus sign is required in Eq. (7.4-7).

We need to supplement Eq. (7.4-7) with two "boundary conditions": $g(0) = 0$ and $g'(0) = 1$. The surprise is the fact that there is an exact solution to this renormalization equation. The solution is

$$g(x) = \frac{x}{1-cx} \qquad (7.4\text{-}8)$$

with $\alpha = 2$. For small values of x, Eq. (7.4-8) reduces to Eq. (7.4-2). Arguments similar to those presented in Chapter 5 and Appendix F show that $\delta = 4$. Hu and Rudnick (HUR82) have shown how to extend these arguments to iterated map functions that have different power law dependences near the tangent bifurcation. The results given here apply to the case for which the dependence is quadratic as shown in Eq. (7.4-2). Thus, like the case of period-doubling, we have many universality classes, but we expect the quadratic case to be the most common.

Exercise 7.4-2. (a) Check that Eq. (7.4-6) satisfies the conditions on $n(\mu)$ stated in the text. (b) Verify that Eq. (7.4-8) satisfies the renormalization condition Eq. (7.4-7) with $\alpha = 2$.

7.5 Types of Intermittency and Experimental Observations

In Chapter 4, in the introductory discussion of the routes to chaos, we mentioned four different types of intermittency. We have repeated that classification in Table 7.1. The names given in quotation marks are intended to indicate the type of bifurcation event that accompanies the intermittency.

The four types of intermittency can be distinguished by the behavior of the Floquet multipliers for the Poincaré map function at the bifurcation event or the slopes of the iterated map functions for one-dimensional maps. Recall from Section 4.6 that a limit cycle becomes unstable by having the absolute value of its Floquet multipliers become greater than 1. For intermittency behavior this event can occur in four ways corresponding to the four types of intermittency listed in Table 7.1. Let us discuss each of the types.

In Type I intermittency, the type discussed in the first sections of this chapter, the Floquet multiplier crosses the unit circle (see Fig. 4.7) along the real axis at + 1. As we have seen, this leads to irregularly occurring bursts of periodic and chaotic behavior. However, during these bursts, the amplitudes of the motion (going back to the full state space description) are stable (on the average). We call this (perhaps, oxymoronically) *stable intermittency* or *tangent bifurcation intermittency* since the bifurcation event is a tangent bifurcation or a saddle-node bifurcation. This type of intermittency has been seen in many experiments, particularly in systems that also show the period-doubling route to chaos.

If the two Floquet multipliers form a complex conjugate pair, then the imaginary part indicates the presence of a second frequency in the behavior of the system. (The first frequency corresponds to the original limit cycle, which disappears at the bifurcation event.) At the bifurcation event, the limit cycle associated with the second frequency becomes unstable, and we observe bursts of two-frequency behavior mixed with intervals of chaotic behavior. Thus, Type II intermittency is a type of Hopf bifurcation event. Type II intermittency has been observed, to date, in only a few experimental studies (HUK87, SEG89).

If the Floquet multiplier is negative and becomes more negative than −1, then a type of period-doubling bifurcation event takes place. The amplitude of the subharmonic behavior created at the bifurcation point grows, while the amplitude of the motion associated with the original period decreases. This periodic behavior, however, is interrupted by bursts of chaotic behavior. Hence, we call this period-doubling intermittency since the Floquet multipliers change as they do for period-doubling, but after the bifurcation event, the period-doubled behavior is not stable.

Table 7.1

Types of Intermittency
1. Type I ("tangent bifurcation intermittency")
2. Type II ("Hopf-bifurcation intermittency")
3. Type III ("period-doubling intermittency")
4. On-off Intermittency

Type III intermittency has been observed by Dubois, Rubio, and Berg, in an experiment on Rayleigh-Bénard convection (DRB83), by Tang, Pujol, and Weiss in an ammonia ring laser (TPW91) and by Kim, Yim, Ryu, and Park (KYR98) in a diode-inductor circuit.

The fourth type of intermittency is called *on-off intermittency* because the behavior of the system seems to alternate between very quiescent behavior and chaotic bursts. This type of intermittency can be viewed as a variant of Type III intermittency with the "new frequency" essentially at zero, that is we get approximately "steady-state" behavior for some period of time. To explain this connection, we need to imagine that the dynamical variable being measured (call it x_m) lies in part of the state space of the system which has a fixed point at (or near) $x_m = 0$. More generally, there may be a hyperplane in the state space with $x_1 = 0$, $x_2 = 0, \ldots x_k = 0$ such that a trajectory starting on that plane stays on the plane. If the observed variable is one of the $\{x_i\}$ associated with this plane or some linear combination of these state space variables, then the system will appear to be "quiescent" when a trajectory gets near to that plane. If the trajectory stays near the plane for some reasonable period of time, then we get the appearance that the system is "off." Of course, the trajectory will eventually leave the neighborhood of that plane. If it then wanders through a chaotic region of state space, we can have a burst of chaotic behavior (the "on" behavior). We can think of the behavior of the system as behavior on a k-dimensional plane perturbed by the degrees of freedom not associated with the plane. If the behavior of the other degrees of freedom is chaotic, then the amount of time spent near the quiescent hyperplane is unpredictable and we get seemingly random intervals of "off" behavior punctuated by seemingly random "on" behavior.

To observe on-off intermittency, the observed variables must correspond to some of those associated with the special hyperplane. It is not obvious that we can always pick out the right variables. Some differential equation models and iterated map models displaying on-off intermittency are discussed in PST93. HPH94 and RCB95 report on experimental systems that exhibit on-off intermittency. There has been some conjecture that on-off intermittency may play a role in a variety of natural phenomena including "bursting" in fluid flow, sunspot activity, and reversals of the Earth's magnetic field.

7.6 Crises

A crisis is a bifurcation event in which a chaotic attractor (and its basin of attraction) disappears or suddenly expands in size (GOY82, GOY83). The former type of crisis is often called a *boundary crisis* for reasons that will become obvious. The sudden expansion (or contraction) of a chaotic attractor is called an *interior crisis*. In both cases, the crisis occurs because an unstable fixed point or an unstable limit cycle "collides" with the chaotic attractor as some control parameter of the system is changed. A *metamorphosis* (GOY87)(ATY91), another type of crisis event, is the appearance or sudden enlargement of fractal structure in a basin boundary.

To illustrate the behavior of a system at a crisis event, let us look at two bifurcation diagrams presented earlier. In Fig. 1.14 the bifurcation diagram for the logistic map suddenly ends at $A = 4$. The chaotic attractor, which is present for A values just below 4, disappears in a boundary crisis event. In Fig. 1.8, the bifurcation diagram for the diode circuit shows the three chaotic bands, which occur at about $V = 0.4$ volts (about 2/3 of the way across the diagram), suddenly expanding into a single chaotic band. This is an example of an interior crisis event. Similar interior crises can be seen in the bifurcation diagram of the Gaussian map in Figs. 5.15 and 5.17. Let us now explore how these crisis events occur.

Boundary Crisis

First, we will discuss the boundary crisis for the logistic map at $A = 4$. Recall from the discussion of Chapter 5 that the logistic map has a fixed point at $x = 0$ and that this fixed point becomes unstable for $A > 1$. As A approaches 4, the chaotic attractor gradually expands in size until at $A = 4$ it touches (collides with) the unstable fixed point at $x = 0$. For $A > 4$, almost all initial points lead to trajectories that end up at $x = -\infty$. If the trajectory starts off between 0 and 1, however, the region in which the chaotic attractor had existed for $A < 4$, the trajectory will wander chaotically around the old attractor region before leaving for the new attractor at $-\infty$. For $A > 4$, the chaotic attractor and its basin of attraction have disappeared and are not replaced by a new attractor in the same region of state space.

> **Exercise 7.6-1.** Investigate the behavior of the quadratic map introduced in Exercise 5.4-6 and show that it has a boundary crisis at $C = 2$. Relate that boundary crisis to the behavior of the unstable fixed point that is "born" at $C = -1/4$.

We can understand some of the dynamics of the behavior for A just greater than 4 by applying the graphic iteration technique. Figure 7.5 shows the logistic map for an A value just greater than 4. Centered on $x = 1/2$ is a narrow range of x values called the "escape region." When a trajectory lands in the escape region, the next iteration will lead to an x value greater than 1; subsequent iterations then rapidly take the trajectory off toward $-\infty$. By finding the x values that lead to $x = 1$, we find the boundaries of this loss region. Straightforward algebra leads to

$$x = \frac{1}{2} \pm \frac{1}{2A^{1/2}}\sqrt{A-4} \qquad (7.6\text{-}1)$$

Thus, we see that the length of this loss region, given by the difference between the two boundary values, varies as $(A-4)^{1/2}$. Since a trajectory starting in the interval between 0 and 1 can escape only if it gets into the escape region, the average time the chaotic transient lasts before escaping to $-\infty$ varies as $(A-4)^{-1/2}$. This

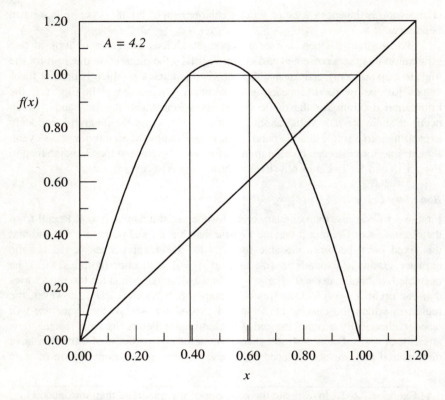

Fig. 7.5. A plot of the logistic map for $A = 4.2$. For x values inside the "escape region" bounded by the vertical lines near $x = 1/2$, trajectories quickly escape to $x = -\infty$. The boundaries of the escape regions are those x values that map to $x = 1$. Trajectories starting at other x values between 0 and 1 will wander chaotically in the interval until they hit the escape region.

behavior has been verified in numerical experiments on various map functions (GOY83). We would expect this behavior to be universal for any map function with quadratic behavior near its maximum value.

Crises may also occur in higher dimensional map functions. In the Hénon map, a boundary crisis occurs as shown in Fig. 5.21. Near $C = 1.08$, the chaotic part of the six-piece attractor collides with a saddle point on the boundary separating the two basins of attraction, and the six-piece attractor (and its basin of attraction) suddenly vanish (GOY83).

Interior Crisis

Let us now turn to the other type of crisis—the interior crisis. For this type of bifurcation event, an unstable fixed point or an unstable limit cycle that exists <u>within</u> the basin of attraction of a chaotic attractor, collides with the chaotic

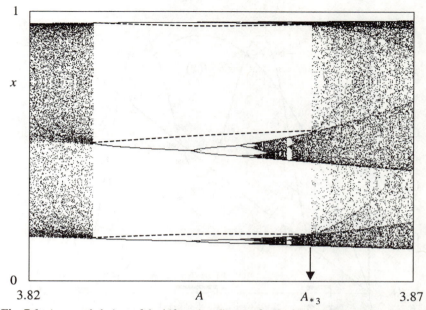

Fig. 7.6. A expanded view of the bifurcation diagram for the logistic map in the vicinity of the period-three interior crisis. At $A = 1 + \sqrt{8}$, a tangent bifurcation gives rise to the period-3 window. (As an aside, note that the effects of intermittency can be seen in the clustering of trajectory points in three bands just to the left of the tangent bifurcation.) Three unstable fixed points are also created. As the control parameter A increases, the stable fixed points undergo a period-doubling sequence to form chaotic bands. The locations of the unstable period-3 fixed points are indicated by the dashed lines. With further increase of A, the unstable fixed points come into contact with the chaotic bands and the chaotic attractor suddenly increases in size.

attractor as some control parameter is varied. When the collision occurs, the chaotic attractor suddenly expands in size. This expansion occurs because trajectories on the chaotic attractor eventually come close to the unstable fixed point (or unstable limit cycle) and then are repelled by the fixed point into regions of the state space that were not visited before the collision took place. The net effect is the expansion of the region of state space over which the trajectories wander.

From the discussion of intermittency and tangent bifurcations, we can see that an interior crisis is a likely partner of a tangent bifurcation that produces a periodic window because unstable as well as stable fixed points come into existence at a tangent bifurcation. For example, Fig. 2.4, which shows the period-5 window for the logistic map, contains an interior crisis to the right of the period-5 period-doubling cascade. The five chaotic bands suddenly expand in size to cover the full range of x values spanned by the period-five trajectories. Before the crisis, the (long-term) trajectories were restricted to the five relatively narrow bands. Figure

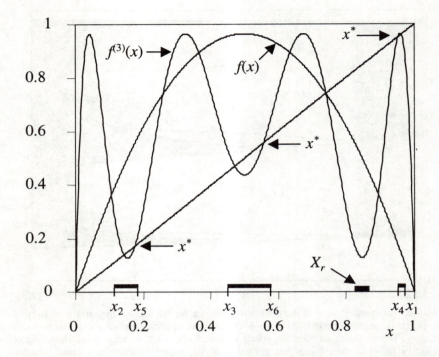

Fig. 7.7. A graph of the logistic map function and its third iterate for $A = 3.86$, just larger than A_{*3}. The unstable fixed points labeled x^* are the three fixed points that have moved into the three chaotic bands indicated by the heavy lines labeled with numbers along the x axis. The region labeled X_r is the "re-injection region." Trajectories landing in X_r are re-injected into the narrow band regions. The points labeled with numbers are the trajectory points for successive iterations starting with $x = 1/2$.

7.6 shows an interior crisis for the period-3 window of the logistic map function. Before the crisis, the attractor consists of three distinct bands, which have developed from the period-three period-doubling cascade. At $A = A_{*3}$, an interior crisis occurs and the attractor suddenly expands to fill the previously avoided regions between the bands.

The behavior of trajectories near an interior crisis shows several universal features (CHW81, GOY83). First, for parameter values just beyond the crisis value A_{*3} for the period-3 interior crisis of the logistic map, the average time trajectories spend in the narrow band regions (the regions in which the trajectories moved before the crisis) varies as $(A - A_{*3})^{-1/2}$. The reasoning here is identical to that used for the boundary crisis: There exists a "loss region" near $x = 1/2$ (for the logistic map) that allows trajectories to get into the previously "forbidden" range. The size of this loss region grows as $(A - A_{*3})^{+1/2}$ and hence the average time, call it τ, spent before the escape occurs varies as $(A - A_{*3})^{-1/2}$. This behavior has been verified in numerical computations (GOY83) and in experiments on semiconductor diodes

(ROH84). For systems that are not well modeled by one-dimensional maps with a quadratic maximum, the exponent in the power-law behavior may not be equal to 1/2 (GOR87). In fact, this type of behavior is sometimes called **crisis-induced intermittency** (GOR87) since the behavior switches intermittently from being confined to the original chaotic bands to wandering over a larger region of state space.

The second type of universal behavior focuses on the fraction of time the trajectories spend in the previously forbidden regions of state space. We want to present the argument leading to this predicted behavior because it illustrates rather nicely how (relatively) simple geometric arguments can lead to detailed predictions about the dynamics of the system. We will use the logistic map function as an example, but it should become obvious that the results apply to any one-dimensional map function with a quadratic maximum.

To simplify notation, let $a = A - A_{*3}$ denote how far the parameter A has increased beyond the crisis value. Next, we note that for small values of a, the fraction of the time the trajectories spend in the previously forbidden region can be expressed as $F = t_n/t_0$, where t_n is the average time spent in the new (previously forbidden region) and t_0 is the average time spent in the old, narrow-band region. (We assume $t_n \ll t_0$.) In the previous paragraph, we argued that t_0 is proportional to $a^{-1/2}$; therefore, we now need to calculate t_n.

To find t_n, let us look at Fig. 7.7. The narrow chaotic bands are indicated by the heavy lines along the x axis. The edges of the chaotic bands are the points labeled by the numbers 1 through 6. These points are the results of applying f n-times to $x = 1/2$, that is, $x_n = f^{(n)}(1/2)$. (Recall from our discussion in Chapter 5 that successive iterates of $x = 1/2$, the critical point, mark off the extreme values of x for a particular trajectory.) For $A > A_{*3}$ the unstable period-three fixed points indicated by x^* have moved into the chaotic bands. A trajectory getting near one of these fixed points will be repelled into the formerly forbidden region. In one of the formerly forbidden regions is an interval marked X_r and called the "re-injection region." If the trajectory lands in this region, it will be re-injected into the narrow band regions. Thus, t_n is determined by how long it takes a trajectory, on the average once it is in the previously forbidden region, to land in X_r.

Exercise 7.6-2. Show that the period-3 interior crisis for the logistic map occurs for $A = 3.8568$. Hint: When $A = A_{*3}$, the period-3 unstable fixed points have just moved into the chaotic bands indicated in Fig. 7.7. Hence, the point labeled x_4 is the same as the point that would be labeled x_7. (x_7 is the result of applying the map function to the point x_6.) Hence, for $A = A_{*3}$, we must have $f^{(4)}(1/2) = f^{(7)}(1/2)$.

Consider a trajectory starting near $x = 1/2$. After six applications of the map function (or two applications of $f^{(3)}$), the trajectory lands near the point labeled x_6 in Fig. 7.7. Note that x_6 is close to one of the unstable fixed points. In fact, the distance $x_6 - x^*$ is proportional to a, at least for small values of a (see Exercise 7.6-3). Since the trajectory is near an unstable fixed point, upon subsequent iteration of

$f^{(3)}$, the distance between the trajectory point and the unstable fixed point will increase by a factor M, the characteristic multiplier for $f^{(3)}$ at the crisis point. ($M = e^{\lambda}$ where λ is the Lyapunov exponent for $f^{(3)}$ at the crisis point.) Hence, it will take n iterations of $f^{(3)}$ for the trajectory to cover the distance d from x^* to X_r, where n satisfies the equation

$$d = kaM^n \qquad (7.6\text{-}2)$$

and k is a constant. By taking the natural log of both sides of Eq. (7.6-2), we can solve for n:

$$n = \frac{\ln d}{\ln M} - \frac{\ln(ka)}{\ln M} \qquad (7.6\text{-}3)$$

Note that the first term on the right-hand side of Eq. (7.6-3) is approximately independent of a for a particular map function.

Since the number of iterations n is proportional to the time spent in the previously forbidden region, we have almost solved the problem. There is, however, a complication: Suppose the nth iterate of $f^{(3)}$ brings the trajectory considered into X_r for a particular value of A. With a further increase of A, there may then be a range of A for which the nth iteration brings the trajectory outside X_r, while the $(n-1)$th iteration has not yet moved into X_r. The trajectory will then not be re-injected, and the time spent in the previously forbidden region will increase. When A is increased yet further, the $(n-1)$th iteration will finally hit X_r and the time spent in the previously forbidden region will decrease. Hence, we expect t_n to be a periodic function of $\ln a$.

To see why the dependence is on $\ln a$ and in fact to find the period length, let a_n be the a value for which the nth iteration brings the trajectory to X_r and a_n-1 be the value that brings the $(n-1)$th iteration to X_r. These two values must satisfy

$$a_n M^n = a_{n-1} M^{n-1} \qquad 7.6\text{-}4)$$

If we take the natural log of both sides of Eq. (7.6-4), we find

$$\ln a_{n-1} - \ln a_n = \ln M \qquad (7.6\text{-}5)$$

From Eq. (7.6-5), we see that this cycle of behavior depends on the difference of the natural logarithms of the a values and that the periodicity of this behavior is $\ln M$. Note, however, that our argument does not tell us specifically what this periodic function is; it need not be sinusoidal.

Putting all of the pieces together, we find that the fraction of time spent in the previously forbidden regions varies with the parameter difference a as:

$$F(a) = K\sqrt{a}\left[constant + P(\ln a) - \frac{\ln a}{\ln M}\right] \qquad (7.6\text{-}6)$$

where P is the unspecified periodic function with period $\ln M$, and K is a constant, independent of a. This rather complicated result has been verified in numerical computations on the quadratic map (GOY83) and in experiments on semiconductor diodes (HIL85). What should be surprising here is that relatively simple geometric arguments give us rather detailed quantitative predictions about the trajectories' behavior near the interior crisis event.

> **Exercise 7.6-3.** In the derivation presented earlier, we assumed that the distance between x_6 and x^* was proportional to the parameter difference a. Suppose that distance was proportional to a^p, where p is some exponent. Follow through the steps of the derivation with this new dependence and show that the result is essentially the same as Eq. (7.6-6).

Noise-Induced Crisis

In the previous section, we had assumed (implicitly) that we were dealing with a perfectly deterministic system; there was no external noise. If noise is present, then a system that is close to, but not yet in a crisis region can be "bumped" into and out of the crisis region by the noise. The average time τ between excursions into the avoided regions between the chaotic bands is described by a scaling law (SOG91)

$$\tau = K\sigma^{-\gamma} g\left(\frac{|A - A_*|}{\sigma} \right) \tag{7.6-7}$$

where σ is a measure of the "strength" of the noise (usually taken to be the standard deviation of the noise signal); A_* is the parameter value at which the crisis occurs in the absence of noise; γ is the scaling exponent for τ for the crisis in the absence of noise; g is some function that depends on the system being studied and the characteristic frequency distribution of the noise. This relationship has been verified in an experiment (SDG91) on a nonlinear driven oscillator system.

Double Crises

If two parameters of a system are varied simultaneously, it is possible to have more complicated crisis events that might involve both an interior crisis and a boundary crisis or an interior crisis and a metamorphosis of a basin boundary, for example. These more complicated situations are discussed in GGY93 and SUG95

7.7 Some Conclusions

In this chapter, we investigated several routes to chaos and several bifurcation events involving chaotic attractors. Intermittency and crisis events are as common as period-doubling. The well-armed chaologist should be able to recognize these events when they occur. In all cases, we have seen that simple geometric and renormalization arguments give us powerful universal predictions for the behavior of systems whose dynamics can be modeled by one-dimensional iterated maps. The extension of these ideas to higher-dimensional maps and to more general types of dynamical systems is still in its infancy.

7.8 Further Reading

The intermittency route to chaos was introduced by P. Manneville and Y. Pomeau, "Intermittency and the Lorenz Model," *Phys. Lett. A* **75**, 1–2 (1979) and Y. Pomeau and P. Manneville, "Intermittent Transition to Turbulence in Dissipative Dynamical Systems," *Commun. Math. Phys.* **74**, 189–97 (1980) (reprinted in [Hao, 1984], and [Cvitanovic, 1984]).

Intermittency is treated in some detail in the texts [Berge, Pomeau and Vidal, 1986], [Thompson and Stewart, 1986], and [Schuster, 1995].

B. Hu and J. Rudnick, "Exact Solutions to the Feigenbaum Renormalization Equations for Intermittency," *Phys. Rev. Lett.* **48**, 1645–48 (1982). This paper applies the renormalization method to intermittency.

1/f Noise and Intermittency

P. Manneville, "Intermittency, Self-Similarity and 1/f-Spectrum in Dissipative Dynamical Systems," *J. Phys.* (Paris) **41**, 1235 (1980).

I. Procaccia and H. G. Schuster, "Functional Renormalization Group Theory of Universal 1/f-noise in Dynamical Systems," *Phys. Rev. A* **28**, 1210–12 (1983).

E. G. Gwinn and R. M. Westervelt, "Intermittent Chaos and Low-Frequency Noise in the Driven Damped Pendulum" *Phys. Rev. Lett.* **54**, 1613–16 (1985). A numerical study of intermittency in the driven damped pendulum.

Observations of Intermittency in Experiments

Type I Intermittency

C. Jeffries and J. Perez, "Observation of a Pomeau–Manneville Intermittent Route to Chaos in a Nonlinear Oscillator," *Phys. Rev. A* **26**, 2117–22 (1982).

W. J. Yeh and Y. H. Kao, "Universal Scaling and Chaotic Behavior of a Josephson-Junction Analog," *Phys. Rev. Lett.* **49**, 1888–91 (1982).

H, Hayashi, S. Ishizuka, and K. Hirakawa, "Transition to Chaos Via Intermittency in the Onchidium Pacemaker Neuron," *Phys. Lett. A* **98**, 474–76 (1983).

R. W. Rollins and E. R. Hunt, "Intermittent Transient Chaos at Interior Crises in the Diode Resonator," *Phys. Rev. A* **29**, 3327–34 (1984).

Type II Intermittency

J.-Y. Huang and J.-J. Kim, "Type-II Intermittency in a Coupled Nonlinear Oscillator: Experimental Observation," *Phys. Rev. A* **36**, 1495–97 (1987).

J. Sacher, W. Elsässer, and E. Göbel, "Intermittency in the Coherence Collapse of a Semiconductor Laser with External Feedback," *Phys. Rev. Lett.* **63**, 2224–27 (1989).

Type III Intermittency

M. Dubois, M. A. Rubio, and P. Berge, "Experimental Evidence of Intermittency Associated with a Subharmonic Bifurcation," *Phys. Rev. Lett.* **51**, 1446–49 (1983).

D. Y. Tang, J. Pujol, and C. O. Weiss, "Type III intermittency of a laser," *Phys. Rev. A* **44**, 35–38 (1991).

C.-M. Kim, G.-S. Yim, J.-W. Ryu, and Y.-J. Park, "Characteristic Relations of Type-III Intermittency in an Electronic Circuit," *Phys. Rev. Lett.* **80**, 5317–20 (1998).

On-Off Intermittency

N. Platt, E. A. Spiegel, and C. Tresser, "On-Off Intermittency: A Mechanism for Bursting," *Phys. Rev. Lett.* **70**, 279–82 (1993).

P. W. Hammer, N. Platt, S. M. Hammel, J. F. Heagy, and B. D. Lee, "Experimental Observation of On-Off Intermittency," *Phys. Rev. Lett.* **73**, 1095–98 (1994). On-off intermittency in an electronic circuit.

F. Rödelsperger, A. Cenys, and H. Benner, "On-Off Intermittency in Spin-Wave Instabilities," *Phys. Rev. Lett.* **75**, 2594–97 (1995). On-off intermittency in ferromagnetic resonance.

Crises

The concept of a crisis in nonlinear dynamics was introduced by C. Grebogi, E. Ott, and J. A. Yorke, "Chaotic Attractors in Crisis," *Phys. Rev. Lett.* **48**, 1507–10 (1982).

C. Grebogi, and E. Ott, and J. A. Yorke, "Crises, Sudden Changes in Chaotic Attractors and Transient Chaos," *Physica D* **7**, 181–200 (1983).

Crisis-induced intermittency is discussed by C. Grebogi, E. Ott, F. Romeiras, and J. A. Yorke, "Critical Exponents for Crisis-induced Intermittency," *Phys. Rev. A* **36**, 5365–80 (1987).

The metamorphosis of a basin boundary is discussed in the following two references:

C. Grebogi, E. Ott, and J. A. Yorke, "Basin Boundary Metamorphoses: Changes in Accessible Boundary Orbits," *Physica D* **24**, 243–62 (1987).

K. T. Alligood, L. Tedeschi-Lalli, and J. A. Yorke, "Metamorphoses: Sudden Jumps in Basin Boundaries," *Commun. Math. Phys.* **141**, 1–8 (1991).

Double crises that can occur when two parameters are varied simultaneously are analyzed in

J. A. C. Gallas, C. Grebogi, and J. A. Yorke, "Vertices in Parameter Space: Double Crises Which Destroy Chaotic Attractors," *Phys. Rev. Lett.* **71**, 1359–1362 (1993).

H. B. Stewart, Y. Ueda, C. Grebogi, and J. A. Yorke, "Double Crises in Two-Parameter Dynamical Systems," *Phys. Rev. Lett.* **75**, 2478–2481 (1995).

Experimental Observations of Crises

R. C. Hilborn, "Quantitative Measurement of the Parameter Dependence of the Onset of a Crisis in a Driven Nonlinear Oscillator," *Phys. Rev. A* **31**, 378–82 (1985).

W. L. Ditto, S. Rauseo, R. Cawley, C. Grebogi, G.-H. Hsu, E. Kostelich, E. Ott, H. T. Savage, R. Segnan, M. L. Spano, and J. A. Yorke, "Experimental Observation of Crisis-Induced Intermittency and Its Critical Exponent," *Phys. Rev. Lett.* **63**, 923–26 (1989).

D. Dangoisse, P. Glorieux, and D. Hennequin, "Laser Chaotic Attractors in Crisis," *Phys. Rev. Lett.* **57**, 2657–60 (1986).

J. C. Sommerer, W. L. Ditto, C. Grebogi, E. Ott, and M. L. Spano, "Experimental Confirmation of the Scaling Theory of Noise-Induced Crises," *Phys. Rev. Lett.* **66**, 1947–50 (1991).

M. Finardi, L. Flepp, J. Parisi, R. Holzner, R. Badii, and E. Brun, "Topological and Metric Analysis of Heteroclinic Crisis in Laser Chaos," *Phys. Rev. Lett.* **68**, 2989–2991 (1992).

Other Crisis Issues

J. A. Yorke, C. Grebogi, E. Ott, and L. Tedeschini-Lalli, "Scaling Behavior of Windows in Dissipative Dynamical Systems," *Phys. Rev. Lett.* **54**, 1095–98 (1985). The scaling of periodic-windows that end in crises.

J. C. Sommer, E. Ott, and C. Grebogi, "Scaling Law for Characteristic Times of Noise-induced Crises," *Phys. Rev. A* **43**, 1754–69 (1991). The theory of so-called noise-induced crises.

7.9 Computer Exercises

CE7-1. Explore intermittency in iterations of the logistic map function in both the period-5 region near $A = 3.7375$ and the period-3 region near $A = 3.8319$ using a program that plots the x value as a function of iteration number.

CE7-2. A more difficult problem: Verify numerically for the logistic map the scaling prediction for the average length of the periodic "bursts" given in Eq. (7.4-7). Try both period-5 and period-3 intermittency.

CE7-3. Explore the crisis regions for the logistic map near $A = 3.86$ and for the Gaussian map function used in Chapter 5 by plotting bifurcation diagrams for those regions.

CE7-4. Study the transient behavior of iterates of the logistic map for trajectories that start near the fixed point at $1 - 1/A$ for $A > 4$. Can you determine how the average length of the transients depends on the parameter difference $A - 4$?

8

Hamiltonian Systems

"Physics is Where the Action Is" or "Minding Your Ps and Qs"

8.1 Introduction

In our discussions of nonlinear dynamics up to this point, we have dealt only with dissipative systems. The crucial feature of a dissipative system from the state space point of view is the "collapse" of a volume of initial conditions in state space. For most purposes, we can focus our attention on the attractor (or attractors, in general) in state space—those "areas" to which trajectories from a range of initial conditions are attracted. That is, we need consider only the attractors to understand the long-term dynamics of the system.

What happens if the amount of dissipation becomes smaller and smaller? In that case the system obviously takes longer and longer for trajectories that start away from the attractor to approach the attractor; it takes longer for a volume of initial conditions to collapse onto the attractor. In the limit in which there is no dissipation at all, we would expect that a volume of initial conditions would remain constant for all time and that there would exist no attractors for the trajectories.

Systems (or models) with no dissipation are called *conservative systems*, or equivalently, *Hamiltonian systems*. The term conservative means that certain physical properties of the system (such as the total mechanical energy, the angular momentum, etc.) remain constant in time. We say that these quantities are *conserved* quantities. If the system starts with a certain amount of energy, then that amount of energy stays the same over time. The name Hamiltonian is applied to these systems because their time evolution can be described by the so-called Hamilton's equations (after Sir William Hamilton, 1805–1865, a noted Scottish mathematician). We shall discuss these equations in the next section.

Do conservative systems occur in nature? In principle, the answer to this depends on our "level of description." Since we know that the total energy of an isolated system is conserved, though the energy may change form, we might conclude that Hamiltonian models are the only appropriate models. However, in practice, this full description is often too complex, and we instead focus our attention on one particular subsystem; the remaining part of the system acts as a source or sink of energy (i.e., as a source of dissipation). In that case, a dissipative model is appropriate.

In practice, many real systems are nearly conservative. The most famous (almost) conservative system is the solar system. In fact, it was a consideration of the dynamics of the solar system that led Poincaré to introduce many of the methods already described for dealing with nonlinear dynamics. Over the time periods of concern, which for the solar system are millions and billions of years, we can neglect most aspects of dissipation. There are, however, dissipation effects in the solar system such as tidal forces, which act on planets and moons, and the drag effects of the "solar wind," streams of particles emitted by the Sun. For example, dissipative tidal forces are responsible for the locking of the Moon's rotation rate to its orbital period so that the same side of the Moon always faces the Earth, as mentioned in Chapter 6. To a high degree of approximation, however, these dissipative effects can be neglected if we limit ourselves to time periods of a few million years. Based on these considerations, we can model the dynamics of the solar system with a dissipation-free, conservative (Hamiltonian) model.

Hamiltonian models are also important in the study of proton beams in high-energy particle accelerations, in quantum mechanics (more on this in Chapter 12), and as a branch of applied mathematics, for which there is now a vast literature. In this chapter, we will describe how chaos appears in Hamiltonian systems, and we will severely limit the amount of mathematical detail so that we can focus on how Hamiltonian systems differ from dissipative systems in some respects but are similar in others. By looking at a model with a variable amount of dissipation, we shall see how the two types of systems are connected. Most of the important theoretical results will simply be stated with only a sketch of a plausibility argument. The goal is to give you an intuitive picture of the rather complex behavior exhibited by Hamiltonian systems. Once the overall picture is in hand, the mathematically inclined reader can use the references at the end of the chapter for more detailed treatments.

We will first introduce some of the basic notions of Hamiltonian systems including the state-space description. We will then discuss an important, but limited, subclass of Hamiltonian systems—those called *integrable*. However, integrable systems cannot show chaotic behavior; therefore, we must explore what happens when a Hamiltonian system becomes nonintegrable. The chapter concludes with a brief description of some applications.

8.2 Hamilton's Equations and the Hamiltonian

Although we shall not make much direct use of Hamilton's equations, it will be helpful to introduce them briefly, both for the following discussion and for the chance to become familiar with some of the specialized jargon used in the study of Hamiltonian systems. In the Hamilton formulation of classical (Newtonian) mechanics, the time evolution of a system is described in terms of a set of dynamical variables, which give the positions (coordinates) and the momenta of the particles of the system. Traditionally, the coordinates are indicated with the symbols q_i and the momenta by p_i. The subscript notation is used to pick out a particular particle and a particular component of the position vector and momentum vector for that particle. If the system consists of N point particles, each of which has three components for its position vector and three components for its momentum vector, the subscript i will run from 1 to $3N$. For example, we might have $q_1 = (\vec{r}_1)_x$ and $p_1 = (\vec{p}_1)_x$. Here, q_1 represents the x component of the position vector for particle number 1, and p_1 the corresponding x component of the particle's momentum vector. Each pair q_i, p_i corresponds to a "degree-of-freedom" for the Hamiltonian system. (Recall the discussion in Section 3.2 about different uses of the term degree-of-freedom.)

The evolution of the Hamiltonian system is completely described if the time dependence of the qs and ps is known. That is, if we know $q_i(t)$ and $p_i(t)$ for all t and for all i, then we know everything there is to know about the time behavior of the system. In the Hamilton formulation, the time-dependence of the qs and ps is determined by solutions of Hamilton's equations, which are written in terms of the derivatives of the **Hamiltonian function** (or just **Hamiltonian**, for short) $H(q,p)$, where the unadorned symbols q and p mean that H depends, in general, on all the q_i and p_i. For the simplest cases, the Hamiltonian is just the total mechanical energy (kinetic energy plus potential energy) of the system, written as a function of the qs and ps. In any case, Hamilton's equations are a set of $2N$ coupled differential equations (for a system of N degrees of freedom)

$$\frac{dq_i}{dt} = \frac{\partial H(q,p)}{\partial p_i}$$

$$\frac{dp_i}{dt} = -\frac{\partial H(q,p)}{\partial q_i} \quad i = 1, \ldots, N \tag{8.2-1}$$

Exercise 8.2-1. Suppose a single particle with a mass m is constrained to move along the x axis. Its Hamiltonian $H = p_x^2/2m + U(q_x)$ is the sum of a kinetic energy term and a potential energy U. Show that Hamilton's equations are equivalent to Newton's Laws of Motion for the system. Hint: The x component of the force acting on the system is given by $F_x = - dU/dx$, the negative gradient (in the x direction) of the potential energy function.

Note that Hamilton's equations are similar in form to the standard first-order differential equations we have been using to describe dynamics in state space for a variety of systems. The similarity can be made more obvious by identifying the state space variables $x_1 = q_1$, $x_2 = p_1$, $x_3 = q_2$, and so on. For a Hamiltonian system, the functions (analogous to the fs in our previous treatment) that give the time dependence of the state space variables can be written as (partial) derivatives of some common function, namely, the Hamiltonian. As we shall see in the next section, that crucial feature embodies the special nature of Hamiltonian systems. The special linkage between the qs and ps and the partial derivatives of the Hamiltonian function give Hamiltonian mechanics a special mathematical form called a *symplectic* structure, which can be exploited to give elegant proofs of many features of the time behavior (see, for example, [Goldstein, 1980]).

An important consequence follows from Hamilton's equations: The value of the Hamiltonian itself represents a conserved quantity; it does not vary in time. We prove this by using the chain rule of differentiation:

$$\frac{dH(q, p)}{dt} = \sum_i \left\{ \frac{\partial H}{\partial p_i} \frac{dp_i}{dt} + \frac{\partial H}{\partial q_i} \frac{dq_i}{dt} \right\} + \frac{\partial H}{\partial t} \qquad (8.2\text{-}2)$$

The terms inside the braces of Eq. (8.2-2) tell us how H depends on time because H depends on the qs and ps and the qs and ps (in general) depend on time. This part describes the so-called *implicit* time-dependence of H. The last term in Eq. (8.2-2) tells us how H depends on time if the time variable appears *explicitly* in H. Explicit time dependence occurs if the system is subject to an externally applied time-dependent force, for example. We will not consider such cases; therefore, we will assume that the last term is 0.

If we now use Hamilton's equations (8.2-1) in Eq. (8.2-2), we find for each term in the sum

$$\frac{\partial H}{\partial p_i} \left(-\frac{\partial H}{\partial q_i} \right) + \frac{\partial H}{\partial q_i} \frac{\partial H}{\partial p_i} = 0 \qquad (8.2\text{-}3)$$

So, we see that the time derivative of H is 0 (if H does not depend explicitly on time). Hence, H represents a conserved quantity. If H represents the total energy of the system (as it usually does), then we say that the total energy is conserved for a Hamiltonian system. Alternatively, we say that the total energy is a "constant of the motion."

Exercise 8.2-2. In simple cases the Hamiltonian of a system can be written as the sum of the kinetic energy and the potential energy of the system, written as a function of momentum and position. For a point particle with mass m moving along the x axis, the kinetic energy, $(1/2)mv_x^2$, can be written in terms of the momentum in the x direction as $p_x^2/2m$ using $p_x = mv_x$, where as usual v_x is the x component of the particle's velocity. For a point mass oscillating under the influence of an ideal spring (for which the force is described by Hooke's law $F_x = -kx$, where x is the displacement of the particle from equilibrium and k is the so-called spring constant), show that the Hamiltonian is $H(q, p) = p^2/2m + (1/2)kq^2$, where $q = x$ is the relevant coordinate and $p = p_x$ is the corresponding momentum. Use Hamilton's equations to find the time evolution equations for this simple harmonic oscillator system.

8.3 Phase Space

We again find that a geometric state space description is useful, if not essential, for understanding how chaos occurs in Hamiltonian systems. State space for a Hamiltonian system is traditionally called *phase space*, and the axes of phase space give the values of the qs and ps. Hence, if we have N degrees of freedom (in the Hamiltonian sense of that phrase), we have N pairs of qs and ps, and the phase space will have $2N$ dimensions. Thus, for a Hamiltonian system (with no explicit time dependence in H), the phase space always has an even number of dimensions. Even for simple cases we will have difficulty visualizing these multi-dimensional phase spaces; therefore, we can anticipate using projections and Poincaré sections to simplify the description.

Since the Hamiltonian function value (usually the energy of the system) is a constant of the motion, a trajectory for a Hamiltonian system cannot go just anywhere in phase space. It can go only to regions of (q, p) space that have the same energy value as the initial point of the trajectory. Thus, we say that trajectories in phase space are confined to a $2N - 1$ dimensional *constant energy surface*. (Of course, this "surface" may be a multidimensional geometric object in general.)

Using techniques similar to those employed for dissipative systems, we can now show that a volume in phase space occupied by a set of initial conditions remains the same as the Hamiltonian system evolves. In Chapter 3, we established that the time-dependence of a small volume V occupied by a set of initial conditions in state space is given by

$$\frac{1}{V}\frac{dV}{dt} = \sum_i \frac{\partial f_i}{\partial x_i} \tag{8.3-1}$$

To show that this time derivative is 0 for a Hamiltonian system, we first need to translate Eq. (8.3-1) into the language of Hamiltonian dynamics. It will be useful in our proof to identify, as we did earlier, x_1 with q_1, x_2 with the corresponding p_1, x_3 with q_2, x_4 with p_2, and so on. We also recall that the time dependence of the xs is given by an equation of the form

$$\dot{x}_i = f_i(x_1, x_2, \ldots) \tag{8.3-2}$$

If we have $2N$ dimensions in the phase space, the index i will run from 1 to $2N$. Then, using the correspondence between the xs and the qs and ps given earlier, we see that

$$\dot{q}_1 = \dot{x}_1 = f_1(x_1, \ldots) = \frac{\partial H}{\partial p_1} \tag{8.3-3}$$

$$\dot{p}_1 = \dot{x}_2 = f_2(x_1,\ldots) = -\frac{\partial H}{\partial q_1} \qquad (8.3\text{-}4)$$

where the last equality in the two previous equations follows from Hamilton's equations.

Let us now examine the terms that appear in the sum on the right-hand side of Eq. (8.3-1). In particular let us look at the first two terms and insert the results of Eqs. (8.3-3) and (8.3-4):

$$\begin{aligned}
\frac{\partial f_1}{\partial x_1} + \frac{\partial f_2}{\partial x_2} &= \frac{\partial}{\partial q_1}\left(\frac{\partial H}{\partial p_1}\right) + \frac{\partial}{\partial p_1}\left(-\frac{\partial H}{\partial q_1}\right) \\
&= \frac{\partial^2 H}{\partial q_1 \partial p_1} - \frac{\partial^2 H}{\partial p_1 \partial q_1} \\
&= 0
\end{aligned} \qquad (8.3\text{-}5)$$

The final equality of Eq. (8.3-5) follows from the fact that the order of differentiation does not change the result for these second "cross" partial derivatives (unless H is an unphysically bizarre function of q and p).

This cancellation of terms continues pairwise for all the qs and ps. Hence, we conclude that for a Hamiltonian system, the volume occupied by a set of initial conditions does not change in time as the system evolves. The practical consequence of this unchanging volume is the fact that Hamiltonian systems do not have phase space attractors in the way dissipative systems do. As we shall see, this lack of attractors is both a simplification and a complication. Since we have no attractors, we do not need to worry about transients; that is, we do not need to let the trajectory run for some time so that it settles onto the appropriate attractor. This usually simplifies the process of finding the appropriate solution for the trajectories. On the other hand, we shall see that the lack of attractors means that trajectories starting with different initial conditions may behave quite differently as time goes on; there is no common attractor onto which they settle.

Liouville's Theorem and Phase Space Distributions

We want to generalize the discussion of phase space volume and its time evolution. This discussion will be important when we take up the issue of chaos in quantum mechanics in Chapter 12. Suppose we specify a *distribution* of initial conditions in phase space by means of a probability density function $\rho(q,p)$, where again q and p stand for the set of q_is and p_is. The probability function is defined so that the probability of finding a trajectory in a small volume dV of phase space centered at the values q and p is given by

$$probability = \rho(q,p)dV \qquad (8.3\text{-}6)$$

Furthermore, suppose that the initial conditions are confined to some volume V in phase space bounded by a "surface" of initial condition points. As the system evolves, the volume V will move through phase space. Its shape will generally be distorted, but the total volume occupied will remain constant. How does the probability density evolve? We can follow the argument used for the time dependence of H given earlier:

$$\frac{d\rho}{dt} = \sum_i \left\{ \frac{\partial\rho}{\partial q_i}\frac{dq_i}{dt} + \frac{\partial\rho}{\partial p_i}\frac{dp_i}{dt} \right\} + \frac{\partial\rho}{\partial t}$$

$$= \sum_i \left\{ \frac{\partial\rho}{\partial q_i}\frac{\partial H}{\partial p_i} + \frac{\partial\rho}{\partial p_i}\left(-\frac{\partial H}{\partial q_i} \right) \right\} + \frac{\partial\rho}{\partial t} \qquad (8.3\text{-}7)$$

$$\equiv \{\rho, H\} + \frac{\partial\rho}{\partial t}$$

The last equality defines the quantity $\{\rho, H\}$ known as the **Poisson bracket**. Again, the terms in the braces tell us how ρ varies because it depends on q and p. (We can think of looking at the time variation of ρ as we "ride" along a trajectory.) The last term tells us how ρ varies at a fixed location in phase space. In some ways, the motion of this probability density in the multidimensional phase space is like the motion of an incompressible fluid. In fluid mechanics, the so-called **Lagrangian** picture follows the trajectory of a small chunk of fluid and focuses attention on how the properties associated with the chunk change as it moves. The total derivative $d\rho/dt$ is then called the **material derivative** or (equivalently) the **hydrodynamic derivative**. The second point of view, watching the time behavior at a fixed point in (phase) space, is called the **Eulerian** picture. We shall exploit this analogy with fluid flow in Chapter 11.

Liouville's Theorem (see, for example, [Goldstein, 1980]) tells us that the total time derivative $d\rho/dt = 0$ and the phase space density around any evolving trajectory point remains constant. As the trajectories evolve, the phase space density in the neighborhood of any trajectory does not change.

The theorem follows from a simple argument: For a Hamiltonian system, by definition, the volume of phase space delimited by some "surface of initial conditions" remains constant in time. Suppose we have M initial points inside that surface distributed with some probability distribution $\rho(q, p)$. As time goes on, the number of trajectory points inside the surface must remain fixed because if M were to change, a trajectory must cross the bounding surface and that cannot occur due to the No-Intersection Theorem. (The surface evolves following the trajectories that arise from its initial conditions.) Both the fixed volume and constant M results apply to very small volumes dV and small numbers dM in phase space. Hence, the ratio $dM/dV = \rho$ cannot change. This establishes Liouville's theorem.

As an aside we want to point out that Liouville's Theorem can be generalized to describe dissipative systems as well. A rather elaborate, but straightforward calculation [Jackson, 1989, pp. 44–45 and Appendix C] shows that $\rho(q, p)$ satisfies the differential equation

$$\frac{\partial \rho}{\partial t} + \sum_i \frac{\partial}{\partial q_i} [\rho(q, p, t) f_i(q, p)] = 0 \qquad (8.3\text{-}8)$$

where the f_is give the time-dependence of the system as shown in Eqs. (8.3-3) and (8.3-4). For Hamiltonian systems, the "divergence requirement" is

$$\sum_i \frac{\partial f_i}{\partial x_i} = 0 \qquad (8.3\text{-}9)$$

so that for Hamiltonian systems Eq. (8.3.-8) reduces to

$$\frac{\partial \rho}{\partial t} + \sum_i f_i(q, p) \frac{\partial \rho}{\partial x_i} = 0 \qquad (8.3\text{-}10)$$

Using the correspondence between the x_is and the qs and ps shows that Eq. (8.3-10) is the same as Eq. (8.3-7).

The main point of the discussion is to note that both Eq. (8.3-6) and (8.3-7) are linear in ρ. Hence, by the general arguments presented earlier, the time evolution of ρ cannot be chaotic. This point is worth reflecting on for a moment. Even though the individual trajectories may show chaotic behavior, the probability distribution of trajectory points evolves in a nonchaotic fashion. More specifically, if we start with a slightly different probability distribution and watch it evolve in time, the difference between its evolution and the evolution of the original distribution will not grow exponentially. (We obviously need to define this "difference" appropriately.) We do not have sensitive dependence on initial conditions for the probability distribution. In other words, if we want to observe chaotic behavior, we cannot look at the probability distribution and its evolution. We must (apparently) look at the individual trajectories. We shall return to this point in our discussion on quantum mechanics in Chapter 12. In Section 8.8, the so-called Arnold cat map provides a simple example of the evolution of phase space distributions for chaotic Hamiltonian systems.

8.4 Constants of the Motion and Integrable Hamiltonians

In Section 8.2, we saw that the energy, represented by the Hamiltonian of a system, is conserved if the Hamiltonian does not depend on time explicitly. Let us flesh out some of the consequences of that result. If a trajectory in phase space starts at a point labeled (q_0, p_0), where q and p represent the entire set of $2N$ phase space coordinates, the system's energy is given by $H(q_0, p_0)$. As time goes on, the qs and ps evolve, but at any later time the energy will have the same value, namely

$$H(q(t), p(t)) = H(q_0, p_0) \qquad (8.4\text{-}1)$$

where $(q(t), p(t))$ gives the phase-space trajectory originating from (q_0, p_0). Hence, each possible trajectory for a Hamiltonian system can be labeled by the energy value that "belongs to" that trajectory. Note that the converse statement is not necessarily true: there may be many different trajectories corresponding to the same energy value.

In some Hamiltonian systems there are additional quantities whose values also remain constant as the trajectory evolves. Let us see why this is important by looking at a special case. Suppose one of the ps, say p_j does not change in time:

$$\dot{p}_j = 0 = -\frac{\partial H}{\partial q_j} \tag{8.4-2}$$

The only way the last term in Eq. (8.4-2) can be 0 for all $(q(t), p(t))$ values along the trajectory is to have $H(q, p)$ not depend on q_j at all! We have the general rule: The momentum p_j is a constant of the motion if, and only if, the Hamiltonian for a system does not depend on the corresponding q_j explicitly. In that case, a trajectory can be labeled by its value of $p_j = p_{j0}$ as well as by its energy value $H(q_0, p_0)$. When this occurs, the trajectories are limited not only to those regions of phase space associated with a particular energy value, they are also constrained by the value of p_j. Thus, the trajectories must "live" on a $2N$-k dimensional "surface" in phase space, where k is the number of conserved quantities.

For a special (and very limited, but theoretically important) class of Hamiltonian systems, there are as many constants of the motion as there are degrees of freedom. Such systems are called **integrable**, for reasons that will shortly become obvious. However, in most cases, the constants of the motion are not the ps in terms of which we initially wrote the Hamiltonian. The constants of the motion, however, can always be expressed as functions of the original qs and ps. The constants of the motion are usually called the **action variables** and are commonly written as $J_i(q, p)$, $i = 1, 2, \ldots, N$. For an integrable Hamiltonian system, the phase space trajectories are confined to an N-dimensional surface in phase space.

Associated with each $J_i(q, p)$ is another variable labeled $\theta_i(q, p)$. This new variable is called the corresponding **angle variable**. (In an upcoming example, we shall see why these names are used.) The $J_i s$ and $\theta_i s$ are chosen so that Hamilton's equations, expressed in terms of the $J_i s$ and $\theta_i s$ have the same mathematical form as the original Hamilton's equations expressed in terms of the qs and ps:

$$\dot{\theta}_i = \frac{\partial H(\theta, J)}{\partial J_i}$$

$$\dot{J}_i = -\frac{\partial H(\theta, J)}{\partial \theta_i} \tag{8.4-3}$$

(Since the angle variable is dimensionless, we see that the action has units of energy multiplied by time, or equivalently, momentum multiplied by distance). If Eqs. (8.4-3) are satisfied, we say that the variables (θ, J) are related to the variables (q, p) by a **canonical transformation**. (Here, canonical means "satisfying some canon or rule".) As we shall see, for a periodic trajectory in phase space, for which the trajectory forms a closed curve, the action has a nice geometric interpretation: The action associated with a periodic trajectory is proportional to the phase space area enclosed by the trajectory.

The special case we are interested in is a canonical transformation that leads to a Hamiltonian that depends <u>only</u> on the J_is and <u>not</u> on the θ_i s. In that case, for all $i = 1, 2, \ldots, N$, we have

$$\dot{J}_i = 0 \qquad\qquad (8.4\text{-}4)$$

and the J_is are the N constants of the motion.

A Hamiltonian system that satisfies Eqs. (8.4-3) and (8.4-4) is called (somewhat unfortunately) an ***integrable system***. The term *integrable* comes from the notion that the action J_i can be expressed as an integral over the motion of the system and that the corresponding equation for θ_i can be easily integrated.

The term *integrable* is a bit misleading because it seems to imply that the character of the system depends on our *ability* to find the appropriate canonical transformation or to do the required integral. In fact, one often finds phrases in the literature such as "A system is integrable, if we can find the canonical transformation . . ." In reality, the character of the system, that is the number of constants of the motion, is independent of our ability to find the appropriate canonical transformation.

A minor technical point: For the system to be integrable and have the simple properties described later, the constants of the motion must have vanishing Poisson brackets among themselves: $\left\{ J_i, J_j \right\} = 0$ for all i and j. If this condition is satisfied, the system is said to have N constants of the motion "in involution."

By expressing the desired canonical transformation in terms of a so-called ***Birkhoff Series*** and by examining the convergence properties of that series, one can determine (at least in principle) whether a given Hamiltonian system is integrable or nonintegrable (HEL80). If the system is ***nonintegrable***, it has fewer constants of the motion than degrees of freedom.

We will now list (without proof) some results, which tell us what kinds of Hamiltonian systems are integrable (HEL80).

1. All one-degree-of-freedom Hamiltonian systems, for which H is an infinitely differentiable (that is, "analytic") function of q and p, are integrable and the corresponding action J satisfies $H = \omega\ J$, where $\omega = \partial H / \partial J$.
2. All Hamiltonian systems for which Hamilton's equations are linear in q and p are integrable (via the so-called normal mode transformations).

3. All Hamiltonian systems with nonlinear Hamilton's equations that can be separated into uncoupled one-degree-of-freedom systems are integrable.

Let us now explore the consequences of having an integrable Hamiltonian, for which all the J_is are constants of the motion. In this case, the time dependence of θ_i is easy to find

$$\dot{\theta}_i = \frac{\partial H}{\partial J_i} \equiv \omega_i(J) \qquad (8.4\text{-}5)$$

The right-hand side of the previous equation defines what is called the angular frequency of the motion. For an integrable system, ω_i depends on the values of all the J_is, but because the J_is are independent of time, the ω_is are also independent of time. Thus, we can immediately write

$$\theta_i(t) = \omega_i t + \theta_i(0) \qquad (8.4\text{-}6)$$

Hence, we see that if the system is integrable and if we can find the canonical transformations that give us Eqs. (8.4-3) and (8.4-4), then, amazingly, we have completely solved the dynamics of the system.

If we want to find the behavior of the system in terms of the original ps and qs, we can use the inverse of the canonical transformations to write

$$q_i(t) = f(\theta, J)$$
$$p_i(t) = g(\theta, J) \qquad (8.4\text{-}7)$$

For a system that is bounded spatially, the qs and ps must be periodic functions of the θ_is since, according to Eq. (8.4-6), $\theta_i(t)$ increases without limit as $t \to \infty$.

A general mathematical procedure for finding the action-angle variable transformations is called *canonical perturbation theory*, in which the original qs and ps are written as power series functions of the new variables J and θ. If the series diverges, we recognize that the system is nonintegrable.

We will now study two examples of one-degree-of-freedom Hamiltonian systems and their phase space behavior.

The Simple Harmonic Oscillator

In Exercise 8.2-2, we learned that the Hamiltonian for a one-dimensional simple harmonic oscillator with mass m and spring constant k is

$$H(q, p) = \frac{p^2}{2m} + \tfrac{1}{2}kq^2 \qquad (8.4\text{-}8)$$

where q is the displacement of the oscillator from its equilibrium position. In this case, the numerical value of the Hamiltonian is the total mechanical energy of the system. The corresponding Hamilton's equations for the time evolution are

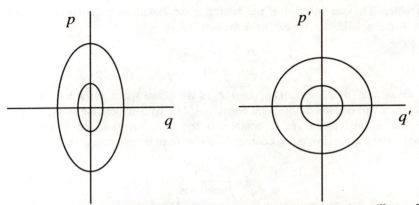

Fig. 8.1. On the left is a phase space trajectory for the simple harmonic oscillator. Each ellipse is associated with a particular value of the energy. A larger ellipse has a larger value of the energy. By rescaling the variables, the trajectories become circles whose radii are equal to $\sqrt{J}$, the square root of the action value associated with that trajectory. The corresponding angle variable θ locates the point on the trajectory.

$$\dot{q} = \frac{\partial H}{\partial p} = \frac{p}{m}$$

$$\dot{p} = -\frac{\partial H}{\partial q} = -kq$$

(8.4-9)

The one (spatial) dimension simple harmonic oscillator model has one degree of freedom and its phase space is two-dimensional. Since the Hamiltonian is independent of time, the phase space trajectories must reside on a $2N-1 = 1$ dimensional "surface" (i.e., on a curve). The trajectories are closed curves because the motion is periodic. Each value of the energy is associated with a unique closed curve.

The phase space trajectories for the simple harmonic oscillator are ellipses with a larger ellipse associated with a larger value of the energy (Hamiltonian) of the system. If the phase space coordinates are suitably rescaled, as shown on the right in Fig. 8.1, then the trajectories become circles. As we shall see, the radius of each circle is equal to the square root of the value of the action associated with that trajectory. The corresponding angle variable gives the location of the trajectory point on the circle.

Note that the simple harmonic oscillator model has only one fixed point, namely ($p = 0$, $q = 0$). In the language of Hamiltonian dynamics this kind of fixed point is called an ***elliptic point*** because the trajectories near the fixed point are ellipses.

For the simple harmonic oscillator, we know that the angular frequency of the oscillatory motion is given by $\omega = \sqrt{k/m}$. Since this is a one-degree-of-freedom system or since Hamilton's equations are linear, we expect that this system is integrable. The one constant of the motion is the Hamiltonian (energy) or some multiple thereof. Hence, we can write the action J as

$$J = \frac{H}{\omega} = \frac{p^2}{2m\omega} + \frac{kq^2}{2\omega} \tag{8.4-10}$$

If we use $p/\sqrt{2m\omega}$ and $q\sqrt{m\omega/2}$ as the phase space variables, then the trajectories will be circles with radii equal to $\sqrt{J}$. To complete the story, we can write the original phase space variables p and q in terms of the action-angle variables (with θ positive going counterclockwise from the positive q axis):

$$p = \sqrt{2m\omega J} \sin\theta$$
$$q = \sqrt{2J/(m\omega)} \cos\theta \tag{8.4-11}$$

Exercise 8.4-1. (a) Show that if the phase space coordinates are transformed as suggested in the text, the phase space trajectories for the simple harmonic oscillator are circles. (b) Show that the radius of the circular trajectory is given by $\sqrt{J}$. (c) Show that in the original phase space the area enclosed by the ellipse is equal to $2\pi J$.

As Exercise 8.4-1 shows, the action associated with a closed trajectory is related to the phase space area enclosed by the trajectory. In general, we may write

$$J = \frac{1}{2\pi} \oint p \, dq \tag{8.4-12}$$

where the symbol $\oint$ means that the integral is taken around the closed path of the trajectory.

The Pendulum

One of the most studied and time-honored examples of a Hamiltonian system is the pendulum. This system consists of a point mass m suspended at the end of a rigid (but massless) rod of length L. The rod is free to pivot about an axis at the other end of the rod. To make the system Hamiltonian, we ignore any dissipation due to friction in the pivot or to air resistance. A picture of this system is shown in Fig. 8.2.

The Hamiltonian for this system is expressed as the sum of kinetic energy of rotation about the pivot point and gravitational potential energy (relative to the equilibrium point when the pendulum mass hangs downward):

$$H = \frac{p_\theta^2}{2mL^2} + mgL(1 - \cos\theta) \tag{8.4-13}$$

where p_θ is the angular momentum associated with the rotation about the axis and g is the acceleration due to gravity. Thus, we see that the pendulum is a one-

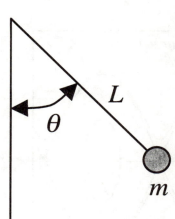

Fig. 8.2. A picture of the pendulum. The angle θ is defined relative to the stable equilibrium position. Gravity acts downward.

degree-of-freedom system (with a two-dimensional phase space). Hence, by the arguments presented earlier, it is an integrable system with one constant of the motion, namely its total mechanical energy E.

For a given value of the energy E, we can use Eq. (8.4-13) to solve for the momentum

$$p_\theta = \pm\sqrt{2mL^2[E - mgL(1 - \cos\theta)]} \qquad (8.4\text{-}14)$$

By convention, the momentum is positive when the pendulum is moving counterclockwise and negative when the pendulum is moving clockwise. From Eq. (8.4-14), we see that the momentum has its largest magnitude when $\theta = 0$, at the bottom of the pendulum's swing. For energies less than $2mgL$, the highest point of the swing occurs when $p_\theta = 0$ or in terms of the angular displacement from the vertical line, when $E = mgL(1 - \cos\theta)$. If we allow the pendulum to swing over the top by giving it sufficient energy (greater than $2mgL$), then the minimum of its momentum magnitude occurs when $\theta = \pi$ at the top of the swing. Eq. (8.4-14) can be used to plot the phase space trajectories as shown in Fig. 8.3.

We can find the corresponding action J for the system by integrating the momentum over one cycle of the motion

$$J = \frac{1}{2\pi}\int d\theta\sqrt{2mL^2[E - mgL(1 - \cos\theta)]} \qquad (8.4\text{-}15)$$

The resulting integral is known as an ***elliptic integral*** and is tabulated numerically in many mathematical handbooks. The important point here is that we can determine the frequency of the motion, numerically, by using Eq. (8.4-15) with Eq. (8.4-5).

The phase space diagram for the pendulum, shown in Fig. 8.3, is typical of the phase space diagrams for many integrable Hamiltonian systems. For relatively small values of the energy, the phase space trajectories are "ellipses" centered on

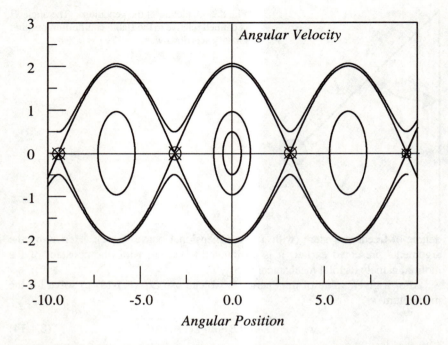

Fig. 8.3. The phase diagram for the pendulum. Angular velocity (vertical axis) is plotted as a function of angular position (horizontal axis). Each trajectory corresponds to a particular value of the total mechanical energy of the system. Elliptic fixed points occur at the origin and at angular positions of $\pm n2\pi$ for positive or negative integer n. Hyperbolic (saddle) points occur at $\theta = \pm\pi$. These are indicated by small circles. The trajectories that join the hyperbolic points are separatrices. Inside the separatrices, the motion is periodic and oscillatory. The motion on trajectories outside the separatrices corresponds to counterclockwise revolutions for $\dot\theta > 0$ and clockwise revolutions for $\dot\theta < 0$.

the origin. At the origin is an elliptic fixed point for the system: If the system starts with $p_\theta = 0$ and $\theta = 0$, then it stays there for all time. These ellipses are the "tori" on which the trajectories live in this two-dimensional phase space.

> **Exercise 8.4-2.** For low energies (small angles of oscillations), we can replace $\sin\theta$ in the pendulum problem with the angle value θ in radians. Use this small angle approximation to evaluate J and then show that the frequency of oscillation is given by $\omega = \sqrt{g/L}$.

There are also fixed points at $p_\theta = 0$ and $\theta = \pm n\pi$, where n is an odd integer. These fixed points correspond to the pendulum's standing straight up with the mass directly above the pivot point. Note that $\theta = \pm\pi$ are physically equivalent points since they both correspond to the vertical position of the pendulum. However, it is occasionally useful to allow the angle to increase or decrease without limit to visualize some aspects of the pendulum's motion. The physical equivalence shows up in the periodicity of the trajectory pictures in state space if you shift along the θ axis by multiples of 2π.

The fixed points corresponding to the inverted vertical position are unstable in the sense that the slightest deviation from those conditions causes the pendulum to swing away from the inverted vertical position. These fixed points are called **hyperbolic points** for Hamiltonian systems because trajectories in their neighborhood look like hyperbolas. The fixed points are, of course, saddle points using the terminology introduced in Chapter 3. Trajectories approach the hyperbolic point in one direction and are repelled in another direction.

The trajectories that lead directly to or directly away from a hyperbolic point are called **separatrices** (plural of **separatrix**) since they separate the phase space into regions of qualitatively different behavior. (The separatrices are the stable and unstable manifolds introduced before.) For the pendulum, the trajectories inside the separatrices correspond to oscillatory motion about the vertically downward equilibrium point. Trajectories outside the separatrices correspond to "running modes" in which the pendulum has sufficient energy to swing over the top. One type of running mode has an angular velocity that is positive (counterclockwise motion); the other type has a negative angular velocity (clockwise motion). In both cases, the magnitude of the angle θ continues to increase with time.

> This organization of the phase space by elliptic points surrounded by the separatrices associated with hyperbolic points is typical for integrable Hamiltonian systems. The separatrices segregate phase space regions that correspond to qualitatively different kinds of motion.

Systems with N Degrees of Freedom

If we compare Eqs. (8.4-5) and (8.4-6) with the results of the simple harmonic oscillator example, we see that an integrable system with N degrees of freedom is equivalent, in terms of action-angle variables, to a set of N uncoupled oscillators. (The oscillators are simple harmonic if the ω_i are independent of the value of the Js. They are otherwise nonlinear oscillators for which ω depends on J.) This connection explains why so much attention is paid to oscillating systems in the study of dynamics.

Since there are N constants of the motion for an integrable system of N degrees of freedom, the trajectories in state space are highly constrained. For example, an integrable system with two degrees of freedom has trajectories confined to a two-dimensional surface in phase space. This surface, in general, is the surface of a torus residing in the original four-dimensional phase space. Like the quasi-periodic motion studied in Chapter 6, the trajectories are characterized by the two frequencies

$$\omega_1 = \frac{\partial H}{\partial J_1} \qquad \omega_2 = \frac{\partial H}{\partial J_2} \qquad\qquad (8.4\text{-}16)$$

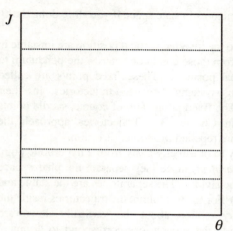

Fig. 8.4. The action-angle diagram for an integrable single degree of freedom Hamiltonian system. Each trajectory consists of a constant action and hence "resides" on a horizontal line in action-angle space. A periodic trajectory of period-n would consist of n points on a horizontal line. A quasi-periodic trajectory would eventually fill in completely a horizontal line.

More generally, we say that a trajectory for an integrable system with N degrees of freedom is constrained to the N-dimensional surface of a torus (which resides in the original $2N$-dimensional phase space). These tori are often called *invariant tori* since the motion is confined to these surfaces for all time.

If the various frequencies ω_i are incommensurate and the motion is quasi-periodic, then the trajectory eventually visits all parts of the torus surface. Such a system is said to be *ergodic* because one could compute the average value of any quantity for that system either by following the time behavior and averaging over time (usually hard to do) or by averaging over the q, p values on the surface of the torus in phase space (usually much easier to do).

Action-Angle Space

Instead of the usual pq phase space, an alternative state space description makes use of the action-angle variables. The motivation for this is threefold. First, for an integrable system, each trajectory is characterized by a fixed value for each of the action variables. For example, for the simple harmonic oscillator, each elliptical trajectory in pq phase space corresponds to a fixed action as shown in Eq. (8.4-10).

In action-angle space, the trajectories of an integrable system reside on horizontal lines of constant action. Each horizontal line in Fig. 8.4 corresponds to a "torus" in the original pq phase space. (we may think of cutting the torus around its outer circumference and then spreading the "surface" flat. The horizontal line corresponds to viewing the surface edge on.)

The second motivation for this kind of diagram comes from the study of nonintegrable systems. As we shall see in the next section, when a Hamiltonian system becomes nonintegrable, the action associated with a trajectory is no longer constant (in general). This fact shows up most obviously in an action-angle space diagram as the trajectory points wander vertically in that diagram.

The third reason for introducing action-angle space is related to the importance of action in quantum mechanics. As we shall see in Chapter 12, each

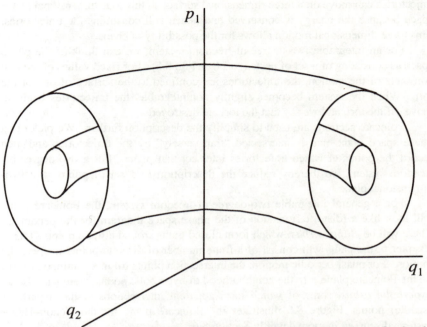

Fig. 8.5. For an integrable two-degree-of-freedom system, the trajectories are confined to the surfaces of a set of nested tori. Each surface corresponds to a different set of values of the two constants of the motion. If the system becomes nonintegrable, the trajectories can move off the tori.

allowable quantum state is associated with integer multiples (in most cases) of a fundamental unit of action. Thus, the allowed quantum states correspond to trajectories equally spaced in the action variable in an action-angle diagram.

8.5 Nonintegrable Systems, the KAM Theorem, and Period-Doubling

Since the behavior of an integrable Hamiltonian system is always periodic or quasi-periodic, an integrable system cannot display chaotic behavior. We have spent some time describing integrable systems because much of the literature on the chaotic behavior of Hamiltonian systems has focused on systems that are, in some sense, just slightly nonintegrable. We can then ask how the behavior of the system deviates from that of an integrable system as the amount of nonintegrability increases.

We are immediately faced with the problem of visualizing the trajectories for nonintegrable systems because, as we learned in the last section, a nonintegrable system must have at least two degrees of freedom. If the system were integrable, then the trajectories would move on the two-dimensional surface of a torus and be either periodic or quasi-periodic. However, if the system is nonintegrable, then the trajectories can move on a three-dimensional surface in this four-dimensional phase space because the energy is conserved and hence still constrains the trajectories. This three-dimensional motion allows for the possibility of chaos.

For an integrable two-degree-of-freedom system, we can think of the phase space as consisting of a set of nested tori (see Fig. 8.5). For fixed values of the two constants of the motion, the trajectories are confined to the surface of one of the tori. When the system becomes slightly nonintegrable, the trajectories begin to move off the tori, and we say that the tori are destroyed.

Poincaré sections are used to simplify the description further. We pick out a phase space plane that is intersected "transversely" by the trajectories and then record the points at which trajectories intersect that plane. For a two-degree-of-freedom system, we thereby reduce the description to a set of points in a two-dimensional plane.

For a general integrable two-degree-of-freedom system, the Poincaré plane will look like a (distorted) version of the phase space diagram for the pendulum: There will be *elliptic orbits*, which form closed paths around elliptic points. (As in Chapter 6, the paths will consist of a finite number of discrete points for periodic motion. For quasi-periodic motion, the intersection points fill in a continuous curve on the Poincaré plane.) In the neighborhood of hyperbolic points, there will also be *hyperbolic orbits*, some of which form apparent intersections at the hyperbolic (saddle) points. Figure 8.6 illustrates the Poincaré plane for the Hénon–Heiles system, discussed in more detail in Section 8.6.

If the system becomes nonintegrable, constraints are removed from the trajectories, and they can begin to move more freely through phase space. Hence, loosely speaking, we expect the highly organized pattern of the integrable system's Poincaré section to "dissolve." Does the entire Poincaré section, however, dissolve simultaneously leaving only a random scattering of points? The answer to this question is provided by the famous ***Kolmogorov–Arnold–Moser (KAM) Theorem*** [Arnold, 1978].

The KAM theorem states that (under various technical conditions that need not concern us here) some phase space tori, in particular those associated with quasi-periodic motion with an irrational winding number, survive (but may be slightly deformed) if a previously integrable system is made slightly nonintegrable. This result is stated more formally as follows: The originally integrable system's Hamiltonian can be written as a function of the action variables alone: $H_0(J)$. We now make the system nonintegrable by adding to $H_0(J)$ a second term, which renders the overall system nonintegrable. The full Hamiltonian is then

$$H(J,\theta) = H_o(J) + \varepsilon H_1(J,\theta) \qquad (8.5\text{-}1)$$

where ε is a parameter that controls the relative size of the nonintegrability term. The second term in (8.5-1) is sometimes called a "perturbation" of the original Hamiltonian, and "perturbation theory" is used to evaluate the effects of this term on the trajectories. The KAM Theorem states that for $\varepsilon \ll 1$ (so the system is almost integrable), the tori with irrational ratios of the frequencies associated with

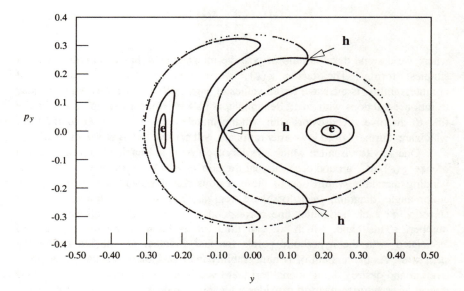

Fig. 8.6. A Poincaré section for a two-degree-of-freedom (almost) integrable system (the Hénon–Heiles system of Section 8.6) with a fixed value of the energy. Each trajectory corresponds to a different value of the other constant of the motion. Two elliptic points, labeled by **e**, are surrounded by "elliptical" orbits. Three hyperbolic points, marked by **h**, are connected by separatrices that divide the plane into well-defined regions.

the actions will survive. These are called *KAM tori*. As ε increases, the tori dissolve one by one with the last survivor being the one with winding number equal to our old friend the Golden Mean, the "most irrational" of the irrational numbers.

As soon as ε increases above 0, the phase space tori associated with rational winding numbers break up. In a Poincaré section representation, the points begin to scatter around the Poincaré plane. This fast break-up can be explained as the kind of resonance effect discussed in Chapter 6. The nonintegrable part of the Hamiltonian essentially couples together what had been independent oscillations in the integrable case. When the frequency ratio for a torus is rational, there is considerable overlap of the harmonics associated with each oscillation. This overlap creates a "strong resonance" condition leading (usually) to a rapid growth of the amplitude of the motion in phase space and a rapid flight from the torus surface to which the trajectories had been confined in the integrable case. When the frequency ratio is irrational, however, there is no overlap in harmonics and we might expect the corresponding torus to survive for larger values of ε.

The KAM theorem states that the tori that survive for a given amount of nonintegrability have winding numbers w that satisfy the following inequality [Schuster, 1995]

$$\left| w - \frac{m}{n} \right| > \frac{g(\varepsilon)}{n^{2.5}} \tag{8.5-2}$$

where we assume that the ratio of positive integers m/n has been reduced to its simplest form. The factor $g(\varepsilon)$, which increases with the amount of nonintegrability, is the same for all values of m/n. We see that tori with w close to rational fractions with small denominators (1/4 or 1/3, for example) will be the first to dissolve. Those tori with winding numbers, such as the Golden Mean, which are "further" from low-order rationals, will survive to larger values of ε.

One way to visualize which tori dissolve first is to imagine building a region of size $g(\varepsilon)/n^{2.5}$ around each rational fraction m/n. Any torus with an irrational winding that falls within one of these regions dissolves. In the corresponding action-angle diagram, each of these regions consists of a horizontal band surrounding each horizontal line corresponding to a J with a rational winding number. This band, which we can think of as bounded by the separatrices associated with the hyperbolic points for that J value, is called a **resonance structure**. As the amount of nonintegrability increases, these resonances begin to overlap and destroy the irrational tori that lived in the region between them. This notion of **resonance overlap** provides a means to approximate for a given system the amount of nonintegrability required to destroy all such tori and to induce complete chaotic behavior (CHI79).

What is the dynamical importance of the KAM tori? In the integrable case, we argued that phase space trajectories are confined to the surfaces of tori in phase space. As the system becomes nonintegrable, trajectories are able to move off these tori. However, the surviving KAM tori still have trajectories associated with their surfaces. In low-dimensional phase spaces, the surviving KAM tori can prevent a trajectory that has moved off a dissolving torus from ranging throughout the allowed energy region of phase space. In a sense, the KAM tori continue to provide some organization for the trajectories in phase space.

Let us see how this organizational ability depends on the dimensionality of the phase space (that is, on the number of degrees of freedom for the system). In a $2N$-dimensional phase space, the constant energy "surface" has $2N - 1$ dimensions. As we argued earlier, the tori for an integrable system have a dimensionality of N. Thus, for the tori to partition phase space, we can have either $N = 1$ or $N = 2$. In other words, the tori can segregate regions of phase space only in systems with one or two degrees of freedom. In higher-dimensionality systems, when the tori begin to dissolve as the system becomes nonintegrable, a so-called **stochastic web** forms. In that case, trajectories may wander over large portions of the allowed energy region of state space. [Zaslevsky, Sagdeev, Usikov, and Chernikov, 1991] gives a very complete description of the formation of this stochastic web.

Poincaré–Birkhoff Theorem

What happens to the phase space trajectories when the rational-ratio tori break-up as the Hamiltonian becomes nonintegrable? This question is answered by the

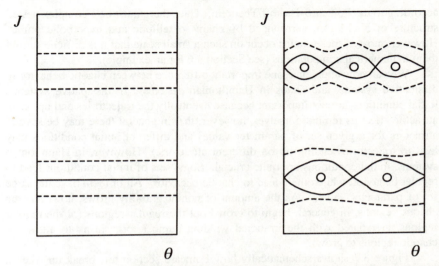

Fig. 8.7. A schematic action-angle diagram of the break-up of a rational winding number torus (solid line on the left) when an integrable system (on the left) becomes slightly nonintegrable (on the right). The elliptic points are indicted by circles. They are separated by hyperbolic points, whose in-sets and out-sets are sketched with solid curves. The in-sets and out-sets of the newly created hyperbolic points must avoid the remaining irrational winding number (KAM) tori (dashed curves). Each set of elliptic points with the corresponding hyperbolic points and their in-sets and out-sets constitute a "resonance."

famous Poincaré–Birkoff Theorem (BIR35) (HEL80). When the system is integrable, each point on a torus corresponding to a rational winding number m/n is part of a periodic orbit. As we saw in Chapter 6, each orbit on that torus leads to n points in the Poincaré section of that torus. According to the Poincaré-Birkhoff Theorem, when the system becomes nonintegrable, the torus breaks up into an alternating sequence of n elliptic points and n hyperbolic points. Around each elliptic point will be a series of elliptic orbits. Associated with the hyperbolic points will be a homoclinic orbit connecting stable and unstable manifolds of the saddle points. The event is shown schematically in Fig. 8.7. These hyperbolic points are important because an in-set (also called the stable manifold) and an out-set (the unstable manifold) is associated with each of them, as we discussed in Chapter 4 for dissipative systems. These in-sets and out-sets, however, must avoid intersecting the remaining tori. Hence, we can see that it is not unreasonable for these in-sets and out-sets to get entangled. Since nearby trajectories feel the influence of these in-sets and out-sets, trajectories near them might be very complicated. From this argument, we expect that the decay of rational winding number tori lead to homoclinic and heteroclinic tangles, which, as we saw in Chapter 4, are a "cause" of chaotic behavior.

The detailed structure, however, is even more complicated. Close to each of the newly formed elliptical orbits must be a chain of elliptic and hyperbolic points according to the Poincaré–Birkoff Theorem. Thus, there must be a complex nested structure of KAM tori, surrounded by chains of elliptic and hyperbolic points. However, most of these chains occur on such a small scale that it is difficult to find them in numerical computations (see Section 8.6 for an example).

We can now appreciate one important difference between chaotic behavior in dissipative systems and chaos in Hamiltonian systems. In dissipative systems, initial conditions are not important because eventually the trajectories end up on an attractor. (Let us remind ourselves, however, that in general there may be several attractors for a given set of parameter values and different initial conditions may lead to trajectories ending up on different attractors.) However, in Hamiltonian systems, initial conditions are quite crucial. Some sets of initial conditions lead to regular behavior, while others lead to chaotic behavior. All of them have the same set of parameter values. As the amount of nonintegrability grows, however, the chaotic regions, in general, begin to crowd out the regular regions (or the regular regions, associated with the irrational winding number tori, shrink to allow the chaotic regions to grow).

Figure 8.7 shows schematically how Poincaré section tori break up when a system becomes nonintegrable. The irrational winding number tori, indicated by dashed curves, survive, but rational winding number tori, indicated by the solid lines, break up into elliptic points and hyperbolic points. The in-sets and out-sets of the hyperbolic points must wind around the newly created ellipses and avoid crossing the still existing KAM tori.

If the original integrable system has some hyperbolic points, the in-sets and out-sets of those hyperbolic points seem to dissolve when the system becomes slightly nonintegrable. According to (HEL80, p. 433), these in-sets and out-sets develop homoclinic and heteroclinic tangles; these tangles cause chaotic behavior for the nonintegrable system.

Chaotic Behavior and Phase Space Mixing

For a two-degree-of-freedom system that becomes nonintegrable, the chaotic behavior is associated with the homoclinic tangles that develop as rational winding number tori break up. However, the associated chaotic trajectories are constrained to relatively small regions of phase space by the surviving KAM tori. In fact, for small amounts of nonintegrability, the chaotic behavior, though present in principle, may not be noticeable at a practical level. With an increase in the amount of nonintegrability, enough KAM tori dissolve to allow the chaotic behavior to extend over a noticeable region of phase space. As the amount of nonintegrability grows further, the remaining KAM tori dissolve and eventually a single chaotic trajectory can roam through most of the allowed region of phase space.

If the system has three (or more) degrees of freedom and becomes nonintegrable, the $(2N - 1)$-dimensional manifolds, which are the surfaces of the KAM tori, can no longer act as boundaries for chaotic trajectories, and a given chaotic trajectory can roam through large regions of phase space. This roaming through phase space is called *Arnold Diffusion*. It represents a kind of statistical mixing due to chaos. However, for two-degree-of-freedom systems, the KAM tori still protect some regions of phase space until the tori dissolve as the amount of nonintegrability increases.

Lyapunov Exponents

How do we know that chaotic behavior is present when a Hamiltonian system becomes nonintegrable? We can again turn to Lyapunov exponents. There are as many Lyapunov exponents as there are dimensions in phase space. We can calculate the Lyapunov exponents for the system's trajectories in a straightforward fashion by techniques to be discussed in detail in Chapters 9 and 10. We find that at least one of the Lyapunov exponents is positive for trajectories associated with the fuzzy regions that develop around hyperbolic points for a nonintegrable system. For a Hamiltonian system, the sum of the Lyapunov exponents must be 0 (since phase space volume is conserved). For a two-degree-of-freedom system (the minimum number required to support chaotic behavior), there are four Lyapunov exponents. For chaotic trajectories, one of these is positive, two are 0, and one is negative.

In many cases, we are interested in the behavior of trajectories that are close to periodic trajectories (some of which may be stable and some unstable). If we use the symbol $\vec{z}(t)$ to represent a trajectory near a periodic orbit (in phase space) and $\delta\vec{z}(t)$ to represent the "distance" between the actual trajectory and the periodic orbit, the evolution of $\delta\vec{z}(t)$ is described by the so-called ***monodromy matrix*** $\bar{\bar{M}}$:

$$\delta\vec{z}(t) = \bar{\bar{M}}\delta\vec{z}(0) \qquad (8.5\text{-}3)$$

The eigenvalues of the monodromy matrix come in pairs λ and $1/\lambda$ since the product of the eigenvalues must equal 1 due to the preservation of phase space area as the system evolves.

Period-Doubling for Hamiltonian Systems

When an integrable system becomes nonintegrable, according to the Poincaré–Birkhoff Theorem, a Poincaré section ellipse corresponding to a torus with winding number m/n breaks up into an equal number of smaller ellipses and hyperbolic points. These smaller ellipses correspond to tori with winding number $2m/2n$, so we see twice as many points in the Poincaré section. As the amount of nonintegrability is further increased, each of these ellipses breaks up into pairs of ellipses (and hyperbolic points) in a sequence reminiscent of period-doubling bifurcations in dissipative systems. In fact, Feigenbaum numbers (BCG80) have been worked out using renormalization type arguments. For Hamiltonian systems the parameter difference scaling δ_H and the size scaling parameters α_H are

$$\delta_H = 8.721097 \ldots \qquad \alpha_H = -4.01807 \ldots \qquad (8.5\text{-}4)$$

where the subscript H reminds us that these numbers apply to Hamiltonian systems. This kind of period-doubling has been seen in many mathematical models (see

Sections 8.6 and 8.7), but we will not pursue the subject further since it has not yet been applied to any actual physical systems.

Period-n-tuplings, with $n > 2$, are seen frequently in Hamiltonian systems. For example, a cycle of period-2 points can undergo a period-quintupling and split into a period-10 cycle. In simple terms, these higher-order period multiplications come about due to various "resonances" among the nonlinear oscillators that make up the system (see, for example, GMV81).

There is some evidence that bifurcations of periodic orbits in Hamiltonian systems often occur in organized groups. The method of "normal-form theory" (see, Appendix B) provides a method for investigating this organization (SSD95). Meyer's theorem [Meyer and Hall, 1992] asserts that the periodic orbits usually undergo only five types of bifurcations.

For the sake of completeness, we would like to mention that some Hamiltonian systems have trajectories that show chaotic behavior, not because of nonintegrability in the sense introduced here, but because the Hamiltonian has "singularities," points at which it is not differentiable. These singularities arise in models of collisions of rigid billiard balls for example. In a rough sense these models are analogous to the piece-wise linear iterated maps discussed in Chapter 5 and piece-wise linear differential equation models to be mentioned briefly in Chapter 12.

The discussion of Hamiltonian systems has been rather abstract and general. In the next two sections, we shall present two model Hamiltonian systems, which illustrate the general features described earlier.

8.6 The Hénon–Heiles Hamiltonian

In this section we will explore the properties of a particular model Hamiltonian to illustrate the dynamics of Hamiltonian systems. The model was first introduced by Hénon and Heiles (HEH64) as a model for the motion of a star inside a galaxy. The Hamiltonian has two degrees of freedom (two pairs of ps and qs) and takes the form

$$H = \tfrac{1}{2}p_1^2 + \tfrac{1}{2}q_1^2 + \tfrac{1}{2}p_2^2 + \tfrac{1}{2}q_2^2 + \left[q_1^2 q_2 - \tfrac{1}{3}q_2^3 \right] \tag{8.6-1}$$

This Hamiltonian represents two simple harmonic oscillators (compare Exercise 8.2-2) coupled by a cubic term, which makes the Hamiltonian nonintegrable. If we let $q_1 = x$, $q_2 = y$, $p_1 = p_x$, and $p_2 = p_y$, then the Hamiltonian can also be interpreted as a model for a single particle moving in two dimensions under the action of a force described by a potential energy function

$$V(x, y) = \tfrac{1}{2}x^2 + \tfrac{1}{2}y^2 + x^2 y - \tfrac{1}{3}y^3 \tag{8.6-2}$$

This potential energy function has a local minimum at the origin ($x = 0$, $y = 0$). A three-dimensional plot of this potential energy function is shown on the left in Fig.

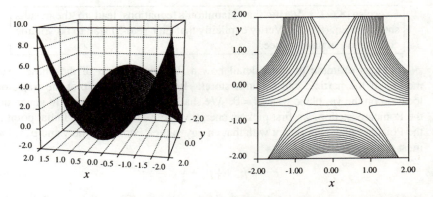

Fig. 8.8. On the left is a three-dimensional plot of the potential energy function for the Hénon–Heiles model. On the right is a contour plot of the same function. We will be concerned with a particle moving in the slight depression near the origin. If the particle's energy is less than 1/6, the particle will be trapped in the triangular region near the origin. For higher energies the particle can escape the local minimum of the potential energy.

8.8. A contour plot of the same potential energy function is shown on the right of Fig. 8.8.

If we start the particle near the origin with an energy value less than 1/6, it will stay in an "orbit" near the origin for all time. If the energy is greater than 1/6, the particle can escape the local minimum of the potential energy and go off to infinity. If the energy is very small, the particle stays close to the origin and the trajectories look much like the periodic motion of a particle in a two-dimensional simple harmonic potential.

Hamilton's equations for this system lead to the following equations for the dynamics of the system:

$$\dot{x} = \frac{\partial H}{\partial p_x} = p_x$$

$$\dot{y} = \frac{\partial H}{\partial p_y} = p_y$$

$$\dot{p}_x = -\frac{\partial H}{\partial x} = -x - 2xy \qquad (8.6\text{-}3)$$

$$\dot{p}_y = -\frac{\partial H}{\partial y} = -y - x^2 + y^2$$

We see that the system lives in a four-dimensional phase space. However, since the system is Hamiltonian, the energy conservation constraint means that the trajectories must live in a three-dimensional volume in this four-dimensional space. Again, we will use the Poincaré section technique to reduce the recorded trajectory points to a two-dimensional plane.

Exercise 8.6-1. Verify that Hamilton's equations lead to the results shown in Eq. (8.6-3). Verify explicitly that Eqs. (8.6-3) lead to no volume contraction in phase space.

Let us examine in some detail how a Poincaré section of the phase space motion of the particle can be understood. It is traditional to plot the trajectory location on the yp_y plane when $x = 0$. We shall follow that tradition. In generating the Poincaré section, we first pick an energy value E and then some initial point on the Poincaré plane consistent with that energy value. For $x = 0$, the y and p_y values must satisfy

$$E = \tfrac{1}{2} p_x^2 + \tfrac{1}{2} p_y^2 + \tfrac{1}{2} y^2 - \tfrac{1}{3} y^3 \qquad (8.6-4)$$

Hence, for a fixed energy and a particular initial value for p_x, there is a finite range on the $y\, p_y$ plane within which the Poincaré section points must fall. The time evolution equations (8.6-3) are then integrated and successive Poincaré section points are generated. Figure 8.9 shows one such orbit in the xy (real space) plane on the left. On the right is the corresponding Poincaré section. The Poincaré section points fall on two "ellipses" that are formed by the intersection of a surface of a three-dimensional torus with the Poincaré plane. (Note that the cross section of the torus is distorted and the part of the torus intersecting the plane for negative values of y has a shape different from the part intersecting at positive values of y.) Thus, we conclude that this particular orbit corresponds to a periodic or quasi-periodic orbit. Near the middle of each of the ellipses is an elliptic point, not shown in Fig. 8.9.

Exercise 8.6-2. Explain in detail the connection between the xy trajectories and the Poincaré section shown in Fig. 8.9.

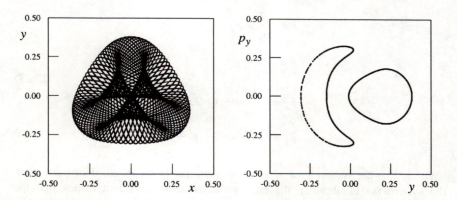

Fig. 8.9. On the left is the xy (real space) trajectory of a particle moving in the Hénon–Heiles potential with $E = 0.06$. The orbit started with $x = 0$, $y = -0.1475$, $p_x = 0.3101$, and $p_y = 0$. On the right is the corresponding p_y–y Poincaré section with $x = 0$. The "ellipses" are formed by the intersection of the surface of a three–dimensional torus with the Poincaré plane.

According to the KAM Theorem, if the Hamiltonian is nonintegrable, the only surviving nonchaotic orbits should be quasi-periodic orbits (with irrational winding numbers). Hence, in the Poincaré section we should see smooth curves where these orbits intersect the Poincaré plane. However, Figs. 8.9 and 8.10 show intersections that appear to be made up of a finite number of points, which we might interpret as due to periodic orbits. The finite number of points is due to two artifacts: (1) The trajectories have been followed for only a finite amount of time. If the irrational winding number for a particular orbit is close to a low-order rational number (e.g., 1/4), it may take a long time to "fill in" the curve on the Poincaré plane. (2) Numerical errors and round-off in the computer's numerical integration of trajectories may lead to apparently periodic orbits.

Note that the trajectory shown in Fig. 8.9 is just one of many trajectories possible for the given energy value. To fill out the Poincaré section, we need to choose a variety of initial conditions consistent with the same energy value. Figure 8.10 shows another orbit (in the xy plane) and its corresponding Poincaré section for the same energy value used in Fig. 8.9. This orbit approaches and is then repelled by three hyperbolic points located near the regions of apparent intersection. Near those hyperbolic points, the trajectory points are smeared and indicate (tentatively) that the behavior is chaotic. However, the chaotic behavior is confined to very small regions of the Poincaré plane. Thus, we see that chaotic orbits and quasi-periodic orbits coexist for the same energy value for Hamiltonian systems. Some initial conditions lead to chaotic orbits, while some lead to quasi-periodic orbits.

Exercise 8.6-3.   Explain in detail the connection between the xy trajectories and the Poincaré section shown in Fig. 8.10.

Figure 8.11 shows a more complete Poincaré section with several initial conditions used to generate a variety of trajectories, all with $E = 0.06$. Note that

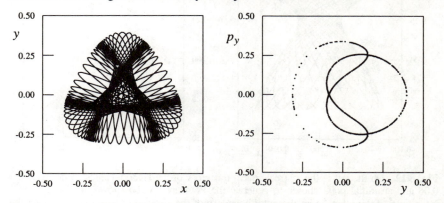

Fig. 8.10. On the left is another orbit of the Hénon–Heiles potential for $E = 0.06$, but with initial conditions different from those in Fig. 8.9. On the right is the corresponding $y\,p_y$ Poincaré section.

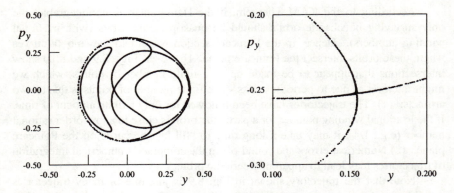

Fig. 8.11. On the left is an $x = 0$, yp_y Poincaré section for the Hénon–Heiles system with $E = 0.06$. On the right is a magnified view of one of the regions near a hyperbolic point. The (slight) smear of intersection points is a symptom of a chaotic orbit.

there is an outer boundary for the allowed intersection points in the yp_y plane (with $x = 0$). Points outside this boundary correspond to trajectories associated with energy values different from $E = 0.06$. The right-hand side of Fig. 8.11 shows a magnified view of the region near the lower hyperbolic point. The chaotic behavior of the intersection points is more obvious. The chaotic regions associated with these hyperbolic points are sometimes called *stochastic layers* or *stochastic webs*. These layers are due to homoclinic and heteroclinic tangles that develop from the stable and unstable manifolds associated with the hyperbolic (saddle) points.

The orbits that come close to the hyperbolic points are close to the separatrices associated with those points. For most Hamiltonian systems, those

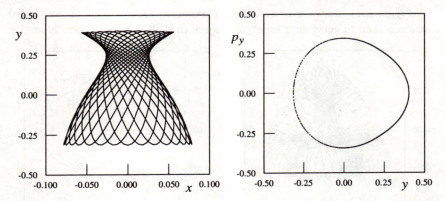

Fig. 8.12. On the left is the xy (real space) trajectory of the Hénon–Heiles model with $E = 0.06$ and an initial point chosen to generate the Poincaré section shown on the right with intersection points that lie on the outer boundary of the allowed energy region. Since this Poincaré section curve is outside the separatrices associated with the hyperbolic points (see Fig. 8.10), the xy trajectory is qualitatively different from the trajectory (shown in Fig. 8.9) for Poincaré curves inside the separatrices.

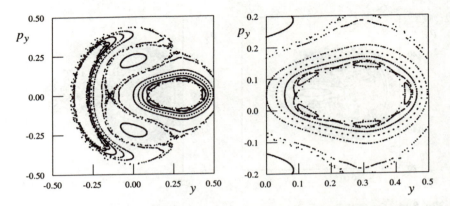

Fig. 8.13. On the left is the Poincaré section for the Hénon–Heiles model with $E = 0.10$. On the right is a magnified view of one of the archipelago island chains of elliptic and hyperbolic points that form from a KAM torus. Surrounding this chain are other surviving KAM tori.

separatrices segregate regions of qualitatively different behavior. In the Hénon–Heiles system, trajectories associated with Poincaré section "curves" that lie outside the separatrices correspond to motion that lies close to the y axis for the real space trajectories. Figure 8.12 shows the trajectory associated with the Poincaré section curve that bounds the allowed region. The vase-shaped real space trajectory is qualitatively different from the trajectories associated with the inner ellipses in the Poincaré section (shown in Figs. 8.9 and 8.10). Since the orbits close to the separatrices are on the border between the two types of behavior, they are quite sensitive to perturbations, and they are the first to show signs of chaotic behavior when the system becomes nonintegrable.

Let us now increase the energy of the particle and see how the Poincaré section changes. For larger values of the energy, we expect the particle to roam over a wider range of xy values and hence the cubic potential term that causes the nonintegrability should become more important.

In Fig. 8.13, we have plotted the yp_y Poincaré section (again with $x = 0$) for $E = 0.10$. The Poincaré section has the same general structure seen in Fig. 8.11: There are two clusters of ellipses around the two elliptic points and an intertwining trajectory that gets near the three hyperbolic points; however, here the orbit associated with the hyperbolic points is more obviously chaotic. In fact, the entire chaotic set of points was generated from a <u>single</u> trajectory launched near one of the hyperbolic points.

A new feature, however, appears as well. On the left in Fig. 8.13, an elliptical band around each of the elliptic points seems to be smeared. On the right of Fig. 8.13, a magnified view of one of these bands shows that the band is actually a cluster (an "archipelago") of five elliptical curves interlaced with an orbit that gets near to five hyperbolic points. You should note that the five elliptical curves were generated by a <u>single</u> trajectory; therefore, these curves should be thought of as

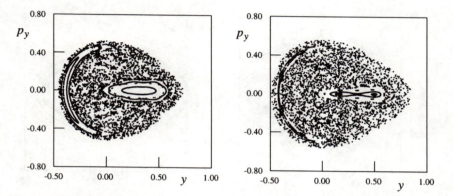

Fig. 8.14. On the left is the Poincaré section (for $x = 0$) yp_y plane for the Hénon–Heiles model with $E = 0.14$. On the right we have $E = 0.16$. In both cases the scattered points were all produced by launching a single trajectory that wanders chaotically through the allowed region of phase space.

cross sections of a "snake" tube that wraps around the main "inner" elliptical tube five times. Similarly, the "necklace" associated with the hyperbolic points is the trace of a single trajectory.

This archipelago and necklace formation is just the structure expected from the Poincaré–Birkoff Theorem. As the energy of the system has increased, the nonintegrable part of the Hamiltonian becomes more important, and the KAM tori corresponding to irrational winding numbers begin to dissolve. Each one dissolves by breaking up into a series of elliptical islands interlaced with a (chaotic) trajectory associated with the hyperbolic points that are "born" when the islands form.

The chaotic trajectory associated with one archipelago, however, is not connected to the chaotic trajectories associated with other clusters of hyperbolic points. In a sense, the remaining KAM tori act as barriers and keep the chaotic trajectories, which would like to roam throughout phase space, confined to certain regions. (Again, we should remind ourselves that this is a feature unique to systems with two degrees of freedom.)

However, if the energy is increased further, the KAM tori continue to dissolve and a single chaotic trajectory eventually wanders throughout almost the entire allowed region of the Poincaré section (consistent with the conservation of energy). Figure 8.14 shows Poincaré sections with $E = 0.14$ (on the left) and $E = 0.16$ (on the right). The scattered dots were all produced from one trajectory that now wanders considerably through the phase space. Some vestiges of KAM tori can still be seen, but they occupy a considerably smaller region of phase space. Calculation of the Lyapunov exponents for the chaotic trajectory shows that one exponent is positive, two are 0, and one is negative, as expected.

Let us summarize what we have seen with the Hénon–Heiles model. For low values of the energy, most of the trajectories are associated with quasi-periodic trajectories (KAM tori). Chaotic behavior is present, but it is barely noticeable

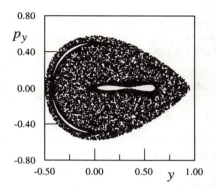

Fig. 8.15. A Poincaré section ($x = 0$) yp_y plane for the Hénon–Heiles model with $E = 0.16666$. All of the KAM tori have dissolved and a single chaotic trajectory wanders throughout almost all the allowed region of phase space.

because it is confined to very small regions of phase space. As the energy increases, the KAM tori begin to dissolve via archipelago formation. The chaotic regions begin to expand. However, for a two-degree-of-freedom system, the remaining KAM tori prevent a given chaotic trajectory from wandering over the entire allowed region of phase space. After the last KAM torus (associated with the Golden Mean winding number) has disappeared, a single chaotic trajectory covers almost the entire allowed region of phase space as shown in Fig. 8.15.

Exercise 8.6-4. Using the methods illustrated in this section, explore the dynamics of trajectories of a particle subject to the following potential

$$V(x, y) = \tfrac{1}{2}(y - 2x^2)^2 + \tfrac{1}{2}x^2$$

Explore similarities and differences with the Hénon–Heiles model.

8.7 The Chirikov Standard Map

Many of the theoretical results for universal behavior of nonintegrable Hamiltonian systems have come from the study of area-preserving iterated map functions. These map functions have been developed to model the behavior of Poincaré section points for Hamiltonian systems. In this section we will discuss one such map, first introduced by B. V. Chirikov (CHI79). That map has become so widely used that it is called the *Standard Map*. For our purposes, we can view the Standard Map as a two-dimensional generalization of the circle maps studied in Chapter 6. We might expect such a map to be relevant for Hamiltonian systems because we expect the trajectories associated with periodic and quasi-periodic motion to reside on the surfaces of tori whose intersection with a Poincaré plane give rise to elliptical "curves." The phase-space-volume conservation property of Hamiltonian systems becomes an area-preserving property of the corresponding two-dimensional maps. In a later section, we will add a dissipative term to the Standard Map to examine the connection between Hamiltonian (conservative) systems and dissipative systems.

The Chirikov Standard Map function is usually written as a function of two variables, r and θ, which can be interpreted as the polar coordinates of a trajectory intersection point on a two-dimensional Poincaré plane:

$$r_{n+1} = r_n - \frac{K}{2\pi} \sin 2\pi\theta_n \ \text{mod} \ [1]$$

$$\theta_{n+1} = \theta_n + r_{n+1} \ \text{mod} \ [1]$$

(8.7-1)

For the Standard Map, the angle θ and the variable r are defined to be in the range [0,1], just as for circle maps. K is a positive nonlinearity parameter.

In the literature on nonlinear dynamics, the Standard Map is often defined with a plus sign in place of the minus sign in Eq. (8.7-1). This is equivalent to shifting the value of θ by 0.5 (π in radian units) and has no fundamental significance. We will use the minus sign to make more obvious the connection to the sine-circle map.

If we fix r and require that the angle variable satisfy

$$\theta_{n+1} = \theta_n + g(r)$$

(8.7-2)

then the Standard Map reduces to what is called the ***Moser Twist Map*** [Moser, 1973]. If $g(r)$ is a rational number, then the trajectories of the twist map are periodic. If $g(r)$ is irrational, then the trajectories are circles (tori) in the xy plane with radius r. If we set $g(r) = r = \Omega$, then the twist map reduces to the $K = 0$ sine-circle map of Chapter 6.

It is also common to replace the variable r_n with the variable J_n to use the action-angle notation of Section 8.4. Then a plot of trajectory points of the Standard Map corresponds to the Poincaré section of a Hamiltonian system plotted in action-angle variables.

> **Exercise 8.7-1.** Verify that the Standard Map corresponds to a nondissipative (area-preserving) map function. Hint: Refer to Section 4.6.

First let us find the fixed points of the Standard Map. Substituting $J = r$ in Eq. (8.7-1), we see that these occur for $\theta = 0$ and 0.5 and $J = 0$ or 1. (Recall that $J = 0$ and $J = 1$ are equivalent under the mod [1] operation.) Using the results of Section 4.6, it is easy to see that the fixed point at the origin is a stable fixed point for $K < 4$, but the one at $\theta = 0.5$ is unstable for any $K > 0$.

> **Exercise 8.7-2.** Consider the Standard Map with the nonlinearity parameter $K = 0$. (a) Show that the iterates of this $K = 0$ Standard Map are just horizontal lines in the θ–J plane. Explain the connection between this diagram and the behavior of an integrable Hamiltonian system in action-angle space. (b) Show that the trajectories correspond to quasi-periodic behavior if J is an irrational number (between 0 and 1) and periodic behavior if J is a rational number (between 0 and 1).

> **Exercise 8.7-2** (continued) (c) Show that with $K = 0$, any point in the θ–J plane with $J = 1/2$ is a period-2 fixed point. (d) Find the period-3 fixed points when $K = 0$. *N.B.* Understanding the $K = 0$ behavior of the Standard Map will guide our thinking about the behavior when K is not equal to 0.

> **Exercise 8.7-3.** Show that the fixed points of the Standard Map have the stability properties stated in the previous paragraph.

> **Exercise 8.7-4.** Show that for a two-dimensional area-preserving map, the stability condition for a fixed point can be expressed as $|TrJ| < 2$, where TrJ is the trace of the Jacobian (or Floquet) matrix of derivatives.

To illustrate the behavior of the trajectories of the Standard Map, we have plotted in Fig. 8.16 the trajectory points generated with $K = 0.2$ by taking 50 initial conditions with $\theta = 0$ and J ranging between 0 and 1 in 50 equal steps. For each initial point, the Standard Map has been iterated 500 times and each of the resulting trajectory points (taken to fall in the interval [0,1]) has been plotted in the θ–J plane.

Let us look at the details of Fig. 8.16. Surrounding the stable fixed point at $J = 0$ and $\theta = 0$ are a series of elliptic orbits. (Recall that the variables are taken mod [1]; therefore, the portions of the ellipses seen near $J = 0$ and 1 and $\theta = 0$ and 1 are all part of the same orbits.) Near $\theta = 0.5$ and $J = 0$ and 1 are hyperbolic (saddle) points. Threading horizontally across the plot are some dotted curves that correspond to quasi-periodic orbits (the apparently continuous lines) and some that appear to correspond to periodic orbits (the ones made up of a few points). Recall that in action-angle space, the surfaces of phase space tori show up as more or less horizontal curves. Approaching and leaving the unstable fixed point at $J = 0$ (or 1) and $\theta = 0.5$ are parts of a hyperbolic orbit.

The full complexity of the behavior of the trajectories is not so obvious in Fig. 8.16. Looked at in finer detail, the map shows small islands interlaced with trajectories associated with hyperbolic points. (The lesson here is that you can be misled easily by the finite resolution of your computations.) Figure 8.17 shows a magnified view of the middle region of Fig. 8.16. Near the center of the plot is a single point that corresponds to a fixed point of the second-iterate of the map function. (The other second-iterate fixed point occurs at $\theta = 0$ near $J = 0.5$.) Around the second-iterate fixed points are elliptical orbits which are part of period-2 trajectories in the sense that trajectory points alternate between an ellipse near $J = 0.5$ and $\theta = 0.5$ and the ellipse near $\theta = 0$ (or 1).

> **Exercise 8.7-5.** Use your knowledge of the $K = 0$ Standard Map to locate approximately the period-3 points and ellipses in Fig. 8.16.

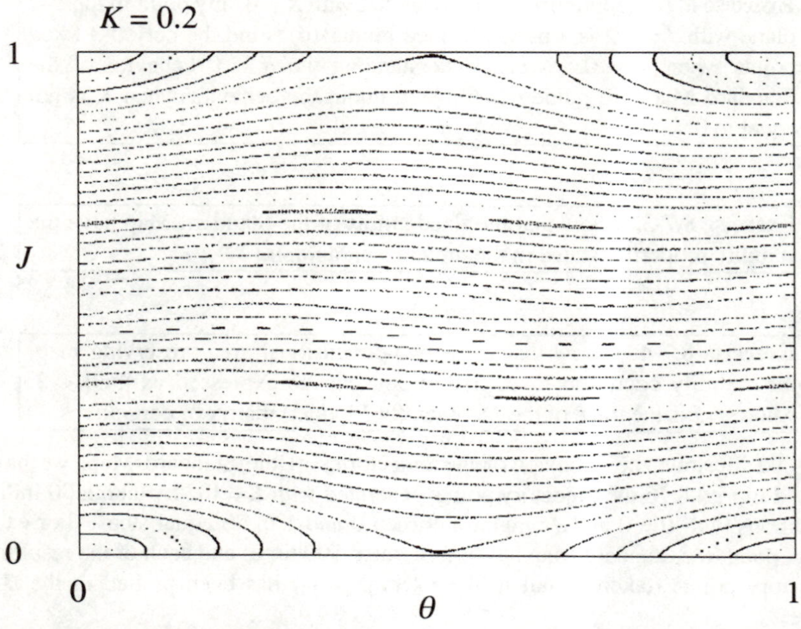

$K = 0.2$

Fig. 8.16. Trajectories for the Standard Map with $K = 0.2$. Fifty trajectories, each consisting of 500 points, were plotted starting with $\theta = 0$ and J, ranging in 50 equal steps between 0 and 1.

If we now increase the parameter K, the chaotic behavior becomes more obvious. Figure 8.18 shows iterates of the Standard Map with $K = 0.26$ on the left and $K = 0.28$ on the right. Near the center of the plot on the right is a series of "islands" surrounding period-two points. (There are still stable period-two points at $J = 0.5$ and $\theta = 0$ and 0.5.) Between the period-2 islands are period-2 hyperbolic points and a thin stochastic band associated with the orbits that come close to those hyperbolic points. By comparing these two figures, we see that the island structure seems to form when the bands corresponding to the separatrices for different periodicities touch.

What is happening is that KAM tori are disappearing as the nonlinearity becomes stronger. Hence, the chaotic regions associated with hyperbolic points can expand and be seen more easily on a particular scale of presentation. The Chirikov resonance overlap criterion (CHI79) can be used to estimate the value of K for which the resonance structures will overlap to destroy the KAM tori that lay between them.

In summary, we note that the Standard Map exhibits many of the same features as the Hénon–Heiles system as the amount of nonlinearity increases. Initially, for small values of K we see mostly simple elliptic orbits around stable fixed points and approximately horizontal tori in action-angle space. As the amount of nonlinearity increases, island structures begin to form around various

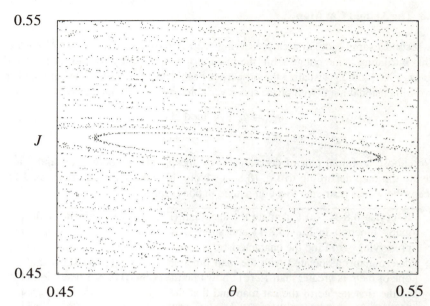

Fig. 8.17. A magnified view of the Standard Map iterates with $K = 0.2$. A period-2 fixed point occurs near the center at about $J = 0.5$ and $\theta = 0.5$. Period-2 elliptical orbits surround the fixed point. The scattered appearance of the diagram is due to the finite number of points used to plot the diagram. Some chaotic behavior is also present, however, but on a small scale.

periodic points and some of the tori are destroyed. As the nonlinearity increases further, resonance structures begin to overlap and "kill off" the KAM tori that lay between them. All the tori are eventually destroyed and a single trajectory will wander over nearly the entire allowed region of phase space.

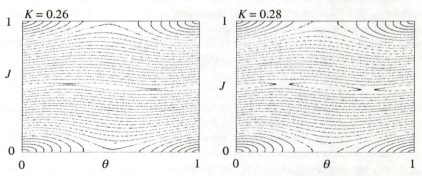

Fig. 8.18. On the left are shown trajectory points for the standard map for $K = 0.26$. On the right $K = 0.28$. We see that the island structures form when separatrix trajectories touch.

8.8 The Arnold Cat Map

A second example of an area-preserving map is the so-called *Arnold Cat Map*. This intriguing name comes from the picture of a cat used by Arnold [Arnold and Avez, 1968] to help visualize the properties of this map function. The cat map is a two-dimensional map (with phase space variables p and q) defined as

$$q_{n+1} = q_n + p_n \ \text{mod} \ [1]$$
$$p_{n+1} = q_n + 2p_n \ \text{mod} \ [1]$$

(8.7-1)

We choose an initial pair of coordinates, q_0 and p_0, that lie between 0 and 1 and iterate the map according to Eq. (8.8-1). Note that taking the variables mod [1] keeps the values restricted to a unit square in phase space.

Exercise 8.8-1. (a) Calculate the Jacobian (Floquet) determinant of the map function in Eq. (8.8-1) and show that the map function is area-preserving. (b) Find the characteristic values (eigenvalues of the Jacobian determinant) associated with the map function Eq. (8.8-1) and show that one of them is greater than 1. (c) Show that the point (0,0) is a fixed point of the first iterate of the cat map and that the points (2/5, 1/5) and (3/5, 4/5) are fixed points of the second iterate of the Cat Map. (d) Show that the fixed points are unstable fixed points.

As Exercise 8.8-1 hints, the fixed points of the cat map are associated with initial points whose coordinates are rational fractions. Initial points whose coordinates are irrational numbers (between 0 and 1) lead to chaotic trajectories. Thus, we see that "most" initial conditions (in the sense that there are mainly irrational numbers, rather than rational numbers in the interval between 0 and 1) lead to chaotic orbits. Thus, the cat map, like the tent map of Chapter 5, provides a model system for which most of the trajectories are chaotic. With the cat map, however, the phase space area of a cluster of initial conditions is conserved.

The cat map can be used to illustrate relatively easily the evolution of probability distributions in chaotic Hamiltonian systems (FOR88). Suppose we specify some initial probability distribution $\rho(q_0, p_0, t = 0)$ for the cat map. Since the cat map is area-preserving, this distribution is simply "dragged along" with the trajectory points. More formally, we write

$$\rho(q_n, p_n, t = n) = \rho(q_0, p_0, t = 0)$$

(8.8-2)

For an iterated map system, we think of time as moving along in steps, so we have used $t = n$ on the left-hand side of Eq. (8.8-2).

To describe the detailed spatial behavior of the evolution of the probability distribution, we will make use of a Fourier spatial analysis (see Appendix A). We can do this because the mod [1] function means we can treat the distribution ρ as a periodic function of p and q with a spatial period equal to unity. The Fourier spatial expansion for $t = n$ and $t = n+1$ takes the form

$$\rho(q_n, p_n, n) = \sum_{j,k} A_{j,k}^{(n)} e^{2\pi i(jq_n + kp_n)}$$

$$\rho(q_{n+1}, p_{n+1}, n+1) = \sum_{j,k} A_{j,k}^{(n+1)} e^{2\pi i(jq_{n+1} + kp_{n+1})}$$

(8.8-3)

The Fourier coefficients $A_{j,k}^{(n)}$ generally depend on the time value n. The positive integer subscripts j and k label the various Fourier modes. They are sometimes called the "mode numbers."

Since the phase space distribution must also satisfy Eq. (8.8-2), we can equate the two sums in Eq. (8.8-3). If we then use the map function given in Eq. (8.8-1) to relate the qs and ps, we find that the equality of the phase space distributions requires that the coefficients satisfy the following relationship

$$A_{j+k,j+2k}^{(n+1)} = A_{j,k}^{(n)}$$

(8.8-4)

This last equation actually tells us a great deal about the evolution of the phase space distribution. To see what is going on, let us consider a special (and not very realistic) case: Suppose that at $t = 0$, only one Fourier amplitude with nonzero j and k is important. Then, only one Fourier amplitude is present upon each iteration of the map function, but according to Eq. (8.8-4) the mode numbers (the subscripts) increase rapidly. Since a large mode number means that the quantity is oscillating rapidly with position, we see that the probability distribution for the cat map quickly becomes a rapidly varying function of position. If we start with a more realistic distribution, which would be described by a range of mode number values, upon iteration of the cat map, the low mode numbers quickly become unimportant and the distribution takes on a very complicated and rapidly varying spatial appearance.

> **Exercise 8.8-2.** Work through the algebra leading from Eqs. (8.8-3) to Eq. (8.8-4).

The crucial point here is that the chaotic nature of the trajectories for the cat map shows up as an evolution of a relatively smooth probability distribution into a highly convoluted probability distribution. However, if we start with just a slightly different probability distribution, manifested by slightly different Fourier coefficients, those Fourier coefficients evolve in time according to Eq. (8.8-3) to values that are not much different from the values for the initial distribution. In short, the probability distributions do not show sensitive dependence on initial conditions. This point will be important in our discussion of chaos in quantum mechanics in Chapter 12.

8.9 The Dissipative Standard Map

Now that we have explored some of the behavior of Hamiltonian systems with chaos, we ought to look at the connection between their behavior and the behavior of the dissipative systems we have studied previously. A useful vehicle for exploring this connection is another iterated map function called the ***Dissipative Standard Map*** (SCW85)(SCH88). This map function is almost like the Standard Map of Section 8.7, but it has built in an adjustable dissipation factor. By changing the dissipation factor we can interpolate smoothly between a dissipative system and a Hamiltonian (conservative) system. The Dissipative Standard Map is given by the following set of equations

$$r_{n+1} = J_D\, r_n - \frac{K}{2\pi} \sin 2\pi\theta_n \ \text{mod} \ [1]$$

$$\theta_{n+1} = \theta_n + r_{n+1} \ \text{mod} \ [1]$$

(8.9-1)

where J_D is the value of the Jacobian determinant for the mapping functions. If $J_D = 1$, then the map functions reduce to the Standard Map. If $J_D = 0$, then we have a one-dimensional iterated map system equivalent to the sine-circle map (discussed in Chapter 6) with $\Omega = 0$.

With $J_D = 0$, we vary the parameter K and observe the usual period-doubling sequence leading to chaos, chaotic bands that remerge as a function of K with various periodic windows as discussed in Chapter 5 for the logistic map function. Schmidt and Wang (SCW85) have studied what happens to these features as J_D increases toward 1. What they found is illustrated in Fig. 8.19. If we proceed up the extreme right-hand side of the figure, where $J_D = 0$, we see the usual period-doubling cascade with period-1 followed by period-2 culminating with an infinite period at $K = K_\infty$. Beyond $K = K_\infty$, there are periodic bands and periodic windows (labeled $p2'$ and $p3'$ in the figure).

As J_D increases toward 1, the amount of dissipation decreases, and it takes many more iterations for the map variables to settle onto an attractor. The basic features of the system remain, but their locations in the $J_D K$ plane change. At $J_D = 1$, many of the "channels" overlap, telling us that different initial conditions lead to different kinds of orbits for an area-preserving map (or Hamiltonian system). We see the following correlations: The bifurcated orbits (those that arise as a breakup of lower period orbits) with period 2^n for the Hamiltonian system, correlate with the 2^n stable orbits of the period-doubling sequence for the sine-circle map. The other periodic orbits of the Hamiltonian system correlate with the periodic windows, which occur for $K > K_\infty$ for the sine-circle map. Between the two extreme cases, for $0 < J_D < 1$, there are several coexisting attractors for a given J_D–K pair, each with its own basin of attraction.

The 2^n chaotic bands, which exist for the $J_D = 0$ case, gradually disappear as $J_D \to 1$. There is some evidence (SCH88) that J_D values at which particular chaotic bands disappear are universal numbers for a wide variety of systems.

The lesson we learn from our quick look at the Dissipative Standard Map is that although there are many differences between the behavior of Hamiltonian and dissipative systems, we can see how the different behaviors are correlated, at least in some cases. The multiplicity of behaviors that occur for a Hamiltonian system

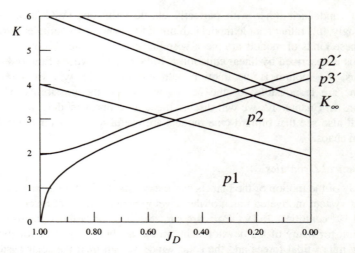

Fig. 8.19. A plot indicating the behavior of the dissipative standard map as a function of K, the nonlinearity parameter, and J_D, the dissipation parameter. For $J_D = 1$, the map function is area-preserving and equivalent to a Hamiltonian system. For $J_D = 0$, the behavior corresponds to that of a one-dimensional iterated map. $p1$ and $p2$ indicate period-1 and period-2 orbits, respectively. K_∞ is the K value where the period-doubling sequence culminates in chaos. $p3'$ indicates a period-3 window (Redrawn from Fig. 1 of SCH88).

with a single set of parameter values eventually turns into a multiplicity of attractors for different parameter values in the one-dimensional dissipative $J_D = 0$ extreme. Indeed there seems to be some universality in the behavior of the transition from strong dissipation to Hamiltonian conditions.

8.10 Applications of Hamiltonian Dynamics

Billiards and Other Games

We begin with a class of Hamiltonian models that make use of perfectly elastic collisions of an object with either boundary walls or with other objects. These models are important in the theory of Hamiltonian systems. One class of such models describes the motion of a ball (usually modeled as a point object) moving horizontally within some confined two-dimensional region, in effect, an idealized game of billiards. If the confining region is rectangular or circular, then it turns out that all the orbits are periodic or quasi-periodic. However, if the boundary is shaped like a stadium, with straight side walls and semi-circular ends, or if a round obstacle is placed inside a rectangular boundary (Sinai Billiards), then the motion can be chaotic, at least for some trajectories. (In almost all cases, it seems possible to find some periodic orbits, but they may be unstable.)

Another model in this general category is one consisting of two balls constrained to move vertically under the action of gravity and interacting with the "ground" and each other with perfectly elastic collisions (WGC90). Perhaps surprisingly, this rather simple model exhibits a full panoply of complex behavior.

These kinds of models are piece-wise linear because the motion between the collisions is described by linear equations. In fact, for the case of the billiard-type problems, the motion is just motion with constant velocity. In a sense, the collisions are responsible for the nonlinearities and the possibility of chaotic behavior. In Chapter 12, we will discuss piece-wise linear models in more detail. We shall also see that billiard-type models are useful test beds for the notions of quantum chaos.

Astronomical Dynamics

The study of the motion of the planets and the question of the long-term stability of the solar system motivated much of the development of classical mechanics in the 18th and 19th centuries. In fact, Poincaré's interest in solar system dynamics was the inspiration for many of his developments in the qualitative analysis of dynamics. Ignoring minor tidal forces and the solar wind, we can treat the solar system as a conservative, Hamiltonian system. There is some evidence (WIS87, KER88) that the orbits of Pluto and some of the asteroids may be chaotic.

Particle Accelerator Dynamics

In contemporary elementary particle physics, the high energy accelerator has become the predominant tool for the exploration of the fundamental structure of matter at the sub-atomic level. Accelerating particles, however, such as protons and electrons to energies hundreds or even thousands times larger than their rest mass energies in a controllable fashion is no trivial task because the orbits of the particles within these accelerators are subject to many perturbations leading to possibly unstable conditions. To a reasonable approximation, dissipation can be neglected in considering these orbits, and the methods of Hamiltonian dynamics come into play. Nonlinear dynamics provides a vocabulary for understanding the possible instabilities of particle orbits and gives us the tools needed to design accelerators that can avoid unstable orbits.

Bulk Superconductivity

When a superconducting material is exposed to an external magnetic field, the circulation of electrons is organized into vortex lattices. These vortices can move in a variety of ways under the action of external forces, and it has been found that phase-locking, Arnold tongues, Farey trees and the devil's staircase can be used to characterize this behavior (REN99).

Optics

When light travels through a uniform dielectric material, the ray dynamics is the same as the Hamiltonian dynamics of a point mass traveling freely within a three-dimensional enclosure. This analogy can be used to understand the behavior of light inside small dielectric spheres, which have been used to produce miniature "whispering gallery" lasers. When the sphere is slightly deformed, the ray trajectories become chaotic (MNC95).

8.11 Further Reading

General Treatments of Hamiltonian Mechanics and Chaos

H. Goldstein, *Classical Mechanics*, 2nd ed. (Addison–Wesley, Reading, MA, 1980). Undoubtedly the best traditional introduction to classical Newtonian mechanics in its many manifestations at the upper undergraduate and graduate level.

R. H. G. Helleman, "Self-Generated Chaotic Behavior in Nonlinear Mechanics," in *Fundamental Problems in Statistical Mechanics*, Vol 5. (E. G. D. Cohen, ed.) (North-Holland, Amsterdam, 1980), pp. 165-233. Reprinted in [Cvitanovic, 1984]. A superb introduction to the chaotic behavior of nonintegrable Hamiltonian systems. Highly recommended.

A. Chernikov, R. Sagdeev, and G. Zaslavsky, "Chaos: How Regular Can It Be?" *Physics Today* **41** (11), 27–35 (November 1988). A nice overview of the question of mixing and chaotic behavior in Hamiltonian systems.

Treatments of chaos in Hamiltonian systems at the level of this book are found in

[Schuster, 1995].

A. J. Lichtenberg and M. A. Liebermann, *Regular and Chaotic Dynamics*, 2nd ed. (Springer, New York, Heidelberg, Berlin, 1992).

R. S. Mackay and J. D. Meiss, *Hamiltonian Dynamical Systems* (Adam Hilger, Bristol, 1987).

G. M. Zaslavsky, R. Z. Sagdeev, D. A. Usikov, and A. A. Chernikov, *Weak Chaos and Quasi-Regular Patterns* (Cambridge University Press, Cambridge, 1991). In addition to providing an excellent introduction to stochastic layers and diffusion in phase space of Hamiltonian systems, this book attempts to link the behavior of Hamiltonian systems to pattern formation.

K. R. Meyer and G. R. Hall, *Introduction to Hamiltonian Dynamical Systems and the N-body Problem* (Springer, New York, 1992).

G. M. Zaslavsky, *Physics of Chaos in Hamiltonian Systems* (Imperial College Press, London, 1998).

Fluid flow techniques provide means for seeing directly state space tori, island chains, and so on. See, for example, G. O. Fountain, D. V. Khakhar, and J. M. Ottino, "Visualization of Three-Dimensional Chaos," *Science* **281**, 683–86 (1998).

A brief but insightful introduction to Hamiltonian chaos for someone with an undergraduate background in classical mechanics is provided by

N. Srivastava, C. Kaufman, and G. Müller, "Hamiltonian Chaos," *Computers in Physics* **4**, 549–53 (1990), "Hamiltonian Chaos II," *Computers in Physics* **5**, 239–43 (1991), "Hamiltonian Chaos III," *Computers in Physics* **6**, 84 (1991). N. Regez, W. Breymann, S. Weigert C. Kaufman, and G. Müller, "Hamiltonian Chaos IV," *Computers in Physics* **10**, 39–45 (1996). The latter two papers focus on connections with quantum chaos (discussed in Chapter 12).

KAM Theorem and Related Topics

V. I. Arnold, *Mathematical Methods in Classical Mechanics* (Springer, New York, 1978).

[Guckenheimer and Holmes, 1990].

[Jackson, Vol. 2, 1991] Appendix L.

Period-Doubling in Hamiltonian Systems

G. Benettin, C. Cercignanni, L. Galgani, and A. Giorgilli, "Universal Properties in Conservative Dynamical Systems," *Lett. Nouvo. Cim.* **28**, 1–4 (1980).

J. M. Greene, R. S. MacKay, F. Vivaldi, and M. J. Feigenbaum, "Universal Behavior in Families of Area-Preserving Maps," *Physica D* **3**, 468–86 (1981).

T. C. Bountis, "Period Doubling Bifurcations and Universality in Conservative Systems," *Physica D* **3**, 577–89 (1981).

D. A. Sadovskif, J. A. Shaw, and J. B. Delos, "Organization of Sequences of Bifurcations of Periodic Orbits," *Phys. Rev. Lett.* **75**, 2120–23 (1995).

Hénon–Heiles Hamiltonian

M. Hénon and C. Heiles, "The Applicability of the Third Integral of Motion: Some Numerical Experiments," *Astrophys. J.* **69**, 73–79 (1964).

[Jackson, Vol. 2, 1991], p. 74.

Chirikov Standard Map

B. V. Chirikov, "A Universal Instability of Many Dimensional Oscillator Systems," *Physics Reports* **52**, 263–379 (1979). The Chirikov Standard Map is discussed in this lengthy review of Hamiltonian dynamics as manifest in nonlinear oscillators.

Moser Twist Map

J. Moser, *Stable and Random Motions in Dynamical Systems* (Princeton University Press, Princeton, NJ, 1973).

Arnold Cat Map

V. I. Arnold and A. Avez, *Ergodic Problems of Classical Mechanics* (Benjamin, New York, 1968).

J. Ford, "Quantum Chaos, Is There Any?" in [Hao, 1988], pp.128–47.

Dissipative Standard Map

G. Schmidt and B. H. Wang, "Dissipative Standard Map," *Phys. Rev. A* **32**, 2994 –99 (1985).

G. Schmidt, "Universality of Dissipative Systems," in [Hao, 1988], pp. 1– 15.

Applications of Hamiltonian Dynamics

R. Helleman, *op. cit.*, has many references to applications of Hamiltonian dynamics in astronomy and accelerator physics.

Billiard and Bouncing Ball Models

G. Benettin and J.-M. Strelcyn, "Numerical Experiments on the Free Motion of a Point Mass Moving in a Plane Convex Region: Stochastic Transition and Entropy," *Phys. Rev. A* **17**, 773–85 (1978).

N. D. Whelan, D. A. Goodings, and J. K. Cannizzo, "Two Balls in One Dimension with Gravity," *Phys. Rev. A* **42**, 742–54 (1990).

Astronomical Dynamics

M. Hénon and C. Heiles, "The Applicability of the Third Integral of Motion: Some Numerical Experiments," *Astron. J.* **69**, 73–79 (1964).

J. Wisdom, "Chaotic Behavior in the Solar System," in *Dynamical Chaos*, (M. Berry, I. Percival, and N. Weiss, eds.) (Princeton University Press, Princeton, NJ, 1987). First published in *Proc. Roy. Soc. Lond.* **A413**, 1–199 (1987).

R. A. Kerr, "Pluto's Orbital Motion Looks Chaotic," *Science* **240**, 986–87 (1988).

Accelerator Dynamics

Helleman, *op. cit.*

A. Chao et al., "Experimental Investigation of Nonlinear Dynamics in the Fermilab Tevatron," *Phys. Rev. Lett.* **61**, 2752–55 (1988).

Superconductivity

C. Reichardt and F. Nori, "Phase Locking, Devil's Staircases, Farey Trees, and Arnold Tongues in Driven Vortex Lattices with Periodic Pinning," *Phys. Rev. Lett.* **82**, 414–17 (1999).

Ray Optics

A. Mekis, J. U. Nöckel, G. Chen, A. D. Stone, and R. K. Chang, "Ray Chaos and Q Spoiling in Lasing Droplets," *Phys. Rev. Lett.* **75**, 2682–85 (1995).

8.12 Computer Exercises

CE8-1. Use *Chaos Demonstrations* Model 10, Chirikov Map, to explore the behavior of the Chirikov Standard Map for various values of K, the nonlinearity parameter. Observe island chains, the break up of tori, and the other features discussed in this chapter.

CE8-2. Use *Chaos Demonstrations* Model 5, The Three-Body Problem, to explore the restricted three-body problem. You can use several "planets" starting from nearby locations to see the exponential divergence of nearby trajectories.

CE8-3. Use *Chaotic Dynamics Workbench* to study the Hénon–Heiles system. Look for island chains, breakup of tori, and other features of nonintegrable Hamiltonian systems. Try several different Poincaré sections to try to visualize the full phase space behavior.

CE8-4. Try out some of the programming exercises given in the articles by Srivastava, Kaufman, and Müller cited in Section 8.11.

III

MEASURES OF CHAOS

9

Quantifying Chaos

Let chaos storm, Let cloud shapes swarm! I wait for form. Robert Frost,
Pertinax

9.1 Introduction

How chaotic is a system's chaotic behavior? In this chapter we shall discuss
several ways to give a quantitative answer to that question. Before we get
immersed in the details of these answers, we should ask *why* we might want to
quantify chaos. One answer lies in a desire to be able to specify quantitatively
whether or not a system's apparently erratic behavior is indeed chaotic. As we
have seen chaotic behavior generates a kind of randomness and loss of information
about initial conditions, which might explain complex behavior (or at least some of
the complex behavior) in real systems. We would like to have some definitive,
quantitative way of recognizing chaos and sorting out "true" chaos from just noisy
behavior or erratic behavior due to complexity (that is, due to a large number of
degrees of freedom). Second, as we shall see in the next chapter, some of these
quantifiers can give us an estimate of the number of (active) degrees of freedom for
the system. A third reason for quantifying chaotic behavior is that we might
anticipate, based on our experience with the universality of the scenarios
connecting regular behavior to chaotic behavior, that there are analogous universal
features, perhaps both qualitative and quantitative, that describe a system's
behavior and changes of its behavior within its chaotic regime as parameters of the
system are changed. We will see that indeed some such universal features have
been discovered and that they seem to describe accurately the behavior of actual
systems. Finally, (although this is rarely possible today), we would hope to be able
to correlate changes in the quantifiers of chaotic behavior with changes in the
physical behavior of a system. For example, is there some quantifier whose
changes are linked to the onset of fibrillation in heartbeats or the beginnings of
turbulence in a fluid or noisy behavior in a semiconductor circuit?

In addition to calculating values for particular quantifiers for chaotic systems,
we need to be able to estimate uncertainties associated with those quantifiers.
Without those uncertainties, it is impossible to make meaningful comparisons
between experimentally measured and theoretically calculated values or to compare
results from different experiments. We will suggest several ways of estimating
these uncertainties in our discussion.

To summarize, here are some reasons for quantifying chaotic behavior:

1. The quantifiers may help distinguish chaotic behavior from
    noisy behavior.

2. The quantifiers may help us determine how many variables
    are needed to model the dynamics of the system.

3. The quantifiers may help us sort systems into universality
    classes.

4. Changes in the quantifiers may be linked to important
    changes in the dynamical behavior of the system.

9.2 Time-Series of Dynamical Variables

The key theoretical tool used for quantifying chaotic behavior is the notion of a
time-series of data for the system. We met up with this idea in Chapter 1 in the
form of a stroboscopic portrait of the current in the semiconductor diode circuit and
later in the more general form of Poincaré sections in state space. In this chapter
we will focus on using a time-sequence of values of a single system variable, say
$x(t)$, to determine the quantitative measures of the system's (possibly) chaotic
behavior. We will assume that we have recorded a sequence of values $x(t_0)$, $x(t_1)$,
$x(t_2)$, ... with $t_0 < t_1 < t_2$, and so on, as illustrated in Fig. 9.1. This could be a series
of time-sampled values of some variable, where the time values are fairly close
together, or it could be a series of Poincaré section values for some variable at fairly
widely separated time values.

It is not obvious that such a set of sampled values of just one variable should
be sufficient to capture the features we want to describe. In fact, as we shall argue
in the next chapter, if the sampling is carried out at appropriate time intervals
(which we shall need to specify) and if the sequence is used cleverly, then we can
indeed "reconstruct" the essential features of the dynamics in state space. We will
show in Chapter 10 that we can often determine the number of state variables
needed to specify the state of the system from the time record of just one variable.

Of course, we need to say what we mean by essential. Sampled values of one
variable will clearly not (or, in general, cannot) tell us what the other variables are
doing (unless we happen to have a complete theory for the system). If we limit our
goals, however, to recognizing bifurcations in the system's behavior and
determining if the behavior is chaotic and if so, how chaotic, then it turns out that
this single variable sequence is sufficient (with some qualifications, of course).

One further comment on measuring a single variable is in order. In almost all
measurements, our instruments measure the dynamical variables indirectly. For
example, if we are interested in temperature, we may actually measure the voltage
produced by a thermocouple in contact with our system. We generally assume that
our "measurement function" provides a fairly straightforward representation of the

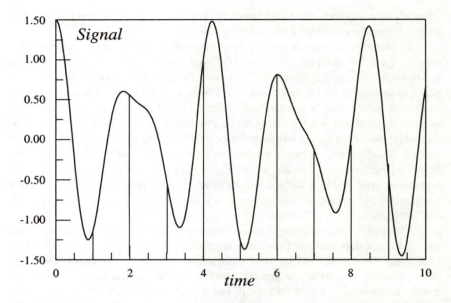

Fig. 9.1. A sketch of the sampling of a dynamical variable. The recorded values form the time series for data analysis. Here the sampling is done at t = 0, 1, 2, . . . and so on.

actual quantity we want to monitor. Strictly speaking, however, we are monitoring the dynamics of our measurement function, not the system directly.

It might be tempting to base our analysis of the system's behavior on continuous time trajectories, given symbolically as $\vec{x}(t)$, where the vector quantity represents a complete set of dynamical variables for the system. (A complete set is the minimum number of variables needed to specify uniquely the state of the system.) In this kind of analysis, the value of $\vec{x}$ is available for any value of the time parameter. However, real experiments always involve discrete time sampling of the variables, and numerical calculations, which we must use for most nonlinear systems, always have discrete time steps. Since both real experiments and actual computer calculations always give the variable values in discrete time steps, we make a virtue of necessity and base our entire discussion on these discrete time sequences.

The problem of choosing the appropriate time between samples (that is, choosing $t_1 - t_0$, $t_2 - t_1$, etc.) is a delicate one. If an infinite amount of noise-free data is available, then almost any set of time intervals will do. However, for more realistic situations, with a finite amount of data contaminated by some noise, we must proceed very cautiously. In the next chapter, we shall develop some "rules of thumb" for selecting time sample intervals and other features of the data. The reader who wants to undertake this kind of analysis for her or his data should consult Chapter 10 and the references at the end of this chapter for more details on sampling time intervals and related matters. For now, we will assume that we have adopted some "reasonable" time-sampling scheme.

In characterizing chaos quantitatively, we will make use of two different, but related, types of description. The first type emphasizes the *dynamics* (time dependence) of chaotic behavior. The now familiar Lyapunov exponent is an example of this type of descriptor. We will also introduce various kinds of "entropy," which play a similar role. These quantifiers tell us how the system evolves in time and what happens to nearby trajectories as time goes on. The second type of quantifier emphasizes the *geometric* nature of the trajectories in state space. We allow the system to evolve for a (reasonably) long time, and then we examine the geometry of the resulting trajectories in state space. Are the trajectories confined to a surface in state space? If so, do they cover that surface completely, and so on? Within this geometric approach we will meet the important and intriguing concept of fractals.

These two types of description are complementary. In the first, we emphasize the actual time dependence of diverging trajectories, for example. In the second we look at the "footprints" left by these trajectories. Intuitively, we expect these two descriptions to be related. At present, however, the theoretical links are weak and mostly conjectural. While numerical and experimental evidence in most cases seems to support these conjectures, there remains much work to be done on the theory linking the dynamical and geometric descriptions.

In the next few sections, we shall introduce four quantifiers of chaos: (1) Lyapunov exponents, (2) Kolmogorov entropy, (3) fractal dimension, and (4) correlation dimension. In the following chapter, we shall discuss ways these notions can be generalized to give more detailed information about the system's behavior. Throughout this chapter, we shall assume, in the spirit of Chapter 5, that we are dealing with an effectively "one-dimensional" system. In the next chapter, we shall show that it is easy, in principle, to relax that restriction to treat multidimensional systems and that, in fact, we can get multidimensional information from the time series of a single variable by a clever "embedding" or "reconstruction" scheme.

We will be limiting our attention to dissipative systems (i.e., to systems for which the effects of transients associated with initial conditions die away and the long-term behavior is restricted to some attracting region or regions in state space). As we proceed with our quantification of the behavior, we shall deal only with trajectories that are assumed to be on attractors. For nondissipative systems (the Hamiltonian systems of Chapter 8), there are no attractors, and we must live with the complication that some trajectories may be periodic and some may be chaotic for the same set of parameter values. Some of the quantifiers discussed in this chapter can be applied to a description of trajectories of Hamiltonian systems, but nearby trajectories in state space may have a completely different character. Of course, dissipative systems with more than one attractor (for a given set of parameter values) have the same complication.

In principle, transients (trajectories starting away from an attractor but evolving toward it), can give useful information about the dynamics of the system. Since transients are often difficult to handle both experimentally and computationally, we shall confine the discussion to long-term behavior, which for a dissipative system, means looking only at trajectories confined to an attractor in state space. A method for applying time-series analysis to chaotic transients is given in JAT94.

9.3 Lyapunov Exponents

In Chapter 4, we introduced the (average) Lyapunov exponent as a measure of the divergence of nearby trajectories. We argued that a system's behavior is chaotic if its average Lyapunov exponent is a positive number. In Chapter 5, we showed how to calculate the Lyapunov exponent for a one-dimensional iterated map function. In this section, we will describe the calculation of the Lyapunov exponent from a one-dimensional time-series of data. We shall label the series $x(t_0), x(t_1), x(t_2), \ldots$ as $x_0, x_1, x_2, \ldots$. For the sake of simplicity, we will assume, as is usually the case, that the time intervals between samples are all equal; therefore, we can write

$$t_n - t_0 = n\tau \qquad (9.3\text{-}1)$$

where τ is the time interval between samples.

If the system is behaving chaotically, the divergence of nearby trajectories will manifest itself in the following way: if we select some value from the sequences of xs, say x_i, and then search the sequence for another x value, say x_j, that is close to x_i, then the sequence of <u>differences</u>

$$d_0 = \left| x_j - x_i \right|$$
$$d_1 = \left| x_{j+1} - x_{i+1} \right|$$
$$d_2 = \left| x_{j+2} - x_{i+2} \right| \qquad (9.3\text{-}2)$$
$$\vdots$$
$$d_n = \left| x_{j+n} - x_{i+n} \right|$$

is assumed to increase <u>exponentially</u>, at least on the average, as n increases. More formally, we assume that

$$d_n = d_0 e^{\lambda n} \qquad (9.3\text{-}3)$$

or, after taking logarithms

$$\lambda = \frac{1}{n} \ln \frac{d_n}{d_0} \qquad (9.3\text{-}4)$$

In practice, we take Eq. (9.3-4) as the definition of the Lyapunov exponent λ. If λ is positive, the behavior is chaotic. In this method of finding λ, we are essentially locating two nearby trajectory points in state space and then following the differences between the two trajectories that follow each of these "initial" points.

Some Technical Details

In principle, the determination of the Lyapunov exponent from the time-series data is straightforward. Several comments, however, are in order:

   1. We have <u>assumed</u> an exponential rate of separation of the two trajectories. For a given time-series, we need to examine the validity of that assumption carefully. This can be done by plotting the (natural) logarithm of the difference d_m as a function of the index m. If the divergence is exponential, the points will fall on (or close to) a straight line, the slope of which is the Lyapunov exponent. A least-squares straight-line fit to that data will give a measure of the goodness of that fit. If the data do not fall close to a straight line on a semi-log plot, then the quoted Lyapunov exponent is meaningless.

   2. The value of λ may (and, in general, does) depend on the value of x_i chosen as the initial value. Hence, we really should write $\lambda(x_i)$. To characterize the attractor, we usually want an average value for λ. We find an average value by calculating $\lambda(x_i)$ for a large number N (say, 30 or 40 in practice) of initial values distributed over the attractor. The average Lyapunov exponent for the attractor is then found from

$$\lambda = \frac{1}{N} \sum_{i=1}^{N} \lambda(x_i) \qquad (9.3-5)$$

From this set of $\lambda(x_i)$ we can also calculate a standard deviation, which can be used to provide an estimate of the uncertainty to be associated with the average value. In the next chapter, we shall see that there may be some interesting universal features associated with the fluctuations of the λs about the average value.

   N.B.: We want the value of λ to reflect the fact that some x_i values occur more frequently than others: trajectories visit some parts of state space more frequently than others. The easiest way to implement this requirement is to choose a large number of initial points according to their subscript indices i. Then those ranges of x values that occur more frequently in the time-series will show up more frequently as initial points.

   3. For the bounded systems with which we are concerned, the number of time steps n used in Eq. (9.3-2) for the determination of λ cannot be too large. Since the xs are limited in size for a bounded system, the differences d_i cannot be larger than the difference between the largest and smallest values of x. Hence, the exponential growth in d cannot go on forever, and we must limit n. To some extent, we must look at the sequence of d_is for a given system to see how large n can be (see Fig. 9.2). The size of n also depends on the value of λ and on d_0. If d_0 is made smaller by requiring that x_j be closer in value to x_i, then the exponential divergence will continue for larger values of n.

4. If the sequence of x values corresponds to periodic behavior, the d values ought to be very small or 0 because the trajectory returns to exactly the same set of values. Hence, this trajectory method would give $\lambda = 0$ reflecting the fact that the ds neither increase nor decrease in size. This result tells us that trajectory points on a periodic orbit neither converge or diverge (on the average). For the state space direction transverse to a stable periodic trajectory, the Lyapunov exponent ought to be negative, indicating that nearby trajectories are attracted to the stable orbit. However, the time-series of values from the trajectory itself cannot tell us how nearby trajectories approach the attractor. More general methods, such as those described in Chapter 5 or in the references at the end of this chapter must be used to determine the Lyapunov exponent when it is negative.

5. There are restrictions on the value of j that we should use for a given x_i. If the time-series results from a closely spaced sampling of some smoothly varying quantity, say the current in the diode circuit in Chapter 1, then we should not choose j too close to i because those two values occur close in time. If the two values are close in time, we expect the behavior to remain close, and we would end up with an anomalously small value for $\lambda(x_i)$. We can avoid this problem by insisting that x_j not follow x_i too closely in time in the sequence. Various criteria have been proposed for choosing a minimum time separation, some using concepts such as "correlation times," which involve technical details we want to avoid here. Generally, a plot of the x_is as a function of time will allow you to determine, at least approximately, what the minimum time delay should be. Of course, if the data are Poincaré section records, which are already widely separated in time, then no such problem arises.

6. A comment on units: Some authors prefer to define λ using

$$d_n = d_0 2^{n\lambda_2}$$

and then interpret λ_2 as a divergence rate in "bits per unit time." By using 2 as the base for the exponential function, we have an exponent that gives the rate of divergence of the sequence of x values written in binary number form (a sequence of bits, 0s or 1s) in analogy with the Bernoulli shift map of Chapter 5. λ_2 is a measure of the rate (in bits per unit time) at which we lose information contained in the initial value of x (expressed in binary form).

Exercise 9.3-1. Show that $\lambda_2 = \lambda / \ln 2$.

---

7. There are practical limits on how small d_0 can be. Since the xs have been either computed with a finite number of decimal places or recorded from an experiment with a finite precision, the number of decimal places produces a lower limit on how small a meaningful difference can be. For example, if the data were recorded with only three decimal places, then it would be meaningless to ask for a difference smaller than 0.001. Of course, if we have taken enough data, eventually,

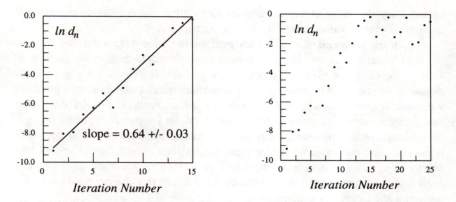

Fig. 9.2. A plot of the logarithm of the trajectory differences for the logistic map as a function of iteration number with $A = 3.99$ and $x_i = 0.1$. The value of d_0 was 0.0001. On the left the trajectory differences are followed for 15 iterations. On the right for 25 iterations. We expect a straight line for exponential divergence of nearby trajectories. The slope of the fitted straight-line on the left gives the Lyapunov exponent. The value for the data on the left is close to the value ln 2 expected for the logistic map with $A = 4$. The data on the right show a saturation and folding when the trajectory differences become close to the overall size of the attractor.

even if the system is chaotic, it will have some repeating values, simply due to the finite digitization accuracy. We could then find two values that are exactly the same. These coincidences will be rare, however, and for a chaotic system, they are an artifact of the finite precision.

Example

To illustrate the time-series method of finding Lyapunov exponents, we have worked out an example. Figure 9.2 shows the trajectory differences as a function of iteration number from the logistic map for two different numbers of iterations each starting from the same x_i. Several features are important in this example: There is a scatter of points about the straight lines in the semilog plots. Thus, we can say that on the average the divergence of nearby trajectories is exponential but there may be considerable fluctuations about that average. Also, for sufficiently long times, the divergence is no longer exponential. This "saturation" occurs whenever the size of the difference between the two trajectories increases to about the size of the attractor. Since the system is bounded, the difference between the trajectories cannot exceed this size, and the semi-log plot of the differences levels off. (The difference may in fact decrease.) Thus, in practice, we must limit the range of time (or equivalently, the range of iteration numbers) for the straight-line fit.

We should also mention an alternative method of computing the Lyapunov exponent from time-series data. If the data are effectively one-dimensional, as explained in Chapter 5, then a plot of x_{n+1} versus x_n should give a sequence of points, which could be connected by a smooth curve. Various "curve-fitting" routines may be used to find an approximate analytic representation of that curve. Given that analytic approximation, we may use the derivative method, Eq. (5.4-12), to find the average Lyapunov exponent. The references at the end of the chapter provide details on other procedures for finding Lyapunov exponents.

Exercise 9.3-2. Use the data in Table 1.2 to plot the natural logarithm of the trajectory differences as a function of iteration number (which is analogous to time). Draw a straight line through those plotted points and from its slope determine the Lyapunov exponent. Compare your value with the value shown in Fig. 9.2. Alternatively, write a computer program to generate the data and to find the Lyapunov exponent.

9.4 Universal Scaling of the Lyapunov Exponent

While a determination of the Lyapunov exponent as described in the previous section can tell us whether or not a system is behaving chaotically, we can go further. If a dissipative system becomes chaotic via the period-doubling route, then we can predict how the Lyapunov exponent will change as the control parameter is varied, driving the system further into the chaotic regime. Figure 9.3 shows the average Lyapunov exponent for the logistic map plotted as a function of the parameter A. For $A < 3.5699 \ldots = A_\infty$, the Lyapunov exponent is negative except at the bifurcation points, where period-doubling occurs. At those points the Lyapunov exponent is equal to 0. For $A > A_\infty$, the Lyapunov exponent is positive but with occasional dips below 0 whenever a periodic window occurs. If we ignore the dips due to the periodic windows, then we see that the Lyapunov exponent grows smoothly as A increases beyond A_∞. We say the system becomes *more chaotic* as A increases, where the "degree of chaoticity" (as measured by the divergence of nearby trajectories) increases with A.

In 1980, Huberman and Rudnick (HUR80) argued that there should be a universal expression for the parameter dependence of the Lyapunov exponent as the system becomes more chaotic following a sequence of period-doubling bifurcations. Their prediction can be written in the following form for $A > A_\infty$, where A_∞ is the parameter value for the accumulation point of the period-doubling sequence:

$$\lambda(A) = \lambda_0 (A - A_\infty)^{\frac{\ln 2}{\ln \delta}}$$
$$= \lambda_0 (A - A_\infty)^{0.445\ldots} \tag{9.4-1}$$

In the previous equation λ_0 is a constant, whose value we shall find later, and δ is once again (!) the Feigenbaum $\delta = 4.669\ldots$ (As we saw before, other values of δ apply if the map function has other than quadratic dependence near its maximum value.)

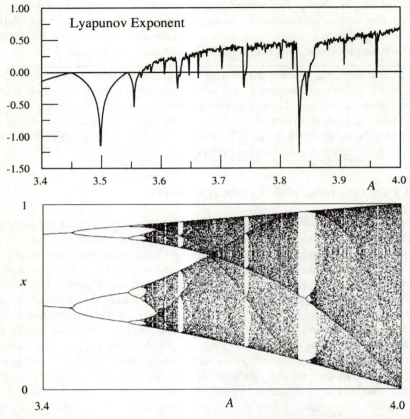

Fig. 9.3. The top figure shows the Lyapunov exponent (calculated using the derivative method in Eq. 5.4-12) for the logistic map function plotted as a function of the control parameter A. The bottom figure shows the corresponding bifurcation diagram. When λ is negative, the behavior is periodic. When the exponent is 0, a bifurcation occurs. When the exponent is positive, the behavior is chaotic.

Eq. (9.4-1) gives us important predictive capabilities: If we see a system become chaotic through a sequence of period-doublings, then we can predict how chaotic it will be (in terms of the average Lyapunov exponent) as a function of the control parameter. Experiments on the semiconductor diode circuit described in Chapter 1 have shown that this scaling law works quite well in describing the behavior of the Lyapunov exponent calculated from actual experimental data (JOH88) for parameter values close to the period-doubling accumulation point.

We should also point out that the scaling law expressed in Eq. (9.4-1) has the same form as scaling laws describing the behavior of physical properties near a second-order phase transition in thermal physics, such as the onset of magnetization in a ferromagnet. The Lyapunov exponent plays the role of the so-called *order parameter*, while $A - A_\infty$ is analogous to the difference between the actual temperature and the critical temperature at which the phase transition occurs. The temperature dependence of these order parameters is described by the same sort of power law with universal classes of scaling exponents. Some other analogies between statistical mechanics and dynamics are explored in Chapter 10.

*Derivation of the Universal Scaling Law for the Lyapunov Exponent

The scaling law in Eq.(9.4-1) can be derived from the following facts: (1) The sequence of chaotic bands, which exist for $A > A_\infty$ (see Fig. 9.3, Fig. 2.1, and Fig. 5.9, for example), merge in a sequence of bifurcations as the control parameter A continues to increase. The bands merge in a kind of "period-undoubling": Eight bands merge to give four; the four merge to give two, and finally the two bands merge to form a single band, as A increases. (2) The chaotic-band-merging sequence is described for A values near A_∞ by a convergence ratio that has the same numerical value as the Feigenbaum convergence ratio δ for the period-doubling sequence leading up to chaos. Hence, there is a relationship between the number of chaotic bands present for a particular parameter value and how far the parameter is from the period-doubling accumulation point value. To be specific, we shall denote the parameter value where 2^m bands merge to give 2^{m-1} bands as $\underline{A}_m$. For example, at the value $\underline{A}_2$ four bands merge to give two bands as A increases. (The underline reminds us that these A values are analogous to, but different from, the A values for the period-doubling sequence.)

If we start two trajectories separated by the distance d_0 within one of the chaotic bands, the trajectory separation increases exponentially. Since the behavior is supposed to be chaotic, we have

$$d_n = d_0 e^{\lambda n} \qquad (9.4-2)$$

where n is the number of iterations. Let us consider the case of the trajectory's cycling among 2^m bands. If the two trajectories start off in one band, after 2^m iterations they will be back in the original band. Then they will be separated by the amount

$$d_{2^m} = d_0 e^{\lambda 2^m} \qquad (9.4-3)$$

Now comes the crucial step: We could view this divergence as due to one iteration of the function $f^{(2^m)}$ with an effective Lyapunov exponent λ_0, which, in some sense, is a characteristic of a single chaotic band. Thus, we could write

$$d_{2^m} = d_0 e^{\lambda 2^m} \equiv d_0 e^{\lambda_0} \qquad (9.4-4)$$

If we *assume* that λ_0 is a constant (that is, the same for all the chaotic bands), we can write from Eq. (9.4-4)

$$\lambda(\underline{A}_m) = \frac{\lambda_0}{2^m} \qquad (9.4-5)$$

$\underline{A}_m$ means we have an A value that gives rise to 2^m chaotic bands.

The index m can be used to tell us how far we are from A_∞. We use the result of Exercise 2.4-1 in the form

$$\underline{A}_m - A_\infty = \left(\frac{1}{\delta}\right)^m \Delta A \qquad (9.4\text{-}6)$$

where

$$\Delta A = \frac{\delta^2}{\delta - 1}(\underline{A}_1 - \underline{A}_2) \qquad (9.4\text{-}7)$$

Essentially, we are assuming that the Feigenbaum scaling for the period-undoublings extends all the way to the single chaotic band. We now solve Eq. (9.4-6) for m

$$m = -\frac{\ln\left[(\underline{A}_m - A_\infty)/\Delta A\right]}{\ln \delta} \qquad (9.4\text{-}8)$$

Using this result in Eq. (9.4-5) yields

$$\lambda(A) = \lambda_0 2^\beta \qquad (9.4\text{-}9)$$

with

$$\beta = \frac{\ln\left[(A - A_\infty)/\Delta A\right]}{\ln \delta} \qquad (9.4\text{-}10)$$

Eq. (9.4-9) is equivalent to the scaling law stated in Eq. (9.4-1).

Exercise 9.4-1. Show that Eq. (9.4-9) is equivalent to Eq. (9.4-1). Hint: Take the natural logarithm of both equations.

The important point here is that the universal scaling law for the Lyapunov exponent follows from the scaling of the chaotic band mergings and the assumed constancy of the effective Lyapunov exponent λ_0 for each chaotic band.

9.5 Invariant Measure

Another important method of characterizing an attractor makes use of a probability distribution function. This notion becomes particularly important as the number of state space dimensions increases. As we have seen, for a larger number of state space dimensions, we have more and more geometric possibilities for attractors. For higher-dimensional state spaces, we need more abstract and less geometric methods of characterizing the attractor. Various kinds of probability distributions are useful in this case. In general terms, we ask what is the probability that a given trajectory point of the dynamical system falls within some particular region of state space.

Definition of Probability

We need to be careful about what we mean by *probability*. We will use the term to mean the **relative frequency** of actual occurrences given a large number of repetitions of the "experiment." That is, if we have an experiment with N possible results (outcomes), and we run the experiment M times (M is usually a large number) and of the M "trials" we find m_i give the ith result (where the index i runs from $i = 1$ to $i = $ N), then we define the probability of getting the ith result as

$$p_i = \frac{m_i}{M} \qquad (9.5\text{-}1)$$

p_i is just the relative fraction of the total number of events that give the ith result. Of course, we must account for all the trials, so we have

$$\sum_{i=1}^{N} m_i = M \qquad (9.5\text{-}2)$$

In addition, if M, the number of trials, is large enough, we expect that the p_i will be reasonably good <u>predictors</u> of the relative number of events in any large sample of <u>future</u> trials carried out under the same conditions.

When we talk about probabilities in state space, we cannot generally ask for the probability that a trajectory lands precisely on some point in state space. Why not? If we use the relative frequency definition of probability, then the number of possible outcomes is infinite if we ask for precise points (assuming that we can specify the coordinates to an arbitrarily large degree of precision). Thus, the probability of getting any *one* precise numerical value is some finite number (the number of trials that have yielded that result) divided by infinity: All the probabilities are 0.

In practice we avoid this problem by dividing the numerical range of outcomes into some finite (but usually large) number of intervals (or "bins" as the statisticians would call them) and asking for the probability of finding a result within a particular interval. This method has the virtue of automatically recognizing the limited precision of any actual measurement or calculation.

Invariant Measure

For our state space probabilities, we divide the state space region occupied by the attractor into a set of intervals (in one dimension) or cells or "boxes" (in two or more dimensions) and ask for the probability that a trajectory visits a particular interval or cell. If we use M intervals and find that the trajectory visits the ith interval m_i times, then we associate the probability $p_i = m_i/M$ with the ith interval.

A graph of p_i as a function of i gives us the probability distribution for that attractor. In many chaotic systems, the calculated p_is do not depend on where we

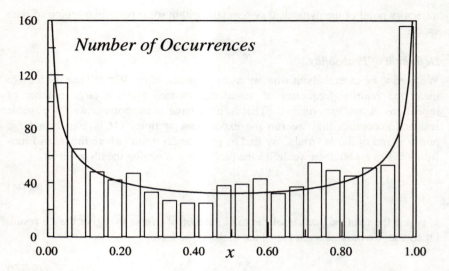

Fig. 9.4. A plot of the invariant distribution for the logistic map with $A = 4.0$. The vertical bars are histogram values for 1024 trajectory points sorted into 20 bins that divide the x axis equally. The solid line is the theoretical probability distribution based on Eq. (9.5-5).

start the trajectory on the attractor as long as we let the trajectory run long enough. In those cases we call the set of p_is a **natural probability measure**. (There are obvious exceptions: If the trajectory starts on an unstable periodic orbit, the measure will be different from that obtained by an initial point that leads to a chaotic trajectory.) The term *measure* is used in the sense of *weight* or *emphasis*. The more often a cell is visited, the larger its measure or weight.

For a one-dimensional system, we express the probability measure as $\mu(x)$. If the dynamics of the system is given by a map function $f(x)$ and if $\mu(x) = \mu(f(x))$, we say that $\mu(x)$ is an **invariant probability measure**. Here the term *invariant* means that the resulting distribution is unchanged under the dynamics of the system.

We can sometimes introduce a continuous function to characterize the probabilities. If we think of dividing state space, which so far is just the x axis, into small intervals or cells, then we can say that the probability of finding the trajectory in the ith cell is given by the integral of some continuous probability *distribution function* $p(x)$ over that cell:

$$p_i = \mu(x_i) = \int_{cell\ i} p(x)dx \qquad (9.5\text{-}3)$$

where x_i labels the location of the ith cell. We interpret $p(x)dx$ as the probability that the trajectory visits the interval between x and $x + dx$.

Another way of visualizing the meaning of p_i is the following: Suppose we assign a mass M to the attractor in state space. We assume that there is a mass density function $\rho(x)$ defined so that $\rho(x)dx$ is equal to the mass of the attractor that lies between x and $x + dx$. Then we say that $m_i = p_i M$ is the mass associated with the ith cell, and p_i is the fraction of the attractor's mass found in cell i.

In Fig. 9.4, we have plotted a histogram of trajectory values for the logistic map with $A = 4.0$. The data values have been sorted into 20 "bins" of equal size lying between $x = 0$ and $x = 1$. As you can see the histogram is relatively smooth at this level of resolution, with the probability tending to be larger for x values near 0 and 1.

For most dynamical systems, we must find the probability distribution numerically by computing the actual trajectories. But for the special case of the logistic map with $A = 4$, we can calculate the invariant probability distribution exactly. To do this, we make use of the connection, introduced in Section 5.8, between the logistic map with $A = 4$ and the Bernoulli shift map. By general arguments for probability distributions, we must have the following relationship between the distributions $p(x)$ for the logistic map (again for $A = 4$) and $P(\theta)$, the distribution given as a function of the variable θ introduced in Eq. (5.8-3):

$$p(x)dx = P(\theta)d\theta \qquad (9.5\text{-}4)$$

The previous equation tells us that the probability of finding x between x and $x + dx$ for the logistic map must be the same as finding θ between θ and $\theta + d\theta$ for the corresponding value of θ for the Bernoulli map. For the Bernoulli shift map, the trajectory values are distributed uniformly over the range $0 \le \theta \le 1$. Thus, we must have $P(\theta) = 1$. Using Eq. (5.8-3), we can relate dx to $d\theta$ to find

$$p(x) = \frac{1}{\pi\sqrt{x(1-x)}} \qquad (9.5\text{-}5)$$

The solid curve in Fig. 9.4 shows this distribution (appropriately normalized). The actual histogram values fall quite close to this curve. (Of course, we expect some deviation between the two since the histogram was generated from only 1024 data points grouped into 20 bins.)

Exercise 9.5-1. Follow through the details of the calculation to obtain the result stated in Eq. (9.5-5).

Ergodic Behavior

One of the reasons for focusing attention on the invariant distribution, if it exists, is that it gives an alternative (and often simpler) way of calculating average properties of a system. For example, if we consider some property B of the system, which depends on the value of the state space variable x, we can define the time average value of B as

$$\langle B \rangle_t = \frac{1}{T}\int_0^T B(x(t))dt \qquad (9.5\text{-}6a)$$

$$\langle B \rangle_t = \frac{1}{N} \sum_{i=1}^{N} B(x(t_i)) \qquad (9.5\text{-}6b)$$

Eq. (9.5-6) gives the time average in two forms. The integral form is useful if we have solved for $x(t)$ as a continuous function of time (an analytic result). The sum form is useful for a discrete time-series of values. In the latter case, $t_1 = T/N$ and $t_N = T$. We follow B in both cases as a function of time for a time interval T and compute the average value of B over that interval. Of course, we usually want the time interval T to be long enough for us to sample the full range of behavior of the system. Thus, we often add the limit $T \to \infty$.

Alternatively, we could evaluate the average of B by finding B as a function of x and multiplying that value by the probability that the system visits the interval $[x, x + dx]$

$$\begin{aligned}\langle B \rangle_p &= \int B(x) p(x) dx \\ &= \sum_{i=1}^{M} B(x_i) p_i\end{aligned} \qquad (9.5\text{-}7)$$

In the integral form, we assume that we have evaluated the (continuous) probability distribution $p(x)$. In the sum form, we sum over the M intervals (or boxes) that divide up the attractor region in state space. (The continuous probability distribution may not exist mathematically in some cases. See [Ott, 1993, pp. 51-55].)

If the average in Eq. (9.5-6) equals the average in Eq. (9.5-7), then we say that the system is **ergodic** (i.e., time averages are the same as state space averages, where the state space average is weighted by the probability that a trajectory visits a particular portion of state space). In traditional statistical mechanics, one often assumes this equivalence in the so-called **ergodic hypothesis**. (You may recall that the issue of ergodicity also appeared in Chapter 8 in connection with Hamiltonian systems.)

As an example of this kind of calculation, let us compute the average Lyapunov exponent for the logistic map function for the parameter value $A = 4$. For this value of A, the probability density is given by Eq. (9.5-5) for x values between 0 and 1. According to Eq. (5.4-11), the "local" value of the Lyapunov exponent is given by

$$\lambda(x) = \ln|f'(x)| \qquad (9.5\text{-}8)$$

We can compute the average Lyapunov exponent by making use of the probability distribution

$$\lambda = \int_0^1 p(x) \ln|f'(x)| dx = \frac{1}{\pi} \int_0^1 \frac{\ln|4(1-2x)|}{\sqrt{x(1-x)}} dx \qquad (9.5\text{-}9)$$

If we now make the (not-so-obvious) substitution $x = \sin^2(\pi y/2)$, we find

$$\lambda = \int_0^1 \ln\left|4\cos(\pi y)\right| dy \qquad (9.5\text{-}10)$$
$$= \ln 2$$

The definite integral in Eq. (9.5-10) can be found in standard integral tables. The main point here is that the Lyapunov exponent is positive, and, in fact, equal to ln 2, the same value obtained for the Bernoulli shift map and for the tent map, Eq. (5.9-1), with $r = 1$.

> Finding the same Lyapunov exponent value for the logistic map with $A = 4$, for the Bernoulli shift map and for the tent map with $r = 1$ may seem quite remarkable. This result tells us that the Lyapunov exponent is *independent of a change in variables* for the map function. (Any one of these three map functions can be transformed into one of the others by a change of variable.) Why is that independence important? When we characterize actual experimental systems, the signal we record is often one that is not directly the dynamical variable being characterized. For example, if we are monitoring the temperature in a fluid for a Lorenz-type system, then we might record the electrical signal from a temperature probe such as a thermistor or a thermocouple. We can use the recorded values of the electrical signal directly to compute the Lyapunov exponent for the system because that exponent is independent of the change in variables (as long as it is one-to-one) in converting from temperature to electrical voltage.

> **Exercise 9.5-2.** Starting from the definition of Lyapunov exponent in Eq. (9.3-4), show explicitly that a linear transformation of variables $u = ax + b$ does not change the Lyapunov exponent. More challenging: what general classes of variable transformations lead to no change in λ ?

9.6 Kolmogorov–Sinai Entropy

In this section we will introduce several related ways of describing chaotic behavior based on notions that are (at least formally) related to the concept of entropy from thermodynamics and statistical mechanics. As we shall see in Chapter 10, these entropy measures can be generalized to give a very powerful formalism for describing dynamical systems.

A Brief Review of Entropy

We will first review the concept of entropy from the point of view of statistical mechanics. From this perspective, the most fundamental definition of entropy is given by counting the number of "accessible states" for the system under consideration. The crucial idea is that statistical mechanics is concerned with relating macroscopic (large-scale) properties of a system, such as pressure, volume occupied, and temperature for a gas, to the microscopic description in terms of positions and velocities of the atoms or molecules that make up that gas. In almost all cases, there is a vast number of microscopic states (for each of which we specify the positions and velocities of the individual atoms or molecules as functions of time) that correspond to the same macroscopic state (e.g., a state of the gas with a particular temperature and pressure). We assume that an isolated system of atoms and molecules in thermal equilibrium visits equally all of these microstates compatible with this set of properties "microstate democratic egalitarianism"). Then we define the entropy S of the system as

$$S = k \ln N \tag{9.6-1}$$

where N is the number of microstates compatible with the assumed macroscopic conditions; k is called Boltzmann's constant and determines the units in which entropy is measured (in the SI system $k = 1.38 \ldots \times 10^{-23}$ Joule/K).

The Second Law of Thermodynamics, one of the most fundamental laws in all of physics, states that in a spontaneous thermal process the entropy of an isolated system will either stay the same or increase. From our microscopic point of view, the Second Law tells us that a system will tend to evolve toward that set of conditions that has the largest number of accessible states compatible with the prescribed macroscopic conditions.

You may have noted that we dodged the difficult question of how we actually count the number of accessible states. In the application of entropy to nonlinear dynamics, we will give one possible scheme for doing this. In quantum statistical mechanics, in which we apply quantum mechanics to describe the microscopic system, the counting procedure is well defined since quantum mechanics automatically leads to a set of discrete states for a system occupying a bounded region of space. In the meantime, we will simply assume that we have adopted some counting procedure.

If the system is not isolated or if it is not in thermal equilibrium, not all the accessible states are equally likely to be occupied. We can generalize the entropy definition to cover this case as well. To see how this comes about let us first rewrite the equally likely case Eq. (9.6-1) in terms of the probability p_i that the system is in the ith of the accessible states. For the equally-likely case $p_i = p = 1/N$ for all the accessible states. Strictly speaking, we should be thinking of an ensemble of systems (i.e., of a mentally constructed large collection of identical systems), each with the same values for the macroscopic properties. We then ask what is the probability that a member of the ensemble is in one particular microstate. Written in terms of these probabilities, the entropy expression becomes

$$S = -k \ln p \tag{9.6-2}$$

Note that S is still a nonnegative quantity since $p \le 1$.

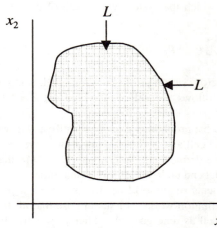

Fig. 9.5. A two-dimensional state space region has been divided up into small cells, each side of which has length L.

If the probabilities are <u>not</u> all the same, then the "obvious" generalization of Eq. (9.6-2) is

$$S = -k \sum_r p_r \ln p_r \qquad (9.6\text{-}3)$$

where p_r is the probability (in the ensemble sense described earlier) for the system to be in the rth microstate. (For a more formal justification of Eq. (9.6-3) see, for example, [Chandler, 1987].) The sum can be taken over *all* the states of the system (assuming that there is a finite number of states). If a state is not accessible, then $p_r = 0$, and the state makes no contribution to the entropy.

Exercise 9.6-1. (a) Show that $p_r \ln p_r \to 0$ as $p_r \to 0$. (b) Show that Eq. (9.6-3) reduces to Eq. (9.6-2) when $p_r = 1/N$ for the N accessible states and $p_r = 0$ for any inaccessible state.

Entropy for State Space Dynamics

We shall now apply these entropy ideas to a description of the state space behavior of a dynamical system. First, we need to decide what "counting states" means in this context. We do this counting by dividing the state space of the system into cells, usually all of the same size. For a dissipative system (for which an attractor exists) we need divide up only the region containing the attractor. (For a conservative system, we divide up the entire range through which the trajectories may wander.) Figure 9.5 shows this division for a two-dimensional state space.

We then start the time evolution of the system with an ensemble (collection) of initial conditions, usually all located within one cell. As the system evolves in time, the trajectories will generally spread out over a larger number of cells in the state space. After n units of time (each of length τ) have gone by, we calculate the relative frequency (probability) p_r with which the system visits each of the cells. We then define the entropy S_n to be

$$S_n = -k \sum_r p_r \ln p_r \qquad (9.6\text{-}4)$$

Note that p_r is the probability that a trajectory starting from our initial cell, is in the rth cell after n units of time. (In what follows, we will set the constant $k = 1$ to simplify notation.)

We again need to be careful about what probability means. Specifically, if we start with M trajectory points in the initial cell and if after n units of time, we have M_r trajectory points in the rth cell, then we define the probability as $p_r = M_r/M$, that is, p_r is the fraction of trajectory points that end up in the rth cell at that time.

Before introducing further refinements of these ideas, let us try out a few simple cases to see how this entropy definition works. First, suppose that all the trajectories move together from cell to cell as time goes on. Then $p_r = 1$ for the occupied cell and is 0 for the unoccupied cells. In that case $S_n = 0$ for all n. Thus, we see that constant entropy corresponds to what we would interpret as regular motion (say, along a limit cycle). For a second example, suppose that the number N_n of occupied cells increases with time and that (for our idealized example) all the occupied cells have the same probability, namely, $1/N_n$. Then the entropy is

$$S_n = +\ln N_n \qquad (9.6\text{-}5)$$

We see that the entropy increases as the natural logarithm of the number of occupied cells.

For a completely random system, each of the M trajectory points would jump to its own cell (assuming that the cells are small enough to distinguish the different trajectory points). In that case, $S_n = \ln M$. The important point is that this number grows without limit as M becomes very large. On the other hand, for a regular or a chaotic deterministic system with a small number of state space variables (i.e., systems that are not completely random), the entropy becomes independent of M for large M.

One important notion should emerge from these examples: We are really concerned with <u>changes</u> in entropy, not in the entropy value itself. For example, if we choose a set of initial conditions in two cells and if the motion is regular, then the entropy value would not be 0, but it would remain <u>constant</u> as the system evolves. The *change* in entropy is characterized by the Kolmogorov–Sinai entropy *rate* (sometimes called the K–S entropy), which describes the *rate of change* of entropy as the system evolves. First we shall give a rough definition of the K–S entropy (rate); then we shall refine it. The K–S entropy K_n after n units of time is defined to be

$$K_n = \frac{1}{\tau}(S_{n+1} - S_n) \qquad (9.6\text{-}6)$$

that is, K_n is the rate of change of the entropy in going from $t = n\tau$ to $t = (n+1)\tau$.

What we actually want is the average of this K–S entropy over the entire attractor, in order to characterize the attractor as a whole. Thus we define the average quantity K

$$K = \lim_{N \to \infty} \frac{1}{N\tau} \sum_{n=0}^{N-1} (S_{n+1} - S_n)$$

$$= \lim_{N \to \infty} \frac{1}{N\tau} [S_N - S_0] \tag{9.6-7}$$

Letting N get very large corresponds to allowing the trajectories to evolve for a long time, hence covering (presumably) all of the attractor.

We will now introduce two further limits: One limit takes the cell size to 0; therefore we use finer and finer divisions of state space. This procedure should make K independent of the details of how we divide up the state space. The second limit takes the time interval τ to 0; therefore, we use smaller time increments and hence a finer description of the dynamics. Putting all of these limits together (and probably pushing the limits of the reader's patience as well), we have the complete definition of the K–S entropy:

$$K = \lim_{\tau \to 0} \lim_{L \to 0} \lim_{N \to \infty} \frac{1}{N\tau} [S_N - S_0] \tag{9.6-8}$$

When the K–S entropy is applied to Poincaré sections and to iterated maps, we set $\tau = 1$ and drop the limit $\tau \to 0$.

Let us try another example. Suppose that the number of occupied cells N_n grows exponentially with time

$$N_n = N_0 e^{\lambda n \tau} \tag{9.6-9}$$

and that all the occupied cells have the same probability $p_r = 1/N_n$. The K–S entropy is then equal to λ, the parameter characterizing the exponential growth of the number of occupied cells. The number of occupied cells, however, is proportional to the distance between the trajectories that all initially occupied one cell. Thus, we see that λ is just the (average) Lyapunov exponent for the system. For exponential growth of the number of occupied cells and for equal probabilities for the occupied cells, the K–S entropy and the Lyapunov exponent are the same. When there is more than one Lyapunov exponent, the K–S entropy turns out to be equal to sum of the positive Lyapunov exponents, a result known as the Pesin Identity (PES77)(PPV86)(GER90).

Exercise 9.6-1. Show that $K_n = \lambda$ for the conditions stated in the previous paragraph.

Alternative Definition of K–S entropy

In the literature on nonlinear dynamics, a slightly different definition of the K–S entropy is sometimes used (see, for example, [Schuster, 1995]. This definition will turn out to be useful in Chapter 10. In this alternative definition, we let a single trajectory run for a long time to map out an attractor. We then cover the attractor region of state space with cells. Next we start a trajectory in one of the cells and label that cell $b(0)$. At a time τ later, the trajectory point will be in cell $b(1)$. At $t = 2\tau$, the trajectory is in cell $b(2)$, and so on, up to time $t = N\tau$, thereby recording a particular sequence of cell labels: $b(0), b(1), ..., b(N)$.

We then start off a second trajectory from the same initial cell. Because the exact initial conditions are slightly different, however, we will generally get a different sequence of cell labels for the second sequence. We repeat this process many times, thereby generating a large number of sequences.

Next, we calculate the relative number of times a particular sequence of N cell labels occurs. Let us call that relative number $p(i)$ for the ith sequence. We then define the entropy S_N to be

$$S_n = < -\sum_i p(i) \ln p(i) > \qquad (9.6\text{-}10)$$

where the sum is taken over all sequences of N cell labels that start with $b(0)$. The brackets $< >$ mean that we average the sum over all starting cells on the attractor.

Finally, we define the K–S entropy to be the average rate of increase of the entropy with respect to sequence length:

$$K = \lim_{N \to \infty} \frac{1}{N} (S_N - S_0) \qquad (9.6\text{-}11)$$

To gain some familiarity with this definition, let us try it out for some special cases. First, let us assume that all the sequences starting from the same initial cell are the same; that is, that we have regular motion: All the trajectories starting from the same cell track each other as time goes on. Then $S_N = 0$ for all N and $K = 0$ since S does not change. At the other extreme, let us assume that each of the M sequences occurs only once (i.e., the system is "purely random"). In that case $p(i) = 1/M$ and $S_N = \ln M$ and grows without limit as M increases. Thus, we see that the K–S entropy increases as the number of sample trajectories increases.

Let us now assume that the number of distinct trajectory cell sequences M_{seq} increases exponentially (on the average) with the length N of the sequences: As the sequences become longer, they split apart due to the distinct behavior of trajectories with (slightly) different initial conditions. In more formal terms, we assume that

$$M_{seq} = e^{\lambda N} \qquad (9.6\text{-}12)$$

Obviously, λ is the Lyapunov exponent for the system. If we further assume that each distinct trajectory sequence occurs with the same probability $p(j) = 1/M_{seq}$, then it is easy to see that the K–S entropy is just $K = \lambda$. So, once again, we get an equality between the K–S entropy and the (positive) Lyapunov exponent.

The relationships among the K–S entropy, the Lyapunov exponent, and the traditional thermodynamic entropy have been explored for a variety of thermodynamic systems. The conclusion is that under many circumstances, they are all proportional to one another. See, for example, BLR95, DAV98, and LAB99.

We shall postpone actual calculations of the K–S entropy to Chapter 10, where we shall see how to find the K–S entropy as a special case of a much more general and powerful calculational scheme.

9.7 Fractal Dimension(s)

The two methods of quantifying chaos described in the previous sections both emphasize the dynamical (time-dependent) aspects of the trajectories. A second category of quantifiers focuses on the geometric aspects of the attractors. In practice, we let the trajectories run for a long time and collect a long time series of data. We can then ask geometric questions about how this series of points is distributed in state space. Perhaps unexpectedly, this geometry provides important clues about the nature of the trajectory dynamics.

A common question is: What is the (effective) dimensionality of the attractor? For example, if the attractor is a fixed point, we say that the dimensionality is equal to 0 because a point is a 0-dimensional object in geometry. If the attractor is a line or a simple closed curve, we say that the dimensionality is equal to 1 because a line or a curve is a one-dimensional object. Similarly, a surface has a dimensionality of 2, a solid volume a dimensionality of 3. Furthermore, we can talk about "hypervolumes" of yet higher dimensions if we wish.

Why is dimensionality important? As we saw in Chapters 3, 4, and 8, the dimensionality of the state space is closely related to dynamics. The dimensionality is important in determining the range of possible dynamical behavior. Similarly, the dimensionality of an attractor tells about the actual long-term dynamics. For example, the dimensionality of an attractor gives us an estimate of the number of active degrees of freedom for the system.

Two points to note: (1) For a dissipative dynamical system (the type of system we are considering here), the dimensionality, D, of the attractor must be less than the dimensionality, call it d, of the full state space because we know that a d-dimensional state space volume of initial conditions must collapse to 0. (You should recall that the dimensionality d of the state space is determined by the minimum number of variables needed to describe the state of the system.) The attractor might be a "surface" of dimension $d-1$ or some other lower-dimensionality object. All we require is that the attractor occupy 0 volume in state space. (2) If we are examining Poincaré section data, the dimensionality D^* of the Poincaré section of the attractor will be one less than the dimensionality D of the full state space attractor ($D = D^* + 1$) since the Poincaré section technique takes out one of the state space dimensions.

For a Hamiltonian system, the dimensionality of the set of points generated by a trajectory must be no larger than $d - 1$, since, as we discussed in Chapter 8, the trajectories are confined to a constant energy "surface" in the state space. If we use a Poincaré section, then the largest possible dimensionality is further reduced. In systems, with additional constants of the motion, the dimensionality is yet smaller.

As an example of the power of dimensionality arguments, let us consider a three-dimensional state space. Let $\vec{f}$ represent the set of time evolution functions for the system (see Section 4.4). If $div\,\vec{f} < 0$ for all points in the state space, then the attractor must shrink to a point or a curve. For such a system, the long-term behavior cannot be quasi-periodic because quasi-periodic trajectories "live" on the surface of a torus. Why does this follow? If we consider a set of initial points distributed through a volume of the torus and if $div\,\vec{f} < 0$ everywhere, then the volume occupied by the initial points inside the torus must shrink to 0, and the torus must disappear. This argument tells us that the Lorenz model described in Chapter 1 cannot have quasi-periodic solutions since the model has $div\,\vec{f} < 0$ for all state space points.

What has come as a surprise to most scientists and mathematicians is that geometric objects with dimensionalities that are not integers play a fundamental role in the dynamics of chaotic systems. These geometric objects have been named *fractals* [Mandlebrot, 1982] because their dimensionality is not an integer. To be able to talk about such fractional dimensions, we need to establish a general means of determining quantitatively the dimensionality.

If an attractor for a dissipative system has a noninteger dimension, then we say that the system has a *strange attractor*.

Unfortunately for the novice in nonlinear dynamics, many apparently different definitions of dimensionality are currently in use. In general these may all give different numerical values for the dimensionality, although in some cases the numbers are close. To exacerbate the difficulties, the name *fractal dimension* is used rather indiscriminately. It is best to recognize that there is a host of fractal dimension measures, none of which can legitimately claim to be the fractal dimension. We will confine our discussion to those measures of dimension that are relatively straightforward to implement for the kind of data generated from the study of a dynamical system. A thorough discussion of the definitions of the various measures of dimension can be found in FOY83. Several examples that yield different values for different dimension methods are treated in ESN90.

We will begin our discussion with a measure called the *box-counting dimension* (often called the *capacity dimension*) because a set of boxes (or cells) is used in the calculation. This particular measure is relatively easy to understand, but it turns out not to be so useful for dimension determinations in higher-dimensionality state spaces. It was first applied to dynamics by Kolmogorov (KOL58).

The box-counting dimension D_b of a geometric object is determined by the following: Construct "boxes" of side length R to cover the space occupied by the geometric object under consideration. For a one-dimensional set (such as the string of x values for a one-dimensional state space trajectory), the boxes are actually line segments of length R. In two-dimensions, they would be two-dimensional squares. In three dimensions they would be cubes, and so on. We then count the <u>minimum</u> number of boxes, $N(R)$, needed to contain all the points of the geometric object. As we let the size of each box get smaller, we expect $N(R)$ to increase as R decreases because we will need a larger number of the smaller boxes to cover all the points of the object. The box-counting dimension D_b is defined to be the number that satisfies

$$N(R) = \lim_{R \to 0} kR^{-D_b} \tag{9.7-1}$$

where k is a proportionality constant. In practice, we find D_b by taking the logarithm of both sides of Eq. (9.7-1) (before taking the limit) to find

$$D_b = \lim_{R \to 0} \left\{ -\frac{\log N(R)}{\log R} + \frac{\log k}{\log R} \right\} \tag{9.7-2}$$

As R becomes very small, the last term in Eq. (9.7-2) goes to 0, and we may define

$$D_b = -\lim_{R \to 0} \frac{\log N(R)}{\log R} \tag{9.7-3}$$

To gain some confidence that Eq. (9.7-3) gives a reasonable definition of dimension, let us apply it to some simple examples. First, consider a two-dimensional space and let the geometric object be a point. In this case, the box is just a square of side R. Only one box is needed to contain the point; therefore, we have $N(R) = 1$ for all values of R. Using this result in Eq. (9.7-3) gives $D_b = 0$, just as we would expect for a point.

What happens if the object consists of a number of <u>isolated</u> points? The answer is that D_b is still equal to 0. To see how this comes about, let N be the number of isolated points. When R is small enough (smaller than the smallest distance between neighboring points), we will have one box around each point. When R gets smaller than this value, the numerator in Eq. (9.7-3) stays fixed while the denominator grows (more negative) without limit. So again we have $D_b = 0$.

Exercise 9.7-1. You might worry that Eq. (9.7-3) involves the logarithm of a length, which carries some units, such as meters or furlongs. Show that the choice of units for R makes no difference in the numerical value of D_b.

As a second example, let us use a line segment of length L as the geometric object. In this case we need $N(R) = L/R$ boxes to cover the segment. [When N is sufficiently large, we can safely ignore any fraction of a box in counting $N(R)$.] We now use this value in Eq. (9.7-3) to find

$$D_b = -\lim_{R \to 0} \frac{\log(L/R)}{\log R} = -\lim_{R \to 0} \frac{\log L - \log R}{\log R} = 1 \qquad (9.7\text{-}4)$$

As we expect, the box-counting dimension of a line segment is equal to 1.

Exercise 9.7-2. Show that $D_b = 1$ for an object consisting of a finite number of isolated line segments.

Exercise 9.7-3. Show that $D_b = 2$ if the geometric object is a surface.

Now that we are convinced that Eq. (9.7-3) gives a reasonable definition of dimension, let us apply it to a case that gives a noninteger result. As a first example, we will discuss the construction of the famous *Cantor set*. [The German mathematician Georg Cantor (1845–1918) introduced this set long before the term *fractal* was invented.] The Cantor Set is constructed in stages by starting with a line segment of length 1. For the first stage of the construction, we delete the middle third of that segment. This leaves two segments, each of length 1/3. For the second stage, we delete the middle third of each of those segments, resulting in four segments, each of length 1/9. For the Mth stage, we remove the middle third of each of remaining segments to produce 2^M segments, each of length $(1/3)^M$. If we continue this process as $M \to \infty$, the Cantor set is left. This process is illustrated in Fig. 9.6.

Let us now calculate the box-counting dimension of this set. We need to proceed cautiously. If we stop at any finite state of deletion, we are left with just a series of 2^M line segments whose $D_b = 1$. But if we let $M \to \infty$, then we seem to be left with just a series of points, and we might expect to find $D_b = 0$. Therefore, we must let $M \to \infty$ and $R \to 0$ simultaneously. We do this by making $N(R)$ the <u>minimum</u> number of boxes required to cover the object at a given stage of construction. We then determine how $N(R)$ and R depend on M. Then, as M gets very large, R will get very small, and we can take both limits simultaneously.

For the Cantor set construction, at the Mth stage of construction, we need a minimum of 2^M boxes with $R = (1/3)^M$. If we use those values in Eq. (9.7-3), we find (leaving off the limit notation)

$$D_b = -\frac{\log 2^M}{\log(1/3)^M} = \frac{\log 2}{\log 3} = 0.63\ldots \qquad (9.7\text{-}5)$$

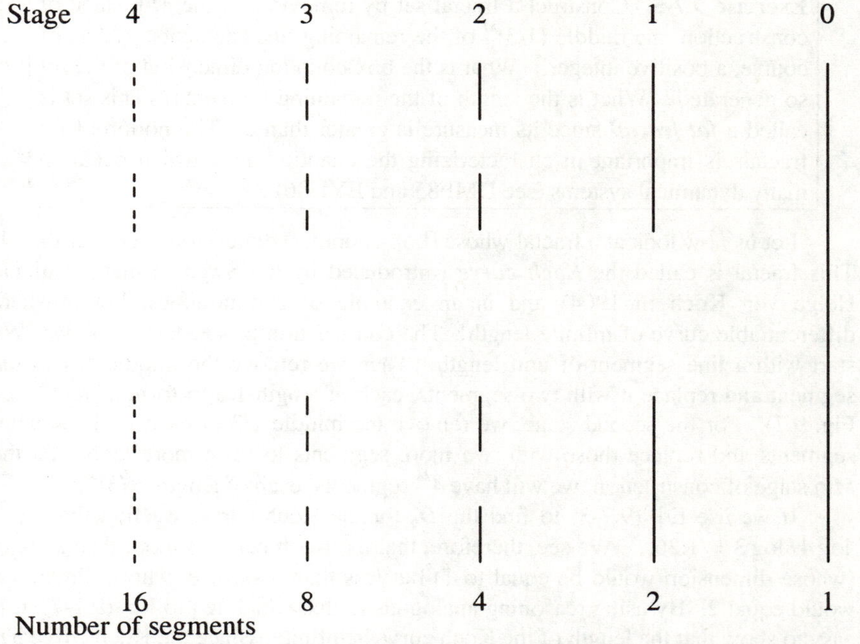

Stage   4   3   2   1   0

    16   8   4   2   1
    Number of segments

Fig. 9.6. An illustration of the first four construction stages of the Cantor set by the removal of middle thirds. At the Mth stage of construction, there are 2^M segments each of length $(1/3)^M$.

We see that D_b for the Cantor set is a noninteger number between 0 and 1. In rough terms, the Cantor set is more than a collection of points but less than a line segment. It is a fractal object.

Exercise 9.7-4. Find D_b for a Cantor set constructed by removing the middle $1/n$ of a line segment (with $n > 1$).

How much of the line segment is left after the Cantor construction process as $M \to \infty$? This length is called the **measure** of the Cantor set. We can compute that measure by noting that at the Mth stage of construction, the length of the line segments remaining is given by

$$length = 1 - 1\left(\tfrac{1}{3}\right) - 2\left(\tfrac{1}{3}\right)^2 - 2^2\left(\tfrac{1}{3}\right)^3 - \ldots - 2^{M-1}\left(\tfrac{1}{3}\right)^M \qquad (9.7\text{-}6)$$

In the limit, the amount left is

$$length = 1 - \frac{1}{3}\left[\sum_{i=0}^{\infty}\left(\tfrac{2}{3}\right)^i\right] \qquad (9.7\text{-}7)$$

The sum in the previous equation is a simple geometric series whose value is 3. Thus, we see that the measure of the Cantor set is 0.

Exercise 9.7-5. Construct a fractal set by removing, at the Mth stage of construction, the middle $(1/3)^M$ of the remaining line segments. (M is, of course, a positive integer.) What is the box-counting dimension of the set so generated? What is the length of the remaining segments? This set is called a ***fat fractal*** since its measure is greater than 0. The notion of fat fractals is important in characterizing the chaotic bands which occur in many dynamical systems (see UMF85 and EYU86).

Let us now look at a fractal whose (box-counting) dimension is greater than 1. This fractal is called the ***Koch curve*** (introduced by the Swedish mathematician Helge von Koch in 1904) and is an example of a continuous, but nowhere differentiable curve of infinite length! The construction proceeds as follows: We start with a line segment of unit length. Then we remove the middle 1/3 of the segment and replace it with two segments, each of length 1/3 to form a "tent" (see Fig. 9.7). For the second stage, we remove the middle 1/3 of each of the smaller segments and replace those with two more segments to form more tents. At the Mth stage of construction, we will have 4^M segments, each of length $(1/3)^M$.

If we use Eq. (9.7-3) to find the D_b for the Koch curve, we find that $D_b = \log 4/\log 3 = 1.26...$ We see, therefore, that the Koch curve is more than a curve (whose dimension would be equal to 1) but less than a surface, whose dimension would equal 2. By using reasoning analogous to that which led to Eq. (9.7-7), it is easy to show that the length of the Koch curve is infinite. Since the Koch curve has an infinity of abrupt changes in slope, it is nowhere differentiable.

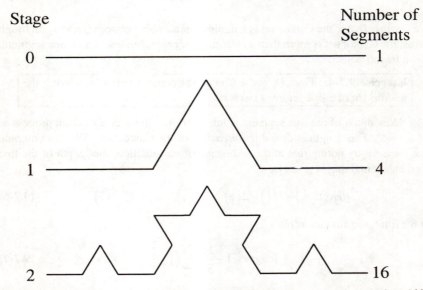

Fig. 9.7. The first few stages of construction of the Koch curve. At each stage the middle 1/3 of each straight segment is removed and replaced with two other segments to form a tent.

> **Exercise 9.7-6.** Work through the details of the calculations that show that $D_b = 1.26\ldots$ for the Koch curve. Show that the Koch curve has infinite length.

> **Exercise 9.7-7.** Start with an equilateral triangle and construct a Koch curve on each side of the triangle. The result is called the **Koch Snowflake**. Show that the boundary has infinite length but that it encloses a finite area.

The fractal objects defined earlier all have the property called **self-similarity** (i.e., a small section of the object, suitably magnified, is identical to the original object). These self-similar objects form a particularly simple class of fractals, but not all fractals are self-similar. Another class of fractals is called **self-affine** (MAN85). Their "self-similarity" is apparent only if different magnification factors are used for different directions. Finally, we distinguish fractals such as the Cantor set generated by a deterministic rule from **random fractals** [Mandlebrot, 1982] generated by stochastic processes. The self-similarity of random fractals requires a statistical description. We shall meet several random fractals in Chapter 11. Random fractals are useful in understanding some aspects of noise, music and various shapes found in nature (VOS89).

Fractal Dimensions of State Space Attractors

Let us now turn our attention to the geometric characterization of attractors of dynamical systems. If we apply the box-counting procedure to the trajectory data from some dynamical system, either experimental or theoretical, we can determine the box-counting dimension for the system's attractor. If the attractor's dimension is not an integer, we say that we have a *strange attractor* for the system. However, the determination of the dimension is not as straightforward as you might imagine. For example, in practice, you cannot take the $R \to 0$ limit because of the finite precision of the data. In Section 9.8, we shall introduce another dimension, which is usually easier to compute from actual data. There we will discuss the computation problems in more detail.

The box-counting dimension has been computed (GRA81) for the logistic map at the period-doubling accumulation point A_∞ by means of renormalization arguments like those given in Chapter 5. The numerical value of the box-counting dimension is 0.5388..., which, according to the renormalization calculation, should be a universal number for any one-dimension unimodal iterated map function with a quadratic maximum.

We can make a rough calculation of the box-counting dimension for the logistic map function by considering the splitting chaotic bands that occur as $A \to A_\infty$ from above (see Figs. 2.1 and 9.3, for example). For a given parameter value, we can think of these bands as constituting line segments (that is, the probability distribution of points within each band is approximately uniform and continuous). When a set of bands splits into twice as many bands, as A decreases toward A_∞, we can think of this as a Cantor set construction that removes some interior piece of each of the chaotic bands. The size of the bands is described by the same Feigenbaum size scaling that holds for the period-doubling sequence for $A < A_\infty$ (see Fig. 2.3), where one section is a factor of $1/\alpha$ smaller than the previous section and the other section is a factor of $(1/\alpha)^2$ smaller.

Suppose we need 2^n segments of minimum length R to cover the 2^n bands for a particular parameter value. Then, we need 2^{n+1} segments of length R/α to cover the segments constructed in the next stage. We can estimate D_b with

$$D_b = -\frac{\log(2^n/2^{n+1})}{\log \alpha} = 0.4498\ldots \tag{9.7-8}$$

which result is somewhat smaller than the result (GRA81) quoted above.

We get a better estimate of D_b by generating a somewhat more complicated Cantor set, which more closely resembles the pattern of band splittings of the logistic map. In this new procedure we start with two segments of lengths 1 and $1/\alpha$ and remove a segment of length $1/\alpha$ from one and $(1/\alpha)^2$ from the other. At the nth stage of construction, we need 2^n segments of length R_n to cover the bands. Then at the $(n+1)$th stage, we will need 2^{n+1} segments of average length

$$R_{n+1} = \frac{R_n}{2}\left[\frac{1}{\alpha} + \frac{1}{\alpha^2}\right] \tag{9.7-9}$$

for the next stage. D_b is found by using this average length (in an admittedly ad hoc manner) in Eq. (9.7-3) to give

$$D_b = -\frac{\log 2}{\log\left(\frac{1}{2}\left[1/\alpha + 1/\alpha^2\right]\right)} = 0.543 \tag{9.7-10}$$

closer to the value cited earlier.

The lesson to be learned here is that when the fractal set does not have a simple self-similar structure, then the value of the box-counting dimension depends on the box covering procedure used. We suspect that for these more general objects we need more than one measure of dimensionality to characterize them. In Chapter 10, we will learn about an infinity of dimensional measures needed to characterize these more complex objects.

* The Similarity Dimension for Nonuniform Fractals

As we saw in the example in the previous subsection, when the fractal set in question does not have a uniform structure, the box-counting dimension value depends upon the covering used for the set. In this section we explore how to generalize the box-counting dimension for this kind of situation. We shall call the new dimension the *similarity dimension* D_s. We shall see that the generalization gives us a useful dimension for the logistic map attractor at the period-doubling accumulation point and for quasi-periodic behavior in the sine-circle map.

For a nonuniform fractal, we are concerned with covering the object with boxes of <u>different</u> sizes r_i. Suppose at a certain stage of construction of the fractal set, we use M_i boxes of size r_i with N_i points in each box. In this method of bookkeeping, different boxes of the same size r that have <u>differing</u> numbers of points in them will be given different indices. Let $p_i = N_i/N$ be the relative number of points in one of the boxes labeled by i. Then we have

$$\sum_i M_i p_i = 1 \qquad (9.7\text{-}11)$$

As we go from one stage of construction to the next, we assume that there is a scaling factor s_i which sets the size r_i of the ith set of boxes relative to some length scale R. That is, $r_i = R/s_i$, where $s_i > 1$. If this construction is being carried out in a d-dimensional state space, then the volume of each of the ith set of boxes is $(R/s_i)^d$ and the total volume of the boxes used must satisfy

$$\sum_i M_i \left(\frac{R}{s_i} \right)^d < R^d \qquad (9.7\text{-}12)$$

or equivalently

$$\sum_i M_i s_i^{-d} < 1 \qquad (9.7\text{-}13)$$

since the volume of the fractal set we are trying to describe is 0 in the full d-dimensional space. The similarity dimension D_s is defined to be the number that satisfies

$$\boxed{\sum_i M_i s_i^{-D_s} = 1 \qquad (9.7\text{-}14)}$$

We can see from Eq. (9.7-14) that we must have $D_s < d$. This dimension is related to the Hausdorff dimension (see later), which uses boxes of varying sizes and asks for the greatest lower bound D that gives a sum greater than 0. Here, however, we are assuming that the fractal satisfies similarity scaling; therefore, we have a definite relationship between the scale factors and the number of boxes of a particular size. This assumption, then, leads to the implicit expression Eq. (9.7-14) for the similarity dimension D_s.

> **Exercise 9.7-8.** Show that if there is only one scale factor, say s_1, and that if we require M_1 boxes to cover the fractal, then the similarity dimension as given by Eq. (9.7-14) is the same as the box-counting dimension D_b.

Before applying this result to some attractors, let us see how this definition of D_s can be developed by using arguments analogous to those used for the box-counting dimension. Suppose that the entire fractal object fits inside a box of volume R^d, where, again, d is the dimensionality of the full space. Suppose we cover the object with boxes of size r. Each box has a volume r^d. Then the number of boxes n_b of size r required to cover the attractor at a particular stage of construction depends on the scaling factors s_i, the overall size of the attractor R and the box size r; therefore, we write

$$n_b(R,r) = \sum_i M_i n_b\left(\frac{R}{s_i},r\right) \tag{9.7-15}$$

Now we invoke two important scaling relations. First, because of the assumed self-similarity of the fractal object, we must have

$$n_b\left(\frac{R}{s_i},r\right) = n_b(R,s_i r) \tag{9.7-16}$$

Second, for self-similar fractals, the box-counting dimension satisfies

$$n_b(R,r) = n_b(R)r^{-D} \tag{9.7-17}$$

The primary assertion is that it is meaningful to use the same D, which we then call D_s, for all the scale factors. Now we use Eqs. (9.7-16) and (9.7-17) in Eq. (9.7-15) to obtain

$$n_b(R)r^{-D_s} = \sum_i M_i n_b(R)s_i^{-D_s} r^{-D_s} \tag{9.7-18}$$

Cancellation of $n_b(R)r^{-D_s}$ from both sides of the previous equation yields Eq. (9.7-14).

Let us now apply Eq. (9.7-14) to the logistic map attractor at the period-doubling accumulation point. First, we need to find the appropriate scaling factors. For the logistic map near the period-doubling accumulation point, we have argued that there are two scaling factors. We can find these factors using the universal function $g(y)$ introduced in Chapter 5 and Appendix F. Let us follow several successive iterations of $g(y)$ starting with $y = 0$. [Recall that the maximum of the function $g(y)$ occurs at $y = 0$.] Then, as we saw in Section 5.5, these trajectory points mark the boundaries of the regions within which all the attractor points reside. For the universal function $g(y)$, we have

$$\begin{aligned} y_1 &= g(0) = 1 \\ y_2 &= g(g(0)) = -1/\alpha \\ y_3 &= g(-1/\alpha) \\ y_4 &= 1/\alpha^2 \end{aligned} \tag{9.7-19}$$

Thus, the lengths of the two intervals in which the attractor resides are given by $y_1 - y_3 = 1 - g(-1/\alpha)$ and $y_4 - y_2 = 1/\alpha^2 + 1/\alpha$. The overall size of the attractor is given by $y_1 - y_2 = 1 + 1/\alpha$. Hence, the two scaling factors are

$$s_1 = \frac{1 + 1/\alpha}{1/\alpha^2 + 1/\alpha} = \alpha = 2.503\ldots$$

$$s_2 = \frac{1 + 1/\alpha}{1 - g(-1/\alpha)} = 5.805\ldots$$

(9.7-20)

We then determine the dimension D by numerically finding the value that satisfies

$$(2.503\ldots)^{-D_s} + (5.805\ldots)^{-D_s} = 1 \qquad (9.7\text{-}21)$$

which yields $D_s = 0.537$. We should point out that the Feigenbaum attractor does not exhibit exactly the rescaling properties required by our calculation of D_s. However, the deviations from rescaling are small, and the method gives a value in close agreement with that calculated by much more complex methods (GRA81).

From this argument, we see that the two scaling factors are α and a number that is almost, but not quite, equal to α^2. Note that our ad hoc use of an average length in the previous calculation of D for the logistic map gives a result for the dimension numerically close to the more rigorous value, but we must admit that the agreement is accidental. There is no theoretical justification for the use of an average length in the dimension calculation.

> **Exercise 9.7-9.** Using the polynomial form of the universal function given in Eq. (F.1-21) of Appendix F, carry out the first four iterations of $g(y)$ starting with $y_0 = 0$. Then verify the numerical values of the scaling factors used in the text. Finally, check the numerical value of D_s determined from Eq. (9.7-21).

We can also use Eq. (9.7-14) to get an approximate value of the similarity dimension associated with the sine-circle map. As described in Chapter 6, at the critical value of the nonlinearity parameter $K = 1$, the frequency-locking intervals supposedly fill (in a rough sense) the entire Ω axis. However, quasi-periodicity still occurs; therefore, we can ask for the value of the similarity dimension of the quasi-periodic intervals. This dimension can be computed to surprising accuracy by using Eq. (9.7-14) with only two scaling numbers based on the sizes of the gaps between related frequency-locking tongues. In particular, we use the gaps between the $p{:}q$, $p'{:}q'$, and $(p+p'){:}(q+q')$ tongues as shown in Fig. 6.14. For example, we can use the distance between the 0:1 tongue at $1/(2\pi)$ and the left-edge of the 1:2 tongue at 0.464 to set the overall size scale. We then use the edges of the 1:3 tongue, which extends from $\Omega = 0.337$ to 0.367 (with an uncertainty of $\pm$ 0.0005), to find $s_1 = 1.71$ and $s_2 = 3.14$. With these results Eq. (9.7-14) gives $D_s = 0.857 \pm 0.001$. This result, even though it uses only two tongues to compute the dimension, is in fairly good agreement with the value 0.87 determined by Jensen, Bak, and Bohr (JBB83).

> **Exercise 9.7-10.** Use the numerical values cited in the previous paragraph to check the values of D_s for the sine-circle map.

The dimension D_s of the frequency-locking intervals has been found from experimental data for systems that can be modeled by the sine-circle map. If a tunnel diode circuit is driven sinusoidally, competition occurs between the driving frequency and the natural oscillation frequency of the tunnel diode. Frequency-locking and quasi-periodic behavior can occur. The similarity dimension has been determined (TES85) for such a system with a value in good agreement with the value stated above. However, an experiment with a driven relaxation oscillator (CUL87) set at the apparent critical value of its control parameter gave a value for the similarity dimension of about 0.795 ± 0.005, not in agreement with the value 0.87. This disagreement was explained by the use of an "integrate-and-fire" model (ACL88) for the relaxation oscillator, which showed that the dimension value found in the experiment was not expected to be universal. The important lesson here is that there are different classes of quasi-periodic systems. Not all quasi-periodic systems belong to the sine-circle map class.

Fractal Dimensions and Basin Boundaries

Fractal dimensions show up in yet another aspect of nonlinear dynamics. As we have mentioned earlier, many nonlinear systems show a *sensitivity to initial conditions* in the sense that trajectories that are initially nearby in state space may evolve, for dissipative systems, to very different *attractors*. In some cases these attractors correspond to fixed points or limit cycles. In other cases the attractors may be chaotic attractors. As we have learned, the set of initial conditions that gives rise to trajectories ending on a particular attractor constitutes the basin of attraction for that attractor. For many nonlinear systems the boundaries of these basins of attraction are rather complex geometric objects, best characterized with fractal dimensions. It is the highly convoluted nature of these basin boundaries that leads to the sensitivity to initial conditions: A slight change in initial conditions could shift the trajectory in an essentially unpredictable way from one basin of attraction to another.

We will give just two brief examples of fractal basin boundaries. The first example is the now familiar driven, damped pendulum. For large enough driving torques, the two predominant modes of motion of the pendulum are rotations that are either clockwise or counterclockwise (on the average). We can ask a very simple question: Which initial conditions lead to clockwise rotations and which to counterclockwise? The answer to this simple question turns out to be very complicated because the boundary between the two basins of attraction is a fractal object (GWW85, GWW86).

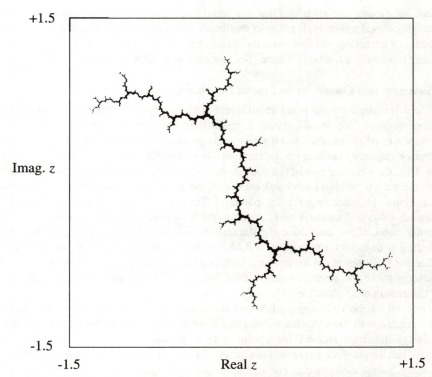

Fig. 9.8. The Julia set for the value of $C = 0 + i\, 1.0$. The central "dendritic" points correspond to initial conditions that give rise to trajectories that remain bounded. All other points correspond to orbits that escape to infinity.

The second example of a fractal basin boundary is the famous *Julia set*, named after the French mathematician Gaston Julia. The Julia set constitutes the boundary between initial conditions of a two-dimensional iterated map function leading to trajectories that escape to infinity and the set of conditions leading to trajectories that remain bounded within some finite state space region. The usual Julia set is based on the iterated map function of a complex variable z:

$$z_{n+1} = z_n^2 + C \qquad\qquad (9.7\text{-}22)$$

where C is a complex control parameter. Figure 9.8 shows the Julia set for the case $C = 0 + i\, 1.0$.

The related and equally famous *Mandlebrot set* gives a similar boundary, but in parameter space: The values of C (a complex number) that give rise to bounded orbits starting from $z = 0$. For more information on Julia and Mandlebrot sets, the reader is referred to the list of books and articles at the end of this chapter.

In addition to basin boundaries being fractal, we can have situations in which the basins associated with different attractors are intermingled. These so-called *riddled basins* are such that for (at least) one of the basins of attraction, any neighborhood about each point in that basin contains points belonging to another basin of attraction. In that case, the basin of attraction (in contrast to the attractor itself) can have a fractal structure. See, for example, (SOO93a) and (LAW94).

Summary and Comments on Fractal Dimensions

We will make a crucial point about terminology to conclude this section on fractal dimensions. We should make a distinction between the apparent geometric properties of an attractor and the geometric properties of a finite data set generated from a trajectory running on the attractor. For example, for the logistic map with $A = 3.58$, the attractor consists of four chaotic bands occupying four distinct intervals of the x axis. Within each band, almost all the points are attractor points (with the exception of unstable periodic points). Thus, we are tempted to say that the attractor has a dimension of 1, which is the dimension of a line interval. On the other hand, if we determine the dimension of a set of data points generated by letting a trajectory run with $A = 3.58$, we find a dimension value (as shown in Section 9.8) that is not an integer. Unless we specifically state otherwise, we will always mean the geometry determined from the data set when talking about the "dimension of an attractor."

All of the definitions of fractal dimension discussed previously should be contrasted with the so-called *topological dimension*. The topological dimension is always an integer and is 0 for a point or set of points, 1 for a line or curve, and so on. The topological dimension is always less than or equal to the box-counting (or other) fractal dimension. For example, the topological dimension of the Koch curve is 1 because it is a curve (or a group of line segments). On the other hand, we have seen that the box-counting dimension for the Koch curve is greater than 1. The box-counting dimension reflects the folded and twisted nature of the Koch curve that makes it, in a sense, more than a curve but less than a surface area.

As an aside, we should point out that another dimension, the Hausdorff dimension D_H, is defined in a fashion quite similar to that for the similarity dimension. The Hausdorff dimension, however, allows us to use variable size boxes to cover the set without the restriction on scaling imposed for the similarity dimension. The length size R in the dimension calculation is the largest size box used to cover the set. We raise the lengths to the power D and sum over all the lengths to find the "measure" of the set. D_H is then defined to be the greatest lower bound on the power D that yields a measure of 0 as $R \to 0$. For details, see FOY83 and ESN90.

9.8 Correlation Dimension and a Computational Case History

Although the box-counting procedure developed in the previous section is conceptually straight-forward, its application to actual data, particularly for higher-dimensional state spaces, is fraught with difficulties. The number of computations required for the box-counting procedure increases exponentially with the state space dimension. Moreover, the box-counting scheme requires us to partition the state space with boxes and then to locate the trajectory points within the boxes, a time-consuming process. To provide a computationally simpler dimension for an attractor, Grassberger and Procaccia (GRP83a) introduced a dimension based on the behavior of a so-called *correlation sum* (or *correlation integral*). This dimension is called the *correlation dimension* D_c and has been widely used to characterize chaotic attractors. It has a computational advantage because it uses the trajectory points directly and does not require a separate partitioning of the state space.

In this section we shall introduce the correlation dimension for one-dimensional data. (In Chapter 10, we shall show how to extend the definition to higher-dimensional systems.) We shall then provide a computational case history by calculating the correlation dimension for various data sets. As we have tried to emphasize throughout this chapter, it is important to have some intuition for the reliability of the numerical results for any quantitative measure of chaotic behavior. We need an estimate of reliability (or uncertainty) both to compare our results with those from other experiments or computations and to monitor the behavior of the system. We have chosen to look at the correlation dimension as a detailed example because it is relatively straightforward to compute and the difficulties it presents are similar to those encountered in the computation of any of the other quantifiers.

To define the correlation dimension, we first let a trajectory (on an attractor) evolve for a long time, and we collect as data the values of N trajectory points. Then for each point i on the trajectory, we ask for the number of trajectory points lying within the distance R of the point i, excluding the point i itself. Call this number $N_i(R)$. Next, we define $p_i(R)$ to be the relative number of points within the distance R of the ith point: $p_i(R) = N_i/(N-1)$. (We divide by $N-1$ because there are at most $N-1$ other points in the neighborhood besides the point i.) Finally, we compute the correlation sum $C(R)$:

$$C(R) = \frac{1}{N} \sum_{i=1}^{N} p_i(R) \qquad (9.8\text{-}1)$$

Note that $C(R)$ is defined such that $C(R) = 1$ if all the data points fall within the distance R of each other. If R is smaller than the smallest distance between trajectory points, then $p_i = 0$ for all i, and $C(R) = 0$.

The smallest nonzero value for $C(R)$ would be $2/[N(N-1)]$ if only two points are within the distance R of each other. If some of the data points happen to have the same numerical value (due to computer round-off, for example), then the smallest value of $C(R)$ is $N^*(N^*-1)/[N(N-1)]$, where N^* is the number of trajectory points that have that value (assuming all other trajectory points are distinct).

As an aside, we point out that if one were to include the ith point itself in counting points within the distance R of point i, then the correlation sum

$C(R)$ would be equal to $1/N$ for sufficiently small R. As we shall see, there is a numerical advantage to excluding the ith point.

The relative number p_i itself can be written in more formal terms by introducing the **Heaviside step function** Θ:

$$\Theta(x) = 0 \quad \text{if} \quad x < 0$$
$$\Theta(x) = 1 \quad \text{if} \quad x \geq 0 \tag{9.8-2}$$

Using this function, we can write

$$p_i(R) = \frac{1}{N-1} \sum_{j=1, j\neq i}^{N} \Theta(R - |x_i - x_j|) \tag{9.8-3}$$

In Eq. (9.8-3), the Heaviside function contributes 1 to the sum for each x_j within the distance R of the point x_i (excluding $j = i$); otherwise, it contributes 0. In terms of the Heaviside function the correlation sum can be written

$$C(R) = \frac{1}{N(N-1)} \sum_{i=1}^{N} \sum_{j=1, j\neq i}^{N} \Theta(R - |x_i - x_j|) \tag{9.8-4}$$

Often the limit $N \to \infty$ is added to assure that we characterize the entire attractor. The correlation dimension D_c is then defined to be the number that satisfies

$$C(R) = \lim_{R \to 0} kR^{D_c} \tag{9.8-5}$$

or after taking logarithms

$$D_c = \lim_{R \to 0} \frac{\log C(R)}{\log R} \tag{9.8-6}$$

For convenience of interpretation, we shall use logarithms to the base 10, though some other workers prefer to use base 2 logarithms.

If the terminology were being defined from scratch, we would prefer to call D_c the **correlation scaling index** (rather than the correlation dimension) because it is the number that tells us how $C(R)$ scales with R. However, the name correlation dimension seems to be firmly entrenched in the literature. The relationship between D_c and other "dimensions" will be discussed in Chapter 10.

There is one obvious difficulty in using Eq. (9.8-6) to determine the correlation dimension: it is not possible to take the limit $R \to 0$. Any real data set consists of a finite number of points. Hence, there is some minimum distance between trajectory points. When R is less than that minimum distance, the correlation sum is equal to 0 and no longer scales with R as Eq. (9.8-5) requires.

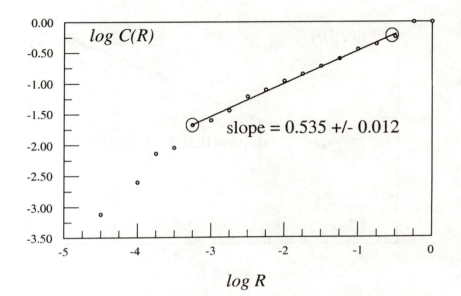

Fig. 9.9. A plot of log $C(R)$ as a function of log R for the logistic map trajectories with A = 3.56995. One hundred data points were used in the analysis. For large values of R the finite size of the attractor makes $C(R)$ "saturate" at 1. For small values of R, the finite number of data points causes ln $C(R)$ to become very negative, which occurs at an R value given roughly by the size of the attractor (given by the range of x values) divided by the number of data points. The intermediate region, in which the curve is approximately a straight line (bounded in the figure by the circled points), is called the *scaling region*. The slope in the scaling region gives the correlation dimension D_c.

What is done in practice is to compute $C(R)$ for some range of R values and then to plot log $C(R)$ as a function of log R as shown in Fig. 9.9.

Let us examine the features of Fig. 9.9, which shows the results of computing $C(R)$ for trajectory data generated from the logistic map function with parameter A = 3.56995, close to the period-doubling accumulation point value. For large values, R is larger than the size of the attractor and all points are within R of each other. For the logistic map data used in that figure, the data values lie between 0.90 and 0.33, approximately. Thus, once R is larger than 0.57, we should have $C(R) = 1$. We see this occurring in the upper right-hand portion of the graph. For R larger than this value, $C(R)$ is equal to 1, independent of R.

At the other extreme, for small values of R, the correlation sum is equal to 0, independent of R. (Those values are not plotted in Fig. 9.9 for obvious reasons.) For Fig. 9.9, only 100 data points were used. From the argument presented earlier, we would expect the smallest nonzero value of $C(R)$ to be about 2/104 (with a corresponding log value of about –3.7). If these data points were spread uniformly over the region of the attractor, then the average spacing would be about 0.005. Thus, we would expect that log $C(R) = -3.7$ when R is about equal to 0.005 (with a

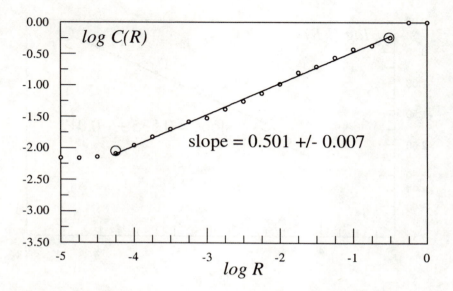

Fig. 9.10. The same calculation as shown in Fig. 9.9 but with 1,000 data points used for calculating the correlation sum. The probabilities for each value of R were averaged over 40 randomly chosen points. The effects due to the finite number of data points are pushed to smaller values of R. The slope is much closer to that found by GRP83.

corresponding log value of -2.3). We see that the graph does reach this level, but for smaller values of R, because the points are clustered on the attractor, not spread uniformly. Note that there is a steeper fall-off for small values of R due to the sparseness of the data.

We see that only in some intermediate R region does $C(R)$ obey the scaling law expressed in Eq. (9.8-5). Hence, this region is called the **scaling region**, and the slope in that region determines the correlation dimension from Eq. (9.8-6), without the limit $R \rightarrow 0$. For the data shown in Fig. 9.9, we find $D_c = 0.535 \pm 0.012$, which is close to, but not quite in agreement with the value found by other authors (GRP83a). The uncertainty was determined from a least-squares fit of a straight line to the scaling region data of Fig. 9.9. Note that the quoted uncertainty is a *statistical* uncertainty based on fitting a straight line in the scaling region. As we shall see, there are also many possible *systematic effects*, which can cause the computed dimension to differ from the "true" value.

Unfortunately, for most sets of experimental data, the determination of the scaling region is rather subjective since in many cases the log $C(R)$ versus log R curve is not a straight line over a very large range of R. You can estimate an uncertainty to be associated with D_c by varying the range of R that you call the scaling region. In other situations, the log $C(R)$ versus log R graph is not straight over any significant range of R. In those cases, the meaning of D_c is at best dubious.

Figure 9.10 shows a similar graph, but 1,000 data points have now been used in the analysis. We see that the scaling region now extends over a larger range of R values. The resulting value for the correlation dimension is now in excellent agreement with the results of GRP83a.

Obviously, for ideal (noise-free) data, the scaling region can be extended to smaller values of R by using a larger number of data points as shown in Fig. 9.10. However, using a larger number of data points considerably increases the computational time. In fact, if Eq. (9.8-4) is used to compute the correlation sum, then the number of comparisons increases as N^2. In practice, one can use a few thousand trajectory points for N, but compute $p_i(R)$ for only a sample, say, 50, of those points. If the points i are chosen "at random" from the larger sample, then the resulting correlation sum seems to be nearly equal to the full correlation sum. This latter result is called the ***pointwise dimension*** (FOY83 and [Moon, 1992]).

We shall now consider several factors that affect the computation of the correlation dimension.

Finite Number of Data Points

The difference between the two calculations in Figs. 9.9 and 9.10 lies solely in the number of data points used in the analysis. The number of data points sets an important **upper limit** on the value of D_c computed from that data set. It is easy to understand why (RUE90), and this lesson provides our first cautionary tale. (We suggest that this limit be called the ***Ruelle limit***.) To see how an upper limit arises, let us imagine calculating the slope of the log $C(R)$ versus log R plot by taking "rise over run":

$$D_c = \frac{\log C(R_2) - \log C(R_1)}{\log R_2 - \log R_1} \qquad (9.8-7)$$

As we have seen, the largest $C(R)$ is 1. The smallest non-0 value of $C(R)$ is $2/N(N-1)$, where N is the number of data points. Let us suppose we have data that stretch over q decades of R values so that the denominator of Eq. (9.8-7) is just q. Thus, the largest value that we could obtain for D_c for this data set is

$$D_c \le \frac{2\log N}{q} \qquad (9.8-8)$$

where we have assumed that $N \gg 1$.

The crucial point here is that the calculated value of D_c has an upper bound limited by the number of data points. If we have too few data points and a potentially large value of D_c, then our value calculated from the data set may be too small. For the case of Fig. 9.9, we have $2 \log N = 4$ and $q \approx 4$, and our upper limit is 1. We are fortunate in this case that the actual correlation dimension is near 0.5; so, the upper limit is not very stringent. The "bottom line," however, is that we need to be very skeptical of correlation dimension values calculated from small data

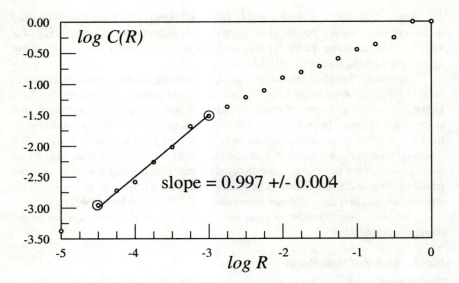

Fig. 9.11. A plot of log $C(R)$ versus log R for logistic map data with artificial noise added. A = 3.56995. The slope is distinctly larger in the noise dominated region below log $R = -3$. The noise was chosen to be a random number between +0.001 and –0.001. One thousand data points were used in the analysis.

sets. This upper limit will become more critical in Chapter 10, where we turn out attention to higher-dimensional systems.

If we use the "pointwise" method of selective averaging to estimate the correlation dimension, then the upper limit on the correlation dimension is given approximately by

$$D_c \leq \frac{\log N + \log N_p}{q} \qquad (9.8\text{-}9)$$

where N_p is the number of points used in computing the average of $p(R)$. We see that the upper limit is smaller in this case because $N_p < N$.

Effects of Noise

The correlation dimension calculation can also be affected by the presence of noise—either "real" noise in experimental data or round-off noise in a numerical computation. If the average "size" of the noise is R_n, then we expect the noise to dominate the structure of the attractor for $R < R_n$. Since noise is supposedly random, the noise-dominated data will tend to be spread out uniformly in the state space, and we would expect to find $D_c = d$, the state space dimension for small values of R. Fig. 9.11 shows a log $C(R)$ versus log R plot for data generated from the logistic map with artificial noise added (from the computer's random number generator) with $R_n = 0.001$. The change in the slope of the curve near log $R = -3$, indicates the presence of noise. The slope in the noise region is 0.997 ± 0.040 in agreement with the value of 1 expected for a one-dimensional system such as the logistic map.

Since the determination of $C(R)$ is affected by noise and choice of scaling region, it is important to test any computational scheme with data sets whose properties are known. Note that noise tends to make the slope of log $C(R)$ versus log R larger for small values of R while the finite number of data points tends to make the slope smaller. Some fortuitous cancellation might occur and produce an artificial scaling region. For test data, you might use the logistic map at $A = 3.56995$ with various amounts of noise added, as we have done here. You might also use a periodic signal with noise added. By seeing how your computational scheme is affected by noise and by investigating its behavior as a function of the number of trajectory points used in calculating $C(R)$, you can put better confidence limits on the correlation dimension (or any other dimension) computed from actual data.

Finite Digitization Precision

Most experimental data sets and essentially all numerical calculations presently result in numerical data stored in a digital computer. Since the computer represents the numbers with only a finite number of bits, we ought to worry about the possible effects of this digitization on computed quantities such as the correlation dimension.

As a concrete example let us consider the data from the semiconductor diode circuit introduced in Chapter 1. These data were logged with a so-called 12-bit analog-to-digital converter. This device samples a signal in the circuit and converts the result into an integer number lying between 0 and 4095 ($4096 = 2^{12}$). Obviously any two circuit readings whose difference is less than 1 in these units will result in the same data number, even though the actual voltages may be different. What effect does this digitization round-off have on the computation of the correlation dimension?

If two circuit readings lie within 1 of each other, then the computer thinks that they are identical values. In essence, having a trajectory separation value R less than 1 gives us no new information about the scaling of the correlation sum for small R. Hence, the main effects of the digitization are to put a lower limit on the range of R that we can use in determining scaling and to set a lower limit on the numerical value of the correlation sum. Alternatively (THE90), if we plot log $C(R)$ as a function of log $(R + s/2)$, where s is the discretization step size, then most of the effects of the step size are removed.

Periodic Data

What is the correlation dimension of a time-series of periodic data points? To illustrate this case, we have plotted in Fig. 9.12 log $C(R)$ versus log R for the logistic map data for $A = 3.5$, a value for which the behavior is period-4. As you can

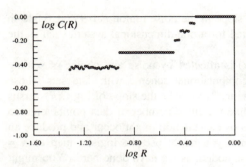

Fig. 9.12. A plot of correlation sum data for a periodic signal from the logistic map with $A = 3.5$ for which the behavior is period-4. The plot shows six distinct steps due to the distinct spacings between the period-4 data values.

see, the plot shows a series of steps. The explanation is fairly straightforward. If R is smaller than the smallest spacing between the four periodic points, then log $C(R)$ = log $(1/4)$ = –0.602... since one-fourth of the data points lie on top of one another. As R increases, it will eventually be large enough so that two of the four periodic points lie within R of each other. (For the logistic map data, the four periodic points are not equally spaced.) For that value of R, we should have log $C(R)$ = log $(1/2)$. As R increases further, we will have further step-wise increases in $C(R)$ until R is large enough to encompass all four numerical values. We will then have log $C(R)$ = 0. For the logistic map with $A = 3.5$, there are six distinct interpoint distances resulting in the six steps seen in Fig. 9.12.

Exercise 9.8-1. Use the numerical data given in Section 1.4 (or compute your own data for $A = 3.5$) for the period-4 behavior of the logistic map function to verify that the steps in Fig. 9.12 occur at the approximately correct values of R.

Data with Gaps

In many situations the string of data values used in the analysis is broken up into distinct ranges. For example, if the data were trajectory data from the logistic map with A near 3.6 (see Fig. 9.3), then the data fall into two ranges corresponding to the two chaotic bands. Hence, there are gaps in the data values. What effects do these gaps have on the determination of D_c? For small values of R, we will be sampling data points all of which lie within the same band. If the two bands each have the same correlation dimension, then the correlation sum should scale as usual with R. (If the correlation dimension is different for the two bands, then the standard correlation dimension procedure yields an average correlation dimension.) Once R is large enough to encompass both bands, there may be a sudden change in the slope of log $C(R)$ versus log R. The details of the changes depend on the size of the gaps compared to the size of the bands. See Fig. 9.13.

Exercise 9.8-2. Compute the correlation sum for the logistic map with A = 3.6 as a function of R as shown in Fig. 9.13. Can you observe effects due to the two-band structure of the trajectories?

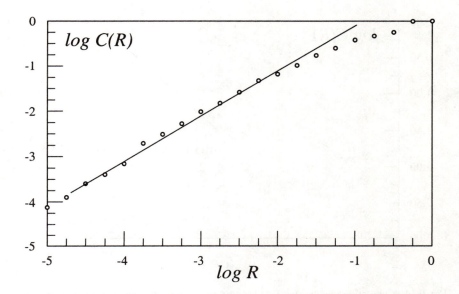

Fig. 9.13. A plot of the correlation sum for logistic map trajectory data with $A = 3.6$ in the region of two chaotic bands. For large values of R the correlation sum plot changes slope when R is big enough to encompass the two bands. One thousand data points were used in the analysis.

Random Data

What is the correlation dimension for purely "random" data? In Fig. 9.14, we have plotted log $C(R)$ versus log R for a data set generated by the random number generator of the computer language QuickBASIC. The data values ranged in size from −0.1 to +0.1 with a uniform distribution between those values.

The plot looks qualitatively like that for the logistic map data. We recognize a saturation effect at the large R end when all of the data points are within R of each other. At the small R end, $C(R)$ goes to 0 when R is less than the range of the data divided by the number of data points. Between these extremes, we see a region that appears to be a scaling region. For purely random data, we would expect the correlation dimension to be equal to the dimension of the state space, which is 1 in the case at hand.

A least-squares fit to the data in the scaling region of Fig. 9.14 gives a correlation dimension value of 0.914 ± 0.033, which is less than 1 within the statistical uncertainty of the fit. Larger data sets give dimension values closer to 1 (see Fig. 9.15, where the same analysis was done with 1,000 trajectory points). Thus, it seems clear that the correlation dimension does not discriminate between random data sets and chaotic data sets whose correlation dimension is close to the dimensionality of the state space unless a very large data set is used. In the example given here, the Ruelle limit for the analysis used in Fig. 9.14 is just less

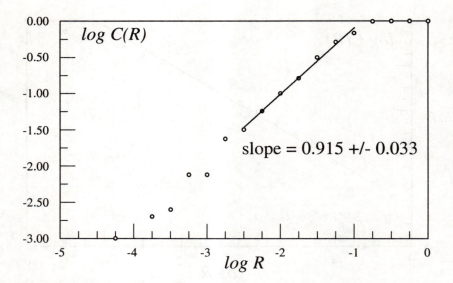

Fig. 9.14. A plot of correlation sum data for a set of random data values lying between –0.1 and 0.1. One hundred trajectory points were used in the data sample. Twenty points were chosen at random to calculate the correlation sum. The plot is qualitatively like that for the logistic map data, but the correlation dimension, given by the slope of the plot in the scaling region, is somewhat less than 1, the dimensionality of the state space.

than 1, and our computed value for the correlation dimension is limited from above by that value. For the data shown in Fig. 9.15, the Ruelle limit is just larger than 1, and the computed correlation dimension agrees with the expected value of 1.

Continuous Data

As a final set of examples, let us apply the correlation dimension calculation to data generated from some continuous (nonchaotic) functions. In Fig. 9.16, we have plotted the correlation sum as a function of R for a data set generated by recording values of $0.5\sin(2\pi n/N)+0.5$, where N is the number of data points and $n = 1$, 2,...N).

There is an apparent scaling region with a slope of 0.876 ± 0.021, significantly less than the value of 1 expected for a purely continuous signal. However, we must remember that we are using a finite data set representation of that continuous signal. For the sine function, the values of the independent variable are uniformly distributed, but the resulting computed values of the function, of course, are not uniformly distributed; they tend to clump near 0 and 1 for the function used. If more data points are used, then the slope of the scaling region becomes closer to 1.

If we use a linear function $y = kx$ with k equal to the reciprocal of the number of data points, we find that the slope of the correlation sum data in the scaling region is closer to 1 as seen in Fig. 9.17.

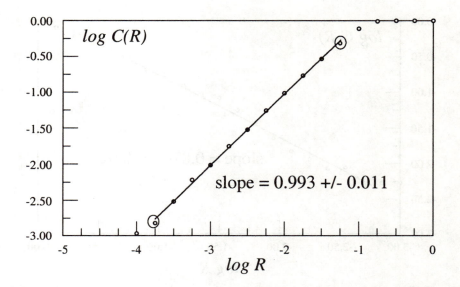

Fig. 9.15. The correlation sum was calculated with the same random noise signal used in Fig. 9.14, but here 1,000 data points were used in the analysis. Forty randomly chosen points were used to calculate the correlation sum. The slope is now, within the computed uncertainty, equal to the expected value of 1.

Effects of Filtering

As a final word of warning, we need to point out that many experimental signals are recorded with some kind of filtering process. For example, many electrical measurements are made with devices called "low-pass filters" to reduce the amount of electrical noise in the signal. These filters are electrical devices that attenuate or enhance selected ranges of frequencies, acting much like the "equalizer" controls on an audio system. If you have what you believe is a slowly varying signal from the system under study, for example, then any rapidly changing signal can be attributed to noise. Filters are used to remove (or at least decrease) the contribution of the rapidly changing noise to the recorded signal. What is the effect of the filtering process on the numerical results for fractal dimensions or scaling exponents? Does filtering artificially raise or lower the calculated dimension value? This problem has only recently been studied; therefore, a definitive statement is impossible to make. Some types of filtering seem to raise the computed dimension value; others tend to lower the value. The general trend seems to be that most filtering processes lower the numerical value of the fractal dimension. This result is consistent with our previous observation that the presence of noise increases the dimension value. If filtering reduces the amount of noise in the signal, it should therefore lower the effective dimension value.

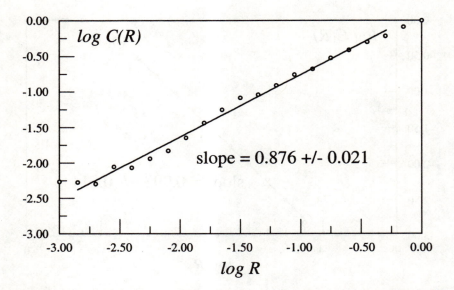

Fig. 9.16. Correlation sum data for a set of points generated from a continuous function, in this case, the sine function $0.5 \sin(2\pi x) + 0.5$ for x from 1/200 to 1. Only 200 data points were used in the analysis. Forty randomly chosen points were used to compute the correlation sum. The computed slope is significantly less than the value of 1 expected for continuous data.

One model of filtering (BBD88) adds another degree of freedom to the dynamical system to account for the filter. For a low-pass filter, as long as the range of frequencies passed is large compared to the system's inherent frequencies, as given by the Lyapunov exponents, for example, then the filtered signal is a faithful reproduction of the original signal and there is no effect on the dimension calculation. However, if the range of frequencies passed is made smaller, then the dynamics of the filter itself become part of the dynamical system and the presence of an extra degree of freedom tends to raise the effective dimension value. In rough terms, the presence of the filter slows the signal response; therefore, the effective signal does not collapse as quickly to the attractor and thus occupies a higher dimensional region in state space (ROC94).

To complicate matters further, it is easy to demonstrate that low-pass filtering can make a "signal" consisting purely of a sequence of random numbers look like a signal from a low-dimensional dynamical system. One clearly has to proceed cautiously here. Again, the best advice is to vary, if possible, the parameters of the filtering process to investigate the effect on the computation of fractal dimension or scaling index values.

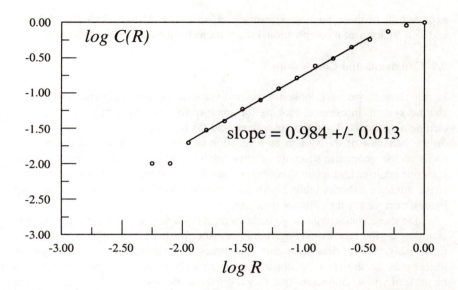

Fig. 9.17. The correlation sum data for a linear function. Just 200 data points were used in the analysis. In this case, the slope in the scaling region is close to the expected value of 1.

Summary

What have we learned from this cautionary tale? The general lesson is that we must be careful in interpreting the numerical value of the correlation dimension (or any other quantifier) when it is computed from actual data. There are important effects due both to the limited number of data points and to the finite resolution and filtering processes used to acquire real data. Furthermore, the presence of noise tends to increase the calculated size of the correlation dimension. Altogether, these processes may influence significantly the actual value of the dimension. Before drawing important conclusions from the numerical value of the correlation dimension, you should carefully investigate how the value depends on the number of data points used, the amount of noise in the data, the arithmetic precision of the data, the data acquisition filtering, and the averaging process used in evaluating the correlation sum.

You should also test the significance of your results by redoing the calculations with surrogate data (TGL91). Surrogate data is data generated by taking your recorded signal and reprocessing it so that it has the same Fourier power spectrum but has lost all of its deterministic character. This method should help you decide whether a model of deterministic chaos or a model of purely stochastic processes best represents your data.

As a "consumer" of results of other scientists' determination of dimensions and exponents, you should be skeptical of results from studies that have not explored all of these possible systematic effects. See ASV88, THE90, CGM90, and TGL91 for useful computational suggestions for the correlation dimension.

9.9 Comments and Conclusions

In this chapter, we have looked at several ways of quantitatively characterizing a chaotic system, in essence, looking for some quantitative way of specifying how chaotic a system is. Lyapunov exponents and the Kolmogorov–Sinai entropy focus on the behavior of the system as a function of time. Various fractal dimensions focus on the geometric structure of attractors in state space. If the system has at least one positive Lyapunov exponent, we say that the system's behavior is *chaotic*. If the attractor exhibits scaling with (in general) a noninteger dimension (scaling index), then we say the attractor is *strange*.

Do these characteristics go hand in hand? That is, if a system's behavior is chaotic, is the corresponding attractor (for a dissipative system) strange? Conversely, if we determine that an attractor is strange, is the behavior of trajectories on the attractor chaotic? Your newly emerging nonlinear intuition might lead you to conjecture that they are linked. For example, the stretching and folding of trajectories, characteristic of chaotic trajectories, would lead you to expect an attractor with strange geometry. Conversely, the complicated geometry of a strange attractor would seem to force the dynamics to be chaotic. However, there are exceptions to this link. For example, at the period-doubling accumulation point of the logistic map function (or any other unimodal map function with a quadratic maximum), the attractor is strange (with $D_c = 0.501 \pm 0.007$ as we have seen in Section 9.8), but the behavior is not chaotic: At the period-doubling accumulation point, the Lyapunov exponent is equal to 0.

In the case of strange attractors that are not chaotic, you might argue that this behavior occurs only for one parameter value (i.e., the period-doubling accumulation point value). Are there attractors that are strange but not chaotic over some range of parameter values? Grebogi, Ott, Pelikan, and Yorke (GOP86) have shown that nonlinear oscillators that are driven externally at two incommensurate frequencies can have attractors that are strange but not chaotic over some finite range of parameter values. They also argue that in general, continuous time systems (that is, systems modeled by differential equations rather than iterated maps) that are not driven with two incommensurate frequencies should not have strange attractors that are not chaotic. Unraveling all the "nots," we would say that for most continuous time systems the attractors are chaotic if they are strange.

We conclude this chapter by pointing out that no single quantifier has emerged as the "best" way to characterize a nonlinear system. All of the quantifiers proposed to date require considerable computational effort to extract their numerical values from the data. All of them require some personal attention to avoid artifacts in their evaluation. A slight preference might be given to Lyapunov exponents, since in a sense, a positive Lyapunov exponent captures the essence of chaotic dynamics: the rapid divergence of nearby trajectories. As we shall see in the next chapter, the Lyapunov exponents, for a multidimensional state space, can be linked to the fractal dimension. However, for higher-dimensional state spaces, the computation of the Lyapunov exponents becomes rather difficult, and most scientists have used some kind of fractal dimension (most commonly the correlation dimension and its generalizations to be discussed in Chapter 10).

9.10 Further Reading

General References on Quantifying Chaos

N. Gershenfeld, "An Experimentalist's Introduction to the Observation of Dynamical Systems." in [Hao, 1988], pp. 310–84. An excellent introduction to many techniques for quantifying chaos with an eye on experimental data.

J.-P. Eckmann and D. Ruelle, "Ergodic Theory of Chaos and Strange Attractors," *Rev. Mod Phys.* **57**, 617–56 (1985). A review article surveying many methods of quantifying chaos at a more sophisticated level.

H. D. I. Abarbanel, R. Brown, J. J. Sidorowich, and L. Sh. Tsimring, "The analysis of observed chaotic data in physical systems," *Rev. Mod. Phys.* **65**, 1331–92 (1993). An excellent and thorough survey of many methods of analyzing data from nonlinear systems.

[Ott, Sauer, and Yorke, 1994]. Contains reprints of 41 papers on the analysis of chaotic data in a variety of fields.

Garnett P. Williams, *Chaos Theory Tamed* (National Academy Press, Washington, DC, 1997). Has nice sections discussing the details of dimension calculations, finding Lyapunov exponents, and so on.

R. Hegger, H. Kantz, and T. Schreiber, "Practical implementation of nonlinear time series methods: The TISEAN package," *Chaos* **9**, 413–35 (1999). Describes a software package TISEAN that implements dimension calculations, estimations of Lyapunov exponents, and so on.

Chaotic Transients

I. M. Jánosi and T. Tél, "Time-Series Analysis of Transient Chaos," *Phys. Rev. E* **49**, 2756–63 (1994).

Lyapunov Exponents

E. N. Lorenz, "The Local Structure of a Chaotic Attractor in Four Dimensions," *Physica D* **13**, 90–104 (1984).

J. Wright, "Method for Calculating a Lyapunov Exponent," *Phys. Rev. A* **29**, 2924–27 (1984).

J.-P. Eckmann, S. O. Kamphorst, D. Ruelle, and S. Ciliberto, "Liapunov Exponents from Time Series," *Phys. Rev. A* **34**, 4971–79 (1986).

A. Wolf, "Quantifying Chaos with Lyapunov Exponents." in [Holden, 1986], pp. 273–90.

S. DeSouza-Machado, R. W. Rollins, D. T. Jacobs, and J.L. Hartman, "Studying Chaotic Systems Using Microcomputer Simulations and Lyapunov Exponents," *Am. J. Phys.* **58**, 321–29 (1990). A good introduction to Lyapunov exponents and their calculation in the case when the time evolution equations are known.

M. T. Rosenstein, J. J. Collins, and C. J. DeLuca, "A practical method for calculating largest Lyapunov exponents from small data sets," *Physica D* **65**, 117–34 (1993).

S. Ellner and P. Turchin, "Chaos in a Noisy World: New Methods and Evidence from Time-Series Analysis," *The American Naturalist* **145**, 343–75 (1995). This article discusses methods for estimating Lyapunov exponents for population dynamics in ecological systems.

Scaling Law for the Average Lyapunov Exponent

B. A. Huberman and J. Rudnick, "Scaling Behavior of Chaotic Flows," *Phys. Rev. Lett.* **45**, 154–56 (1980).

S. C. Johnston and R. C. Hilborn, "Experimental Verification of a Universal Scaling Law for the Lyapunov Exponent of a Chaotic System," *Phys. Rev. A* **37**, 2680–82 (1988).

Invariant Measure

E. Ott, *Chaos in Dynamical Systems* (Cambridge University Press, Cambridge, 1993). Pages 51-55 have a nice discussion of invariant measures and invariant distributions.

K–S entropy

D. Chandler, *Introduction to Modern Statistical Mechanics* (Oxford University Press, New York, 1987). Gives an excellent discussion of entropy within the framework of traditional statistical mechanics.

J. D. Farmer, "Information Dimension and Probabilistic Structure of Chaos," *Z. Naturforsch.* **37a**, 1304 (1982).

P. Grassberger and I. Procaccia, "Estimation of the Kolmogorov Entropy from a Chaotic Signal," *Phys. Rev. A* **28**, 2591–93 (1983).

G. Benettin, L. Galgani, and J.-M. Strelcyn, "Kolmogorov Entropy and Numerical Experiments," *Phys. Rev. A* **14**, 2338–45 (1976). Reprinted in [Hao, 1985]. A nice discussion of the relationship between K–S entropy and Lyapunov exponents.

Ya. B. Pesin, *Russ. Math. Surveys* **32**, 55 (1977). The Kolmogorov entropy equals the sum of the positive Lyapunov exponents.

G. Paladin, L. Peliti, and A. Vulpiani, "Intermittency as multifractality in history space," *J. Phys. A* **19**, L991–6 (1986). More on the Pesin relation.

A. Bonasera, V. Latora, and A. Rapisarda, "Universal Behavior of Lyapunov Exponents in Unstable Systems," *Phys. Rev. Lett.* **75**, 3434–37 (1995).

M. Dzugutov, E. Aurell, and A. Vulpiani, "Universal Relation between the Kolmogorov-Sinai Entropy and the Thermodynamic Entropy in Simple Liquids," *Phys. Rev. Lett.* **81**, 1762–65 (1998).

V. Latora and M. Baranger, "Kolmogorov-Sinai Entropy Rate versus Physical Entropy," *Phys. Rev. Lett.* **82**, 520–23 (1999).

E. Cohen and L. Rondoni, "Comment on 'Universal Relation between the Kolmogorov-Sinai Entropy and the Thermodynamic Entropy in Simple Liquids,' " *Phys. Rev. Lett.* **84**, 394 (2000).

M. Dzugutov, "Dzugtov Replies," *Phys. Rev. Lett.* **84**, 395 (2000).

Fractals

The following books provide excellent introductions to fractal geometric objects:

B. B. Mandelbrot, *The Fractal Geometry of Nature* (W. H. Freeman, San Francisco, 1982). The masterful (but often frustratingly diffusive) account by the inventor of the term fractal.

H.-O. Peitgen and P. H. Richter, *The Beauty of Fractals* (Springer-Verlag, Berlin, Heidelberg, New York, Tokyo, 1986). A book with beautiful pictures and enough detail of the mathematics so you can begin developing your own fractal pictures.

M. Barnsley, *Fractals Everywhere* (Academic Press, San Diego, 1988). A more mathematically sophisticated introduction to fractals.

H. Jurgens, H.-O. Peitgen, and D. Saupe, "The Language of Fractals," *Sci. Am.* **263**, 60–67 (1990). A nice introductory article.

E. Guyon and H. E. Stanley, *Fractal Forms* (Elsevier, New York, 1991).

D. K. Umberger and J. D. Farmer, "Fat Fractals on the Energy Surface", *Phys. Rev. Lett.* **55**, 661–64 (1985). Fat fractals and ways of characterizing them.

R. Eykholt and D. K. Umberger, "Characterization of Fat Fractals in Nonlinear Dynamical Systems," *Phys. Rev. Lett.* **57**, 2333–36 (1986).

B. B. Mandelbrot, "Self-Affine Fractals and Fractal Dimension," *Phys. Scr.* **32**, 257–60 (1985). A discussion of self-affine fractals.

M. A. Rubio, C. A. Edwards, A. Dougherty, and J. P. Gollub, "Self-Affine Fractal Interfaces from Immiscible Displacements in Porous Media," *Phys. Rev. Lett.* **63**, 1685–87 (1989). An experiment analyzed in terms of a self-affine fractal.

Fractal Dimensions

A. N. Kolmogorov, "A New Invariant for Transitive Dynamical Systems," *Dokl. Akad. Nauk. SSSR* **119**, 861–64 (1958).

P. Grassberger, "On the Hausdorff Dimension of Fractal Attractors," *J. Stat. Phys.* **26**, 173–79 (1981).

J. D. Farmer, E. Ott, and J. A. Yorke, "The Dimension of Chaotic Attractors," *Physica D* **7**, 153–80 (1983). Careful definitions of Hausdorff dimension, capacity (box-counting) dimension, and information dimension with various examples.

A good survey and comparison of the various kinds of dimensions and their practical computation at the level of this book are found in Chapter 7 of [Moon, 1992].

D. Ruelle, *Chaotic Evolution and Strange Attractors* (Cambridge University Press, New York, 1989). A somewhat mathematically more sophisticated look at characterizing strange attractors.

H. Hentschel and I. Procaccia, "The Infinite Number of Generalized Dimensions of Fractals of Strange Attractors," *Physica D* **8**, 435–44 (1983). This article discusses the similarity dimension as well as many other dimensions.

C. Essex and M. Nerenberg, "Fractal Dimension: Limit Capacity or Hausdorff Dimension?" *Am. J. Phys.* **58**, 986–88 (1990). Some examples of sets that yield different numerical values for the different definitions of dimensions.

M. H. Jensen, P. Bak, and T. Bohr, "Complete Devil's Staircase, Fractal Dimension and Universality of Mode-Locking Structure in the Circle Map," *Phys. Rev. Lett.* **50**, 1637–39 (1983). The sine-circle map shows a fractal dimension in the structure of Arnold tongues.

J. Testa, "Fractal Dimension at Chaos of a Quasiperiodic Driven Tunnel Diode," *Phys. Lett. A* **111**, 243–45 (1985). An experiment showing agreement with the fractal dimension predicted for the sine-circle map.

A. Cumming and P. S. Linsay, "Deviations from Universality in the Transition from Quasi-Periodicity to Chaos," *Phys. Rev. Lett.* **59**, 1633–36 (1987). An experiment that shows disagreement with the sine-circle map.

P. Alstrom, B. Christiansen, and M. T. Levinsen, "Nonchaotic Transition from Quasi-periodicity to Complete Phase Locking," *Phys. Rev. Lett.* **61**, 1679–82 (1988). This paper explains the disagreement found in the Cumming and Linsay paper.

C. Grebogi, E. Ott, S. Pelikan, and J. A. Yorke, "Strange Attractors that are not Chaotic," *Physica D* **13**, 261–68 (1984). Are all strange attractors chaotic? Apparently not.

R. F. Voss, "Random Fractals, Self-Affinity in Noise, Music, Mountains, and Clouds," *Physica D* **38**, 362–71 (1989).

Fractal Basin Boundaries

E. G. Gwinn and R. M. Westervelt, "Intermittent Chaos and Low-Frequency Noise in the Driven Damped Pendulum," *Phys. Rev. Lett.* **54**, 1613–16 (1985).

E. G. Gwinn and R. M. Westervelt, "Fractal Basin Boundaries and Intermittency in the Driven Damped Pendulum," *Phys. Rev. A* **33**, 4143–55 (1986).

J. C. Sommerer and E. Ott, "A physical system with qualitatively uncertain dynamics," *Nature* **365**, 136–140 (1993).

Y.-C. Lai and R. L. Winslow, "Riddled Parameter Space in Spatiotemporal Chaotic Dynamical Systems," *Phys. Rev. Lett.* **72**, 1640–43 (1994).

Correlation Dimension

P. Grassberger and I. Procaccia, "Characterization of Strange Attractors," *Phys. Rev. Lett.* **50**, 346–49 (1983). The introduction of the correlation dimension.

N. B. Abraham, A. M. Albano, B. Das, G. De Guzman, S. Yong, R. S. Gioggia, G. P. Puccioni, and J. R. Tredicce, "Calculating the Dimension of Attractors from Small Data Sets," *Phys. Lett. A* **114**, 217–21 (1986). Good discussion of the computational pitfalls surrounding correlation dimensions.

H. Atmanspacker, H. Scheingraber, and W. Voges, "Global Scaling Properties of a Chaotic Attractor Reconstructed from Experimental Data," *Phys. Rev. A* **37**, 1314–22 (1988). More hints on finding dimensions from experimental data.

J. W. Havstad and C. L. Ehlers, "Attractor Dimension of Nonstationary Dynamical Systems from Small Data Sets," *Phys. Rev. A* **39**, 845–53 (1989).

J. Theiler, "Estimating Fractal Dimension," *J. Opt. Soc. Am. A* **7**, 1055–73 (1990).

M. Möller, W. Lange, F. Mitschke, N. B. Abraham, and U. Hübner, "Errors from Digitizing and Noise in Estimating Attractor Dimensions," *Phys. Lett. A* **138**, 176–82 (1989). The effect of digitization on dimension calculations.

R. Badii, G. Broggi, B. Derighetti, M. Ravani, S. Ciliberto, A. Politi, and M. A. Rubio, "Dimension Increase in Filtered Chaotic Signals," *Phys. Rev. Lett.* **60**, 979–82 (1988). The effect of filtering on computed dimensions is discussed, with examples.

F. Mitschke, M. Möller, and W. Lange, "Measuring Filtered Chaotic Signals," *Phys. Rev. A* **37**, 4518–21 (1988).

D. Ruelle, "Deterministic Chaos: The Science and the Fiction," *Proc. Roy. Soc. Lond. A* **427**, 241–48 (1990). This essay includes cautions on calculating the correlation dimension with small data sets.

C. Essex and M. A. H. Nerenberg, "Coments on 'Deterministic Chaos: the Science and the Fiction' by D. Ruelle," *Proc. Roy. Soc. London A* **435**, 287–92 (1991). Some comments on Ruelle's proposed limits that in many cases are not sufficiently stringent.

J. Theiler, B. Galdrikian, A. Longtin, S. Eubank, and J. D. Farmer, in *Nonlinear Modeling and Forecasting*, Santa Fe Institute Studies in the Sciences of Complexity, Proc. Vol. XII, pp. 163–88. M. Casdagli and S. Eubank (eds.) (Addison-Wesley, Reading, MA, 1991). Proposes the use of "surrogate data" to test calculations of fractal dimensions.

A. K. Agarwal, K. Banerjee, and J. K. Bhattacharjee, "Universality of Fractal Dimension at the Onset of Period-Doubling Chaos," *Phys. Lett. A* **119**, 280–83 (1986). A scaling law for fractal dimensions at the onset of chaotic behavior.

M. T. Rosenstein and J. J. Collins, "Visualizing the Effects of Filtering Chaotic Signals," *Computers and Graphics* **18**, 587–92 (1994). Examines the effects of filtering on the determination of fractal dimensions.

9.11 Computer Exercises

CE9-1. Compute the average Lyapunov exponent for the logistic map data for $A = 4$ and show that it is equal to ln 2 as calculated analytically in Section 9.5.

CE9-2. Use *Chaos Demonstrations* to plot the Lyapunov exponent for the logistic map as a function of the control parameter L. (Use the Lyapunov "view" in the Logistic Map section.) Check, as best you can, whether the Lyapunov exponent obeys the scaling law given in Section 9.4.

CE9-3. (Challenging, but not too difficult.) Write a computer program to calculate the correlation sum for a time series of data. Have the program plot log $C(R)$ as a function of log R and verify the results given in Section 9.8.

CE9-4. Use the correlation dimension program you wrote for CE9-3 to investigate the conjecture (ABB86) that the dimension D as function of the parameter A (say, for the logistic map) increases as A increases beyond the period-doubling accumulation point:

$$D(A) - D(A_\infty) = k\left|A - A_\infty\right|$$

where k is a constant.

CE9-5. A shareware program FRACTINT provides a wealth of fractal generating programs including those for the famous Mandlebrot set, Julia sets, and many more. A very versatile and sophisticated program. Information on the program is available through FRACTINT Copyright (C) 1990-97 The Stone Soup Group. Primary Authors: Bert Tyler (73477.433@compuserve.com), Timothy Wegner (twegner@phoenix.net), Jonathan Osuch (73277.1432@compuserve.com), Wesley Loewer (loewer@tenet.edu), Mark Peterson (Past Primary Author) or on the FRACTINT homepage (http://spanky.triumf.ca/www/fractint/fractint.html).

CE9-6. Use *Chaos Demonstrations* to explore the Julia and Mandlebrot sets, which are mathematical examples of fractal basin boundaries.

10

Many Dimensions

and

Multifractals

Fate shall yield to fickle Chance, and Chaos judge the strife.   Milton, *Paradise Lost*, ii. 232.

10.1 General Comments and Introduction

In this chapter we shall describe several generalizations of the methods of quantifying chaos that were introduced in Chapter 9. First, we shall show how to extract a multidimensional description of state space dynamics from the time series data of a *single* dynamical variable. The so-called *embedding* (or *reconstruction*) *scheme*, which enables this process, is certainly one of the most important technical contributions to the study of nonlinear dynamics in the last decade; it has become the tool of choice in analyzing nonlinear systems.

With the embedding (reconstruction) method at hand, we shall show how we can generalize in various ways the quantitative measures of chaotic behavior. We will focus on the correlation dimension and its generalizations and on extensions of the concept of Kolmogorov entropy. This presentation should bring the reader very close to the research literature on quantifying chaos.

In the final sections of the chapter we shall describe three powerful, but rather abstract, ways of characterizing dynamics. One method looks at the spectrum or spread of fractal dimensions and Lyapunov exponents across the state space. The second classifies attractors according to the topological properties of the unstable periodic orbits embedded in those attractors. The third uses a formalism much like statistical mechanics to capture the dynamical behavior of the system in one powerful mathematical function.

The reader who is less enamored of mathematical formalism should certainly read Section 10.2 on the embedding technique, but could well skip the remainder of the chapter. On the other hand, for anyone interested in actually applying the analysis of chaos to real data, the material on generalized dimensions and entropies is worth studying in detail.

10.2 Embedding (Reconstruction) Spaces

In Chapter 9, our discussion of Lyapunov exponents, dimensions, and entropies used data consisting of a single variable, say x, which we had sampled at successive times $\{t_i\}$ to generate a set of sampled values $\{x_i\}$. (We use the braces $\{\ \}$ to denote the set or collection of values.) We assumed that the dynamics of the system was well described by this single set of numbers. In essence we took for granted, at least implicitly, a one-dimensional model for the dynamics.

We now want to lift that one-dimensional restriction. Of course, one way to proceed would be to record simultaneously values of all the dynamical variables for the system. For a three-dimensional system, such as the Lorenz model of Chapter 1, we would need to record three values, say $X(t)$, $Y(t)$, and $Z(t)$. As the number of state space dimensions increases, the number of values to be recorded increases. At first sight, this procedure seems straightforward. One problem, however, should be immediately apparent: For many systems, such as a fluid system or a multimode laser, we may not know how many degrees of freedom are active. We do not know in advance the required number of variables. Moreover, in practical terms, some variables are much easier to measure accurately than others. We would prefer to build our analysis on data of accurately measured variables if we can.

Both of these difficulties are circumvented by a technique, which is turning out to be one of the major technical contributions of chaos theory. This technique is based on the notion that even for a multidimensional dynamical system, the time series record of a <u>single</u> variable is often sufficient to determine many of the properties of the full dynamics of the system. We do not need to record values of all of the dynamical variables. Just one will do. In particular we can use this single time series to "reconstruct" the dimensions, entropies, and Lyapunov exponents of the dynamics of the system.

The basic idea is very simple. We use the time series data of a single variable to create a multidimensional *embedding space*. (As we shall see, a better name would be *reconstruction space*, but the term embedding space is more commonly used in nonlinear dynamics.) If the embedding space is generated properly, the behavior of trajectories in this embedding space will have the same geometric and dynamical properties that characterize the actual trajectories in the full multi-dimensional state space for the system. The evolution of the trajectories in the embedding space, in a sense, mimics the behavior of the actual trajectories in the full state space.

The use of a single time series to generate a reconstruction space to characterize nonlinear dynamical systems was suggested in 1980 by Packard, Crutchfield, Farmer and Shaw (PCF80) and was put on a firm theoretical basis by F. Takens (TAK81).

To see how the embedding procedure works, let us suppose that we have recorded a series of X values for some dynamical system. We then want to use the series of values $X_1, X_2, X_3,\ldots$ to reconstruct the full dynamics of the system. We do this by grouping the values to form "vectors." For example, suppose we decide to

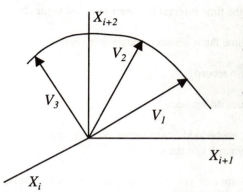

Fig. 10.1. A schematic representation of the use of a single time series to form a three–dimensional embedding space. The tip of each vector marks a trajectory point in the embedding space. The vectors are labeled V_1, V_2, and V_3 to distinguish them from the sampled values.

use a d-dimensional reconstruction. We could then group together d values, for example, $(X_1, X_2,..., X_d)$. Each number in the vector gives the value of the "component" of the vector along one of the axes in this d-dimensional space. Together, the d values give the coordinates of a single point in the d-dimensional space. Thus we say we are embedding the dynamics in a space. (Alternatively, we could say that we are *reconstructing* the d-dimensional behavior from that of a single variable.)

The time evolution of the system is followed by seeing how successive vectors $(X_{n+1}, X_{n+2}, ... X_{n+d})$, where n labels the successive vectors, move in the d-dimensional space. (As we shall see shortly, the time spacing between the vector components and between successive vectors need not be the same.) Our basic assumption is that the geometry and dynamics of the vectors constructed in this way are the same as the geometry and dynamics of the trajectories in the full state space of the system. Figure 10.1 gives a schematic representation of the embedding construction for $d = 3$.

Why does this embedding procedure work? One rough answer is that the character of the overall dynamics of the system must be reflected in the behavior of any one dynamical variable. For example, if the overall behavior is periodic, we would expect the behavior of any one variable to be periodic also. If the overall behavior is chaotic, the accompanying divergence of nearby trajectories should, in general, also show up in the behavior of a single variable. Another way of seeing why this method works is to reverse the procedures we used in Section 3.3 to reduce the dynamical equations to a set of first-order differential equations. If we run the procedure in the opposite way, we may write the dynamics in terms of a higher-order differential equation for <u>one</u> variable. We can then use the time series data of the single variable to calculate approximately the various derivatives required to express that differential equation.

Several questions immediately arise:

    1.   How many dimensions should we use in the embedding space?

2. How should we choose the time interval between sampled values?

3. How do we choose the time lag n between successive vectors?

4. Which variable should we record?

5. How many sampled values do we need?

6. How do we use the embedded trajectories to get quantitative information about the dynamics of the system?

Unfortunately, it is difficult to answer any of these questions definitively, partly because the theoretical underpinnings of the embedding technique are not fully developed at present and partly because the answers depend to some extent on the system being studied, the accuracy of the data, the amount of noise present and so on. In practice, we need to explore how the results for a specific system depend on the embedding dimension, the time between samples, the number of samples, and so on. In all cases, we need to use some scientific judgment to decide what values or range of values give "reasonable" results. We will give some guidelines for choosing these numbers based on commonsense augmented by experience with analyzing actual systems later. At the end of the chapter, you will find references in which more detailed information can be found.

Takens (TAK81) has shown that if the underlying state space of a system has d_s dimensions, and if the embedding space has $2d_s + 1$ dimensions, then we can capture completely the dynamics of the system with the embedding space vectors. Otherwise, the "projection" of the attractor onto a smaller dimensional space might reduce the apparent dimensionality of the attractor. We can understand this result in a simple way: If the state space has d_s dimensions, then we can express the dynamics in terms of a d_s-order differential equation for some variable. To calculate those derivatives numerically, we need two time samples for each derivative. In addition, we need the current value of the variable. Hence, we need a total of $2\,d_s + 1$ values.

In practice, however, as we have seen, for dissipative systems the *effective* dimensionality for the long-term behavior is that of the system's attractor (or attractors). This dimensionality may be considerably smaller than that of the original state space. Thus, we may be able to use embeddings whose dimension is about twice the (possibly fractal) dimension of the *attractor* to mimic the dynamics on the attractor.

Most recent work in characterizing dynamical systems by the embedding technique has focused on the evaluation of fractal dimensions by calculating the Grassberger–Procaccia correlation sum, introduced in Chapter 9. We shall use the correlation sum to illustrate how to implement the embedding technique. Let us begin the discussion by writing the correlation sum in the following form:

$$C^{(d)}(R) = \frac{1}{N(N-1)} \sum_{i,j=1, i \neq j}^{N} \Theta \left[R - \left| \vec{x}_i - \vec{x}_j \right| \right] \qquad (10.2\text{-}1)$$

You should note that Eq. (10.2-1) is almost identical to Eq. (9.8-4), but we have added a superscript to C indicating that the correlation sum may depend on d, the number of embedding dimensions. We have used arrows to indicate that the x values are now considered to be vectors in the embedding space. If we think of the vectors as giving the coordinates of a point in the embedding space (as shown in Fig. 10.1), then the correlation sum tells us the relative number of pairs of points that are located within the distance R of each other in this space.

The labeling of the vectors is a bit messy, but let us see what is involved. A d-dimensional vector is the collection of d components

$$\vec{x}_i = (x_i, x_{i+t_L}, x_{i+2t_L}, \ldots, x_{i+(d-1)t_L}) \qquad (10.2\text{-}2)$$

where t_L is called the **time lag** and represents the time interval between the successively sampled values that we use to construct the vector $\vec{x}_i$. Later we shall discuss how to choose t_L.

The "length" of the difference between two vectors is usually taken to be the "Euclidean length":

$$\left| \vec{x}_i - \vec{x}_j \right| = \sqrt{\sum_{k=0}^{d-1} (x_{i+kt_L} - x_{j+kt_L})^2} \qquad (10.2\text{-}3)$$

However, some authors have advocated the use of the "maximum coordinate difference" as a measure of the length (to save computation time):

$$\left| \vec{x}_i - \vec{x}_j \right| = \text{Max}_k \left| x_{i+kt_L} - x_{j+kt_L} \right| \qquad (10.2\text{-}4)$$

In other words, we find the largest difference between corresponding components and use that difference as a measure of the difference between two vectors. The correlation dimension (scaling exponent) does not seem to depend sensitively on the measure used for calculating the vector differences. The maximum coordinate difference (sometimes called the **maximum norm**) has the advantage that the largest difference will be the same for all embedding dimensions, thus making the comparison of results for different embedding dimensions easier.

We are now ready to find the correlation dimension (scaling exponent) for the system just as we did in the previous chapter. We define $D_c(d)$ to be the number that satisfies

$$C^{(d)}(R) = kR^{-D_c(d)} \qquad (10.2\text{-}5)$$

for some range of R values, which we again call the scaling region. Here we have included a parenthetical d to indicate that the value of D_c may (and in general does) depend upon the dimension of the embedding space.

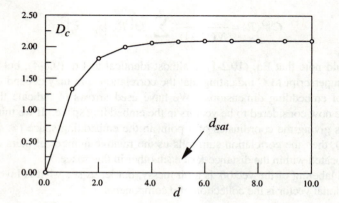

Fig. 10.2. A schematic graph of the correlation dimension plotted as a function of the embedding dimension. When the embedding dimension is equal to or greater than about twice the dimension of the state space attractor (the value labeled as d_{sat}) the correlation dimension $D_c(d)$ becomes independent of d. The correlation dimension of the attractor in this case is about 2.1.

What we do in practice is compute $D_c(d)$ for $d = 1, 2, 3, \dots$ and plot the values of D_c as a function of d. We expect D_c to vary with d until d is equal to or becomes greater than about twice the dimension of the state space attractor for the system. For $d > d_{sat}$ (a "saturation" value), D_c becomes independent of the embedding dimension d. Hence, we expect to see a graph something like that shown in Fig. 10.2. For the data shown in Fig. 10.2, we see that the correlation dimension of the attractor is about 2.1.

From this example, we see the power of the embedding technique. From the data for a single variable, we can find the correlation dimension of the state space attractor. This information can then guide us in constructing mathematical models of the dynamics of the system.

Some caveats, however, are in order. When we set out to find the correlation dimension as a function of embedding dimension, we often find that the size and location of the scaling region (as a function of R) also depends on the embedding dimension. Thus, in many cases, we need to examine the $\ln C(R)$ versus $\ln R$ graph "by hand" for each value of the embedding dimension in order to find the scaling region and the value of the correlation dimension. (In this chapter we shall use natural logarithms, which have some formal advantages as we shall see.)

As an example of this scaling region problem, let us look at Fig. 10.3, which shows a plot of the $\ln C(R)$ versus $\ln R$ calculated from data based on the logistic map function (with $A = 3.99$) for various values of the embedding dimension. We would expect the correlation dimension (as determined by the slope of the graphs in the scaling region) to be the same for any embedding dimension because the system is one-dimensional. This seems to be the case, but we also see that the length of the scaling region depends on the embedding dimension. Hence, we must be cautious in selecting the range of R values chosen to determine the slope. A detailed

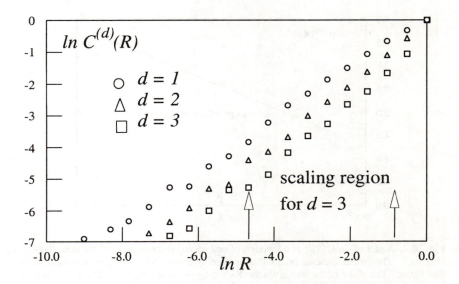

Fig. 10.3. A plot of the natural logarithm of the correlation sum as a function of ln R for various embedding dimensions for data computed from the logistic map function with $A = 3.99$. Two thousand data points were used in the analysis. Note that the slope of the graphs are nearly the same in the scaling regions, but the scaling region length changes with embedding dimension.

detailed analysis of how the estimate of D_c varies with embedding dimension is found in DGO93a and DGO93b. It turns out that one can estimate in advance what the "best" embedding dimension will be (DIH96), but in practice examining the results with the equivalent of Figures 10.2 and 10.3 is very important.

There is yet another method for determining when you have chosen the appropriate embedding dimension. The idea is called "false nearest neighbors" [Abarbanel, 1996]. The basic notion is the following: If you use too small an embedding dimension, then in projecting the state space trajectories onto a lower dimensional "surface," two trajectory points that might be far apart in the full state space end up near each other in the smaller dimensional surface. (As an analogy, think about projecting points on a basketball downward to the floor under the ball. The projection of a point near the top of the ball might end up close to the projection of a point near the bottom of the ball.) As the embedding dimension increases, the number of nearest neighbors to a particular point in the embedding space should decrease until the embedding dimension is sufficiently large to reflect accurately the geometry of the attractor.

We can, however, get yet more information from $C^{(d)}(R)$. As d increases, we expect $C^{(d)}(R)$ to <u>decrease</u> for a fixed value of R. Why? When we compute $C^{(d)}(R)$, we are asking for the probability that trajectory points in the embedding space stay within the distance R of each other. As d increases, we are increasing the length of the trajectory segments being compared. If you recall the discussion of K–S

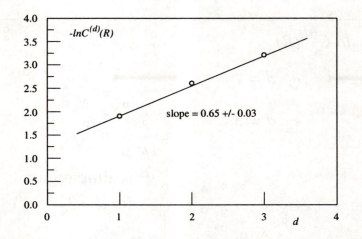

Fig. 10.4. A plot of -ln $C^{(d)}(R)$ as a function of embedding dimension from the data shown in Fig. 10.3. The circles indicate points that are averages over the scaling region indicated in that figure. The slope of the line gives the K–S entropy for the system. The value obtained here is close to the value of ln 2 expected for the logistic map with $A = 4$.

entropy in Chapter 9, you will remember that the K–S entropy measures the rate of change of this probability. Thus, we might expect (we will provide further arguments along this line in a later section of this chapter)

$$C^{(d)}(R) = be^{-dKt_L} \qquad (10.2\text{-}6)$$

where K is the Kolmogorov (K–S) entropy and t_L is the time lag between successive components of an embedding space vector. Hence, from our plot of ln $C^{(d)}(R)$ versus ln R, we can determine the K–S entropy by seeing how ln $C^{(d)}(R)$ changes with d for a fixed value of R. (Of course, we choose a value of R in the scaling regions for all values of d.) These results are illustrated in Fig. 10.4 for the data shown in Fig. 10.3. The value of the slope is close to ln 2, the value expected for the K–S entropy for the logistic map with $A = 4$, where we use $t_L = 1$.

To summarize, we see that the correlation sum calculated by the embedding technique allows us to determine both the correlation dimension of the attractor and the K–S entropy of the dynamics. Let us now turn to a conjecture relating a fractal dimension and the Lyapunov exponents.

The Kaplan–Yorke Conjecture: The Lyapunov Dimension

Kaplan and Yorke (KAY79) have suggested that the dimension of an attractor in a multidimensional state space (or multidimensional reconstruction space) can be defined in terms of the average Lyapunov exponents in that space. Recall that there are as many Lyapunov exponents as there are state space dimensions. Let us rank the Lyapunov exponents from the largest λ_1 to the smallest λ_d for a d-dimensional space producing a *spectrum* of Lyapunov exponents: $\lambda_1 > \lambda_2 > \ldots > \lambda_d$. Let j be the largest integer such that $\lambda_1 + \lambda_2 + \ldots + \lambda_j > 0$. (Recall that the sum of all the Lyapunov exponents is negative for a dissipative system and 0 for a Hamiltonian system.)

The Lyapunov dimension D_L is then defined to be

$$D_L = j + \frac{\sum\limits_{i=1}^{j} \lambda_i}{-\lambda_{j+1}} \qquad (10.2\text{-}7)$$

[Jackson, 1992, Vol 2, pp. 217–222] provides a simple argument for the result given in Eq. (10.2-7). There is some numerical evidence (see, for example, RHO80) that D_L is numerically close to the box-counting dimension and the correlation dimension. Constantin and Foias (COF85) have proved that in general $D_L \geq D_H$, the Hausdorff dimension. The references at the end of the chapter discuss methods for determining the spectrum of Lyapunov exponents from the time-series data. One word of caution: the time-delay embedding technique can lead to "spurious" Lyapunov exponents that are artifacts of the method.

10.3 Practical Considerations for Embedding Calculations

In this section we will discuss in a rather general way some of the practical details of implementing the embedding calculations described in the previous section. We shall try to justify some "rules of thumb" for choosing the number of data points, the time lag between "components" for the embedding vectors, and the time jumps between successive vectors. If we had available extremely fast computers, lots of data, and plenty of time, then we could be profligate and simply use millions of data points and repeat calculations dozens of times with variations in time lags and time jumps. In practice, we need to make a compromise between computation time and the accuracy of our results. We want to do just enough work, but no more, to achieve a specified level of accuracy. Many readers may wish to skip this section until they need to use the embedding scheme to analyze data.

We assume that we are dealing with the time record of a single dynamical variable for the system. To be completely general, let us also assume that the time between successive samples, call it t_s, is a parameter under our control. (If we are using a Poincaré section sampling technique, then t_s is determined by the dynamics. In that case, we ought to use several Poincaré sections for the system to see if our results depend on the "location" of the Poincaré section in state space.)

A fundamental time scale for the system's behavior is a time interval called the *autocorrelation time*, which can be defined in terms of the *autocorrelation function*:

$$g(n) = \frac{\sum\limits_{k} x_k x_{k+n}}{\sum\limits_{k} |x_k|^2} \qquad (10.3\text{-}1)$$

$g(n)$ compares a data point in the series with a data point located n units of time away. If, on the average, they are uncorrelated, then we have $g(n) = 0$. If they are nearly the same, then we find $g(n) = 1$. For data sets from stochastic or chaotic systems, the autocorrelation function is expected to fall off exponentially with time: $g(n) = ae^{-n\tau}$, where τ is called the *autocorrelation time*. This time will set the scale of our choice of several embedding scheme parameters.

When we compute the autocorrelation function, it is traditional first to subtract the mean of the data values from each of the data points. This procedure means that $g(n)$ goes to zero for uncorrelated data. One further practical matter: for real data sets we have to be sure that the index $k + n$ in Eq. (10.3-1) stays within the range of actual data indices.

A second time period under our control is the time period between successive "components" of each of the embedding space vectors. We need not take successive samples, say $(x_4, x_5, x_6, \ldots)$. We could use, if we choose, $(x_4, x_9, x_{14}, \ldots)$, where the successive components are separated by five sample periods. Let us call the time period between successive components t_L, the **time lag**. The time lag can be written as a multiple of the sampling time: $t_L = L\, t_s$, with $L = 1, 2, \ldots$.

Experience with embedding calculations has shown that what seems to be crucial is the overall time span (or "time window") covered by the vector (AMS88 and references cited therein). This time span is given by the product $(d-1)Lt_s$. If the time span is too short or if the time span is too long, the scaling region for the correlation sum seems to get very small. It has been suggested (AMS88) that using a time span about two or three times the autocorrelation time gives the broadest scaling region for the correlation sum.

We can also choose the time interval between successive vectors, which we shall call t_J, the **jump time**. We want t_J small enough so that we can map out most of the details of the attractor geometry. We often choose t_J so that the vectors used are distributed uniformly (in time) throughout the data set.

A fourth time interval we can choose is a minimum time separation between vectors for comparison in the correlation sum. That is, we might choose to compare $\vec{x}_i$ and $\vec{x}_j$ only if the corresponding components are separated in time by some minimum amount, which we shall call the **comparison time** t_c. This minimum separation would avoid "excessive" correlation due solely to the fact the samples occur close in time. (This problem is almost always automatically avoided if we are using Poincaré section data.) (THE86 discusses the problems that arise when the comparison time is too small for data sets with "autocorrelated" data, that is, data points that are nearly the same for some significant stretch of time.) We can avoid these problems by choosing the comparison time to be greater than the autocorrelation time.

Finally, we need to choose N the total number of data points to be used in the analysis. Usually, more is better, but the computation time increases rapidly with N, and we may need to compromise.

To illustrate how we organize the data, we write out a set of d-dimensional vectors, which we now call $\vec{V}$ (to avoid confusion with the data points themselves). The lag time t_L is given as a multiple of the time interval between successive samples: $t_L = L\,t_s$. The jump time t_J between successive vectors is also given as a multiple of t_s: $t_J = J\,t_s$. Both L and J are taken to be positive integers. As before x_n is the variable value recorded at time $t = nt_s$. The set of vectors is

$$\vec{V}_0 = (x_0, x_L, \ldots, x_{(d-1)L})$$
$$\vec{V}_1 = (x_J, x_{J+L}, \ldots, x_{(d-1)(J+L)})$$
$$\vdots$$
$$\vec{V}_p = (x_{pJ}, x_{pJ+L}, \ldots, x_{(d-1)(pJ+L)})$$

(10.3-2)

To analyze our choices of t_s, t_L, t_J, and N, we need at least rough estimates of the characteristics of the attractor. In most cases, these can be estimated (guessed at) from the time series we have recorded. First, we need to know the time interval required for a trajectory to go around the attractor. Since this is roughly equivalent to the time interval between successive Poincaré section crossings if we had used a Poincaré section sampling, let us call this time t_p (the "Poincaré time"). (If the actual data are Poincaré section samples, then we already have $t_s = t_p$). Next, we need an estimate of the "size" of the attractor. From the record of the single variable x, we can find the range of x values visited by the data. Let us use A to denote the size of the attractor. Then in a d-dimensional embedding, the "volume" occupied by the attractor will be approximately A^d for $d < D_c$, the correlation dimension of the attractor. (For $d > D_c$, the actual volume of the attractor is determined by D_c.) Finally, we need an estimate of the autocorrelation time and the divergence rate of nearby trajectory points. This rate can be estimated by finding the Lyapunov exponent for the one-dimensional data set $\{x_i\}$ using the methods discussed in Chapter 9.

In addition to the estimate of the attractor characteristics given earlier, we need to take into account the precision with which the data are recorded. We usually specify this precision as the number of binary bits used to represent the data. In an experiment, this number might be determined by a computer-driven sampling device (usually an "analog-to-digital converter"). If we are dealing with data generated by a computer's numerical solution of some equations, then the number of bits is determined by the software. (For typical experiments, data are often recorded with at least 8-bit resolution. Sometimes 14-bit or 16-bit resolution is used for high precision measurements. For computations, "single-precision" on a PC-type computer uses seven decimal digits, while "double precision" uses 15 or 16 decimal digits.)

The fundamental considerations in applying the embedding technique are to use enough data to characterize the attractor reasonably while choosing the time lag and the jump time to avoid spurious correlations. To be more specific, we want to have enough data so that in the neighborhood of each vector in a d-dimensional

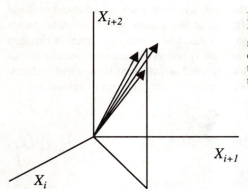

Fig. 10.5. A schematic representation of vectors in a 3–dimensional embedding space. If the time lag between successive components is too small, the vectors all tend to cluster around a 45° line passing through the origin.

embedding space, there will be a sufficient number of vectors to get an accurate estimate of the correlation sum. We want those vectors, however, to consist of data points reasonably separated in time so that we are comparing trajectory points that have passed through a particular section of state space at quite different times.

Let us consider the sampling time question first. Generally, we want to have the time between successive samples be much less than t_p, the time to go around the attractor. We will then have samples over nearly the complete range of the attractor. (Obviously, this point is moot if we are using a Poincaré sampling scheme.) However, in many cases, the sampling time has been fixed by some experimental or computational criterion, and we must simply work with what data we have.

What about the choice of time lag t_L between successive components of a vector? We would like to choose the time lag to be large enough so that the successive components are "independent." The time required for independence is given roughly by the reciprocal of the (positive) Lyapunov exponent λ for the data. As a rough rule of thumb, we choose the lag time to be such that the total time span represented by a given d-dimensional vector $(d-1)\, t_L$ is greater than say $3\lambda^{-1}$. (For at least some systems, the autocorrelation time τ is roughly the same as λ^{-1}.) If we choose $(d-1)\, t_L < \lambda^{-1}$ (or less than the autocorrelation time), then the embedding vectors will tend to cluster around a "45° line" passing through the origin of the embedding space, as shown in Fig. 10.5, because all the components will tend to have nearly the same numerical value. In this case, the vectors do not give a very good representation of the geometry of the attractor in the original state space. (If we had, unrealistically, a large amount of extremely high precision data, we could still extract correlation dimensions from these vectors. In practice, however, experimental noise or computer round-off can mask fine details.) A detailed analysis of time lag considerations is given in RCD94.

Let us now think about choosing t_J, the minimum jump time between successive embedding vectors. We want to have t_J small enough so that we map out the trajectory as it proceeds around the attractor. If we make t_J too small, however, then, as we shall see, the total number of data points required becomes very large. Furthermore, as we argued earlier, if the vectors used are closer in time than roughly the autocorrelation time, then we learn nothing new by comparing the vectors because they are nearly identical. Thus, we generally choose t_J to be several times larger than the autocorrelation time.

How do we choose the total number of data points needed? As we mentioned earlier we want enough vectors in the neighborhood of each individual vector to give a reasonable characterization of the attractor. We can estimate the number of data points as follows. For a d-dimensional embedding, the attractor volume is approximately A^d, where A, as before, is an estimate of the "radius" of the attractor. If we use a total of M vectors, then we will have $m = M/A^d$ vectors (or trajectory points) per unit volume of embedding space. In a sphere of radius R (used in the calculation of the correlation sum) around a given trajectory point, we expect, on the average $m\,R^d = (R/A)^d M$ trajectory points. If we have collected data with b bit resolution, then the smallest R we want to use in the correlation sum is $R_{min} = A/2b$. (In practice, we may use a somewhat larger R_{min}, say $R_{min} = A / (2b-2)$ to avoid effects due to digital round-off.) Now suppose for R_{min} we want to have at least N points in the neighborhood to get a reasonable estimate of the correlation sum. We must therefore choose M to satisfy

$$N = \left(\frac{R_{min}}{A} \right)^d M = \left(\frac{1}{2^{b-2}} \right)^d M \qquad (10.3\text{-}3)$$

To get a feeling for what is required, let us find M for the minimal case with $N = 2$ (only two points in the neighborhood), $b = 8$ (8 bit resolution), and $d = 3$. We find we need $M = 524,288$. Thus, even for this modest case, we find that we need many data points. What is worse, the required number of data points goes up exponentially with the embedding dimension.

Another possible consideration in the choice of the number of data points is to make M large enough so that there are nearly as many trajectory points "transverse" to a given trajectory as there are points along the trajectory. This assures us that there will be enough trajectory points in the neighborhood of a given trajectory point so that we can calculate the geometric properties of the attractor reasonably well (see AMS88 for details).

Generally speaking, we would like to have as many data points as possible to compute dimensions, entropies, and so on. In practice, the number of data points may be limited either by the data-taking capabilities of our equipment (some digital oscilloscopes, for example, can record only 1,000 or 2,000 successive samples) or by the computation time required to calculate the quantities of interest. If we need to do the computation only once, then the computation time would not be so important. In many cases, however, we want to see how the correlation dimension, for example, changes as a control parameter is changed, or in a clinical setting, we may want to use some quantitative characteristic of an EKG signal to monitor in more or less real time the status of a patient. In those cases, we must strike some compromise between accuracy (requiring a large number of data points) and computation time (which limits the number of data points we can use). In practice, data sets of a few thousand data points seem to be sufficient for "reasonable" estimates of the correlation dimension (AAD86) for systems exhibiting low-dimensional dynamics. If you anticipate a high dimensional system (many active degrees of freedom), normalizing the correlation sum in terms of the sum for a "random" distribution of points on the attractor seems to help produce results that are useful in discriminating between deterministic chaos and random noise (BHM93).

When the system is driven by an external periodic force, the period of the external drive establishes a natural period for sampling the dynamics of the system. In that case, as we have mentioned several times before, we sample some variable of the system at a particular phase of the external force to form a Poincaré section. In this case the sampling time t_s is the period of the external force and is equal to the Poincaré time. It is then reasonable to use a time lag for vector components and a time jump between successive vectors both equal to t_s since the variable samples are already reasonably separated in time. In such a situation, we need assure only that we have a sufficient number of data points and we need not worry about choosing the time lag and the time jump. However, the analysis should be repeated for different phases of the sampling to see if the results depend on the phase. Different phases put the Poincaré section in different parts of the attractor and there is no general reason to believe that the correlation dimension, for example, should be exactly the same everywhere on the attractor.

Summary of the use of the Embedding Technique

1. From the time series data of a single variable, estimate the values of A (the attractor size), λ (the largest positive Lyapunov exponent), τ (the autocorrelation time), and t_p (the Poincaré time, the time to cycle around the attractor).

2. Choose the sample time $t_s \ll t_p$. (For Poincaré section data, we have $t_s = t_p$. See item 6.)

3. Choose t_L (the time lag between successive components of the embedding space vector) such that the time span of a vector $(d-1)t_L$ is about three times λ^{-1} or about three times τ, the autocorrelation time.

4. Choose the jump time t_J (the time between successive vectors) sufficiently small compared to the Poincaré time t_p so the vectors cover the attractor with small enough spacing to pick out details of the attractor. (For Poincaré section data, we usually choose $t_J = t_p$.)

5. Choose the total number of data points to satisfy Eq. (10.3-3).

6. Test your results for dependence on t_s, t_J, t_L, and the total number of data points. If you are using Poincaré section data, repeat the calculations for other positions of the Poincaré section in state space. This is obviously a time-consuming process, but it is needed to be able to provide estimates of the uncertainties to be associated with the results.

7. Test your results by running the analysis on a data set modified to remove any deterministic evolution. This technique uses *surrogate data* (TEL92, PRT94, SCS96, [Abarbanel, 1996], CAR97) generated, as mentioned at the end of Chapter 9, by taking the Fourier transform of your original data, randomizing the phases, and then regenerating the surrogate data with an inverse Fourier transform.

The references at the end of the chapter include several experiments analyzed by the embedding dimension technique. Before embarking on an analysis of your own data, you should read several case histories of embedding analysis. A cautionary tale is provided in JBT94.

10.4 Generalized Dimensions and Generalized Correlation Sums

In Chapter 9, we introduced several different measures of the "dimension" of a state space attractor. All of these measures were averages over the attractor. However, as you might expect, the local contributions from different parts of the state space are, in general, different; stating just an average value does not acknowledge all of the complexity of the attractor's geometry. In this section we will introduce a set of so-called *generalized dimensions*, the purpose of which is to provide more detailed information about the geometry of the attractor.

We can understand the motivation behind these generalized dimensions by considering two statistical distributions of some quantity (family income, for example). Two different communities might have the same *average* income but nevertheless have very different distributions of income as illustrated in Fig. 10.6.

What kind of measures can be used to distinguish the two distributions? Obviously, if we know the two distributions in detail, we can simply display the distributions and note the differences. However, it is common in statistics to distinguish the distributions quantitatively by calculating the so-called *moments of the distribution*, which are defined as the average difference between the individual data values and the mean value, with the difference raised to some power. More formally, we define the qth moment of the distribution to be

$$M_q \overset{def}{=} \frac{1}{N} \sum_{i=1}^{N} (x_i - <x>)^q \qquad (10.4\text{-}1)$$

where, as usual, $<x>$ is the mean (average) of the x values. The various moments give a way of quantitatively specifying the difference between two distributions. M_2 is the square of the well-known standard deviation. For the data shown in Fig. 10.6, the standard deviation for the Figville data is smaller than that for Tree City.

A similar scheme can be applied to characterize the geometry of an attractor. Let us begin by following a procedure similar to that used for the box-counting definition in Chapter 9. Suppose we have N trajectory points on an attractor. As before, we divide the attractor region of state space into cells of size R labeled $i = 1$,

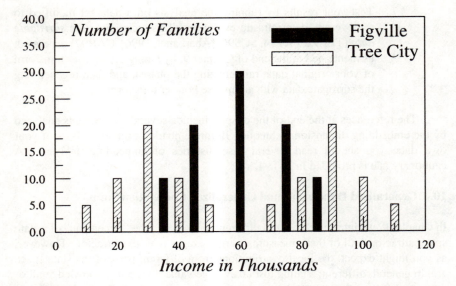

Fig. 10.6. Two graphs of the hypothetical distribution of family income in two communities. The average income is the same, but the distributions differ in detail.

2, 3, ..., $N(R)$. In general, $N \neq N(R)$. (Recall that in general the number of cells depends on R.) For a Hamiltonian system, we use the full region of state space visited by the trajectory.

The number of trajectory points in the ith cell is denoted N_i. We define the probability p_i to be the relative number of trajectory points in the ith cell: $p_i = N_i/N$. The generalized (box-counting) dimension D_q is then defined as

$$D_q = \lim_{R \to 0} \frac{1}{q-1} \frac{\ln \sum_{i=1}^{N(R)} p_i^q}{\ln R} \qquad (10.4\text{-}2)$$

Note that the generalized dimension D_q involves the probabilities raised to the qth power. You might expect that we use only $q = 0, 1, 2$, and so on, but the definition of D_q can be extended to apply to any (real) value of q. (As we mentioned in Chapter 9, we cannot in practice take the $R \to 0$ limit. Instead, we look for the slope of the graph of the numerator of Eq. (10.4-2) plotted as a function of ln R.)

The factor of $q - 1$ has been included in the denominator so that for $q = 0$, we have

$$D_0 = -\lim_{R \to 0} \frac{\ln N(R)}{\ln R} \qquad (10.4\text{-}3)$$

the right-hand side of which we recognize as Eq. (9.7-3); therefore, D_0 is the same as the box-counting dimension.

Let us now look at some general properties of these dimensions. As $q \to \infty$, the largest probability value, call it p_{max}, will dominate the sum, and we have

$$D_\infty = \lim_{R \to 0} \frac{\ln p_{max}}{\ln R} \qquad (10.4\text{-}4)$$

At the other extreme, as $q \to -\infty$, the smallest probability value, p_{min}, will dominate the sum, and we find

$$D_{-\infty} = \lim_{R \to 0} \frac{\ln p_{min}}{\ln R} \qquad (10.4\text{-}5)$$

Hence, we see that D_∞ is associated with the most densely occupied region of the attractor while $D_{-\infty}$ is associated with the most rarefied (least populated) region of the attractor.

We also see that $D_{-\infty} \geq D_\infty$ and that in general $D_q \geq D_{q'}$ for $q < q'$. For a self-similar fractal with equal probabilities for all the cells, $p_i = 1/N(R)$, and we have $D_q = D_0$ for all q.

Exercise 10.4-1. Starting with Eq. (10.4-2), verify that $D_q = D_0$ for all q for a self-similar fractal.

Let us turn to the $q = 1$ case. This requires some special mathematical attention because $q - 1 = 0$ in this case, and it may appear that we are dividing by 0 in Eq. (10.4-2). However, if we allow q to approach 1, then we can avoid this problem. The gimmick is to introduce a new function $y(q)$:

$$y(q) \stackrel{def}{=} \ln \sum_{i=1}^{N(R)} p_i^q \qquad (10.4\text{-}6)$$

Now we write a Taylor-series expansion of $y(q)$ near $q = 1$:

$$y(q) = y(1) + \frac{dy}{dq}(q-1) + \dots \qquad (10.4\text{-}7)$$

We note that $y(1) = 0$ since the probabilities must add up to 1. We then evaluate the first-derivative term:

$$\frac{dy}{dq} = \frac{\sum_i p_i^q \ln p_i}{\sum_i p_i^q} = \sum_i p_i \ln p_i \qquad (10.4\text{-}8)$$

where we used the elementary derivative formula

$$\frac{da^x}{dx} = a^x \ln a \qquad (10.4\text{-}9)$$

The last equality in Eq. (10.4-8) follows from setting $q = 1$. Assembling all these results in the expression for D_1 gives us as $q \to 1$

$$D_1 = \lim_{R \to 0} \lim_{q \to 1} \frac{q-1}{q-1} \frac{\sum_i p_i \ln p_i}{\ln R} = \lim_{R \to 0} \frac{\sum_i p_i \ln p_i}{\ln R} \qquad (10.4\text{-}10)$$

D_1 is called the **information dimension** because it makes use of the $p \ln p$ form associated with the usual definition of "information" for a probability distribution.

Next, we will show that D_2 is just the correlation dimension. (This argument is based on [Schuster, 1995].) To see this connection, let us look at the probability sum

$$S_q(R) = \sum_{i=1}^{N(R)} p_i^q \qquad (10.4\text{-}11)$$

We should remind ourselves that the sum in Eq. (10.4-11) is over $N(R)$ cells uniformly distributed over the attractor region of state space. We want to write this sum in terms of cells centered on trajectory points. (Recall that we use such cells in the calculation of the correlation sum.) First, we rewrite the probability sum in the following way:

$$S_q(R) = \sum_{i=1}^{N(R)} p_i p_i^{q-1} \qquad (10.4\text{-}12)$$

Note that the probability p_i is 0 if the trajectory does not actually visit cell i. The sum over i, therefore, can be written as a sum over cells actually visited by the trajectory. Furthermore, we write the first factor p_i for the visited cells as $1/N$, where N is the number of trajectory points. This essentially says that we will let the sum run over the trajectory points (labeled by the indices j and k) and each trajectory point contributes $1/N$ to the probability. We will <u>not</u>, however, use $1/N$ for the other probability factor, the one raised to the $q-1$ power. (This is obviously a bit of a dodge, but it makes the connection we want to make. One way of justifying this procedure is to say that the probabilities are *almost* all equal to $1/N$. As we raise the probabilities to various powers, the differences will become more noticeable.) Finally, and this is the crucial step, we replace the probability p_i^{q-1} in Eq. (10.4-12) with the Heaviside step-function sum (used in the correlation sum) raised to the $q-1$ power:

$$p_j^{q-1} = \left[\frac{1}{N-1} \sum_{k=1, k \neq j}^{N} \Theta\left(R - \left| x_j - x_k \right| \right) \right]^{q-1} \qquad (10.4\text{-}13)$$

which then gives us for the probability sum

$$S_q(R) = \frac{1}{N} \sum_{j=1}^{N} \left[\frac{1}{N-1} \sum_{k=1, k \neq j}^{N} \Theta\left(R - \left|x_j - x_k\right|\right) \right]^{q-1} \tag{10.4-14}$$

We then define the **generalized correlation sum** $C_q(R)$:

$$C_q(R) = \sum_{j=1}^{N} p_j p_j^{q-1} = \frac{1}{N} \sum_{j=1}^{N} \left[\frac{1}{N-1} \sum_{k=1, k \neq j}^{N} \Theta\left(R - \left|x_j - x_k\right|\right) \right]^{q-1} \tag{10.4-15}$$

Note that for $q = 1$, $C_1 = 1$ and that for $q = 2$, $C_2(R)$ is the same as the (unsubscripted) correlation sum introduced in Eq. (9.8-4).

From here on, we will use the generalized correlation sum C_q and the sum of the probabilities S_q more or less interchangeably. In general, they are different because S_q is based on the probability that a cell is occupied, while C_q is based on the correlation between trajectory points. These are not necessarily the same, but in what follows, we require only that their dependence on R be the same. A general and rigorous proof that they have the same R dependence is lacking.

Putting Eq. (10.4-14) into Eq. (10.4-2) yields the generalized dimension D_q in terms of the generalized correlation sum

$$D_q = \lim_{R \to 0} \frac{1}{q-1} \frac{\ln C_q(R)}{\ln R} \tag{10.4-16}$$

Hence, we see that $D_2 = D_c$. [As an aside, if we wish to find D_1, it is best to go back to Eq.(10.4-10).]

> **Exercise 10.4-2.** Verify that $C_1 = 1$ and that C_2 is the same as the correlation sum of Chapter 9.

There are three important points to note. First, in practice, almost all calculations of D_q from experimental data use the generalized correlation sum rather than the box-counting method. Hence, we should view Eq. (10.4-16) as defining D_q. Second, to find D_q, we usually plot $1/(q-1) \ln C_q(R)$ as a function of $\ln R$ and look for a scaling region as we did for D_c in Chapter 9. Third, D_q can be applied to higher-dimensional state space data by using the embedding technique introduced in Section 10.2. Formally, we replace x_j and x_k in Eq. (10.4-15) with their d-dimensional vector equivalents. This extension will be discussed in Section 10.6.

10.5 Multifractals and the Spectrum of Scaling Indices $f(\alpha)$

As we mentioned in the previous section, we generally expect that different parts of an attractor may be characterized by different values of the fractal dimensions. In such a situation, a single value of some fractal dimension is not sufficient to characterize the attractor adequately. Two quite different attractors might have the same correlation dimension, for example, but still differ widely in their "appearance." An object with a multiplicity of fractal dimensions is called a *multifractal*. We can visualize this object as a collection of overlapping fractal objects, each with its own fractal dimension. In fact, the multifractal description seems to be the appropriate description for many objects in nature (not just for attractors in state space). To some extent it has been the extension of the notion of fractals to multifractals that has led to the wide range of applications of fractals in almost all areas of science. Some of these applications will be described in Chapter 11.

A natural question to ask is how many "regions" have a particular value or range of values of the fractal dimension. A very powerful, but somewhat abstract, scheme has been developed to answer that question. That scheme provides a distribution function, usually called $f(\alpha)$, which gives the distribution of "scaling exponents," labeled by α (not to be confused with the Feigenbaum number α).

These notions can be understood by considering, once again, the partition of the attractor region of state space into a group of cells of size R labeled by an index i, with $i = 1, 2..., N(R)$. As before, we let a trajectory run for a long time and ask for the probability that the trajectory points fall in the ith cell. That probability is defined as $p_i(R) = N_i/N$, where N_i is the number of trajectory points in the ith cell and N is the total number of trajectory points. Alternatively, if we are considering an actual geometric object, we "cover" the object with cells of size R and ask what fraction of the mass of the object (call that fraction p_i) falls within cell i.

We now assume that $p_i(R_i)$ satisfies a scaling relation

$$p_i(R_i) = kR_i^{\alpha_i} \tag{10.5-1}$$

where k is some (unimportant) proportionality constant and R_i is the size of the ith cell. α_i is called the scaling index for cell i. As we make R smaller, we increase the number of cells $N(R)$, and we ask: What is the number of cells that have a scaling index in the range between α and $\alpha + d\alpha$? We call that number $n(\alpha)d\alpha$.

The crucial assumption is that we expect the number of cells with α in the range α to $\alpha + d\alpha$ to scale with the size of the cells with a characteristic exponent, which we shall call $f(\alpha)$. This characteristic exponent is formally defined by

$$n(\alpha) = KR^{-f(\alpha)} \tag{10.5-2}$$

Recalling our discussion in Chapter 9, we see that $f(\alpha)$ plays the role of a fractal dimension. We may interpret $f(\alpha)$ as the fractal dimension of the set of points with scaling index α.

If we plot $f(\alpha)$ as a function of α, we get, for a multifractal, a graph like that shown in Fig. 10.7. The maximum value of $f(\alpha)$ corresponds to the (average) box-counting dimension of the object. For a truly one-dimensional attractor, such as the attractor for the logistic map at the period-doubling accumulation point, we would expect $f(\alpha)$ to be less than 1 since we know that the attractor is a fractal. For data

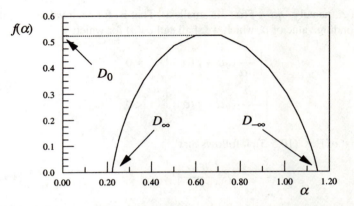

Fig. 10.7. A plot of $f(\alpha)$ as a function of α for a two-scale Cantor (multifractal) set. (See Fig. 10.8 and the text following Exercise 10.5-4.) The maximum value of $f(\alpha)$ is equal to the (average) box-counting dimension for the set. Here, $p_1 = 0.7$, $L_1 = 0.2$, $p_2 = 0.3$, and $L_2 = 0.35$.

from a two-dimensional mapping such as the Hénon map, we might expect $f(\alpha)$ to be less than 2. We will treat the particular example illustrated in Fig. 10.7 in more detail after introducing a few more formalities.

Both the generalized dimensions introduced in the previous section and $f(\alpha)$ describe properties of the multifractal. We might reasonably expect that those quantities are related. We will now show what that relationship is. In fact, in some cases, we compute $f(\alpha)$ by first finding the generalized dimensions and then following the procedure to be presented shortly.

The connection between $f(\alpha)$ and D_q is most easily established by looking at the probability sum, repeated from Eq. (10.4-11):

$$S_q(R) = \sum_{i=1}^{N(R)} p_i^q \tag{10.5-3}$$

where the sum is over cells labeled by the index i. $S_q(R)$ is sometimes called the qth-order **partition function**, in analogy with partition functions used in statistical mechanics. (In Section 10.8, we will exploit this similarity more formally.)

In order to make the connection to $f(\alpha)$, we write the probabilities in terms of α and then integrate over the distribution of α values to get the partition function

$$S_q(R) = C \int d\alpha R^{-f(\alpha)} R^{q\alpha} \tag{10.5-4}$$

where C is an unimportant proportionality constant. The first factor in the integrand in Eq. (10.5-4) tells us how many cells have scaling index α while the second factor is the qth power of the probability associated with the index α.

We now use the following argument to evaluate the integral: Since R is supposed to be small, the numerical value of the integrand is largest when the exponent of R, namely, $q\alpha - f(\alpha)$, is smallest. That is, for each value of q, there is a corresponding value of α, which we shall call $\alpha_*(q)$ for which

$$\left[\frac{d}{d\alpha}(q\alpha - f(\alpha))\right]_{\alpha=\alpha_*} = 0$$

$$\left[\frac{d^2}{d\alpha^2}(q\alpha - f(\alpha))\right]_{\alpha=\alpha_*} < 0 \qquad (10.5\text{-}5)$$

From the first of Eqs. (10.5-5), it follows that

$$q = \frac{df(\alpha)}{d\alpha}\bigg|_{\alpha=\alpha_*} \qquad (10.5\text{-}6)$$

which tells us that the slope of the $f(\alpha)$ curve is equal to 1 for the α value corresponding to $q = 1$. We also see that

$$\frac{d^2 f(\alpha)}{d\alpha^2}\bigg|_{\alpha=\alpha_*} < 0 \qquad (10.5\text{-}7)$$

which tells us that the $f(\alpha)$ curve must be concave downward.

Returning to the evaluation of the integral, we note that the integral value must be approximately the value of the integrand where it is a maximum multiplied by some proportionality constant. Thus, we write

$$S_q(R) = C'R^{q\alpha_* - f(\alpha_*)} \qquad (10.5\text{-}8)$$

If we now recall the definition of D_q in Eq. (10.4-2) and compare it to Eqs. (10.5-3) and (10.5-8), we see that

$$\boxed{(q-1)D_q = q\alpha - f(\alpha) \qquad (10.5\text{-}9)}$$

where we have dropped the subscript * on the variable α.

Exercise 10.5-1. Provide the details of the argument leading to Eq. (10.5-9).

In practice, for data generated by the trajectory of some model dynamical system or for data from some experiment, we often find D_q by the methods described in Section 10.4. Then we carry out various manipulations to find for each q the corresponding values of α and $f(\alpha)$. To see how this works out, we first differentiate Eq. (10.5-9) with respect to q:

$$\frac{d}{dq}[(q-1)D_q] = \alpha + q\frac{d\alpha}{dq} - \frac{df}{d\alpha}\frac{d\alpha}{dq} \qquad (10.5\text{-}10)$$

which, together with Eq. (10.5-6), yields

$$\alpha(q) = \frac{d}{dq}[(q-1)D_q] \qquad (10.5\text{-}11)$$

thus giving us a value of α. We then solve Eq. (10.5-9) for $f(\alpha)$:

$$f(\alpha) = q\frac{d}{dq}[(q-1)D_q] - (q-1)D_q \qquad (10.5\text{-}12)$$

To summarize, we see that once we have found D_q as a function of q, we can compute $f(\alpha)$ and α from Eq. (10.5-11) and (10.5-12). This change from the variables q and D_q to α and $f(\alpha)$ is an example of a *Legendre transformation*, commonly used in the formalism of thermodynamics. See, for example, [Morse, 1964] or [Chandler, 1987].

Since $f(\alpha)$ and α are based on the D_qs, why should we go to the extra computational trouble? The answer is twofold: First, there is a relatively simple interpretation of various aspects of the $f(\alpha)$ curve, as we shall see later. Second, this curve displays in a straightforward fashion some expected universal features.

To get some feeling for $f(\alpha)$, let us first compute this quantity for a simple probability distribution. We can do this by using Eq. (10.5-1) with a specified distribution function. Specifically, let us assume that the probability distribution function for x between 0 and 1 is given by

$$\rho(x) = kx^{-1/2} \qquad (10.5\text{-}13)$$

where k is a proportionality constant. The probability of finding the particle in segment of length R ($R \ll 1$) is given by the integral

$$p(R) = \int_{x_0}^{x_0+R} \rho(x)dx \qquad (10.5\text{-}14)$$

which gives

$$\begin{aligned} p(R) &= kR^{1/2} \text{ for } x_0 = 0 \\ p(R) &= kR \text{ for } x_0 > 0 \end{aligned} \qquad (10.5\text{-}15)$$

Since only one point ($x_0 = 0$) has the scaling index $\alpha = 1/2$, the corresponding fractal dimension $f(\alpha = 1/2) = 0$. On the other hand, a continuous one-dimensional segment of x values has the scaling index $\alpha = 1$ with the corresponding fractal dimension $f(\alpha) = 1$.

Generalized Partition Functions and f(α) for Weighted and Asymmetric Cantor Sets

For a variety of interesting examples involving various fractals generated by recursive procedures, it is useful to define a generalized partition function that allows for the possibility of cells of variable size. By evaluating this function for several simple examples, we shall gain considerable insight into the workings of the $f(\alpha)$ curve. This generalized partition function is defined to be

$$\Gamma(q,\tau) = \sum_i \frac{p_i^q}{R_i^\tau} \qquad (10.5\text{-}16)$$

The parameter τ is chosen so that for a fixed value of q, $\Gamma(q,\tau) = 1$. The meaning of the parameter τ becomes apparent if we evaluate Γ for the special case of equal cell sizes $R_i = R$. Then taking the logarithm of both sides of Eq. (10.5-16) yields

$$\tau = \frac{\ln \sum_i p_i^q}{\ln R} \qquad (10.5\text{-}17)$$

Comparing this last equation with Eq. (10.4-16) tells us that

$$\tau = (q-1)D_q \qquad (10.5\text{-}18)$$

Thus, we see that the generalized dimension D_q (or, more specifically, $(q-1)D_q$) is that number which, for a given value of q, makes the generalized partition function equal to 1. [The actual numerical value of the partition function is not so important; for τ not equal to $(q-1)D_q$, the partition function diverges to $\pm\infty$.]

To see how this generalized partition function allows us to find $f(\alpha)$ for a recursively generated fractal, let us look at the case of a "weighted" Cantor set. You will recall from Chapter 9 that the canonical "middle-thirds" Cantor set is generated by starting with a line segment of length 1 and eliminating the middle one-third, leaving two segments each of length 1/3. We can generalize this Cantor set by allowing the two newly generated line segments to have different "weights," say p_1 and p_2, with $p_1 + p_2 = 1$. This weighting means that when we assign points to the two segments, a fraction p_1 of the points go to the segment on the left and p_2 go to the segment on the right at the first level of generation. At the second level, we have four segments: one with weight p_1^2; one with p_2^2, and two with $p_1 p_2$. Figure 10.8 shows the first few levels of recursion for this set. A moment's consideration should convince you that the generalized partition function for the nth generation can be written as

$$\Gamma_n(q,\tau) = \frac{1}{R^{n\tau}}(p_1^q + p_2^q)^n \qquad (10.5\text{-}19)$$

where $R = 1/3$ for the standard Cantor set. Again, for a specified value of q, the parameter τ is chosen so that the generalized partition function is equal to 1.

To evaluate the generalized partition function, we follow an argument similar to that used by HJK86. The binomial expression in Eq. (10.5-19) can be expressed in the standard way as a sum of products of p_1 and p_2 raised to various powers:

$$\Gamma_n(q,\tau) = \frac{1}{R^{n\tau}} \sum_{w=0}^{w=n} \frac{n!}{(n-w)!w!} p_1^{qw} p_2^{q(n-w)} \qquad (10.5\text{-}20)$$

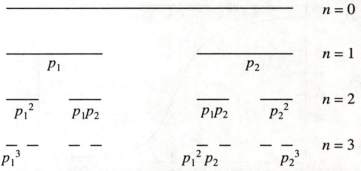

Fig. 10.8. A sketch of the first few generations of the construction of a weighted Cantor set. For the $n = 3$ generation, only a few of the probabilities (weightings) have been indicated explicitly.

where $n!$ (read "n factorial") $= 1 \times 2 \times 3 \times ... \times n$. The factorial combination in the previous equation tells us how many line segments have the weight corresponding to a particular value of w.

Now comes the crucial part of the argument: For large n, there is one term in the sum in Eq. (10.5-20) that is largest and in fact dominates the sum. We can find the corresponding value of w (call it w_*) by differentiating the natural logarithm of the summand with respect to w and setting the resulting derivative equal to 0. (We use the natural logarithm because we can then make use of the Stirling approximation: $\ln n! \approx n \ln n - n$. If the logarithm has a maximum, so does the original function.) When we carry out that differentiation, we find that w_* satisfies

$$\frac{n}{w_*} = 1 + \left(\frac{p_2}{p_1} \right)^q \qquad (10.5\text{-}21)$$

As an aside, we can find the value of τ by requiring that the associated term in the sum satisfy

$$\frac{1}{R^{n\tau}} \frac{n!}{(n - w_*)! w_*!} p_1^{qw_*} p_2^{q(n - w_*)} = 1 \qquad (10.5\text{-}22)$$

This last expression can readily be solved for $\tau = (q - 1)D_q$.

Figure 10.9 shows a plot of D_q as a function of q for the weighted Cantor set with $p_1 = 0.7$ and $p_2 = 0.3$. We shall return to a discussion of the numerical details of that plot in a moment.

We make connection with $f(\alpha)$ by requiring that α be the exponent for the R dependence of our dominant probability (weighting):

$$p_{w_*} = p_1^{w_*} p_2^{(n - w_*)} = R_n^{\alpha} \qquad (10.5\text{-}23)$$

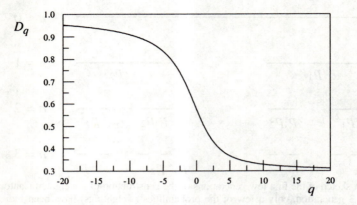

Fig. 10.9. A plot of the generalized dimension D_q as a function of q for the weighted Cantor set with $p_1 = 0.7$ and $p_2 = 0.3$. The numerical values are discussed in the text.

where $R_n = (R = 1/3)^n$ is the length of the segments constructed at the nth generation. Hence, we find

$$\alpha = \frac{w_* \ln p_1 + (n - w_*) \ln p_2}{n \ln R} \tag{10.5-24}$$

Note that α depends only on the ratio n/w_*. Similarly, we find $f(\alpha)$ by asking for the length dependence of the number of segments corresponding to the value w_*:

$$\frac{n!}{(n - w_*)! \, w_* !} = R_n^{-f(\alpha)} \tag{10.5-25}$$

Using the Stirling approximation for the factorials and solving for $f(\alpha)$ yields, after a bit of algebra,

$$f(\alpha) = \frac{\ln\left(1 - \frac{w_*}{n}\right) - \frac{w_*}{n} \ln\left(\frac{w_*}{n} - 1\right)}{\ln R} \tag{10.5-26}$$

which, like α, depends only on the ratio n/w_*.

To determine the entire $f(\alpha)$ curve, for prescribed values of p_1 and p_2, we pick values of q (usually ranging from about -40 to $+40$) and then find n/w_* from Eq. (10.5-21). We then compute α and $f(\alpha)$ from Eqs. (10.5-24) and (10.5-26). The results for $p_1 = 0.7$ and $p_2 = 0.3$ are shown in Fig. 10.10. Let us now try to understand the results shown in Figs. 10.7, 10.9, and 10.10.

When the parameter $q \to \infty$ we see from Eq. (10.5-21) that $w_* \to n$. From Eq. (10.5-26), we see in this situation that $f(\alpha) \to 0$ and we get the smallest value for α, namely $\alpha_{\min} = (\ln p_1)/(\ln R) = 0.324\ldots$ for the case shown in Fig. 10.10. On the other hand, for $q \to -\infty$, we see that $f(\alpha) \to 0$, but we have the largest value for α, namely $\alpha_{\max} = (\ln p_2)/(\ln R) = 1.095\ldots$. The largest value of $f(\alpha)$

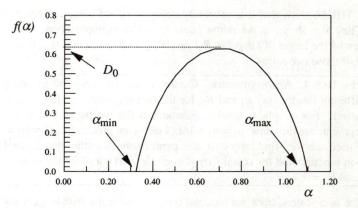

Fig. 10.10. A plot of $f(\alpha)$ for the weighted Cantor set with $p_1 = 0.7$, $p_2 = 0.3$ and $R = 1/3$. The various features of the curve are interpreted in the text.

occurs for $q = 0$ for which we have $n/w_* = 2$. From Eq. (10.5-26), we find $f_{max} = \ln 2/\ln 3$, which is just the box-counting dimension D_0 for the Cantor set with $R = 1/3$.

As we can see from Eq.(10.5-9), the largest value of α is equal to $D_{-\infty}$ and the smallest value of α is equal to D_{∞}. Figure 10.9 shows how D_q approaches these values.

Since $q \to \infty$ emphasizes the largest probability in the generalized partition function, we see that α_{min} corresponds to the most densely populated part of the fractal (the part with the largest probability or weighting). At the other extreme, α_{max} corresponds to $q \to -\infty$ and emphasizes that part of the fractal with the smallest probability, the least densely populated part. This interpretation carries over to all $f(\alpha)$ curves.

Exercise 10.5-2. Work through the algebraic details for the expressions for the weighted Cantor set. Verify the numerical relations for the case given in the text.

Exercise 10.5-3. The $f(\alpha)$ curve can also be calculated for the weighted Cantor set starting with Eq.(10.5-17) which gives $(q-1)D_q$ in terms of the probability sum and then using the Legendre transformation relations in Eqs. (10.5-11) and (10.5-12). Follow through this calculation and compare your results with those given earlier and in Fig.10.10.

Let us now return to the example illustrated in Fig. 10.7. These results were computed for a so-called *two-scale* Cantor set, which is a combination of the asymmetric and weighted Cantor sets introduced earlier. At each stage of construction, we have two different segment lengths R_1 and R_2 and two different weights p_1 and p_2. Using a method similar to that used earlier for the weighted

Cantor set, HJK86 show that it is relatively straightforward to find α and $f(\alpha)$ for this set. They also show, as we might guess from the example given above, that α_{max} is given by the larger of $\{\ln p_1/\ln R_1\}$ or $\{\ln p_2/\ln R_2\}$ and that α_{min} is given by the smaller of those two ratios.

Exercise 10.5-4. An "asymmetric" Cantor set can be generated by using two different lengths, say R_1 and R_2, for the two segments created at each generation. For simplicity's sake, assume that the weights given to the two segments are the same: $p_1 = p_2 = 1/2$. Following procedures similar to those used above, find approximate expressions for the generalized partition function and for α and $f(\alpha)$. Graph the $f(\alpha)$ curve for the case $R_1 = 0.2$ and $R_2 = 0.4$.

For the two-scale Cantor set models, once we have the two lengths and the two probabilities, we can sketch the complete $f(\alpha)$ curve at least approximately: The two lengths and the two probabilities give us the α values for which $f(\alpha)$ goes to 0. We can estimate the peak value of $f(\alpha)$ from the value of the similarity dimension as described in Section 9.7. We also know that the $f(\alpha)$ curve must be concave downward. This information is sufficient to give us at least a rough picture of the entire curve.

Exercise 10.5-5. Using the numerical values listed in the caption of Fig. 10.7, verify the results stated in the previous paragraph.

Let us now look at another example, the $f(\alpha)$ distribution for the logistic map at the period-doubling accumulation point. The results are shown in Fig. 10.11.

The logistic map data illustrate again the basic features of the $f(\alpha)$ distribution. The α values for which $f(\alpha)$ is not 0 lie between 0 and 1. We expect this for the logistic map because, as we argued in Section 9.7, we can treat the attractor at the period-doubling accumulation point approximately as a Cantor set with two length scales ($1/\alpha_F$ and $1/\alpha_F^2$), where α_F is the Feigenbaum $\alpha = 2.502\ldots$ The numerical value of $f(\alpha)$ at its peak is just the box-counting dimension about equal to $0.5388\ldots$.

Exercise 10.5-6. Use the method suggested in the previous paragraph to calculate the values of $D_{-\infty}$ and D_{∞} for the logistic map. Compare your values with the results plotted in Fig. 10.11.

As our final example of multifractal distributions, let us look at the $f(\alpha)$ distribution for data calculated from the sine-circle map [Eq. (6.7-6)] with parameter $K = 1$ and with winding number Ω equal to the golden mean, $G = (\sqrt{5}-1)/2 = 0.6180\ldots$. As we saw in Chapter 6, this set of conditions corresponds to the onset of chaotic behavior. The distribution is plotted in Fig. 10.12.

The general behavior of the distribution is similar to that for the logistic map. The largest value of α can be expressed analytically as

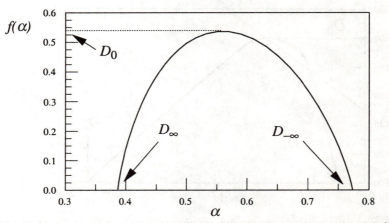

Fig. 10.11. A plot of $f(\alpha)$ distribution for the logistic map attractor at the period-doubling accumulation point. The numerical value of $f(\alpha)$ at the peak of the function corresponds to the box-counting dimension of the attractor.

$$\alpha_{max} = \frac{\ln G}{\ln \beta^{-1}} = 1.8980\ldots \qquad (10.5\text{-}27)$$

where $\beta = 1.2885$, the scaling factor in the neighborhood of $\theta = 0$ for the sine-circle map (HJK86). At the other extreme, the smallest value for α is given by

$$\alpha_{min} = \frac{\ln G}{\ln \beta^{-3}} = 0.6326\ldots \qquad (10.5\text{-}28)$$

The peak of the distribution occurs at $f(\alpha) = 1$, which is to be expected because the iterates of the sine-circle map completely cover the interval $\theta = [0, 2\pi]$ thus giving a box-counting exponent of 1.

A variety of experiments have been analyzed using the $f(\alpha)$ formalism. (Several references are given at the end of the chapter.) The usefulness of this approach is in its ability to recognize universality classes among a diversity of dynamical systems even though the attractors look different to the eye. For example, JKL85 analyzed data from a Rayleigh–Bénard convection experiment and found that its $f(\alpha)$ distribution matched, within experimental uncertainty, the distribution shown in Fig. 10.12 for the sine-circle map. A similar analysis was carried out (SRH87) for a system of coupled semiconductor diode oscillators. In practice, as we have mentioned before, the correlation sum is used to compute the generalized dimensions D_q and the $f(\alpha)$ function is then found by using the Legendre transformations Eqs. (10.5-11) and (10.5-12).

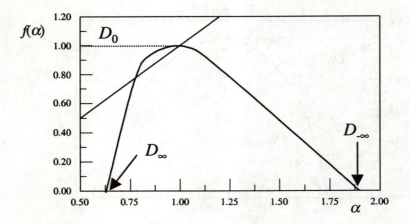

Fig. 10.12. A plot of the $f(\alpha)$ distribution for data from the sine-circle map with $K = 1$ and the winding number equal to the Golden Mean. The α_{max} and α_{min} values can be estimated analytically as described in the text. The 45° line $f(\alpha) = \alpha$ is shown for reference.

10.6 Generalized Entropy and the $g(\Lambda)$ Spectrum

In Section 10.4 we saw how to generalize the notion of dimension to create an entire spectrum of dimensions. Similarly, we can generalize the Kolmogorov entropy to create a series of entropy functions, which, at least in principle, give us more detailed quantitative information about the dynamics of the system.

The easiest starting point is Eq. (9.6-11), the definition of K–S entropy in terms of the sequence probability sum. Let's for the moment restrict ourselves to a one-dimensional system and write the probabilities of the cell sequences in the expanded form $p(i_1, i_2, \ldots, i_d)$ so that N in Eq. (9.6-11) is replaced by d. (Later we will identify d with an embedding space dimension. For now it is just a measure of the length of the cell sequence.) We also set $S_0 = 0$ since all the trajectories start in

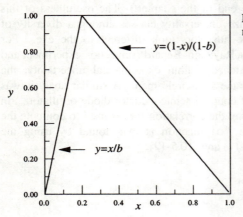

Fig. 10.13. A graph of the tilted tent map function with $b = 0.2$.

the same cell. Then with D_1 in Eq. (10.4-10) in mind, we call the K–S entropy K_1 with

$$K_1 = \frac{-\sum_i p(i_1, i_2, \ldots, i_d) \ln p(i_1, i_2, \ldots, i_d)}{d}$$

We now reverse the argument connecting Eq. (10.4-2) for D_q to Eq. (10.4-10) for D_1 and define a generalized entropy K_q as

$$K_q = \frac{-\ln \sum_i p(i_1, i_2, \ldots, i_d)^q}{(q-1)d} \qquad (10.6\text{-}1)$$

To get a feeling for the K_qs and what the probabilities $p(i_1, i_2, \ldots, i_d)$ mean, let us look at a simple one-dimensional iterated map scheme, called the *tilted tent map*, with the mapping function

$$f_b(x) = \frac{x}{b} \quad \text{for} \quad 0 < x < b$$

$$\qquad (10.6\text{-}2)$$

$$f_b(x) = \frac{1-x}{1-b} \quad \text{for} \quad b < x < 1$$

The map function is plotted in Fig. 10.13.

To evaluate the probabilities, we recall that the probability associated with a trajectory point x landing in a cell can be expressed in terms of the Lyapunov exponent associated with that value of x:

$$p(i_1) = e^{-\lambda(x)} = e^{-\ln|f'(x)|} \qquad (10.6\text{-}3)$$

that is, if the Lyapunov exponent is large and positive, then the probability will be small since trajectories are repelled rapidly from that region of state space. Thus, for the tilted tent map, we argue that the probability associated with trajectory segments with just one step is $p(i_1) = b$ if x is in the interval $[0,b]$ and $p(i_1) = 1-b$ if x is in the interval $[b,1]$. For the sake of simplicity and to develop expressions that are analogous to those used in Section 10.5, let us use $p_1 = b$ and $p_2 = 1-b$. We then have

$$\sum_{i_1} p(i_1)^q \equiv S_q^1 = p_1^q + p_2^q \qquad (10.6\text{-}4)$$

where the middle equality defines the probability sum S_q^1.

For trajectory segments whose length is two steps, we have four possibilities: (1) Both points are in the interval $[0,b]$; (2) both points are in $[b,1]$; (3) the first point is in $[0,b]$ and the second is in $[b,1]$; (4) the first point is in $[b,1]$ and the second is in $[0,b]$. The associated probabilities are $p_1^2, p_2^2, p_1 p_2$, and $p_2 p_1$, respectively. Then the probability sum is

$$\sum_{i_1,i_2} p(i_1,i_2)^q \equiv S_q^2 = (p_1^q + p_2^q)^2 \qquad (10.6\text{-}5)$$

With a quick generalization, we write

$$\sum_{i's} p(i_1,\ldots i_d)^q \equiv S_q^d = (p_1^q + p_2^q)^d \qquad (10.6\text{-}6$$

Using Eq. (10.6-6) in Eq. (10.6-1), the definition of K_q, yields

$$K_q = -\frac{1}{q-1}\ln(p_1^q + p_2^q) \qquad (10.6\text{-}7)$$

With the use of the techniques leading to Eq. (10.4-10), we can show that

$$K_1 = p_1 \ln\left(\frac{1}{p_1}\right) + p_2 \ln\left(\frac{1}{p_2}\right) \qquad (10.6\text{-}8)$$

If we calculate the average Lyapunov exponent for the tilted tent map using the first equality in Eq. (9.5-9) with $p(x) = 1$ for the tilted tent map, we see that K_1 is just the average Lyapunov exponent for the system.

Exercise 10.6-1. Use the first equality in Eq. (9.5-9) with $p(x) = 1$ for the tilted tent map to verify that K_1 is indeed the average Lyapunov exponent for the system.

Exercise 10.6-2. Show that $K_q = K_1 = \ln 2$ for all q if $p_1 = p_2$. In general if all the local Lyapunov exponents for the system are the same then $K_q = K_1$ for all q.

Let us look at a few more special cases. From Eq. (10.6-7), we find that $K_0 = \ln 2$. K_0 is called the topological entropy because it is given by the natural

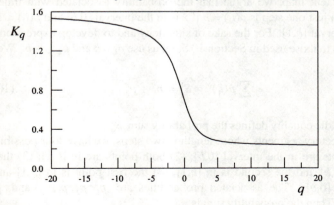

Fig. 10.14. A plot of K_q for the tilted tent map model with $b = 0.2$.

logarithm of the number of fixed points of the system, a topological property. When $q \to \infty$, the larger of p_1 and p_2 dominates in Eq. (10.6-7) and we have $K_\infty = -\ln p_{\max}$. Similarly, when $q \to -\infty$, the smaller of the two probabilities dominates, and we find $K_{-\infty} = -\ln p_{\min}$. Figure 10.14 shows a plot of K_q as a function of q for the tilted tent map with $b = 0.2$.

Exercise 10.6-3. Verify the numerical aspects of Fig. 10.14.

Generalized Entropies and Generalized Correlation Sums

Trying to find the entropies by calculating probabilities is nearly a hopeless task for all but the simplest systems. However, the entropies K_q can be related to the generalized correlation sums C_q whose computation is more straightforward. Let us see how this works for $q = 2$, and then we will generalize the result to other values of q.

K_2 is used because it is particularly easy to compute. Using Eq. (10.6-1), we have

$$K_2 = -\frac{1}{d}\ln \sum_{i_1,i_2,\ldots} p(i_1,i_2,\ldots,i_d)^2 \qquad (10.6\text{-}9)$$

For the case of $d = 1$, following Grassberger and Procaccia (GRA83), we argue that the sum of $p(i_1)^2$ can be expressed as the correlation sum $C_2(R)$ (except for some overall numerical factor that is unimportant when we take logarithms) defined in Eq. (10.4-15). That is, the correlation sum basically involves the probability that two trajectory points fall within a cell of size R. This probability is given approximately by the square of the probability that a single trajectory point falls in that cell (assuming that the "events" are independent).

For $d > 1$, the cell sequence of trajectory points is equivalent to a vector in a d-dimensional embedding space. We relate $p(i_1,\ldots,i_d)^2$ to a generalized correlation sum $C_2^{(d)}(R)$, which can be thought of as a (yet further) generalization of Eq.(10.2-1) and Eq. (10.4-15):

$$\sum_{i's} p(i_1,\ldots,i_d)^2 \approx C_2^{(d)}(R) \qquad (10.6\text{-}10)$$

We see that when $d = 1$, Eq. (10.6-10) reduces to our previous definition of the correlation sum in Eq. (10.4-15).

In practice, K_2 can be computed from the correlation sum $C_2^{(d)}(R)$ by finding how $C_2^{(d)}(R)$ depends on the embedding dimension. Why does this work? Increasing the embedding dimension means that we increase the number of elements of the time-series "vector"; that is, we increase the time length of the sequence of sampled values used in the computation. If the system is behaving chaotically, then we would expect the corresponding trajectory elements to diverge (on the average) as the length of the time sequence increases. The generalized entropy, just like the K–S entropy, measures the <u>rate</u> at which this divergence occurs. Thus, if we determine how the correlation sum decreases with increasing embedding dimension, we can determine the corresponding K_2.

From Eq. (10.6-1), we see that

$$\sum_{i's} p(i_1, \ldots, i_d)^2 = e^{-dK_2} \tag{10.6-11}$$

To complete the connections, we recall that the generalized correlation sum is related to the correlation dimension D_2: $C_2^{(d)}(R) \sim R^{D_2}$. Putting all of this together yields

$$C_2^{(d)}(R) = A R^{D_2} e^{-dK_2} \tag{10.6-12}$$

Taking the natural logarithm of both sides of the previous equation gives us

$$\ln C_2^{(d)}(R) \approx D_2 \ln R - dK_2 \tag{10.6-13}$$

where we have dropped the unimportant $\ln A$ term. From Eq. (10.6-13), we see how to find K_2: With R fixed we plot $\ln C_2^{(d)}(R)$ as a function of d, the embedding dimension; the slope of that curve should be $-K_2$. In practice, we need to be sure to use a value of R that puts the $\ln C_2^{(d)}(R)$ versus $\ln R$ curve in the "scaling region."

We are now ready for the final generalization. Following Pawelzik and Schuster (PAS87), we may define a new correlation sum. To simplify the formalism, let us first introduce a notation for a difference between two of our embedding space "vectors":

$$R_{ij}(d) \equiv \sqrt{\sum_{m=0}^{d-1} (x_{i+m} - x_{j+m})^2} \tag{10.6-14}$$

(As mentioned before, we could also use the maximum component difference as a measure of the difference between two vectors.) In terms of this R_{ij}, we can define a generalized correlation sum for a d-dimensional embedding space as

$$\boxed{\begin{aligned} C_q^{(d)}(R) &= \left\{ \frac{1}{N} \sum_i \left[\frac{1}{N-1} \sum_{j, j \neq i} \Theta(R - R_{ij}) \right]^{q-1} \right\}^{\frac{1}{q-1}} \\ &\approx \sum_{i's} p(i_1, \ldots, i_d)^q \end{aligned} \tag{10.6-15}}$$

Recalling the connection between the generalized dimension D_q and the generalized correlation sum C_q, we have in analogy with Eq. (10.6-13)

$$\boxed{\ln C_q^{(d)}(R) \approx (q-1)D_q \ln R - d(q-1)K_q} \tag{10.6-16}$$

Thus, if we plot $\ln C_q^{(d)}$ as a function of d, we see that the slope (in the scaling region) should be $-(q-1)K_q$.

Let us now look at a few special cases to see what the Cs and the K_qs tell us. For $q = 0$, the entropy K_0, the topological entropy, simply counts the topological properties of the attractor. As we see from the second equality in Eq. (10.6-15), $C_0^{(d)}$ is a sum over all sequences of state space cells for which the probability of visitation is not 0. Thus, K_0 tells us how the logarithm of the number of such sequences grows with embedding dimension d, a topological property of the attractor. We will see later that this number can be related to how the number of (unstable) periodic trajectories increases with the length of the period.

We need a special treatment for $q = 1$. K_1 is sometimes called the *information entropy* in analogy with the information dimension D_1. Using arguments identical to those we used for looking at D_1, we first define a special correlation sum

$$C_1^{(d)}(R) = \frac{1}{N} \sum_i \ln\left[\frac{1}{N-1} \sum_{j, j \neq i} \Theta(R - R_{ij}) \right] \qquad (10.6\text{-}17)$$

If we use Eq. (10.6-1) and the procedures leading to Eq. (10.4-10) we find that K_1 can be written as

$$K_1 = -\frac{1}{d}\frac{1}{N} \sum_i \ln\left[\frac{1}{N-1} \sum_{j, j \neq i} \Theta(R - R_{ij}) \right] \qquad (10.6\text{-}18)$$

so that we have

$$C_1^{(d)}(R) = -dK_1 \qquad (10.6\text{-}19)$$

If we plot $C_1^{(d)}(R)$ as a function of d, the slope should be $-K_1$.

> **Exercise 10.6-4.** Show that Eq. (10.6-18) follows from Eq. (10.6-1) in the limit $q \to 1$.

Dynamical Spectrum $g(\Lambda)$

The generalized entropies suggest that there might be a set of scaling relations that are the analogues of the $f(\alpha)$ spectrum for the generalized dimensions. To see how this comes about, let us recall the definition of the generalized entropies Eq. (10.6-1) and rewrite it using $T = e^{-d}$ to obtain

$$K_q = \frac{1}{q-1}\frac{1}{\ln T} \ln \sum_{i's} p^q(i_1, \ldots, i_d) \qquad (10.6\text{-}20)$$

This form is exactly like that of Eq. (10.4-2) for the generalized dimensions.

Based on the arguments in Section 10.5, we might expect there to be a scaling exponent, usually called $\Lambda(i_1, \ldots, i_d)$, defined in such a way that the probabilities $p(i_1, \ldots, i_d)$ are given by

$$p(i_1,\dots,i_d) = T^{\Lambda(i_1,\dots,i_d)} \tag{10.6-21}$$

Then the number of times we find a value of $\Lambda(i_1,\dots,i_d)$ lying between Λ and $\Lambda + d\Lambda$ is given by

$$n(\Lambda)d\Lambda = T^{-g(\Lambda)}d\Lambda \tag{10.6-22}$$

which defines the spectrum $g(\Lambda)$ of the scaling exponents.

Now the argument proceeds exactly as in Section 10.5 with $f(\alpha)$ replaced by $g(\Lambda)$. In analogy with Eq. (10.5-9), we end up with the relationship

$$K_q = \frac{1}{q-1}[\Lambda q - g(\Lambda)] \tag{10.6-23}$$

In practice, $g(\Lambda)$ is computed by first finding the generalized correlation sum and the generalized entropy. Next, we compute the auxiliary quantity

$$\tau(q) = (q-1)K_q = \Lambda q - g(\Lambda) \tag{10.6-24}$$

Finally, we construct [in analogy to Eqs. (10.5-11) and (10.5-12)] the Legendre transformation relations

$$\Lambda(q) = \frac{d}{dq}[(q-1)K_q]$$
$$g(\Lambda) = q\frac{d}{dq}[(q-1)K_q] - (q-1)K_q \tag{10.6-25}$$

Before examining the results of some computations of $g(\Lambda)$, let us connect Λ to something more familiar, namely, Lyapunov exponents. For a one-dimensional system described by an iterated map function $f(x)$, a cluster of trajectories starting inside a cell of size R located at x_1 will spread to $|f'(x_1)|$ cells after one iteration

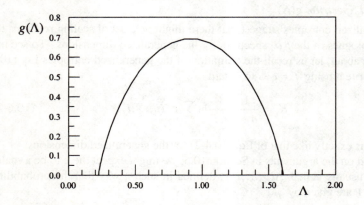

Fig. 10.15. A plot of the $g(\Lambda)$ distribution for the tilted tent map with $b = 0.2$.

step. Thus, the probability of a certain sequence of cells being visited can be expressed as

$$p(i_1,\ldots,i_d) = \frac{p(1)}{|f'(x_1)f'(x_2)\ldots f'(x_{d-1})|} \tag{10.6-26}$$

where $p(1)$ is the probability of being in the first cell. Hence, we find for the logarithm of the probability

$$\ln p(i_1,\ldots,i_d) = \ln p(1) - \sum_{j=1}^{d-1} \ln |f'(x_j)| \tag{10.6-27}$$

For large values of d, we can neglect the $p(1)$ term in comparison to the other term on the right of Eq. (10.6-27), and we find that

$$p(i_1,\ldots,i_d) \approx e^{-\sum_j \ln|f'(x_j)|} = T^{\frac{1}{d}\sum_j \ln|f'(x_j)|} \tag{10.6-28}$$

We immediately (?) recognize the sum of the logarithms of the absolute values of the derivatives of the mapping function as defining the (local) Lyapunov exponent for the particular trajectory segment leading through the sequence of cells $i_1,\ldots,i_d$. Thus, the individual $\Lambda_{i_1,\ldots,i_d}$ is the Lyapunov exponent for that particular sequence. Hence, we see that the $g(\Lambda)$ function describes the distribution of Lyapunov exponents over the attractor.

The actual computation of the $g(\Lambda)$ distribution proceeds in analogy with the $f(\alpha)$ calculations described in Section 10.5. For the tilted tent map, Eq. (10.6-7) gives us an explicit expression for K_q. Using that expression, we invoke the Legendre transformations in Eq. (10.6-25) to find

$$\Lambda(q) = \frac{-p_1^q \ln p_1 - p_2^q \ln p_2}{p_1^q + p_2^q} \tag{10.6-29}$$

$$g(\Lambda(q)) = \ln(p_1^q + p_2^q) + q\Lambda(q)$$

Figure 10.15 shows the $g(\Lambda)$ distribution for the tilted tent map with $b = 0.2$. Note that the largest value of $g(\Lambda)$ corresponds to K_0 and that the end points of the distribution correspond to K_∞ and $K_{-\infty}$ in direct analogy to the $f(\alpha)$ plots.

Exercise 10.6-5. Verify the calculations leading to Eq. (10.6-29) and the numerical values shown in Fig. 10.15.

As another example, Fig. 10.16 displays the results of finding K_q and the resulting $g(\Lambda)$ distribution for data taken from the semiconductor diode circuit described in Chapter 1. Conditions were set so that the trajectories are chaotic with an average Lyapunov exponent of about 0.5 (per cycle of the driving signal). We see that there is a distribution of Λ values. The largest value corresponds to regions of the attractor with the most rapid divergence of nearby trajectories, while

412 Chapter 10

the smallest value of Λ corresponds to that region of the attractor with the least rapid divergence, which for a system modeled by a quadratic maximum map should be close to 0. Other examples of $g(\Lambda)$ distributions can be found in PAS87 and [Schuster, 1995].

> **Exercise 10.6-6.** By using a partition function defined in analogy to Eq. (10.5-16)
>
> $$\Gamma(q,\tau,d) = \frac{\sum_i p^q(i_1,\ldots,i_d)}{T^\tau}$$
>
> with $T = e^{-d}$, use the method outlined in Section 10.5 for the weighted Cantor set to find $g(\Lambda)$ for the tilted tent map. Compare your results with those computed using the Legendre transformation method.

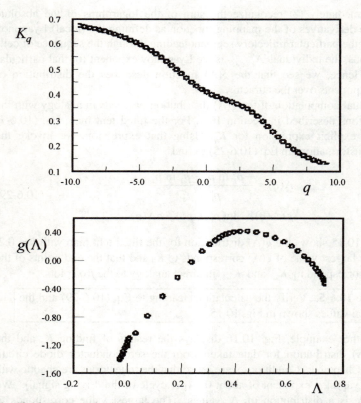

Fig. 10.16. Computed values of K_q and $g(\Lambda)$ for data from the semiconductor diode circuit of Chapter 1. (Data courtesy of Ryan Wallach.)

10.7 Characterizing Chaos via Periodic Orbits

In this section we shall describe yet another method of providing a quantitative description of chaotic behavior. Although this method is fairly new and not as widely tested as some of the methods described earlier, we believe that it provides many insights lacking in the statistical methods. This new analysis may eventually turn out to be more useful.

To introduce this new method, we first recall that the attracting region of state space for chaotic behavior contains, in addition to the chaotic orbit points, a vast number of periodic trajectory points. For example, for the logistic map function for parameter $A > A_\infty$ there is an infinity of periodic points, which developed as a result of the period-doubling bifurcations. However, these periodic trajectory points are generally (except within the periodic windows) unstable: A trajectory in the neighborhood of one of those points is repelled by it. In a sense, a chaotic trajectory is chaotic because it must weave in and around all of these unstable periodic points yet remain in a bounded region of state space. The basic idea of this new method is to characterize the chaotic attractor by means of the properties (for example, the characteristic exponents) of the unstable periodic points.

This idea is not new. For example, the Lorenz model attractor's positive Lyapunov exponent could be estimated from the positive characteristic exponent associated with the unstable fixed points of the Lorenz model. In other words, the exponential divergence of trajectories is dominated, at least for some situations, by the positive characteristic exponent associated with the out-set of the system's saddle points. Also, Poincaré realized the importance of periodic orbits [Poincaré, 1892](translation from DHB92): "What renders these periodic trajectories so precious to us is that they are, so to speak, the only breach through which we might try to penetrate into a stronghold hitherto reputed unassailable."

What is new is the realization that the unstable periodic points, at least for low order periodicities, can be found directly from a time-series record of chaotic trajectories. There is no need to know the underlying differential equations or map function. Furthermore, once the periodic points are found, their characteristic values can be estimated (often fairly accurately) directly from the chaotic trajectory time-series data. (See the references at the end of this chapter.) There are also powerful methods for finding periodic points when the time evolution equations are known (DHB92).

A further advantage of this periodic orbit analysis is that it gives a hierarchical method of characterizing chaotic attractors. Much of our success in understanding in detail the attractors at the period-doubling accumulation point or at the onset of chaotic behavior for quasi-periodicity with the golden-mean winding number is due to our ability to describe these attractors as the limit of a sequence of periodic attractors. The renormalization formalism described in Chapters 5 and 6 allows the determination of the quantitative properties of the attractor. In an analogous fashion, the analysis of a chaotic attractor (perhaps deep within the chaotic regime) can be described in terms of a hierarchy of (unstable) periodic orbits of increasing length.

How do we find those periodic points if we have available only data from a chaotic trajectory, which by definition, is not periodic? The basic idea can be understood for one-dimensional data by considering the Poincaré map generated directly from experimental data. For example, we saw in Chapter 5, Fig. 5.3, that a plot of I_{n+1} as a function of I_n for the semiconductor diode circuit of Chapter 1 gives a one-dimensional curve (if the dissipation is sufficiently high). The intersection of that curve with the $I_{n+1} = I_n$ line gives the location of a period-1 trajectory point. However, that periodic point is unstable. If we plot I_{n+2} as a function of I_n, then the intersection of that curve with the $I_{n+2} = I_n$ line gives the location of unstable period-2 points. In principle, we can continue in this way to locate periodic points of higher and higher order. Furthermore, from the slopes of the curves at the periodic points, we can determine the Lyapunov exponents for those points. Obviously, in practice, we are limited by the precision of the data to finding periodic points of low order.

For higher-dimensional systems, we can proceed in essentially the same way. From a time-series record of a single variable we can construct an embedding (reconstruction) space as described earlier in this chapter. We can then scan through this record to find trajectory points that are within some specified distance of one another after n time units. This is sometimes called the "method of close returns." These points are assumed to be in the neighborhood of the unstable period-n points in the embedding space. Since there is generally more than one period-n cycle for a given set of parameter values, the period-n points are then sorted into the individual period-n cycles. Once the period-n cycles are found, the characteristic values for the cycle points can be estimated by finding how nearby trajectories diverge from the cycle points. We will not go into the details here; the reader is referred to ACE87, CVI88a, and MHS90 for more information. Numerical experiments indicate (ACE87) that finding orbits up to $n = 10$ is sufficient for determining the topological entropy K_0 and the box-counting dimension D_0 of the attractor to within a few percent accuracy.

As an example of this kind of treatment, we quote the relationship between the number of periodic points and the topological entropy K_0 (BIW89):

$$K_0 = \lim_{n \to \infty} \frac{1}{n} \log_2[N(n) - 1] \qquad (10.7\text{-}1)$$

where $N(n)$ is the number of points belonging to periodic orbits of order n and its divisors (including 1 and n).

In addition to providing an alternative method of determining some of the statistical properties of a chaotic attractor, the periodic orbit analysis provides a way of classifying chaotic attractors and of distinguishing among attractors that may have the same statistical description (MHS90). In the statistical characterization of chaotic attractors, we have focused on so-called metric properties such as

Lyapunov exponents and fractal dimensions that do not depend on the coordinates used to describe the system. However, the metric properties do change, as we have seen, with parameter values. Hence, it is difficult to use them as definitive tests of the validity of any particular model. Another type of description emphasizes the topological properties of the periodic orbits (SOG88, CGP88, GLV89). In simple terms, the topological properties specify how the periodic orbits weave in and around each other in the chaotic attracting region of state space. This topological information tells us about the "organization" of the attractor in a way that is independent of both the coordinates used to describe state space and the control parameter values. As control parameters are changed, the periodic orbits may appear or disappear, but the evidence is that the topological properties of the remaining orbits are not affected by this appearance or disappearance.

The topological classification is carried out (MHS90) by finding the relative rotation rates for the periodic orbits. The relative rotation rate specifies how many times one orbit rotates around the other orbit during one period. This rotation rate can be determined by using a Poincaré section of the two periodic orbits: One draws a vector from the intersection point of one orbit with the Poincaré plane to an intersection point of the other orbit with the plane. As the trajectories evolve (between Poincaré section crossings), this vector will rotate. If one orbit has period n_A (that is, its period is n_A times the time between Poincaré section crossings) and the other orbit has period n_B, then the vector will return to its original position after $n_A \times n_B$ periods. The relative rotation rate is the number of full rotations of the vector that occur during $n_A \times n_B$ periods divided by the number of periods.

10.8 *Statistical Mechanical and Thermodynamic Formalism

All of the formalism introduced in this chapter can be brought together in formal structures that resemble the relationships of statistical mechanics and thermodynamics. This connection is perhaps not too surprising since the goal of the generalized dimensions and generalized entropies is to give a detailed statistical description of attractors and trajectories. First, we give a brief summary of the formalism of statistical mechanics and thermodynamics. We will then describe how an analogous formalism can be constructed that embodies all of the results of this chapter. Readers who are not familiar with statistical mechanics may skip this section.

In statistical mechanics, the primary computational tool for a system in thermal equilibrium is the so-called *partition function*, which is generated by summing a Boltzmann-type factor over all the energy states of the system (which, for the sake of simplicity, we take to be a discrete set):

$$Z = \sum_n g_n e^{-\beta E_n} \qquad (10.8\text{-}1)$$

where g_n is the degeneracy factor for the nth energy; that is, g_n specifies how many states have the energy E_n. As usual, β is proportional to the reciprocal of the absolute (Kelvin) temperature for the system. The crucial point is that all the thermodynamic properties of the system can be computed from the partition function.

Let us illustrate how we use the partition function to compute various thermodynamic quantities. For example, the internal energy of the system is obtained from

$$E(\beta) = -\frac{\partial \ln Z}{\partial \beta} \tag{10.8-2}$$

and the Helmholtz free energy (the energy that can be extracted from the system at constant temperature) is given by

$$F(\beta) = -\frac{\ln Z}{\beta} \tag{10.8-3}$$

Finally, the dimensionless entropy for the system is

$$S(\beta) = \beta^2 \frac{\partial F}{\partial \beta} = \beta\{E(\beta) - F(\beta)\} \tag{10.8-4}$$

For dynamical systems, two partition functions, Z_m and $\Gamma(q,\tau)$, can be defined. The first is given in terms of the local Lyapunov exponent associated with all the periodic points of order m; that is, points belonging to a period-m trajectory. (This analogy between energy states and trajectories associated with periodic points will appear again in Chapter 12 for our discussion of the relationship between quantum mechanics and chaotic dynamical systems.) We call this first partition function the *dynamical partition function* since it emphasizes the temporal dynamics of the system as embodied in the Lyapunov exponents. The second type of partition function involves the spatial scaling exponents, and we call it the "structural partition function."

Dynamical Partition Function

Let us begin with the dynamical partition function defined as

$$Z_m = \sum_{x_m} e^{-\beta\lambda(x_m)} \tag{10.8-5}$$

In this expression, β is a parameter that plays the role of inverse temperature. x_m is a trajectory point on a periodic cycle of length m. The sum is over all such points. In this formalism, we are using the (unstable) periodic points to characterize the system. If the system has a multidimensional state space, then $\lambda(x_m)$ is the largest (positive) Lyapunov exponent at the trajectory point x_m.

If the dynamical system can be modeled by a one-dimensional iterated map, we can write the partition function Z_m as follows

$$Z_m = \sum_{x_m} e^{-\beta \ln|f^{(m)'}(x_m)|} \tag{10.8-6}$$

where $f^{(m)'}(x_m)$ is the derivative of the mth iterate of the map function evaluated at the point x_m, one of the fixed points of the mth iterate.

Let us see how this formalism works by applying it to the tilted tent map (see Fig. 10.13), an iterated map system with the mapping function repeated here

$$f_p(x) = \frac{x}{b} \quad \text{for} \quad 0 < x < b$$

$$f_p(x) = \frac{1-x}{1-b} \quad \text{for} \quad b < x < 1 \tag{10.8-7}$$

To evaluate the partition functions we need to know the slope of the function (and its iterates) at its fixed points. From Fig. 10.13 we see that the function itself has two fixed points, one at $x = 0$ with slope $1/b = 1/p_1$ and the other between b and 1 with slope $-1/(1-p_1) = -1/p_2$ (using the notation introduced in Section 10.6). The second iterate has four fixed points, one with slope $(1/p_1)^2$, one with slope $-(1/p_2)^2$ and two with slopes $\pm 1/(p_1 p_2)$.

Let us now evaluate the partition functions Z_1 and Z_2. From Eq. (10.8-6), we write Z_1 directly:

$$Z_1 = e^{-\beta \ln\frac{1}{p_1}} + e^{-\beta \ln\frac{1}{p_2}} \tag{10.8-8}$$

Some straightforward algebra allows us to write the partition function as

$$Z_1 = p_1^\beta + p_2^\beta \tag{10.8-9}$$

We evaluate Z_2 by first noting that $f^{(2)}$ has four fixed points. Hence, we have

$$Z_2 = e^{-\beta \ln\left(\frac{1}{p_1}\right)^2} + e^{-\beta \ln\left(\frac{1}{p_2}\right)^2} + 2e^{-\beta \ln\left(\frac{1}{p_1 p_2}\right)} \tag{10.8-10}$$

which, again after a little algebra gives us

$$Z_2 = (p_1^\beta + p_2^\beta)^2 \tag{10.8-11}$$

We then see that the obvious generalization for Z_m is

$$Z_m = (p_1^\beta + p_2^\beta)^m \tag{10.8-12}$$

(You should note the similarity between this result and the results for the generalized partition function for the Cantor sets discussed in Section 10.5.)

Exercise 10.8-1. Check Eq. (10.8-12) explicitly for the case $m = 3$ by writing out the partition function term by term.

Once we have the partition functions in hand, we can then calculate other thermodynamic properties. The (average) internal energy of the system is given by

$$E(\beta) = -\frac{1}{m}\frac{\partial \ln Z_m}{\partial \beta}$$

$$= -\frac{p_1^\beta \ln p_1 + p_2^\beta \ln p_2}{p_1^\beta + p_2^\beta} \tag{10.8-13}$$

Of particular interest is $E(\beta = 0)$:

$$E(0) = \frac{-\ln p_1 - \ln p_2}{2} \tag{10.8-14}$$

Hence, we see that $E(0)$ is just the average (unweighted) Lyapunov exponent for the system. This result should not be too surprising since our construction of the partition function started with the Lyapunov exponent in the argument of the exponential in the partition function. That observation suggests that we should view

$$\frac{e^{-\lambda(x_m)}}{Z_m} \tag{10.8-15}$$

as a probability associated with the fixed point x_m. The larger the (positive) Lyapunov exponent, the smaller is the probability that a trajectory finds itself near that fixed point. Next, we note that the usual (weighted) average Lyapunov exponent is given by $E(1) = -p_1 \ln p_1 - p_2 \ln p_2$.

Let us now find the Helmholtz free energy for the system:

$$F(\beta) = -\frac{1}{m\beta}\ln Z_m(\beta)$$

$$= -\frac{1}{\beta}\ln(p_1^\beta + p_2^\beta) \tag{10.8-16}$$

Of somewhat more interest are the various dimensionless entropies, which we find according to the standard procedures by taking the derivative of the free energy with respect to β, the inverse temperature parameter:

$$S(\beta) = \beta^2 \frac{\partial F}{\partial \beta} \tag{10.8-17}$$

which for the tilted tent map gives

$$S(\beta) = \frac{(p_1^\beta + p_2^\beta)\ln(p_1^\beta + p_2^\beta) - \beta(p_1^\beta \ln p_1 + p_2^\beta \ln p_2)}{p_1^\beta + p_2^\beta} \tag{10.8-18}$$

We can unravel the meaning of this rather complicated looking expression by evaluating the entropy for specific values of β. For example, setting $\beta = 0$ gives

$$S(0) = \ln 2 \qquad (10.8\text{-}19)$$

which is the map's topological entropy (i.e., the natural logarithm of the number of fixed points for the first iterate of the map function).

For $\beta = 1$, we get from Eq. (10.8-18)

$$S(1) = -p_1 \ln p_1 - p_2 \ln p_2 \qquad (10.8\text{-}20)$$

which we recognize as the K–S entropy K_1 for the system. For the tilted tent map, this is also the average Lyapunov exponent.

Let us summarize what we have learned from this example. For a system such as the tilted tent map for which we can write down explicitly the sum over fixed points in Eq. (10.8-5), we can calculate the generalized partition function Z_m. From that partition function we can then calculate the generalized entropies for the system. What is somewhat surprising is the role played by β, the inverse temperature parameter: β picks out the order q of the generalized entropy.

Exercise 10.8-2. Discuss the relationship between the definition of topological entropy in Eq. (10.8-19) and the one given after Eq. (10.6-7).

Structural Partition Function

The second type of partition function emphasizes the spatial structure of the dynamical system's attractor. Let us call it the **structural partition function**. It was defined in Eq. (10.5-16), but we repeat it here:

$$\Gamma(q,\tau) = \sum_i \frac{p_i^q}{R_i^\tau} = 1 \qquad (10.8\text{-}21)$$

where p_i is the probability that a trajectory lands in cell i. Recall that the parameter $\tau(q)$ is chosen so that the right-hand equality holds, and we established that $\tau = (q-1)D_q$.

To make a connection to a thermodynamic formalism, we note that $\tau(q)$ is proportional to the logarithm of a partition function. Hence, it plays the role of a free energy. From Eq. (10.5-9), we see that $\tau(q)$ is related to the scaling spectrum $f(\alpha)$ by

$$\tau(q) = q\alpha - f(\alpha) \qquad (10.8\text{-}22)$$

If we compare this result with the usual thermodynamic relation among internal energy E, free energy F and entropy S, namely $F = E - S/\beta$, we see that we can identify $f(\alpha)$ with entropy and α with internal energy.

At this point it is worth asking: Why bother with this thermodynamic formal analogy? The answer is that once this identification has been made, we can use the accumulated experience of thermodynamics to help us think about and organize our experiences with dynamical systems characterized either by the dynamical spectrum $g(\Lambda)$ or by the scaling spectrum $f(\alpha)$. For example, some dynamical systems show a discontinuity in their $f(\alpha)$ versus α curves. A discontinuous jump in $f(\alpha)$ is then called a phase transition in analogy with first-order phase transitions in thermodynamics, which have a discontinuous change in entropy. In rough terms, if the $f(\alpha)$ curve shows a jump, we can think of the attractor as being a mixture of two (or sometimes more) distinct phases with different densities, much like a mixture of water droplets and water vapor. Moreover, the many methods for characterizing thermodynamic systems may suggest novel ways of studying dynamical systems. Conversely, we might expect that our knowledge of dynamical systems will enrich our understanding of thermodynamics and statistical mechanics. For a recent review, see ZAS99.

10.9 Wavelet Analysis, q-Calculus and Related Topics

In this section, we mention briefly several other (related) techniques for the analysis of time series data from nonlinear systems. All share the feature of looking at how the properties of the system change under a rescaling of the coordinates, much in the spirit of the multifractal analysis of Section 10.5.

The first of these techniques makes use of a mathematical technique known as *wavelet analysis*. Wavelet analysis can be thought of as a generalization of Fourier analysis. As Appendix A discusses, Fourier analysis tells us how much our data look like sine and cosine functions of different periodicities. Analogously, the wavelet analysis tells us how much our data look like various "wavelets," which are functions peaked at various (spatial or temporal) locations with various "widths" or spreads in space or time. The references at the end of the chapter give a sampling of wavelet analysis applied to nonlinear data. See BON98 in particular.

Another mathematical technique called *q*-calculus or *q*-analysis can be thought of as a generalization of ordinary calculus. Instead of having just first derivatives, second derivatives, and so on, we can define a "fractional" q-derivative:

$$\partial_x^{(q)} f(x, y, \ldots) \equiv \frac{f(qx, y, \ldots) - f(x, y, \ldots)}{(q-1)x} \qquad (10.9\text{-}1)$$

The q-derivative tells us how the function f changes when the x coordinate, for example, is stretched by the factor q. (Recall that the ordinary derivative tells us how the function changes when the x coordinate is shifted by a small amount.) In the limit $q \to 1$, the q-derivative is the same as the ordinary derivative. It turns out that q calculus can be used to characterize multifractal sets (ERE97).

Recently, Tsallis (TSA88) has introduced a new version of thermodynamics, called **nonextensive thermodynamics**, in which entropy, free energy and so on are not proportional to the size of the system (as they are in ordinary thermodynamics). This model then provides a further generalization of the entropy ideas and multifractal distributions discussed in this chapter. Such generalizations may prove useful in characterizing nonlinear systems (CLP97, LYT98, ANT98).

10.10 Summary

Let us summarize what we have seen in this chapter. There have been many so many trees that it may be easy to lose sight of the forest. In brief, we would say that there are now available many different ways of quantitatively describing chaotic systems. Some of these methods emphasize the spatial characteristics of the state space region visited by trajectories in terms of generalized (fractal) dimensions. Other methods, such as the generalized entropies, emphasize the temporal dynamics of the system. Both the generalized dimensions and entropies can be computed from the data from the time series samples of a single dynamical variable by combining the embedding space techniques with the generalized correlation sums.

All of these methods are descriptive. They allow us to provide numerical values of quantities that describe the dynamics of the system. In a few cases there are predictions of universality. For example, for systems that follow the quasi-periodic route to chaos and have dynamics in the same class as the sine-circle map, the $f(\alpha)$ spectrum at the critical point at the onset of chaos is supposed to be universal. Unfortunately, these methods do not generally provide us with any predictive power. Moreover, although much progress has been made in simplifying and speeding up the computation of dimensions and entropies, these calculations still consume an enormous amount of computer time for even relatively small data sets. If we simply want to have some quantitative measure to monitor a chaotic dynamical system, then perhaps finding just the largest Lyapunov exponent or the box-counting or correlation dimension is sufficient. In any case, the goal of computing any of these quantitative measures for a dynamical system should be an understanding of the dynamics of that system. The computation is not a goal in itself.

The topological classification of dynamics in terms of periodic orbits seems to hold great promise in developing our understanding of dynamical systems. However, this field is just in its infancy, and we must wait to see if this young tree will bear useful fruit.

To conclude, we also point out that all of the analysis given in Chapters 9 and 10 has assumed that the system's parameters remained constant. For actual experimental or "real world" systems, you must worry about the validity of this assumption. For the laboratory realization of many chemical and physical systems, this assumption does not pose a significant problem. However, for living biological systems or uncontrolled systems such as variable stars or the stock market, the validity of this assumption must be questioned. We are then faced with a difficult question: What is the significance of some quantifier, such as a correlation dimension, if it is computed from a time-series of data recorded over a period of time during which the system's parameters were changing? Here we are on rather mushy ground. At best, in this kind of situation, any calculated quantity represents an average over parameter values. At worst, they may have no meaning at all because the system has not settled into an attracting region of state space and the computed quantities are some kind of (rather ill-defined) characteristic of the transient behavior of the system. In some cases, however, there is interest in analyzing the transient behavior itself. Time-series methods can be helpful there (JAT94).

Several methods have been proposed for recognizing and dealing with "non-stationary" systems—systems whose parameters are changing with time—since in many cases these are in fact quite interesting systems and in other cases they are unavoidable. These methods have generally focused on identifying *recurrence times*, which serve to measure how often a trajectory revisits a particular neighborhood in the state space. The statistics of these times can help identify non-stationary conditions. For a recent survey, see GAO99.

10.11 Further Reading

Time-Series Embedding

N. H. Packard, J. P. Crutchfield, J. D. Farmer, and R. S. Shaw, "Geometry from A Time Series," *Phys. Rev. Lett.* **45**, 712–15 (1980).

F. Takens in *Dynamical Systems and Turbulence*. Vol. 898 of Lecture Notes in Mathematics, D. A. Rand and L. S. Young, eds. (Springer Verlag, Berlin, 1981).

M. Casdagli, S. Eubank, J. D. Farmer, and J. Gibson, "State Space Reconstruction in the Presence of Noise," *Physica D* **51**, 52–98 (1991). An extensive discussion of state space reconstruction techniques using embedding.

T. Sauer, J. A. Yorke, and M. Casdagli, "Embedology," *J. Stat. Phys.* **65**, 579–616 (1991).

H. D. I. Abarbanel, R. Brown, J. J. Sidorowich, and L. Sh. Tsimring, "The analysis of observed chaotic data in physical systems," *Rev. Mod. Phys.* **65**, 1331–92 (1993).

M. T. Rosenstein, J. J. Collins, and C. J. De Luca, "Reconstruction expansion as a geometry-based framework for choosing proper delay times," *Physica D* **73**, 82–98 (1994).

H. Abarbanel, *Analysis of Observed Chaotic Data* (Springer, New York, 1996). Highly recommended for its encyclopedic and careful coverage of many analysis techniques.

[Williams, 1997]. This book has extensive discussions of time-series analysis using the techniques discussed in this chapter.

R. Hegger, H. Kantz, and T. Schreiber, "Practical implementation of nonlinear time series methods: The TISEAN package," *Chaos* **9**, 413–35 (1999). Describes a software package TISEAN with many features for time-series analysis.

Generalized Dimensions

P. Grassberger, "Generalized Dimensions of Strange Attractors," *Phys. Lett. A* **97**, 227–30 (1983).

H. Hentschel and I. Procaccia, "The Infinite Number of Generalized Dimensions of Fractals and Strange Attractors," *Physica D* **8**, 435–44 (1983).

Kaplan–Yorke Conjecture, the Lyapunov Spectrum, and the Lyapunov Dimension

J. Kaplan and J. A. Yorke, *Springer Lecture Notes in Mathematics* **730**, 204 (1979).

D. A. Russell, J. D. Hansen, and E. Ott, "Dimensions of Strange Attractors," *Phys. Rev. Lett.* **45**, 1175–78 (1980).

P. Constantin and C. Foias, *Commun. Pure Appl. Math.* **38**, 1 (1985). This paper proves that the Lyapunov dimension D_L is always greater than or equal to the Hausdorff dimension D_H.

X. Zeng, R. Eykholt, and R. Pielke, "Estimating the Lyapunov-Exponent Spectrum from Short Time Series of Low Precision," *Phys. Rev. Lett.* **66**, 3229–32 (1991). This paper suggest a method for determining the full set of Lyapunov exponents given only time-series data from a single variable.

J. C. Sommerer and E. Ott, "Particles Floating on a Moving Fluid: A Dynamical and Comprehensible Physical Fractal," *Science* **259**, 335–39 (1993). The information dimension, Lyapunov exponents, and Lyapunov dimension are used to analyze the behavior of fluorescent tracer particles on a fluid surface.

Methods for determining the spectrum of Lyapunov exponents from time series data

U. Parlitz, "Identification of True and Spurious Lyapunov Exponents from Time Series," *Int. J. Bifur. Chaos*, **2**, 155–65 (1992).

Th.-M. Kruel, M. Eiswirth, and F. W. Schneider, "Computation of Lyapunov spectra: Effect of interactive noise and application to a chemical oscillator," *Physica D* **63**, 117–37 (1993).

T. D. Sauer, J. A. Tempkin, and J. A. Yorke, "Spurious Lyapunov Exponents in Attractor Reconstruction," *Phys. Rev. Lett.* **81**, 4341–44 (1998).

R. Hegger, "Estimating the Lyapunov spectrum of time delay feedback systems from scalar time series," *Phys. Rev. E* **60**, 1563–66 (1999).

K. Ramasubramanian and M. S. Sriram, "Alternative algorithm for the computation of Lyapunov spectra of dynamical systems," *Phys. Rev. E* **60**, R1126–29 (1999).

Analyzing Data with Embedding and Correlation Dimensions

See the references at the end of Chapter 9.

N. B. Abraham, A. M. Albano, B. Das, G. De Guzman, S. Yong, R. S. Gioggia, G. P. Puccioni, and J. R. Tredicce, "Calculating the Dimension of Attractors from Small Data Sets," *Phys. Lett A* **114**, 217–21 (1986).

J. W. Havstad and C. L. Ehlers, "Attractor Dimension of Nonstationary Dynamical Systems from Small Data Sets," *Phys. Rev. A* **39**, 845–53 (1989).

A. M. Albano, J. Muench, C. Schwartz, A. I. Mees, and P. E. Rapp, "Singular-Value Decomposition and the Grassberger–Procaccia Algorithm," *Phys. Rev. A* **38**, 3017–26 (1988). This paper suggests the use of the so-called singular-value decomposition to enhance the speed of calculation of the correlation dimension.

J. Theiler, "Spurious Dimension from Correlation Algorithms Applied to Limited Time-Series Data," *Phys. Rev. A* **34**, 2427–32 (1986). Computes the correlation dimension for autocorrelated stochastic data and for numerical data from a delay-differential equation model. Emphasizes the need for a minimum time delay for comparison vectors.

J. P. Eckmann and D. Ruelle, "Ergodic Theory of Chaos and Strange Attractors," *Rev. Mod. Phys.* **57**, 617–56 (1985). A good review of the techniques and theory of the methods of quantifying chaos.

E. N. Lorenz, "Dimension of Weather and Climate Attractors," *Nature* **353**, 241–44 (1991). Correlation dimensions may not always give simple results if the system consists of several loosely coupled parts.

M. Ding, C. Grebogi, E. Ott, T. Sauer, and J. A. Yorke, "Plateau Onset for Correlation Dimension: When Does It Occur?" *Phys. Rev. Lett.* **70**, 3872–75 (1993).

M. Ding, C. Grebogi, E. Ott, T. Sauer, and J. A. Yorke, "Estimating correlation dimension from a chaotic time series: when does plateau onset occur?" *Physica D* **69**, 404–24 (1993). These two papers give a detailed account of finding the correlation dimension D_c from time series data using the embedding (reconstruction) method including an analysis of what can go wrong and how the results may depend on the embedding dimension.

M. Ding and R. C. Hilborn, "Optimal Reconstruction Space for Estimating Correlation Dimension," *Int. J. Bifur. Chaos* **6**, 377-381 (1996). How to estimate the appropriate embedding dimension in advance.

M. Bauer, H. Heng, and W. Martienssen, "Characterizing of Spatiotemporal Chaos from Time Series," *Phys. Rev. Lett.* **71**, 521–24 (1993). Suggests a slight modification of the Grassberger-Proccacia method that allows analysis of high-dimensional systems.

Surrogate Data

See the references at the end of Chapter 9.

J. Theiler, S. Eubank, A. Longtin, B. Galdrikian, and J. D. Farmer, "Testing for nonlinearity in time series: the method of surrogate data," *Physica D* **58**, 77–94 (1992).

D. Prichard and J. Theiler, "Generating Surrogate Data for Time Series with Several Simultaneously Measured Variables," *Phys. Rev. Lett.* **73**, 951–54 (1994).

T. Schreiber and A. Schmitz, "Improved Surrogate Data for Nonlinearity Tests," *Phys. Rev. Lett.* **77**, 635–38 (1996).

C. Cellucci, A. Albano, R. Rapp, R. Pittenger, and R. Josiassen, "Detecting noise in a time series," *Chaos* **7**, 414–22 (1997). Surrogate data methods can help in estimating the extent to which time series data are corrupted with noise.

Multifractals and $f(\alpha)$

T. C. Halsey, M. H. Jensen, L. P. Kadanoff, I. Procaccia, and B. I. Schraiman, "Fractal Measures and Their Singularities: The Characterization of Strange Sets," *Phys. Rev. A* **33**, 1141–51 (1986).

M. H. Jensen, L. P. Kadanoff, A. Libchaber, I. Procaccia, and J. Stavans, "Global Universality at the Onset of Chaos: Results of a Forced Rayleigh-B‚nard Experiment," *Phys. Rev. Lett.* **55**, 2798–801 (1985). The $f(\alpha)$ method is used to describe experimental data from a forced Rayleigh–Bénard experiment.

Z. Su, R. W. Rollins, and E. R. Hunt, "Measurements of $f(\alpha)$ Spectrum of Attractors at Transitions to Chaos in Driven Diode Resonator Systems," *Phys. Rev. A* **36**, 3515–17 (1987).

A. Chhabra and R. V. Jensen, "Direct Determination of the $f(\alpha)$ Singularity Spectrum," *Phys. Rev. Lett.* **62**, 1327–30 (1989). A more direct method of computing the $f(\alpha)$ spectrum.

B. B. Mandelbrot, "A Multifractal Walk down Wall Street," *Scientific American* **280** (2), 70–73 (February, 1999).

B. B. Mandelbrot, *Fractals and Scaling in Finance: Discontinuity, Concentration, Risk* (Springer-Verlag, New York, 1997).

B. B. Mandelbrot, *Multifractals and 1/f Noise: Wild Self-Affinity in Physics* (Springer-Verlag, New York, 1999).

Legendre Transformations

P. M. Morse, *Thermal Physics* (W. A. Benjamin, New York, 1964).

D. Chandler, *Introduction to Modern Statistical Mechanics* (Oxford University Press, New York, 1987).

Generalized Entropies and $g(\Lambda)$

A. Renyi, *Probability Theory* (North-Holland, Amsterdam, 1970).

P. Grassberger and I. Procaccia, "Estimation of the Kolmogorov Entropy from a Chaotic Signal." *Phys. Rev. A* **28**, 2591–93 (1983). The connection between the derivative of the correlation sum with embedding dimension and the K–S entropy.

P. Grassberger and I. Procacci, "Dimensions and Entropies of Strange Attractors from a Fluctuating Dynamics Approach," *Physica D* **13**, 34–54 (1984).

J.-P. Eckmann and I. Procaccia, "Fluctuations of Dynamical Scaling Indices in Nonlinear Systems," *Phys. Rev. A* **34**, 659–61 (1986).

G. Paladin, L. Peliti, and A. Vulpiani, "Intermittency as multifractality in history space," *J. Phys. A* **19**, L991–6 (1986).

G. Paladin and A. Vulpiani, "Intermittency in chaotic systems and Renyi entropies," *J. Phys. A*. **19**, L997–1001 (1986).

T. Tél, "Dynamical Spectrum and Thermodynamic Functions of Strange Sets from an Eigenvalue Problem," *Phys. Rev. A* **36**, 2507–10 (1987).

D. Beigie, A. Leonard, and S. Wiggins, "Statistical Relaxation under Nonturbulent Chaotic Flows: Non-Gaussian High-Stretch Tails of Finite-Time Lyapunov Exponent Distributions," *Phys. Rev. Lett.* **70**, 275–78 (1993). Relates $g(\Lambda)$ to the distribution of stretching rates in fluid flow.

R. Badii, "Generalized entropies of chaotic maps and flows: A unified approach," *Chaos* **7**, 694–700 (1997).

Generalized Correlation Sums

K. Pawelzik and H. G. Schuster, "Generalized Dimensions and Entropies from a Measured Time Series," *Phys. Rev. A* **35**, 481–84 (1987). Discussion of the generalized correlation sums $C_q^{(d)}$ and $g(\Lambda)$ distribution.

Experiments Analyzed Using Embedding Techniques

G. P. Puccioni, A. Poggi, W. Gadomski, J. R. Tredicce, and F. T. Arecchi, "Measurement of the Formation and Evolution of a Strange Attractor in a Laser," *Phys. Rev. Lett.* **55**, 339–41 (1985). The authors calculate the correlation dimension and K–S entropy for a CO_2 laser with modulated losses both for periodic behavior and for chaotic behavior.

A. Brandstater and H. L. Swinney, "Strange Attractors in Weakly Turbulent Couette–Taylor Flow," *Phys. Rev. A* **35**, 2207–20 (1987). This paper analyzes experimental data using the embedding technique and carefully examines how choices of time lags, sampling rates, embedding dimensions and number of data points affect the computed value of the correlation dimension. They also test their computations on sets of random numbers.

J. Kurths and H. Herzel, "An Attractor in a Solar Time Series," *Physica D* **25**, 165–72 (1987). Are the pulsations in radio emissions from the Sun describable by a model with a chaotic attractor? Calculations of the generalized dimensions, Lyapunov exponents, and Kolmogorov entropies do not provide convincing evidence. The paper contains a nice summary of the algorithms for calculating these quantities.

J. K. Cannizzo and D. A. Goodings, "Chaos in SS Cygni?" *Astrophys. J.* **334**, L31–34 (1988). The authors analyze variable light signals from a close binary system and conclude that the signal is not characterized by a low-dimensional strange attractor.

J. K. Cannizzo, D. A. Goodings, and J. A. Mattei, "A Search for Chaotic Behavior in the Light Curves of Three Long-Period Variables," *Astrophys. J.* **357**, 235–42 (1990).

A. Jedynak, M. Bach, and J. Timmer, "Failure of dimension analysis in a simple five-dimensional system," *Phys. Rev. E* **50**, 1170–80 (1994). A detailed examination of how correlation dimension calculations can go wrong.

J. R. Buchler, T. Serre, and Z. Kolláth, "A Chaotic Pulsating Star: The Case of R Scuti," *Phys. Rev. Lett.* **73**, 842–45 (1995).

D. K. Ivanov, H. A. Posch, and Ch. Stumpf, "Statistical measures derived from the correlation integrals of physiological time series," *Chaos*, **6**, 243–53 (1996). Correlation dimension calculations fail to show low-dimensional chaos but are still useful for detecting different states of brain dynamics.

A. Provenzale, E. A. Spiegel, and R. Thieberger, "Cosmic lacunarity," *Chaos* **7**, 82–88 (1997). The distribution of galaxies in space is analyzed using generalized dimensions.

K. Lehnertz and C. E. Elger, "Can Epileptic Seizures be Predicted? Evidence from Nonlinear Time Series Analysis of Brain Electrical Activity," *Phys. Rev. Lett.* **80**, 5019–22 (1998).

Unstable Periodic Orbit Analysis

H. Poincaré, *Les Méthods Nouvelles de la Méchanique Céleste*, Vol. I, Chap. III, Art. 36. (Gauthier-Villars, Paris, 1892). Translated in H. Poincaré, *New Methods of Celestial Mechanics* (American Institute of Physics,Woodbury, NY, 1993).

D. Auerbach, P. Cvitanovic, J.-P. Eckmann, G. Gunaratne, and I. Procaccia, "Exploring Chaotic Motion Through Periodic Orbits," *Phys. Rev. Lett.* **58**, 2387–89 (1987).

P. Cvitanovic, "Invariant Measurement of Strange Sets in Terms of Cycles," *Phys. Rev. Lett.* **61**, 2729–32 (1988).

H. G. Solari and R. Gilmore, "Relative Rotation Rates for Driven Dynamical Systems," *Phys. Rev. A* **37**, 3096–109 (1988).

P. Cvitanovic, "Topological and Metric Properties of Hénon-Type Strange Attractors," *Phys. Rev. A* **38**, 1503–20 (1988).

G. H. Gunaratne, P. S. Linsay, and M. J. Vinson, "Chaos beyond Onset: A Comparison of Theory and Experiment," *Phys. Rev. Lett.* **63**, 1–4 (1989). This paper uses a topological analysis in terms of periodic orbits to characterize the data from an experiment using coupled semiconductor diode oscillators.

O. Biham and W. Wentzel, "Characterization of Unstable Periodic Orbits in Chaotic Attractors and Repellers," *Phys. Rev. Lett.* **63**, 819–22 (1989).

G. B. Mindlin, X.-J. Hou, H. G. Solari, R. Gilmore, and N. B. Tufillaro, "Classification of Strange Attractors by Integers," *Phys. Rev. Lett.* **64**, 2350–3 (1990).

K. Davies, T. Huston, and M. Baranger, "Calculations of Periodic Trajectories for the Hénon–Heiles Hamiltonian Using the Monodromy Method," *Chaos* **2**, 215–24 (1992). Provides a nice introduction to the monodromy matrix method for finding periodic orbits in two-dimensional Hamiltonian systems.

[Tufillaro, Abbott, and Reilly, 1992]. This book has an excellent introduction to topological analysis of attractors.

P. So, E. Ott, S. J. Schiff, D. T. Kaplan, T. Sauer, and C. Grebogi, "Detecting Unstable Periodic Orbits in Chaotic Experimental Data," *Phys. Rev. Lett.* **76**, 4705–8 (1996).

B. R. Hunt and E. Ott, "Optimal Periodic Orbits of Chaotic Attractors," *Phys. Rev. Lett.* **76**, 2254–57 (1996).

P. Schmelcher and F. K. Diakonos, "Detecting Unstable Periodic Orbits of Chaotic Dynamical Systems," *Phys. Rev. Lett.* **78**, 4733–36 (1997).

Y.-C. Lai, Y. Nagai, and C. Grebogi, "Characterization of the Natural Measure by Unstable Periodic Orbits in Chaotic Attractors," *Phys. Rev. Lett.* **79**, 649–52 (1997).

F. K. Diakonos, P. Schmelcher, and O. Biham, "Systematic Computation of the Least Unstable Periodic Orbits in Chaotic Attractors," *Phys. Rev. Lett.* **81**, 4349–52 (1998).

N. S. Simonovic, "Calculations of periodic orbits: The monodromy method and application to regularized systems," *Chaos* **9**, 854–64 (1999).

Thermodynamic Formalism

T. Bohr and T. Tél, "The Thermodynamics of Fractals." In B. Hao (ed.) *Chaos*, Vol. 2 (World Scientific, Singapore, 1988). A thorough discussion of the thermodynamic formalism (but limited to one-dimensional systems).

G. M. Zaslavsky, "Chaotic Dynamics and the Origin of Statistical Laws," *Physics Today* **52** (8), 39–45 (1999).

Wavelet Analysis, q-calculus, and Nonextensive Thermodynamics

J. F. Muzy, E. Bacry, and A. Arneodo, "Wavelets and Multifractal Formalism for Singular Signals: Application to Turbulence Data," *Phys. Rev. Lett.* **67**, 3515–18 (1991).

A. Arneodo, E. Bacry, P. V. Graves, and J. F. Muzy, "Characterizing Long-Range Correlations in DNA Sequences from Wavelet Analysis," *Phys. Rev. Lett.* **74**, 3293–96 (1995).

A. Arneodo, E. Bacry, and J. F. Muzy, "Oscillating Singularities in Locally Self-Similar Functions," *Phys. Rev. Lett.* **74**, 4823–26 (1995).

G. Kaiser, *A Friendly Guide to Wavelets* (Springer, New York, 1994).

C. S. Burrus, R. A. Gopinath, and H. Guo, *Introduction to Wavelets and Transforms: A Primer* (Prentice-Hall, 1997).

C. Bowman and A. C. Newell, "Natural patterns and wavelets," *Rev. Mod. Phys.* **70**, 289–301 (1998). A very nice introduction to pattern formation and the use of the wavelet transform to analyze patterns.

A. Erzan and J.-P. Eckmann, "*q*-analysis of Fractal Sets," *Phys. Rev. Lett.* **78**, 3245–48 (1997).

C. Tsallis, "Possible Generalization of Boltzmann-Gibbs Statistics," *J. Stat. Phys.* **52**, 479–87 (1988).

U. M. S. Costa, M. L. Lyra, A. R. Plastino, and C. Tsallis, "Power-law sensitivity to initial conditions within a logisticlike family of maps: Fractality and nonextensivity," *Phys. Rev. E.* **56**, 245–50 (1997).

M. L. Lyra and C. Tsallis, "Nonextensivity and Multifracticality in Low-Dimensional Dissipative Systems," *Phys. Rev. Lett.* **80**, 53–56 (1998).

C. Anteneodo and C. Tsallis, "Breakdown of Exponential Sensitivity to Initial Conditions: Role of the Range of Interactions," *Phys. Rev. Lett.* **80**, 5313–16 (1998).

Transient Chaos and Recurrence Time Analysis for Non-Stationary Systems

I. M. Jánosi and T. Tél, "Time-series Analysis of Transient Chaos," *Phys. Rev. E* **49**, 2756–63 (1994).

J. B. Gao, "Recurrence Time Statistics for Chaotic Systems and Their Applications," *Phys. Rev. Lett.* **83**, 3178–81 (1999).

D. Yu, W. Lu, and R. G. Harrison, "Detecting dynamical nonstationarity in time series data," *Chaos* **9**, 865–70 (1999). These authors use a "cross-time index" and its distribution to test for nonstationary parameters.

10.12 Computer Exercises

CE10-1. Write a computer program that computes and plots $f(\alpha)$ for the weighted Cantor set as described in Section 10.6.

CE10-2. Write a computer program that computes and plots $f(\alpha)$ for the asymmetric Cantor set of Exercise 10.5-4. Hint: Use Newton's Method to solve the resulting implicit equation for n/w_*.

CE10-3. Read the article HJK86 and implement their analysis to plot $f(\alpha)$ for the two-scale Cantor set. Verify that the results reduce to the special cases, the weighted Cantor set and the asymmetric Cantor set, in the appropriate limits.

CE10-4. Read the article by H. Gould and J. Tobochnik, "More on Fractals and Chaos: Multifractals," *Computers in Physics* **4** (2), 202–7 (March/April 1990) and try some of the program suggestions in that article.

IV

SPECIAL TOPICS

11

Pattern Formation

and

Spatiotemporal Chaos

The next great awakening of human intellect may well produce a method of understanding the qualitative content of equations. Today we cannot. Today we cannot see that the water flow equations contain such things as the barber pole structure of turbulence that one sees between rotating cylinders. Today we cannot see whether Schrödinger's equation contains frogs, musical composers, or morality—or whether it does not. We cannot say whether something beyond it like God is needed, or not. And so we can all hold strong opinions either way. R. P. Feynman in [Feynman, Leighton, Sands, 1964], Vol. II, p. 41–12.

11.1 Introduction

Intricate, lacy ice crystals form from swirling mists; clouds take on regular, periodic patterns that stretch for hundreds of kilometers; a warm pan of tomato soup develops a regular pattern of hexagons on its surface. Order arises from (apparent) disorder. Complexity emerges from uniformity. Nature seems to generate order and spatial patterns, even where we least expect to find them. How does this happen?

All of these questions concern systems with significant spatial variation of their properties. So, we must now consider the spatial dependence of the dynamics of the system, as well as the temporal dynamics, which so far has been our prime concern in this book.

From the fundamental physics point of view, there is a real puzzle here. At the microscopic atomic and molecular level, we know that the significant interactions (or forces) among atoms and molecules extend over rather short distances, something on the order of 10 Å (roughly 5 to 10 atom diameters). Beyond those distances the interactions between atoms become insignificant compared to the random thermal agitation of the environment at all but the lowest temperatures. Given that fact, the puzzle is to explain the emergence of molecular order which extends over very large distances, perhaps a few millimeters in the case of ice crystals or even kilometers in the case of clouds. How do short range interactions conspire to give us long range order?

One possible explanation is to say that there are long-range interactions (not accounted for by the fundamental atomic and molecular forces) that allow the molecules to "communicate" with each other; therefore, a particular water molecule, for example, "knows" where to place itself to form the correct shape for the tip of a snowflake. This long-range communication, however, seems too much like magic and, in any case, is ad hoc. We do not expect new kinds of forces to emerge in this way.

As scientists, we would like to explain the emergence of order on the basis of *known* forces and the appropriate dynamics. That is, we hope that we can explain the production of long-range order "simply" by applying the known laws of physics to the appropriate dynamics for the system. Within the past twenty years or so, we have made some progress in understanding what these appropriate dynamics are. What may be surprising is that many of the concepts that help us understand temporal chaotic behavior seem to be useful in understanding the emergence of order and complexity.

What do these order-producing systems have in common? First, and most importantly, they are systems that are driven away from thermal equilibrium. A system at thermal equilibrium, by definition, has a uniform temperature and no energy flows among its "parts" (assuming that it has distinguishable parts). Energy flows and nonequilibrium conditions are necessary for the emergence of order. Second, dissipation is important in allowing nonequilibrium systems to settle into definite structures, in analogy with the importance of dissipation in allowing a system's dynamical trajectories to settle onto an attractor in state space. The distinguished physicist and Nobel Prize winner Ilya Prigogine has coined the term "dissipative structures" for the ordered spatial patterns that emerge in nonequilibrium dissipative systems.

The third commonality among these pattern-producing systems is our old friend nonlinearity. Without nonlinearity, no patterns are produced. Nonequilibrium, however, is important, too, because for most systems at, or very close to, thermal equilibrium the nonlinearities are "hidden." The effects of nonlinearities become noticeable only when the system is driven sufficiently far from equilibrium.

These statements about nonlinearity are perhaps too strong. Many linear systems such as rays of light passing through slits or gratings or water drops, for example, produce many intriguing patterns. These patterns, however, are determined by the geometry of the slits, gratings, or water drops. The kinds of patterns we will treat in this chapter, generated by nonlinearities in the system, often have geometric shapes that are independent of the shape of their boundaries. In addition, nonlinear patterns often become time-dependent even when all the system parameters and boundaries are time independent. Both of these features—

patterns independent of the geometry of boundaries and inherent time dependence—are signals of what we call *nonlinear pattern formation*.

Another characteristic feature of nonlinear pattern formation is the production of *coherent structures*. These structures can be a pattern of stripes, a localized pulse, a vortex of fluid flow and so on that persist through time and often move around within the medium in which the pattern has formed. These structures are formed by a variety of physical, chemical or biological mechanisms. SHO93 and RTV99 give some flavor of the methods used to attempt to provide a general understanding of these structures.

Although much has been learned about the emergence of spatial patterns, the story of this vast, complex, and important field is far from complete. In this chapter, we shall introduce some of the basic aspects of what is now being called pattern formation and spatiotemporal chaos. (Of course, chaotic behavior is just one feature of the broader landscape of nonlinear dynamics in systems with significant spatial structure, but the same nonlinearities that give rise to the pattern formation can produce chaotic time behavior.) In the subsequent sections we provide a potpourri of examples to give some sense of the wide range of concerns in spatiotemporal nonlinear dynamics. These by no means exhaust the list of fascinating phenomena under active investigation, but they should give you some notion of the basic issues, the concepts and formalism in current use, and the range of application of these phenomena. Wherever possible, we try to show how the concepts we have learned in the study of nonlinear temporal dynamics can carry over to those systems with interesting spatial structure. In fact, we shall see that fractal dimensions, applied here to actual spatial structures, play an important role in characterizing many of these systems.

As we proceed through these sections, we shall introduce some of the basic physics and formalism for the description of fluid flow and transport phenomena. For better or worse, these topics have disappeared from most introductory physics courses, producing an unfortunate lacuna in physicists' educations. Our goal is to provide just enough introduction to these topics so that we can see the connections among the formalisms used to describe the phenomena.

Spatiotemporal patterns are examples of how simple order can arise from complex behavior. Chaos, as we have mentioned before, can be viewed as complex behavior arising from (often) simple systems. Is there a theory (or at least a point-of-view) that links these together and provides an understanding of the dynamics of complex systems, including biological systems, social systems, as well as systems in the physical sciences? As yet there is no simple answer to that question. The references at the end of the chapter give some flavor of what some scientists and mathematicians think a "theory of *complexity*" would be like. What follows in this chapter are several examples of the kinds of problems a theory of complexity would try to solve.

11.2 Two-Dimensional Fluid Flow

As a first example of spatially dependent dynamics, we describe some aspects of two-dimensional fluid flow. The first obvious question is what do we mean by two-dimensional fluid flow? Aren't fluids inherently three dimensional? Two-dimensional fluid flow occurs when the fluid conditions remain uniform in one direction as the fluid (perhaps) swirls and twists in the other two (usually mutually perpendicular) directions. An example is probably helpful. Imagine a fluid contained in a rectangular chamber, as illustrated in Fig. 11.1, whose right and left sides can slide up and down. The sliding will induce fluid flow in the x and y directions (in the plane of the page), but the flow will be the same at any value of z, the coordinate direction perpendicular to the page (except perhaps in a very thin layer immediately at the front and back surfaces). Hence, we say that the fluid motion is (essentially) two-dimensional.

This type of fluid system has been used extensively in recent years to study the conditions for fluid mixing. (See [Ottino, 1989] and other references at the end of this chapter.) Fluid mixing is important in a variety of industrial and natural settings. How can a paint manufacturer assure that pigments are mixed thoroughly into the paint medium? How are gases, such as CO_2, which are dissolved in the surface layers of the ocean, mixed with deeper lying layers? In fact, the question of how to characterize mixing appropriately and its relation to chaos is still unresolved (RKZ00). For our purposes, mixing occurs when a small blob of colored fluid, for example, injected at some point into the main body of fluid, eventually becomes stretched and twisted sufficiently to extend essentially throughout the entire fluid region. (We will ignore the contribution of molecular diffusion to the mixing. We will be concerned only with the mixing induced by the bulk motion of the fluid.) This mixing occurs, as we all know from stirring our morning coffee, when the fluid motion involves some kind of twisting and folding. If the fluid flows "smoothly," then little mixing occurs.

The dynamical analogy to the mixing of fluids is the stretching and folding of a cluster of initial conditions associated with chaotic motion in *state space*, which causes the cluster to be spread throughout a significant region of state space. As we

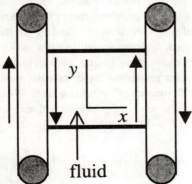

Fig. 11.1. A schematic diagram of a fluid system in which the flow is essentially two–dimensional. The right and left sides of the fluid cell can move up and down vertically, thereby inducing fluid motion in the xy plane.

shall see, this analogy can be formulated mathematically; in fact, the equivalent of the average Lyapunov exponent for the divergence of nearby trajectories can be used to provide a quantitative measure of the amount of mixing [Ottino, 1989, p. 117].

The hydrodynamic theory of two-dimensional fluid motion shows that the velocity of the fluid at a particular location (x,y) can be specified in terms of a so-called *streamfunction* $\Psi(x, y, t)$ [Ottino, 1989] for what is called isochoric flow. From the streamfunction, we compute the velocity components by taking partial derivatives

$$v_x(x, y) = \frac{\partial \Psi(x, y, t)}{\partial y}$$

$$v_y(x, y) = -\frac{\partial \Psi(x, y, t)}{\partial x}$$

(11.2-1)

At a fixed time t, the lines of constant $\Psi(x, y, t)$ map out the so-called *streamlines* of the fluid motion. The streamlines are curves tangent to the fluid velocity vector at that particular point in space.

The crucial point here is that Eqs. (11.2-1) for the velocity components of the fluid flow have the same mathematical structure as Hamilton's equations, introduced in Chapter 8. By comparing Eqs. (11.2-1) with Eq. (8.2-1), we see that the fluid streamfunction plays the role of the Hamiltonian function for Hamiltonian systems (ARE83, ARE84). This realization tells us that the trajectory of a fluid particle in real xy space for two-dimensional fluid flow [that is the solution of the differential equations embodied in Eqs. (11.2-1)] is the same as the state space (pq space) trajectory in the corresponding Hamiltonian system. (Recall that the state space trajectories are the solutions of the Hamiltonian differential equations.) If the streamfunction is independent of time, then a particle of the fluid moves along the curve determined by $\Psi(x, y)$ = a constant. This behavior has a direct analogy in Hamiltonian dynamics: A point in state space for a Hamiltonian system moves along a state space curve for which $H(q,p)$ = a constant if the Hamiltonian is independent of time.

In most of traditional fluid dynamics, finding the streamfunction and streamlines for a given set of conditions is the goal of most calculations. However, for the study of mixing we must go further: We need to find the actual paths followed by particles of the fluid. As we mentioned earlier, mixing is associated with the exponential divergence of two nearby particle paths. Two different (but related) methods are used to characterize particle paths. In the first method, we find the so-called *pathlines*, the actual particle paths. In more formal terms, we want to find the position vector of a particle as a function of time. We write this position vector as

$$\vec{x} = \vec{\Phi}(\vec{x}_0, t)$$

(11.2-2)

where $\vec{x}_0$ is the initial position vector of the particle whose path we are following. If we know the streamfunction for a particular situation, then, in principle, we can find the pathline by solving the differential equation

$$\frac{d\vec{x}}{dt} = \vec{v}(\vec{x}, t) \qquad (11.2\text{-}3)$$

In experiments, we can map out pathlines by injecting a tracer particle or small blob of dye at the position $\vec{x}_0$ at time $t = 0$ and following its path as a function of time.

The second method of following motion in the fluid is to inject, usually at a fixed spatial location, a stream of tracer particles or dye and to trace the path laid out by this stream. Such a path is called a ***streakline***. If a flow is steady in the sense that the fluid velocity at a fixed spatial location is independent of time, then streaklines and pathlines coincide. In the more general and more interesting case of nonsteady flows, streaklines and pathlines are not the same because the velocity at the tracer injection point changes in time.

Let us now return to the analogy between two-dimensional fluid flow and Hamiltonian state space dynamics. Two important results follow from this analogy. First, we may use the actual fluid paths to visualize the analogous state space trajectories in the corresponding Hamiltonian system. By watching the fluid motion, we can develop more intuition about the meaning of folding and stretching in state space dynamics. (This visualization was suggested early in this century by the physicist J. W. Gibbs. See [Gibbs, 1948], Vol. II, pt. 1, pp. 141–156. Second, we can use what we have learned about Hamiltonian dynamics to guide us in our understanding of two-dimensional fluid flow.

As an example of the latter claim, we make the following argument: If the streamfunction is independent of time, and the fluid flow is steady, then the fluid system, as mentioned earlier, is equivalent to a Hamiltonian system with one degree of freedom. Hence, as we saw in Chapter 8, there is no possibility of chaotic behavior. In fluid flow language, we can say that the fluid will not exhibit mixing behavior. The folding and exponential stretching required for mixing cannot occur in a *steady* two-dimensional flow system. On the other hand, if the streamfunction is periodic in time, the fluid flow system is equivalent to a Hamiltonian system with "one-and-a-half degrees of freedom," and chaotic (mixing) behavior is possible. Even in the latter case, however, as we saw in Chapter 8 for chaotic behavior in state space, the mixing behavior, for some range of parameter values, may not extend throughout the fluid; some parts may be mixed, but other parts may show smooth, nonmixing flow.

Figure 11.2 shows some pathlines for fluid flow in the system illustrated in Fig. 11.1. The pathlines are made visible by the injection of a small, colored blob of fluid. For the case shown in Fig. 11.2, the flow was caused by a steady motion of the right and left plates in opposite directions. The fluid pathlines look much like the phase space trajectories of a Hamiltonian system. In fact, elliptic points and hyperbolic points, analogous to those for a Hamiltonian system organize the fluid flow.

If a blob of fluid is located near an elliptic point, it either remains fixed in

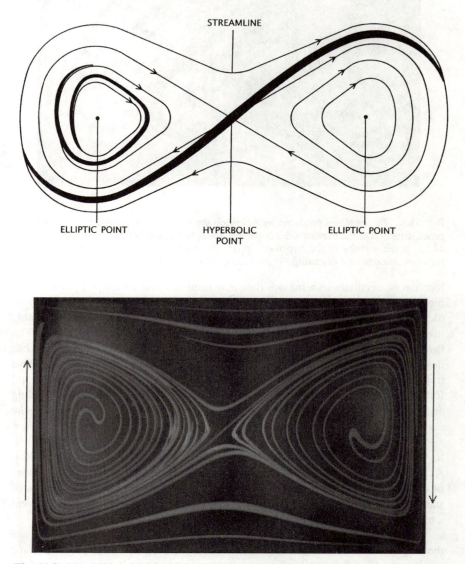

Fig. 11.2. The pathlines produced by following the motion of a colored blob of fluid in the type of system illustrated in Fig. 11.1. In this case the right and left sides were moved in opposite directions with constant speed. The locations of elliptic points and a hyperbolic (saddle) point are indicated. Photograph used by permission from J. Ottino.

Fig. 11.3. The pathlines produced by following the motion of a colored blob of fluid for periodic motion of the moving sides of the apparatus, but rotated 90° from that shown in Fig. 11.1. In the vicinity of the hyperbolic point, the fluid shows the stretching and folding behavior characteristic of mixing. (From [Ottino, 1989], p. 210.)

location or circulates in a smooth (laminar) fashion about the elliptic point. For a blob injected near a hyperbolic point, the fluid is stretched and pushed away from the hyperbolic point along the direction associated with the point's unstable manifold, while the blob is pulled toward the point along the stable manifold. The flow can develop mixing characteristics if the equivalent Hamiltonian system has developed homoclinic or heteroclinic trajectories, as discussed in Chapter 8. These trajectories are possible if the right and left sides of the cavity are moved back and forth periodically (see Fig. 11.3). Then the blob of tracer dye will undergo folding and stretching that mimics the homoclinic or heteroclinic tangles of the corresponding phase space trajectories (see Fig. 11.4).

As one further illustration of the use of the Hamiltonian analogy to fluid flow, let us recall the important role played by KAM surfaces in Hamiltonian dynamics. As we saw in Chapter 8, KAM surfaces form barriers in state space for trajectories of the dynamical system. Trajectories may wander chaotically in some parts of state space, but KAM surfaces, at least for some range of parameter values, prevent a trajectory from visiting all of the allowed region of state space.

Do analogous KAM surfaces restrict fluid mixing? Alternatively, can we use the notion of KAM surfaces to understand why mixing does not occur in some regions of a fluid flow while efficient mixing seems to occur in other regions of the same flow? Experiments (KUS92) (KUO92) (OMT92) seem to indicate that the answer to both questions is yes. Figure 11.5 shows the streaklines of dye injected into the fluid flow through a chamber called a partitioned-pipe mixer. The fluid

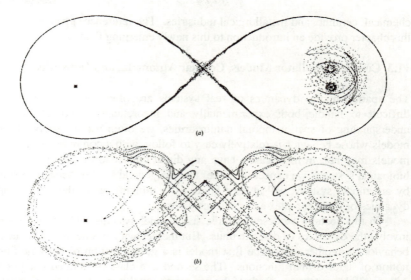

Fig. 11.4. An illustration of how fluid pathlines mimic a homoclinic tangle in a Hamiltonian system. The paths were calculated for a model system called the Tendril–Whorl flow model. (*a*) For lower flow strength, the homoclinic behavior near the saddle point at the center is hidden. (*b*) For higher flow strength, the homoclinic tangle becomes obvious. (From [Ottino, 1989] p. d (Color illustrations).

enters near the top and is removed at the bottom. As you can see, dye injected at one point travels a fairly regular path through the tube and does not mix with its surroundings. That dye is injected in a region inside a KAM surface, a region in which the motion is fairly regular. On the other hand, dye injected near the edge of the tube follows a tortuous path, and the trajectory appears to be chaotic. We see that our knowledge of the behavior of trajectories in state space for a Hamiltonian system can guide us in understanding the complex fluid flows in actual three-dimensional space. This method can also be extended to visualize three-dimensional chaos (FKO98).

Granular Flows

There is yet another type of "fluid" flow whose study has benefited from the ideas of nonlinear dynamics. This is the flow of so-called granular materials such as sand, small beads, and even a mix of cocktail nuts. These materials can flow under appropriate conditions (think of sand pouring out of a child's bucket at the beach) but with properties quite different from ordinary fluids like water or air. Try shaking a container of mixed nuts with cashews, peanuts, and Brazil nuts, for example. You find that, perhaps to your surprise, that the large nuts rise to the top and the small nuts go toward the bottom of the container. The flow and mixing of these granular materials is important in pharmaceuticals, food processing, and the chemical, ceramic, and metallurgical industries. The references given at the end of this chapter provide an introduction to this newly emerging field.

11.3 Coupled-Oscillator Models, Cellular Automata, and Networks

The spatiotemporal dynamics of real systems are often very complicated and difficult to handle both experimentally and theoretically. To develop some understanding of spatiotemporal nonlinearities, we often use a variety of simple models whose behavior is relatively easy to follow with a computer. Although the models may not, in most cases, have any direct connection to actual physical or biological systems, we expect (or hope) that the general kind of behavior exhibited by our model systems will at least guide us in thinking about the spatiotemporal dynamics of more complex and realistic systems.

In this section we describe three such model systems. The three examples involve the dynamics of discrete units. But the behavior of one unit influences the behavior of another unit. The first model is a generalization of the now familiar notion of iterated map functions. The second is a class of iterated computations called cellular automata. Both of these systems involve iterative schemes that lead to the generation of interesting and complex spatial patterns starting from random initial conditions. The third example is a network of units that process an input and provide an output.

Coupled Iterated Map Functions

Let us make a model system consisting of a collection of identical dynamical systems (such as the logistic map function). Each member of the collection is to be associated with a different spatial location. With each spatial location, we also associate a number determined by the dynamical rule that defines the chosen dynamical system. We are interested in how the numbers at different spatial locations get correlated or uncorrelated. In the simplest situation these locations might be along a straight line, so we need worry about only one spatial direction.

Of course, if each member of the overall system simply acts independently, nothing interesting would occur. We need some coupling or interaction among the different spatial locations. We find that as the strength of the coupling between different spatial locations increases the system can spontaneously generate spatial order, that is spatial patterns of correlated changes of the numbers occur. Although this model is not directly related to any real physical system, we might anticipate that real systems with coupling between spatially separated "oscillators" might show some of the same dynamical and spatial features.

Let us consider a specific example: a one (spatial) dimensional model. Suppose we have N dynamical systems at locations labeled by $i = 1, 2, 3, ..., N$. At each location we assign a number $x(i)$ by the following iterative procedure:

$$x_{n+1}(i) = \frac{1}{1+2\varepsilon}[f(x_n(i)) + \varepsilon f(x_n(i-1)) + \varepsilon f(x_n(i+1))] \qquad (11.3\text{-}1)$$

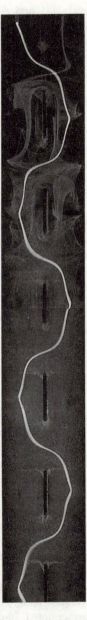

Fig. 11.5. Streaklines of dye injected into the flow through a partitioned-pipe mixer. (The mixer itself is illustrated in the diagram on the left.) On the right, fluid enters at the top of the pipe and is removed at the bottom. The dye injected inside a KAM surface near the top left does not mix with its surroundings. The dye injected at the top right of the tube shows chaotic trajectories and good mixing behavior. (From KUO92)

is determined by the value at i at the nth time step and by the values at the neighboring sites $i-1$ and $i+1$. The parameter ε determines the strength of the coupling to the neighbors. The function f might be any function like the logistic map function. The factor $1/(1+2\varepsilon)$ assures that the x values stay between 0 and 1, for example, if the function f maps the range [0,1] onto itself. To avoid ambiguities at the end points 1 and N, we often identify $x(1) = x(N)$. In general we could use many different map functions, and we might also allow for coupling among more than just the nearest neighbors. For now, let us stay with this simple model.

To explore the dynamics of this kind of system, we "seed" the system with a random selection of x values (usually, between 0 and 1) and then implement the dynamics by applying Eq. (11.3-1) to the x values at all the locations simultaneously. We now have two parameters to vary. One is the parameter associated with the map function (e.g., the parameter A in the logistic map function). The other is ε, the parameter that determines the coupling between neighbors.

The behavior of such a coupled system is quite complex and by no means fully explored. The most interesting effects occur when the A values, assumed to be the same at all spatial locations, are in the region for which each individual map function behaves chaotically in the absence of coupling with its neighbors. As the coupling strength is increased, the overall system develops spatial organization with groups or clusters of numerical values all changing together. The system has essentially generated spatial order from a random initial state.

K. Kaneko (KAN89a,b, KAN90, KAN92) studied a one-dimensional coupled iterated map system in which all the spatial locations are coupled together:

$$x_{n+1}(i) = (1-\varepsilon)f(x_n(i)) + \frac{\varepsilon}{N}\sum_k f(x_n(k)) \qquad (11.3-2)$$

To further simplify the analysis, we can divide the range of x values into two segments, as we did in Chapter 5, and label one range R (right) and the other L (left). Then we look for correlations in the values of R and L at different spatial locations. Kaneko found that the system quite commonly settled into completely coherent oscillations of the entire system: All locations would switch back and forth together between R and L. For some values of ε and the map parameter (say, A for the logistic map), the system would break up into two interlaced groups (that is, the members of one group are not necessarily spatial neighbors) that oscillate between R and L out of phase. The general conclusion is that these coupled map systems seem to generate interesting spatial order as a result of the competition between the chaotic behavior of an individual element and the tendency for uniformity due to the coupling among neighbors. This kind of competition will be seen to be a common feature of many pattern generating systems.

These coupled iterated map models have many surprising features. Additional noise can cause them to become more ordered (SCK99). If the dynamics becomes high-dimensional, the shadowing theorem (Chapter 2) may fail and no mathematical model can produce reasonably long solutions that match those of the system (LAG99). Chaotic coupled maps can also be configured to perform simple computations (SID98). Recently, Egolf (EGO00) has argued that studies of coupled chaotic map models indicate that the tools of equilibrium statistical mechanics can be used to understand far-from-equilibrium, spatially-extended chaotic systems. Such systems have heretofore been difficult, if not impossible, to understand using standard analytic methods developed for equilibrium situations.

Cellular Automata

Cellular automata are simple discrete systems manifesting interesting spatiotemporal patterns. The name comes from the procedure for implementing this kind of system. We divide up some spatial region into discrete cells. A numerical value (or a color or some other attribute) is associated with each cell. The attribute is allowed to evolve in discrete time steps according to a set of rules. In most cases the rules are based on "local" interactions: A given cell changes its attribute depending on its own value and the values possessed by its neighbors. Hence, the evolution is "automatic." A one-dimensional array of coupled iterated maps forms a simple cellular automaton with the evolution rules specified by something like Eq. (11.3-1) or Eq. (11.3-2). However, most cellular automata use two or more spatial dimensions and have attributes with a limited (and usually fairly small) number of values. Often the rules for evolution can be stated in simple verbal form. Most cellular automata are implemented on computers, and their history, in fact, begins with some of the early electronic computers in the 1940s [Toffoli and Margolus, 1987].

Let us make a simple one-dimensional automaton by dividing a one-dimensional region (which might be infinitely long) into cells. We assign a number (usually an integer) to each cell in some initial pattern. Then, in discrete time steps, we change the numerical values in the cells all at the same time (that is, synchronously) to new values based on a set of rules. For example, we might assign a 1 (meaning "alive," to give the model a more interesting flavor) or a 0 (meaning "dead"). The evolution rules might be the following:

1. If a live cell (that is, a cell with a 1 associated with it) has live neighbors on both sides, it dies (the number changes to 0) (because of "overcrowding").
2. If a dead cell (with a 0) has a live neighbor on each side, then the cell becomes alive ("1") on the next time step. (The couple gives birth to a child.)
3. If a live cell has no live neighbors, it dies (changes to 0) (due to "loneliness").
4. If a cell (dead or alive) has one live neighbor and one dead neighbor, it remains unchanged.
5. If a dead cell has two dead neighbors, it remains dead.

If we have a spatial region of finite length, then we must also give rules for what happens at the ends. For now, we will assume we have an infinitely long region.

Next, we assign an initial pattern of 1s and 0s to the cells and let the system evolve according to the rules. For example, we might start with ...0,1,0,..., that is, all 0s and an isolated 1. According to rule (3), this isolated live cell dies on the next time step and the system evolves to all 0s and remains there forever.

Let us try a second initial pattern, say ...0,1,1,0,... (i.e., almost all 0s with an isolated live pair). Since each 1 has a neighboring 1 and 0, it remains unchanged according to rule (4). Thus, this pattern remains unchanged forever. As a third example, let us consider ...0,1,1,1,0,... where again the ellipsis means all 0s except for the 1s shown. This pattern evolves to ...,1,0,1,0,... on the first time step and then to all 0s. Thus, we see that at least two different initial patterns end up evolving to the same final state.

From these simple examples, we see that we should ask the following questions about the evolution of cellular automata:

1. What initial patterns (if any) are unchanged under the rules of evolution?
2. What set of initial patterns leads to complete extinction (all 0s)?
3. What set of initial patterns leads to other patterns that are then unchanged under the rules of evolution?
4. What set of patterns have cyclical (periodic) behavior for the evolution?
5. What kinds of patterns (if any) evolve from "random" initial patterns?
6. Are recurring patterns (or cycles) stable with respect to "perturbations" such as changing the value of one of the cells?
7. What happens to the patterns if we change the rules for evolution?

The similarity between these questions for cellular automata and the questions we have been asking about the temporal dynamics of systems should be obvious. In fact, we can use almost exactly the same language here. For example, a pattern that remains unchanged is a *fixed point* of the evolution. A sequence of repeating patterns constitutes a *limit cycle*. We can consider a pattern (or periodic cycle of patterns) to which a set of other patterns evolve as an *attractor* for the evolution of the system. The set of all patterns that evolve to a particular final pattern (or cycle of patterns) constitutes the *basin of attraction* for that pattern. If two almost identical patterns evolve to very different final patterns (suitably defined), then we can say that the evolution displays *sensitive dependence on initial conditions*. A pattern is said to be stable if a small perturbation of that pattern (for example, changing the numerical value of one cell) evolves back to the original pattern. Otherwise, the pattern is unstable.

Exercise 11.3-1. Try various initial patterns for the one-dimensional automaton example given earlier. Find at least one other recurring pattern. Can you find a periodic cycle of patterns? Are these recurring patterns stable?

> **Exercise 11.3-2.** Change one of the rules for the evolution of the one-dimensional automaton. Can you find recurring patterns and cycles? Explore the stability of these patterns.

> **Exercise 11.3-3.** Construct a one-dimensional automaton with a finite number of cells. Set the boundary conditions so that $x(1) = x(N)$, where N is the number of cells. Using the rules given in the text, explore the attractors, basins of attraction, and so on, for this automaton.

Two-dimensional cellular automata show even more complex possibilities. We shall discuss two examples to give some taste of what can happen. The "Game of Life," introduced by the mathematician John Conway, divides a two-dimensional region into a square array of cells. A "1" ("alive") or "0" ("dead") is assigned to each cell. Each cell has eight neighbors whose states influence that particular cell's evolution. Conway's original rules (GAR70) of evolution were the following:

1. A live cell ("1") stays alive only if it is surrounded by two or three live neighbors.
2. A live cell dies if it has more than 3 live neighbors (overcrowding) or fewer than two live neighbors (loneliness).
3. A dead cell ("0") changes to a 1 if it has exactly three live neighbors. (Obviously, rather complicated social arrangements exist in this make-believe world.)

We can again pose the set of questions listed earlier. Figure 11.6 shows a stable pattern and a periodic pattern for the Game of Life.

Figure 11.7 shows the pattern that emerged from a set of random initial values. Obviously, we have a very rich set of possibilities: We can change the set of evolution rules, and we can extend the automata construction to three or more dimensions. See the readings at the end of this chapter for a taste of this rich and only incompletely mapped out smorgasbord.

A second example of a two-dimensional cellular automaton uses a hexagonal

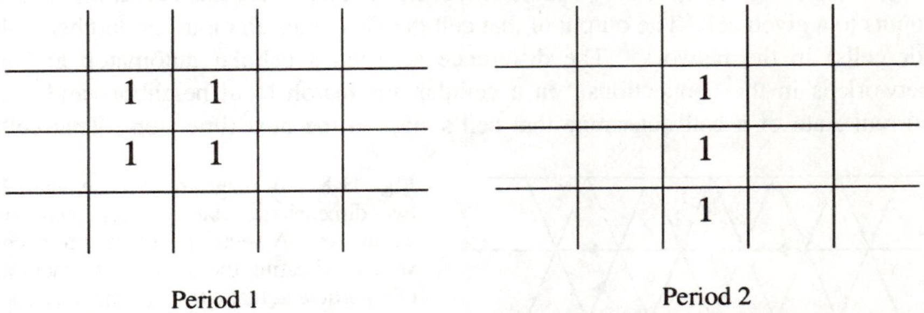

Period 1   Period 2

Fig. 11.6. Some recurring patterns for the Game of Life. On the left is a pattern that remains unchanged under the evolution rules. On the right is a pattern that recurs periodically (with period-2) under the rules.

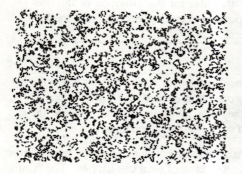

Fig. 11.7. A pattern that evolved for the Game of Life starting with a set of random initial conditions in a rectangular grid. The pattern has not yet reached a steady state, but the emergence of "structures" on all length scales seems apparent. The fractal (box-counting) dimension of the pattern is about 1.7. (See Section 11.6 for other patterns with similar dimensions.)

array of cells (vertices) and assigns a vector (which may point along one of the lines connecting the vertices) to a "particle" that can move through the lattice from one vertex to another. Figure 11.8 illustrates a typical array. (This type of model is often called a "lattice gas model" because the motion of the particles from vertex to vertex is something like the motion of gas particles in a container.) With an appropriate choice of the rules for the behavior of the particles on the lattice, the particles' motion can mimic the flow of a fluid in two-dimensions. See the readings at the end of this chapter for details of this kind of application. Cellular automata models have also been used to study aspects of the human immune system (ZOB98).

Networks

For the last example of discrete spatial systems, we discuss a class of systems with a somewhat different structure from the first two in this section: **networks** (sometimes called **neural networks** because of their similarity to networks of nerve cells in an organism). Networks are arrays (more often conceptual than real) of cells. Each cell is assigned an attribute, usually one of two possible values. A given cell in the network may have one or more connections to other cells in the network. These connections serve as input signals to that cell. The time evolution rules are often given by binary operations, such as those of Boolean algebra, on the inputs to a given cell. The output of that cell then becomes an input for another cell (or cells) in the network. The difference between a cellular automaton and a network is in the connections. In a cellular automaton local neighbors and the current state of a cell determine that cell's state in the next time step. That cell

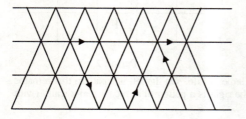

Fig. 11.8. A diagram of a hexagonal two–dimensional lattice gas cellular automaton. A vector is assigned to each vertex indicating the direction of motion of a particle at that site. In some models, more than one particle can reside at a vertex at any one moment.

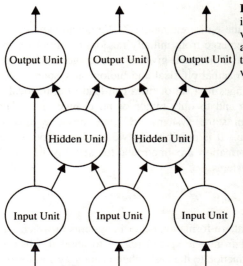

Fig. 11.9. A diagram of a network. The input units or cells are connected to various output units and also, in this case, a layer of hidden units. The weights of the outputs to the various layers can be varied.

affects its neighbors. In a network, say with two input connections per cell, the two cells that provide the input to a particular cell need not be the cells to which the output connection goes. Figure 11.9 shows a simple neural network.

Stuart A. Kaufman (KAU91) has studied Boolean networks extensively. The behavior of such a network depends both on the number of connections that provide input to a particular cell and on the rules that govern how the cell processes the input connections to give an output. Kaufman has shown that in a Boolean network for which the number of cells and the number of inputs per cell are the same, the network behaves chaotically with sensitive dependence on initial conditions. However, the number of possible state cycles and the number of basins of attraction are relatively small. Hence, these networks show a surprising amount of order. In fact, these properties seem to hold even when the number of inputs is reduced to three (KAU91). Networks with only two inputs per cell tend to have quite stable collective behavior; small perturbations will not remove the system from its attracting state. Kaufman and others have suggested that a network residing on the border between order and chaos is a good model for an adaptable system, one that can respond favorably and quickly to changes in its environment.

Although these network models are highly suggestive, whether or not they capture the actual logic of adaptation and evolution remains an open question. See [Kaufman, 1993] and [Kaufman, 1995] for a wider discussion of these issues. On a different note many "naturally occurring" networks such as the World Wide Web, electrical power grids, and groups of actors seem to have dynamics that lead to interesting scaling behavior (BAA99). The common features of these networks are that the networks expand by adding new vertices (or connection points) and that new vertices attach preferentially to other vertices that are already well connected.

Summary

What do we learn from the study of these discrete models? We quickly appreciate that interesting spatial patterns can emerge from initially random conditions even for very simple time evolution rules. This result gives us encouragement to seek simple rules for pattern formation in actual physical and biological systems. We also learn that even very minor changes in a set of rules can lead to dramatically different patterns for some systems and to only minor or no changes for other systems. Some spatial patterns in some systems are quite unstable against perturbations; others are quite stable. It is possible that we can learn from these models what features of pattern formation are important for stability in actual physical, chemical, and biological systems.

11.4 Transport Models

Many examples of spatiotemporal pattern formation occur in systems in which one or more substances are transported from one spatial region to another. In this section we will provide a brief introduction to the basic phenomenology and formal description of transport. We shall, in the following sections, apply these descriptions to a sample of pattern formation problems. These transport models use a differential equations approach, which may be more comfortable and certainly more familiar to most scientists and engineers. We shall eventually see that these models, like the discrete models of the previous section, can lead to interesting pattern formation.

In these transport models, the spatial and time variables are assumed to be continuous. By way of contrast, the discrete models discussed in the previous section use discrete spatial (and sometimes temporal) variables. We met with this distinction between discreteness and continuity in our discussion of temporal dynamics: Ordinary differential equations describe the dynamics of systems with a continuous time variable; iterated maps use discrete time steps. The relationship between spatiotemporal models with discrete variables and those with continuous variables is often not very clear. Rigorous proofs are generally lacking, and we must rely on some mixture of experience and commonsense to guide us in the comparison and applications of these distinct categories of models.

Molecular Diffusion

As a prototype of transport, we examine the spatial transport of molecules via the process of diffusion. At the microscopic level, diffusion is due simply to the random motion of the molecules. As the molecules collide with each other and with the other molecules that make up the medium in which the molecules "live," in their endless dance of random thermal motion, the molecules will tend to spread out from regions of high concentration to regions of low concentration. At the macroscopic level (to which we will confine our discussion), we focus our attention on the net outcome of these microscopic collisions and random motion: The

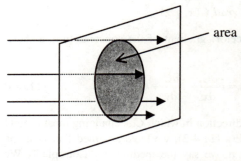

Fig. 11.10. A sketch illustrating the meaning of the current density j_x. The current density multiplied by the cross-sectional area (perpendicular to the flow direction) through which the molecules pass gives the number of molecules passing through that area per unit time.

molecules tend to diffuse from a region of high concentration toward a region of low concentration.

You should ponder, at least for a moment, the jump in level of description: From the microscopic point of view, each molecule is just as likely to move to the left, for example, as it is to move to the right. It is the preponderance of molecules in the high-concentration region, not any inherent property of the molecules themselves, that leads to the "flow" from regions of high concentration to regions of low concentration. If we imagine a thin, but permeable barrier dividing a high-concentration region from a low-concentration region, then we can see that even if there is equal probability for an individual molecule near the barrier to move left or right across the barrier, there can still be a net flow from high concentration to low concentration simply because there are more molecules present on the high-concentration side. At the macroscopic level, we describe this behavior as a flow of a continuous variable, the concentration of molecules.

Let us now turn to a more formal description of diffusion. The flow of the molecules is usually expressed in terms of a ***molecular current density***, the number of molecules per unit area passing a point in space per unit time (see Fig. 11.10). The current density in the x direction, for example, j_x is proportional to the rate of change of concentration with position as expressed in Fick's Law of Diffusion:

$$j_x = -D \frac{\partial C(x, y, z)}{\partial x} \qquad (11.4\text{-}1)$$

where D is called the ***diffusion coefficient*** and $C(x,y,z)$ is the concentration of molecules (the number per unit volume) under consideration. The partial derivative gives us the so-called gradient of the concentration in the x direction. The minus sign in Eq. (11.4-1) tells us that the flow is from high-concentration regions to low-concentration regions. (By definition, the gradient is positive in going from low concentration to high concentration.) The diffusion coefficient D depends on the type of molecule and on the conditions of the environment such as viscosity. A large value of D means that the molecules diffuse relatively rapidly through their environment.

If the concentration of molecules also varies along the y and z directions, then we need to write a vector equation

$$\vec{j} = -D \; grad \; C(x,y,z) \tag{11.4-2}$$

The symbol *grad* $C(x,y,z)$ is the (vector) gradient of the concentration, a short-hand notation for the following vector

$$grad \; C(x,y,z) = \hat{i}\frac{\partial C}{\partial x} + \hat{j}\frac{\partial C}{\partial y} + \hat{k}\frac{\partial C}{\partial z} \tag{11.4-3}$$

which yields a vector pointing in the direction in which C is changing most rapidly with position. Note that in writing Eq. (11.4-2), we have assumed that D is the same for all directions in the medium; we say the medium is "isotropic." We would otherwise have different Ds for different directions.

If the molecules being transported are neither created nor destroyed (due to chemical reactions), we say that the behavior of the molecules is described by a ***conservation law***. In such a situation, if there is a net flow of molecules out of some region, then the concentration of the molecules in that region must decrease. As we saw in Chapter 3, the net flow out of a region is described mathematically by the so-called divergence of the current density vector. Thus, the conservation law (sometimes called the ***continuity equation***) can be written

$$\frac{\partial C(x,y,z,t)}{\partial t} = -div \; \vec{j}(x,y,z,t) \tag{11.4-4}$$

This conservation law tells us that if there is a net flow out of some spatial region, as indicated by a positive value of the divergence of the current density vector, then the concentration in that region will decrease (its time derivative will be negative).

We are now ready to establish our most important equation describing transport by diffusion by combining Eqs. (11.4-4) and (11.4-2):

$$\frac{\partial C(x,y,z,t)}{\partial t} = div[D \; grad \; C(x,y,z,t)] \tag{11.4-5}$$

In many cases, the diffusion coefficient is independent of position, so we may write (dropping the arguments of the concentration function to simplify notation)

$$\frac{\partial C}{\partial t} = D \; div \; grad \; C \tag{11.4-6}$$
$$\equiv D\nabla^2 C$$

where the last equality defines ∇^2, called the ***Laplacian*** after the French mathematician J. P. Laplace, whom we met in Chapter 1. In Cartesian (x,y,z) coordinates, the Laplacian is simply the sum of three second derivatives: $\nabla^2 C = \partial^2 C/\partial x^2 + \partial^2 C/\partial y^2 + \partial^2 C/\partial z^2$. As we shall see, Eq. (11.4-6) and its generalizations can be used to describe many other transport phenomena in addition to diffusion.

We can use Eq. (11.4-6) to define some quantities that will help us get a feeling for various transport phenomena. First, we define a distance d over which the concentration of molecules varies significantly. We also define a characteristic diffusion time τ_D. Specifically, the definitions are

$$d^2 \nabla^2 C \approx C$$
$$\tau_D \frac{\partial C}{\partial t} \approx C \qquad (11.4\text{-}7)$$

Roughly speaking, we say that τ_D is the time required for the concentration to change a substantial amount due to diffusion. The inverse of this characteristic diffusion time is called the "diffusion rate." Using these definitions in Eq. (11.4-6) yields the following relationship:

$$\tau_D = \frac{d^2}{D} \qquad (11.4\text{-}8)$$

which relates the diffusion coefficient D to the characteristic concentration distance d and the characteristic diffusion time τ_D. Note that for small values of D, we have a large value for τ_D (for a given value of d). In that case the molecules diffuse slowly, and it takes a long time for the concentration to change significantly.

Some Refinements

We have focused our attention so far on the concentration of molecules in a single (small) spatial region and we have described how the concentration changes due to diffusion of molecules in and out of that region. In other situations, we may be concerned with the concentration in some region that is moving. The concentration can change both through diffusion and because our region of interest moves, with the fluid flow, into a spatial region where the concentration is different. We take this "bulk flow" (as compared to diffusive flow) of the material medium into account by adding a $\vec{v} \cdot grad\, C$ term to the left-hand side of Eq. (11.4-6):

$$\frac{\partial C}{\partial t} + \vec{v} \cdot grad\, C = D \nabla^2 C \qquad (11.4\text{-}9)$$

Mathematically, we can say that the left-hand-side gives the *total* time derivative of the concentration. The first term tells us how C changes at a fixed spatial location while the second term tells us how C changes due to fluid flow into a region of different concentration. The first term expresses the time derivative from the point of view of an observer at a fixed spatial location. In fluid mechanics, this is called the **Eulerian** point of view. The two terms together give us the **Lagrangian** point of view: the time dependence seen by an observer moving with a volume element of the fluid.

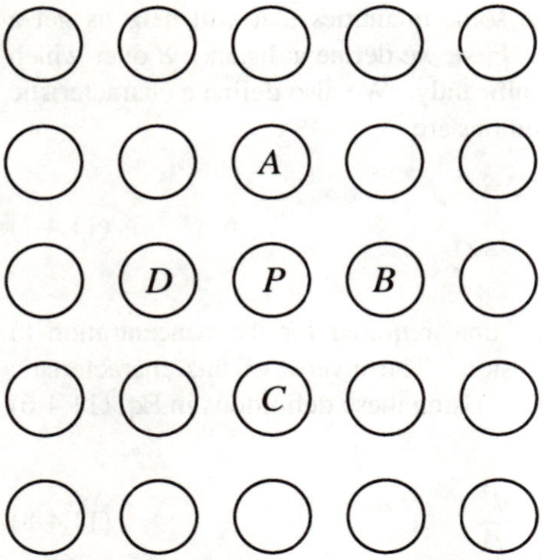

Fig. 11.11. An illustration of the use of the averaging property of the Laplacian to solve (approximately) Laplace's equation. The value of the function at point P is taken as the average of the values at the neighboring points $V(P) = 1/4 \{V(A) + V(B) + V(C) + V(D)\}$.

Exercise 11.4-1. Use the chain rule for partial differentiation to show that the left-hand side of Eq. (11.4-9) is indeed the total time derivative for the concentration.

For our second refinement of the diffusion equation, we can allow for chemical reactions that change the concentration. We take the reactions into account by adding a term f_{source} to the right-hand side of Eq. (11.4-9). (If the reactions remove the molecules of interest, we make f_{source} negative.) Assembling all of these possibilities yields our most general diffusion transport expression

$$\frac{\partial C}{\partial t} + \vec{v} \cdot grad\, C = D\nabla^2 C + f_{source} \qquad (11.4\text{-}10)$$

As we shall see, the Laplacian ∇^2 appears in a wide variety of transport phenomena. Although we will not do much with the formal mathematical properties of the Laplacian, we do want to point out an important feature: The Laplacian expresses a kind of averaging effect. It shows up in the description of phenomena, such as diffusion, which tend to have a smoothing effect. That is, diffusion tends to decrease (smooth out) concentration differences. On the other hand, spatially concentrated chemical reactions tend to produce concentration differences. It is this competition between smoothing due to diffusion and the production of differences due to chemical reactions (or other mechanisms) that leads to interesting spatial patterns.

Because a detailed mathematical proof of the averaging nature of the Laplacian would take us too far afield, we shall refer the interested reader to two books that treat this property in the context of the theory of electrostatic fields [Griffiths, 1981] and [Purcell, 1985]. The essential statement is: If a function, say $C(x,y,z)$, satisfies the equation $\nabla^2 C = 0$ in some region, then the value of C at some point P in that region is equal to the average value of C taken over a sphere (contained in that region) centered on the point P. This averaging property is often used in numerical solutions of ***Laplace's Equation*** $\nabla^2 C = 0$. Figure 11.11 illustrates the method in two-dimensions.

Exercise 11.4-2. We can illustrate the averaging property of the Laplacian with a simple one-dimensional example. Suppose we are concerned with some function $f(x)$ and its average over the interval $[-L/2, L/2]$. The average of $f(x)$ over that interval is defined as

$$< f(x) >= \frac{1}{L} \int_{-L/2}^{L/2} f(x)dx$$

(a) Use a Taylor series expansion about $x = 0$ to show that

$$\frac{d^2 f(x)}{dx^2} = \frac{12}{L^2}[< f(x) > -f(0)]$$

if we neglect all derivatives higher than the second.

(b) Show that if the second derivative is equal to 0 (i.e., f satisfies a one-dimensional version of Laplace's equation), then $f(0) = <f(x)>$; that is, the value of the function at the center of the interval is equal to the average value over the interval.

(c) Extend the argument to three dimensions.

Conductive Thermal Energy Flow and Other Transport Phenomena

Transport equations for many other phenomena take the same form as the equations describing diffusion. We illustrate this point with several examples.

The transport of thermal energy ("heat") due to conduction is driven by a temperature gradient. Thus, we can define a thermal energy current density (so many watts per unit area) $\vec{j}_T$ that is proportional to the temperature gradient:

$$\vec{j}_T = -\sigma_T \, grad \, T(x, y, z) \tag{11.4-11}$$

where T is the temperature of the material and σ_T is called the thermal conductivity of the material. This equation tells us that thermal energy flows from regions of high temperature to regions of low temperature. The equivalent conservation law says that if there are no energy sources or sinks in a region, then as thermal energy leaves that region, there must be a change in the thermal energy density ρ_T associated with that region. The conservation law takes the form

$$\frac{\partial \rho_T}{\partial t} = -div \, \vec{j}_T \tag{11.4-12}$$

A change in thermal energy density, however, means that the temperature of that region will change. The *specific heat capacity* C_T is the quantity that relates a change in thermal energy density to a change in temperature of that region:

$$\Delta\rho_T = C_T\Delta T \qquad (11.4\text{-}13)$$

Using Eq. (11.4-13) and Eq. (11.4-12) allows us to write

$$\frac{\partial T}{\partial t} = \frac{\sigma_T}{C_T}\nabla^2 T \qquad (11.4\text{-}14)$$

$$\equiv D_T\nabla^2 T$$

where the last equality defines D_T, the thermal diffusivity of the material, in analogy with the diffusion coefficient D for molecular diffusion. By comparing Eq. (11.4-14) with Eq. (11.4-6), we see that changes in temperature due to thermal conduction (thermal energy diffusion) are described by an equation identical in form to the equation describing concentration changes due to molecular diffusion.

Charged Particle Motion in a Resistive Medium

If electrically charged particles move through a resistive medium, the electric current density $\vec{j}_e$ is proportional to the gradient of the electric potential function $V(x,y,z)$. (This statement is equivalent to the well-known Ohm's Law.) In mathematical terms, we write

$$\vec{j}_e = -\sigma\ grad\ V(x, y, z) \qquad (11.4\text{-}15)$$

where σ is the *electrical conductivity* of the medium. (The negative gradient of the electric potential function is equal to the electric field.) The corresponding conservation law relates the time derivative of the electric charge density ρ_e to the (negative) divergence of the electric current density:

$$\frac{\partial\rho_e}{\partial t} = -div\ \vec{j}_e \qquad (11.4\text{-}16)$$

Combining Eqs. (11.4-15) and (11.4-16) yields

$$\frac{\partial\rho_e}{\partial t} = \sigma\nabla^2 V \qquad (11.4\text{-}17)$$

This equation has some of the same structure as the previous transport equations, but there are important differences. For charged particles, we do not have a simple relationship between the charge density ρ_e and the electric potential V. Equation (11.4-17), however, will be sufficient for establishing analogies in a later section.

Eq. (11.4-17) may look rather strange to the experienced students of electrostatics. We should point out that the usual Poisson equation for the electric potential function

$$\nabla^2 V = \frac{\rho_e}{\varepsilon_o} \qquad (11.4\text{-}18)$$

relates the Laplacian of the electrical potential function to the charge density that is responsible for the electrical potential (i.e., the sources of the fields). In Eq. (11.4-17), we are relating the change in charge density of charges moving under the influence of a potential function produced by <u>other</u> (usually fixed in position) electrical charges.

Momentum Transport in Fluid Flow

The motion of a fluid can be expressed in terms of the flow of momentum (mass times velocity) using Newton's Second Law (force equals rate of change of momentum). For a fluid, this law is usually expressed in terms of momentum *per unit volume* (mass per unit volume ρ times velocity). Conceptually, using the Lagrangian point of view, we follow a certain spatial region as it moves with the fluid flow. Then, for momentum in the x direction, for example, we write

$$\frac{\partial(\rho v_x)}{\partial t} + \vec{v} \cdot grad\,(\rho v_x) = F_x \qquad (11.4\text{-}19)$$

where F_x is the x component of the force (per unit volume) acting on the packet of fluid in the region we are following.

The first term on the left-hand side of Eq. (11.4-19) tells us that the momentum may change because the velocity at a particular spatial location may change with time (we shall consider only fluids for which the density is essentially constant). The second term tells us that the momentum may change because the fluid packet moves (with velocity $\vec{v}$) to a location with a new momentum value. (Compare the previous discussion about the Eulerian and Lagrangian points of view in fluids.) Note that this second term is *nonlinear* in the velocity—a portent of interesting things to come.

The fluid also satisfies a conservation law. We assume that the fluid is neither created nor destroyed. If the fluid density ρ changes in a certain region, there must be a net flow into or out of that region. Hence, the conservation law takes the familiar form:

$$\frac{\partial \rho}{\partial t} = -div\,(\rho \vec{v}) \qquad (11.4\text{-}20)$$

For fluids such as water, the density is essentially constant (We say that the fluid is *incompressible*.) and the conservation law reduces to

$$div\,\vec{v} = 0 \qquad (11.4\text{-}21)$$

which result we shall invoke in our discussion later.

For the fluids that we shall consider, the important forces will be due to (1) "body forces" (such as gravity, proportional to ρ), (2) pressure gradients, and (3) viscosity. Thus, we write

$$F_x = \rho f_x - \frac{\partial p}{\partial x} + \mu \nabla^2 \upsilon_x \qquad (11.4\text{-}22)$$

where p is the fluid pressure at that location and μ is the fluid viscosity. (A simple development of the viscosity force expression is given at the end of this section.) The body forces (per unit mass density) are represented by the symbol f_x. If we combine Eqs. (11.4-22) and (11.4-19), we arrive at the fundamental fluid transport equation

$$\frac{\partial(\rho\upsilon_x)}{\partial t} + \vec{\upsilon} \cdot grad(\rho\upsilon_x) = \rho f_x - \frac{\partial p}{\partial x} + \mu \nabla^2 \upsilon_x \qquad (11.4\text{-}23)$$

with analogous expressions for υ_y and υ_z. If the density is constant for the fluid of interest, then ρ can be removed from the derivative terms, and we obtain, after dividing through the last equation by ρ,

$$\frac{\partial \upsilon_x}{\partial t} + \vec{\upsilon} \cdot grad\, \upsilon_x = f_x - \frac{1}{\rho}\frac{\partial p}{\partial x} + \nu \nabla^2 \upsilon_x \qquad (11.4\text{-}24)$$

where $\nu = \mu/\rho$ is called the **kinematic viscosity**. Eq. (11.4-24) is the famous **Navier–Stokes equation** for incompressible fluid flow.

Summary

As we have seen, the formal descriptions of many transport phenomena have nearly identical forms: An expression giving the temporal and spatial derivatives of the transported quantity is equal to various "force" terms, many of which can be written in terms of the Laplacian ∇^2. Given these similar formal expressions, we might expect to find similarities in the behaviors of these transport systems.

*Simple Development of the Form of the Viscosity Force

Here we present a simplified formal treatment of viscosity. The goal is to provide some justification for the $\mu \nabla^2 \upsilon_x$ form of the viscosity force used in Eq. (11.4-22) because we will use that form to establish several analogies in sections to follow. It is also used in the development of the Lorenz model equations in Appendix C. More detailed treatments can be found in standard texts on fluid mechanics.

We begin by considering a fluid flowing in the x direction as shown in Fig. 11.12. Let us assume that the magnitude of the velocity increases with distance above the lower plate. Let us now focus our attention on a thin section of fluid. The fluid just above the section is traveling faster than the fluid in that section and tends to drag it forward in the x direction. This drag force is just what we mean by the viscous force acting on the fluid section. Isaac Newton was apparently the first

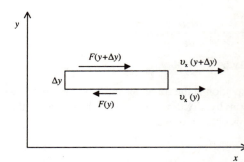

Fig. 11.12. A diagram of a section of fluid moving in the x direction (to the right). The velocity increases in the vertical (y) direction. A thin section of fluid of height Δy is shown. The fluid above drags the rectangular section to the right. The fluid below provides a force acting to the left.

to suggest that this force is proportional to (1) the area ΔA of the top of the section and (2) the change in speed per unit distance in going across the thin section:

$$F_x(y + \Delta y) = \mu \frac{\Delta v_x}{\Delta y} \Delta A \qquad (11.4\text{-}25)$$

where the proportionality constant is defined to be the viscosity of the fluid. The argument of F_x reminds us that this is just the force due to the fluid above the section. In the limit of $\Delta y \to 0$, the ratio in the last term in Eq. (11.4-25) becomes a partial derivative. You should note the important physical point: The force is proportional to a velocity (actually a change in velocity with position), a typical situation for a frictional force in a fluid.

We now need to consider the force exerted on this section by the fluid below it, traveling with a slower speed. That viscous force tends to slow down the section of fluid. In fact, if the force due to the fluid below is equal in magnitude, but opposite in direction, compared to the force due to the fluid above, there is no net force on this section. (There is just a tendency to shear or distort the section.) However, if the gradient at the top of the section is larger than the gradient at the bottom of the section, then there will be a net force to the right. Thus, we can write the net force as

$$
\begin{aligned}
F_x &= F_x(y + \Delta y) - F_x(y) \\
&= \frac{\partial F_x}{\partial y} \Delta y \qquad\qquad (11.4\text{-}26) \\
&= \mu \frac{\partial^2 v_x}{\partial y^2} \Delta y \Delta A
\end{aligned}
$$

where we have replaced the difference by the derivative multiplied by the change in distance over which the difference occurs. Thus, we see that the *net* force in the x direction due to the top and bottom sections is proportional to the <u>second</u> derivative of the x component of the velocity.

Since the x component of the velocity might also vary in the x and z directions, we get two more contributions to the viscous force. Together, they yield

$$F_x = \mu \left\{ \frac{\partial^2 v_x}{\partial x^2} + \frac{\partial^2 v_x}{\partial y^2} + \frac{\partial^2 v_x}{\partial z^2} \right\} \Delta y \Delta A \qquad (11.4\text{-}27)$$

$$\equiv \mu \{\nabla^2 v_x\} \Delta y \Delta A$$

The force we need for the fluid flow equation is the force per unit volume, which is obtained from the previous expression by dividing both sides by $\Delta y \Delta A$, the volume of our section of fluid. With this operation, we find

$$f_x = \mu \nabla^2 v_x \qquad (11.4\text{-}28)$$

which is the desired result.

We should point out that we have dodged some of the more intricate formalism needed to exclude from the viscous force expression the fluid motion that arises from a pure rotation of the fluid, for example, in a bucket of water rotating about a vertical axis through the center of the bucket. Such a pure rotation does not involve the shearing of one layer of water relative to another. That type of motion is excluded by writing the force on the top of the section, for example, as

$$F_{xy} = \mu \left\{ \frac{\partial v_x}{\partial y} + \frac{\partial v_y}{\partial x} \right\} \Delta A \qquad (11.4\text{-}29)$$

where the first subscript on the force term tells us that the force is in the x direction; the second subscript tells us that the force is acting on a plane perpendicular to the y axis. Similar expressions give us the forces acting on the faces in the x and z directions. (Note that using these expressions does not change the simple argument used earlier because there we had velocity only in the x direction.) The calculation proceeds as previously, but we need to invoke the fact that $div \, \vec{v} = 0$ for an incompressible fluid to arrive at the desired result.

> **Exercise 11.4-3.** Carry through the calculation outlined in the previous paragraph to show that Eq. (11.4-28) is the correct expression for the viscous force. Hint: See [Feynman, Leighton, Sands, 1964], Vol. II, p. 41–4, or an introductory book on fluid mechanics.

11.5 Reaction-Diffusion Systems: A Paradigm for Pattern Formation

As an example of a system that can develop interesting spatial patterns from transport phenomena, we discuss so-called *reaction-diffusion* systems. In these systems we focus our attention on the concentrations of two or more chemical species, which both diffuse through the spatial region under consideration and react chemically. This type of system was considered by Turing (TUR52) in a classic paper in theoretical biology. Turing showed that reaction-diffusion systems can lead, via spontaneous symmetry-breaking, to interesting spatial patterns. Such patterns are now called *Turing structures*. Although the theory of these structures has been thoroughly explored, unambiguous experimental evidence of their existence has been available only recently (CDB90) (OUS91). These reaction-diffusion mechanisms are thought to play an important role in pattern formation in biological systems [Murray, 1989]. Here we shall use this model as an example of how a spatially extended system, described by a set of partial differential equations, is analyzed in terms of pattern formation. The important lesson to be learned is that by expressing the behavior of the system in terms of the time behavior of various *spatial modes*, we can reduce the problem of pattern formation to a problem in which the amplitudes for these spatial modes play the role of state space variables. Once that reduction has been made, we can then apply all that we have learned about temporal dynamics to the study of pattern formation.

In a reaction-diffusion system, the concentrations C_i of the chemical species are described by

$$\frac{\partial C_i}{\partial t} = f_i(\ldots C_j \ldots) + D_i \nabla^2 C_i \tag{11.5-2}$$

where D_i is the diffusion coefficient for the ith substance. The function f_i specifies the reaction mechanism, which in general depends nonlinearly on all the concentrations. In addition, we need to specify "boundary conditions." The feed (or input) and removal of reactants can be concentrated in space or time or the reactants can diffuse in and out from boundaries. What is surprising is that these equations and boundary conditions can lead to spatial patterns with an intrinsic length scale (a "wavelength") that depends only on the concentrations or rates of input or removal and <u>not</u> on the geometric size of the system.

In order for spatial patterns to form, several requirements must be met. The reaction kinetics need a positive feedback mechanism for at least one of the species, called the activator, and an inhibitory process. Numerical evidence seems to require that the inhibitor diffuse through the spatial region more rapidly than the activator. Otherwise, the patterns exist only over a rather narrow range of control parameters (rates of feed, rates of diffusion, and so on).

As an example of a reaction-diffusion system, let us consider a two-component model [Murray, 1989, Chapter 14] [Nicolis and Prigogine, 1989, Appendix A]. We assume that the system has only two substances whose concentrations are given by C_1 and C_2. The diffusion constant of substance 1 is D_1; it is D_2 for substance 2. The equations describing this system are

$$\frac{\partial C_1}{\partial t} = f_1(C_1, C_2) + D_1 \nabla^2 C_1$$
$$\frac{\partial C_2}{\partial t} = f_2(C_1, C_2) + D_2 \nabla^2 C_2 \tag{11.5-3}$$

We also need to specify the boundary conditions. Since we are interested in pattern formation without "external forcing," we choose to specify that the flux (flow) of species at the boundary be 0. In addition, we need to specify the initial spatial concentration $C_1(\vec{r}, t = 0)$, $C_2(\vec{r}, t = 0)$. Let us assume that those specifications have been made.

Since the governing equations are partial differential equations, we follow the standard solution procedure, which assumes that the resulting solutions can be written as (combinations of) <u>products</u> of functions, one depending only on time, the other carrying all the spatial dependence. To see how this procedure works, let us first ignore the spatial dependence; that is, we "turn off" diffusion. Then the time-dependent equations are

$$\frac{dC_1}{dt} = f_1(C_1, C_2)$$
$$\frac{dC_2}{dt} = f_2(C_1, C_2)$$

(11.5-3)

These equations have exactly the same form as the general two-dimensional state space equations treated in detail in Chapter 3. To make use of the results developed earlier, we will limit our attention to small deviations from steady-state concentrations C_{1o} and C_{2o}, values that are obtained as the fixed-point values in Eq. (11.5-3). (We assume for now that there is only one fixed point.)

In parallel with our treatment of state-space dynamics in Chapter 3, we define variables that express the deviation from the steady-state values

$$c_1 = C_1 - C_{1o}$$
$$c_2 = C_2 - C_{2o}$$

(11.5-4)

Then, as we saw in Chapter 3, the time dependence near the steady-state is controlled by the characteristic values of the Jacobian matrix of derivatives of the functions f_1 and f_2. Using this expansion about the steady-state values in Eq. (11.5-3), we obtain the time evolution equations

$$\dot{c}_1 = f_{11}c_1 + f_{12}c_2$$
$$\dot{c}_2 = f_{21}c_1 + f_{22}c_2$$

(11.5-5)

where we have used a notation like that in Section 3.11. The results developed in Chapter 3 can be applied immediately to this case. For example, the steady-state will be stable if the real parts of the characteristic values of the Jacobian matrix are negative.

Let us now return to the full space-time problem. We focus our attention on small deviations from the steady-state condition, which we take, for the sake of simplicity, to be a spatially uniform state. Making use of $\nabla^2 C_{io} = 0$ for the spatially uniform steady-state, we find that the cs are described by the following equations:

$$\dot{c}_1 = f_{11}c_1 + f_{12}c_2 + D_1\nabla^2 c_1$$
$$\dot{c}_2 = f_{21}c_1 + f_{22}c_2 + D_2\nabla^2 c_2$$

(11.5-6)

If there were no coupling between the substances (that is, if f_{12} and f_{21} are both 0), we could write c_1 and c_2 as products of functions, one depending on time, the other on spatial variables. For example, $c_1(\vec{r},t) = \phi(t)U(\vec{r})$. When this product form is inserted into Eq. (11.5-6), we find for c_1

$$U(\vec{r})\dot{\phi}(t) = f_{11}\phi(t)U(\vec{r}) + \phi(t)D_1\nabla^2 U(\vec{r}) \qquad (11.5\text{-}7)$$

with an analogous equation for c_2. Rearranging the previous equation and dividing through by the product ϕU gives us

$$\frac{1}{D_1}\left(\frac{\dot{\phi}(t)}{\phi(t)} - f_{11}\right) = \frac{1}{U(\vec{r})}\nabla^2 U(\vec{r}) \qquad (11.5\text{-}8)$$

Now comes a crucial part of the argument: The left-hand side of Eq. (11.5-8) is a function of time alone; the right-hand side is a function of position alone. The only way the two sides can be equal for all combinations of time and position is for each to be equal to a constant (independent of space and time), which we shall call, with some foresight, $-k^2$. Thus we have separated the original partial differential equation into two ordinary differential equations:

$$\dot{\phi}(t) = (f_{11} - k^2 D_1)\phi(t)$$
$$\nabla^2 U(\vec{r}) + k^2 U(\vec{r}) = 0 \qquad (11.5\text{-}9)$$

The first equation in Eq. (11.5-9) tells us that the time dependence can be written as an exponential function of time with a characteristic exponent that depends on both the reaction function derivatives f_{11} and on the constant k. The second equation in Eq. (11.5-9) and the boundary conditions imposed on the system determine the so-called *spatial modes* for the system. For a region of finite spatial extent, it turns out that only certain values of k lead to solutions that satisfy the equation and the boundary conditions. These values of k are called the eigenvalues (or characteristic values) for the spatial modes.

As a simple example of spatial modes, let us consider a one-dimensional problem. We choose the x axis for that one dimension. Let us assume that the boundary conditions are that the flow of the substance is 0 at the boundaries of the region at $x = 0$ and $x = L$. Since the flow is proportional to the gradient (spatial derivative) of the concentration for diffusion problems, the physical boundary condition translates into the mathematical boundary condition of having the derivative of $U(x)$ vanish at $x = 0$ and $x = L$. The set of functions that satisfy the one-dimensional equation

$$\frac{d^2 U(x)}{dx^2} + k^2 U(x) = 0 \qquad (11.5\text{-}10)$$

and these boundary conditions is $U_k(x) = A\cos kx$ with $k = n\pi/L$ and $n = 0, 1, 2, 3,$ ... The subscript on the function U reminds us that the function depends on k. We see that only a discrete set of k values, the set of eigenvalues, leads to functions satisfying both the differential equation and the boundary conditions.

Note that as a function of x, $U_k(x)$ is periodic with spatial period ("wavelength") $2\pi/k$. For higher-dimensional problems, there will generally be a set of ks for each of the spatial dimensions. Hence, k is sometimes called the **wavevector** for the spatial modes. Note that $k = 0$ corresponds to the spatially uniform state.

Exercise 11.5-1. Verify that $U_k(x) = A \cos kx$ with $k = n\pi/L$ with $n = 0$, 1, 2, 3, ... satisfies Eq. (11.5-10) and the stated boundary conditions at $x = 0$ and $x = L$.

Returning to our general, but still uncoupled, problem, let us assume that we have found the spatial mode functions and the allowed values of k for the problem at hand. We then assume that the most general solution for c_1, for example, can be written as a superposition of products of temporal functions and spatial mode functions:

$$c_1(\vec{r},t) = \sum_k a_{1k}(t) U_k(\vec{r}) \tag{11.5-11}$$

The coefficients $a_{1k}(t)$ are called the **spatial mode amplitudes**. The coefficients at $t = 0$, $a_{1k}(t = 0)$, are determined by the initial conditions on the concentration. The expression for c_2 can be written in a similar form.

With these preliminaries in mind, let us get back to the problem of the coupled concentrations. Our basic mathematical assumption is that the solutions to Eqs. (11.5-6) can be written in the form of Eq. (11.5-11) even when the concentrations are coupled. If we insert the superposition form Eq. (11.5-11) into Eqs. (11.5-6) and replace $\nabla^2 U$ with $-k^2 U$ according to Eq. (11.5-9), we find:

$$\sum_k \dot{a}_{1k} U_k(\vec{r}) = f_{11} \sum_k a_{1k}(t) U_k(\vec{r}) + f_{12} \sum_k a_{2k}(t) U_k(\vec{r})$$
$$- D_1 \sum_k a_{1k}(t) k^2 U_k(\vec{r}) \tag{11.5-12}$$

with an analogous equation for the second of Eq. (11.5-6).

Now we come to another second crucial point: We claim that the functions $a_{ik}(t) U_k(\vec{r})$ associated with each value of k must satisfy the equations independently. (In more formal terms, we say the spatial mode functions are "orthogonal" when integrated over the spatial region of interest. We can use that orthogonality to pick out a particular value of k. See the references on mathematical methods at the end of this chapter for more details.) Hence, we find the following equations for the kth spatial mode amplitudes:

$$\dot{a}_{1k}(t) = f_{11} a_{1k}(t) + f_{12} a_{2k}(t) - D_1 a_{1k}(t) k^2$$
$$\dot{a}_{2k}(t) = f_{21} a_{1k}(t) + f_{22} a_{2k}(t) - D_2 a_{2k}(t) k^2 \tag{11.5-13}$$

We see that the spatial mode amplitudes depend on each other; we say we have **coupled modes**.

> Using the spatial mode expansion technique, we have transformed our original set of partial differential equations into a set of coupled ordinary differential equations for the spatial mode amplitudes. These equations are exactly of the form of the equations describing temporal dynamics in state space, where now the spatial mode amplitudes play the role of state space variables. We have one complication: since there are an infinite number of possible k values, the state space has, in principle, an infinite number of dimensions. This result tells us that systems described by partial differential equations are dynamical systems with an infinite number of degrees of freedom. What saves the day and makes some progress possible is that for many cases, only a few spatial modes are "active." For the inactive modes, the time dependence of their amplitudes is an exponential with a characteristic value whose real part is negative. Any initial "excitation" of that inactive mode dies away exponentially with time. An inactive mode plays no role in the long-term dynamics of the system.

Since Eq. (11.5-13) is exactly like the set of equations treated in Section 3.11, we can, without further ado, apply all that we learned in Chapter 3 to analyze the dynamics of the system. In particular, we assume that the temporal behavior of each amplitude can be written as an exponential function of time, as we did in Section 3.11:

$$a_{ik}(t) = b_{ik}e^{\lambda(k)t} \tag{11.5-14}$$

where $b_{ik} = a_{ik}(0)$ is independent of time and $\lambda(k)$ is a characteristic exponent. (Note that λ is <u>not</u> the spatial wavelength.) Using this form in Eq. (11.5-13) yields the following coupled, linear *algebraic* equations

$$\lambda b_{1k} = f_{11}b_{1k} + f_{12}b_{2k} - D_1k^2b_{1k}$$
$$\lambda b_{2k} = f_{21}b_{1k} + f_{22}b_{2k} - D_2k^2b_{2k} \tag{11.5-15}$$

The coupled, linear equations for the modes Eqs.(11.5-15) have "nontrivial" solutions only if the determinant of the coefficients of the b_{ik} is equal to 0. (The "trivial" solutions are $b_{ik} = 0$.) Evaluating that determinant then leads to an expression for the characteristic exponent $\lambda(k)$ in terms of the spatial mode parameter k. For our coupled equations, the resulting equation for λ is

$$\lambda^2 + \lambda(k^2[D_1 + D_2] - f_{11} - f_{22}) + k^2(D_1D_2k^2 - f_{11}D_2 - f_{22}D_1)$$
$$+ f_{11}f_{22} - f_{12}f_{21} = 0 \tag{11.5-16}$$

In general, this equation yields two values of λ for each value of the spatial mode wavevector k. The resulting function $\lambda(k)$ is often called the **dispersion relation** for the system. (The terminology is borrowed from wave optics, where the analogous relation tells us how light waves may be dispersed in traveling through some medium due to the relationship between temporal frequency, here denoted by $\lambda(k)$, and wavevector. This relationship determines the speed of the waves.)

> **Exercise 11.5-2.** (a) Check the calculation leading from Eq. (11.5-15) to Eq. (11.5-16). (b) Check that Eq. (11.5-16) leads to the expected result when $k = 0$. (c) What conditions must hold for the f_{ij}s if the spatially uniform ($k = 0$) solution is to be a stable state?

In order to find where interesting spatial patterns emerge from the initial homogeneous state, we look for the conditions for which the real part of $\lambda(k)$ becomes positive for some value of k. Using our experience with temporal dynamics in state space, we see that when the real part of $\lambda(k)$ becomes positive, the "motion" of the system in state space becomes unstable along that corresponding direction. In the spatial mode case, that means that a pattern corresponding to the spatial mode function $U_k(\vec{r})$ begins to emerge from the spatially uniform initial state. In other words, if we start in the spatially uniform initial state and if the real part of $\lambda(k)$ is positive for $k \neq 0$, then any slight disturbance will push the system away from the unstable fixed point corresponding to the spatially uniform state.

As an example of what happens, let us take the diffusion constant D_2 as the control parameter with all the other parameters held fixed. Consulting Table 3.3, we see that the real part of λ will be positive for some k if

$$D_1 D_2 k^4 - k^2 (D_1 f_{22} + D_2 f_{11}) + f_{11} f_{22} - f_{12} f_{21} < 0 \qquad (11.5\text{-}17)$$

This case has been analyzed in some detail in [Murray, 1989, Chapter 14]. A typical plot of the dispersion relation for different values of the control parameter D_2 is shown in Fig. 11.13. For $D_2 < D_c$, a "critical value," all the λs have <u>negative</u> real parts, and the spatially uniform state is stable. At $D_2 = D_c$, the spatial mode with $k = k_c$ begins to grow. For $D_2 > D_c$ modes associated with several ks may begin to grow. A note of modesty is important here: Our analysis does not tell us what happens to these modes as their amplitudes begin to grow. Our analysis is valid only very close to the initial spatially uniform steady-state.

> **Exercise 11.5-3.** In Exercise 11.5-2, we found certain constraints on the f_{ij} given the assumed stability of the $k = 0$ state. (a) Use those results and Eq. (11.5-17) to show that no spatial mode can be unstable (and hence, no spatial pattern formed) if the two diffusion coefficients are the same (that is, if $D_1 = D_2$). (b) Assume that $f_{11} > 0$ (substance 1 is a self-activator) and $f_{22} < 0$ (substance 2 is a self-inhibitor). Show that we need $D_2 > D_1$ in order to have some spatial mode become unstable. *Hint*: See [Murray, 1989], Chapter 14.

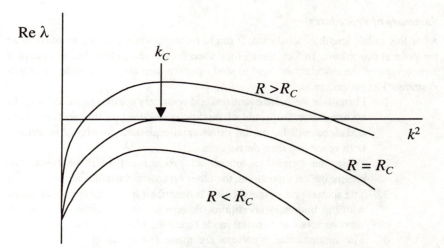

Fig. 11.13. A plot of the real part of typical dispersion relations $\lambda(k)$ for a two-substance coupled reaction-diffusion system. When the real part of $\lambda(k)$ becomes positive, spatial modes corresponding to those values of k can begin to grow in amplitude. $R = D_2/D_1$ is the ratio of the two diffusion coefficients.

Figure 11.14 shows a sketch of the concentration pattern that begins to grow for a one-dimensional situation with $k_c = 8\pi/L$.

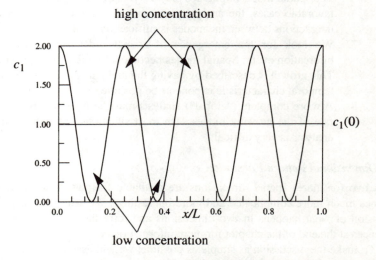

Fig. 11.14. A sketch of the spatial pattern of concentration c_1 that begins to emerge with $k_c = 8\pi/L$. When c_1 is above $c_1(0)$, the concentration of substance 1 is higher than in the spatially uniform state. The pattern is an alternating sequence of high-concentration and low-concentration regions. (The numerical scale for c_1 is arbitrary.)

Summary of Procedures

After this rather lengthy calculation, it might be worthwhile pausing to summarize the general procedure. In fact, the method used for the reaction-diffusion system is a prototype of the calculations used to study pattern formation in a wide variety of systems. The procedure is:

1. Formulate the basic equations and boundary conditions describing the system. For the kinds of models considered in this section, these equations will be partial differential equations involving in general both space and time derivatives.
2. Write the general solutions as products of two functions, one depending on time alone, the other on spatial variables alone.
3. The spatial part of the solution is described by an equation that, along with the boundary conditions, determines the possible spatial mode wavevectors k and spatial mode functions $U_k(\vec{r})$ for the system.
4. The fundamental equations are made linear (using a Taylor series expansion) around a steady state (fixed point) condition. A spatially uniform state is often taken as the steady state.
5. The expansion in terms of (possibly time-dependent) spatial modes is used in the fundamental equations leading to (in general) a set of coupled ordinary differential equations for the time-dependent amplitudes of the spatial modes.
6. The time dependence is then determined by characteristic exponents in direct analogy with analysis of temporal dynamics in state space. The spatial mode amplitudes play the role of state space variables. In favorable cases, the analysis need include only a few modes if the interactions between the modes is sufficiently weak.
7. We look for the emergence of spatial order as signaled by a bifurcation event: Spatial modes become unstable and begin to grow. This growth is described by having the real part of at least one of the temporal characteristic exponents be positive.
8. A more complete ("global") analysis must be done to determine the fate of the growing modes. (In most situations this more complete analysis is very difficult.)

Two-Dimensional Pattern Formation

When two (or more) spatial dimensions are available for pattern formation, life becomes much more complicated and more interesting. We will give only a brief discussion of what happens in two dimensions and refer the reader to the list of references at the end of the chapter for more information.

To make the discussion as simple as possible, we will assume that the pattern is restricted to two spatial dimensions. We focus our attention on a single-component reaction-diffusion system as an example. In that case, Eq. (11.5-1) can be written as

$$\frac{\partial C(x, y)}{\partial t} = f(C) + D\left\{\frac{\partial^2 C(x, y)}{\partial x^2} + \frac{\partial^2 C(x, y)}{\partial y^2}\right\} \quad (11.5\text{-}18)$$

where the reaction term is a nonlinear function of C. (We have an autocatalytic reaction.) We follow the general procedure outlined earlier. Let us suppose that we have found that only one wavevector <u>magnitude</u> satisfies the condition for a pattern to emerge out of the uniform steady-state. However, the stability conditions generally do not pick out a <u>direction</u> for the wavevector. What are the possible patterns?

We assume that there is no preferred direction in our system and there is no preferred point. (We say that the system is *isotropic* and *homogeneous*.) If a pattern forms, it must consist of regular polygons that fit together to cover the plane surface. This regular polygon requirement severely restricts the possible patterns: If the polygon has n sides, then each of the interior angles is $\pi(1 - 2/n)$ radians. In order to cover the plane completely, this angle must divide 2π an integer number of times. Call that integer m. These two conditions can be satisfied simultaneously for the following (n,m) pairs: (3,6), (4,4), (6,3) corresponding to equilateral triangles, squares, and hexagons. (The triangles and hexagons are essentially the same pattern.) Of course, the system may develop other, more complex patterns by superposing, with various amplitudes and phases, the simple polygon patterns. We will assume, however, that we are near the threshold of pattern formation; therefore, we need to consider only the simple polygon solutions.

In many systems, we also see "roll" patterns, which are the two-dimensional generalization of the one-dimensional pattern shown in Fig. 11.14. In a roll pattern, the concentration (or other relevant physical parameter) varies periodically in one direction but is uniform in the orthogonal direction.

In both the roll case and the polygon case, some asymmetry of the system or some random fluctuation must pick out the orientation of the pattern since we have assumed that the basic system is isotropic. Something must break the rotational symmetry of the system.

The reaction function determines which of the possible patterns is stable once it begins to grow. Of course, the reaction function depends on the particular system under study. In most cases, we express the reaction function in terms of $c(x, y) = C(x, y) - C_o$, the deviation of the concentration from a uniform steady-state. A commonly studied reaction term contains a linear term and a cubic term and leads to what is called the ***Ginzburg–Landau equation***:

$$\frac{\partial c(x, y)}{\partial t} = ac(x, y) + bc^3(x, y) + D\nabla^2 c(x, y) \quad (11.5\text{-}19)$$

A quadratic term is missing because we want the equation to be the same if c is replaced by $-c$; that is, deviations above and below the steady-state concentration are assumed to be identical. The Ginzburg–Landau equation is used to model pattern formation in fluids, chemical reactions, laser light intensity patterns, and in

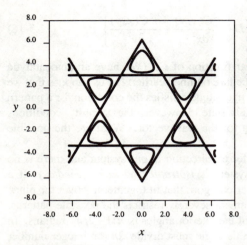

Fig. 11.15. A schematic diagram of a single hexagon cell described by Eq. (11.5-20). Inside the center hexagon, we have $c(x,y) > -1$. The interiors of the rounded triangular contours have $c(x,y) < -1$.

many other areas with appropriate redefinition of the dependent variable. See the references at the end of the chapter. In principle, although it is often difficult in practice, we can use the explicit form of the reaction function to determine which of the possible patterns will be stable for a given set of parameters.

Hexagonal patterns occur surprisingly frequently in nature (see, for example, [Murray, 1989] and [Chandrasekhar, 1981]). A solution of the two-dimensional generalization of Eq.(11.5-10) satisfying no-flux boundary conditions and describing a hexagonal pattern is given in polar coordinates by the following expression [Chandrasekhar, 1981]:

$$c(r,\theta) = \cos\{kr\sin(\theta)\}$$
$$+\cos\{kr\sin(\theta + \pi/3)\} \quad (11.5\text{-}20)$$
$$+\cos\{kr\sin(\theta + 2\pi/3)\}$$

where $k = 4\pi/(3L)$ and L is the length of a side of the hexagon. (The amplitude of the concentration difference is arbitrary. Here, we have $c = 3$ at the origin.) We see again that the stability conditions, not boundary geometry, determine the magnitude of the wavevector and the geometric size of the pattern. A diagram of one of the resulting hexagonal cells is shown in Fig. 11.15.

Pattern Selection and Spatiotemporal "Turbulence"

The process by which a spatially nonuniform pattern emerges from an initially spatially uniform state is called *pattern selection*. In the simplest cases, there is only one spatial mode whose amplitude will grow if the system is "bumped" away from the initial state (say, by a random disturbance). However, in general there may be several unstable modes available for a given set of conditions. Which mode is actually selected is determined by the exact initial conditions and the precise nature of the "random" disturbance. In this general case, successive repetition of the experiment can lead to different patterns because the initial conditions can never be exactly reproduced. Thus, in such cases, even though we may not need to take into account noise and fluctuations in the "final state" of the system, noise and fluctuations often play a critical role in the birth and evolution of these structures.

As we have seen for reaction-diffusion systems, the possible spatial modes of the system determine the kinds of patterns that can emerge as the system moves away from a spatially uniform state. In most realistic situations, we need additional help to sort out the possible modes. If the system and its boundary conditions have some spatial symmetry, then the mathematical theory of symmetry groups can be used to classify the possible patterns (see [Golubitsky, Stewart, Schaeffer, 1988] and GMD90 for further details).

When the emerging spatial patterns show sufficient spatial (or perhaps temporal) complexity, we say that the system exhibits spatiotemporal chaos, or rather loosely, turbulence. (Technically, this situation is called "weak turbulence." The connection to strong turbulence with vortices, and so on, is not yet well established.) Many systems that exhibit spatiotemporal chaos have spatial patterns that are characterized by so-called *defects*. For our purposes, we can think of a defect as boundary between two regions with different spatial structures. For example in one region, the pattern might be spatially periodic with one period. In the adjacent region the pattern might also be spatially periodic with the same period, but there is a phase change at the boundary. Alternatively, the second region might have a spatial periodicity with a different period. Again, the boundary between the two regions constitutes a defect. Eckmann and Procaccia (ECP91) have shown that the stability analysis methods outlined in this section can be extended to show why this defect-mediated complexity appears so commonly.

In much of our discussion, we have assumed that the notion of attractors in a state space of spatial modes is a useful concept in talking about complex spatial patterns. This assumption is analyzed and counterexamples are given in CRK88.

11.6 Diffusion-Limited Aggregation, Dielectric Breakdown and Viscous Fingering: Fractals Revisited

In the previous section, we saw that we could understand the onset of pattern formation (at least for reaction-diffusion systems) by studying the stability and instability of spatial mode amplitudes. However, our analysis, limited to conditions close to steady-state, could only tell us when some spatial mode begins to grow. It could not tell us the final fate of the system. In fact, we know that many systems develop intricate spatial patterns not captured by the analysis of the previous section. In this section, we describe three rather different physical systems, all of which develop similar, intricate spatial patterns.

Our approach will be more descriptive than analytic; to see how these patterns can be understood from mathematical models, we must, in a sense, solve the models exactly. In practice, that means solving or mimicking the system on a

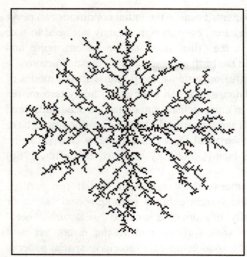

Fig. 11.16. An example of a pattern formed by diffusion limited aggregation in a computer model. Particles diffuse in at random from the border until they hit an occupied site on the cluster, where they then stick. At the upper left is a particle approaching the cluster.

computer. We shall see that these systems are linked both by the kinds of patterns they produce and by the form of the mathematics used to describe them.

Diffusion Limited Aggregation

Diffusion Limited Aggregation (DLA) describes a process in which a substance diffuses rather slowly through a medium until it hits an object to which it can stick. Initially, this might be a piece of dirt on which a cluster of the diffusing substance will start to grow. Once the cluster begins to grow, other molecules of the substance can stick to the cluster and a kind of "crystal" begins to form. In reasonably realistic situations, the diffusing molecules approach the cluster at random but with equal probabilities from different directions. We might expect that under these isotropic condition the cluster that forms would be more or less a sphere. In fact, the clusters that form are anything but spherical. They tend to be lacy, intricate structures, such as the one shown in Fig. 11.16.

How do we describe DLA analytically? We start with the diffusion equation, Eq. (11.4-10) but assume that the diffusion is so slow that we can ignore the time derivative of the concentration and the $\vec{v} \cdot grad$ term. The diffusion equation then reduces to

$$\nabla^2 C(x, y, z) = 0 \tag{11.6-1}$$

As usual, we must also state the boundary conditions for the system. Here we assume that the concentration is constant on the boundaries, both at the periphery of the system and on the surface of the cluster that is being formed. As a reminder, the transport flow equation is Eq. (11.4-2).

It should not be obvious that Eq. (11.6-1) with the stated boundary conditions can lead to the intricate patterns seen in DLA. In fact, Eq. (11.6-1) is the kind of equation solved in undergraduate physics courses to understand electrical potentials

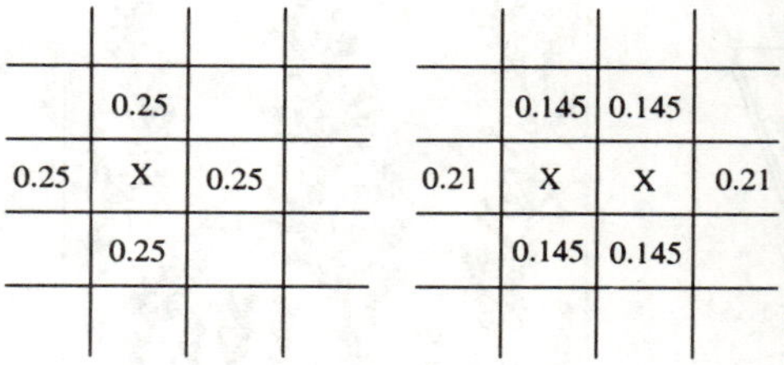

Fig. 11.17. A diffusion limited aggregation lattice. On the left we have one occupied site indicated by X. The probability to land in one of the four sticking sites is 1/4. On the right, we have two neighboring occupied sites that constitute the beginning of a filament. The probability to stick at the end is enhanced over the probability of sticking on the side.

around charged objects. In all those cases, the resulting solutions are rather smooth functions of position, nothing at all like the DLA clusters. After introducing the other two physical systems, we shall return to the mathematical question of relating Eq. (11.6-1) to the observed patterns.

Even without solving any equations, however, we can get some physical intuition about the formation of these patterns. In fact, it is rather simple to program a computer to mimic DLA. You instruct the computer to release a "molecule" at some random location at the edge of the computer terminal screen. The molecule then makes a random walk across the screen until it hits an object (an initial seed, or an occupied site of the cluster) at which point it sticks. Another molecule is then released and the process is repeated. You quickly discover that the resulting cluster shows an intricate, lacy structure similar to that shown in Fig. 11.16.

The basic physical reason for the tendency to grow long filaments (or *dendrites*) is easily understood. On the left Fig. 11.17 shows one occupied site (on the computer screen, for example) and the empty sites next to it. Consider a very simple model in which the incoming molecule can stick only if it lands in a site directly above or below or directly to the right or left of the occupied site. Since the molecule is diffusing isotropically, the probability of hitting one of those sites is 1/4. Now suppose that two adjacent sites are occupied as shown on the right of Fig. 11.17. Again, an incoming molecule will stick if it lands directly above or below or directly to the right or left of an occupied site. It is relatively easy to show that the probabilities are approximately 0.145 for the top and bottom sites, on the sides of our rather short "filament," and 0.21 for the end sites. The important point is there is enhanced probability for the incoming molecule to stick to the ends and hence to make the filament longer.

Exercise 11.6-1. Derive the sticking probabilities illustrated in Fig. 11.6-2. *Hint*: Use symmetry arguments.

A second mechanism also tends to enhance the lacy structure of the cluster. If two filaments begin to grow near each other, they form a "fjord," the interior of which is difficult to reach by a diffusing particle since the particle is more likely to stick to one of the filaments as it diffuses around. Once the molecule sticks to the edge of a filament, it forms the seed of a branch filament that further shields the

Electrodeposition Cell

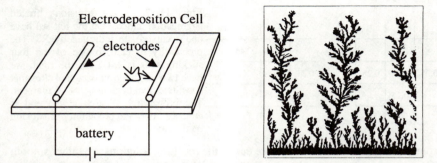

Fig. 11.18. On the left is a sketch of a rectangular electrodeposition cell. An electrical current flows between the two electrodes, each about 0.1 mm in diameter, resting on a glass plate. The electrodes are covered by a second glass plate (not shown) to restrict the current to a two–dimensional region. The space between the plates is filled with a $ZnSO_4$ aqueous solution. On the right is the resulting electrodeposition pattern. (From AAG88.)

interior of the fjord. Thus we see, at least in qualitative terms, why DLA tends to lead to patterns with long thin filaments with many branches.

The DLA model has been used to understand snowflake patterns, electrodeposition structures, and many other physical phenomena. For more information, see the references at the end of the chapter.

Electrodeposition

If electrodes are inserted into a solution containing ionic species and if an electrical potential difference is developed between the electrodes, then an ionic current will flow between the electrodes. If the ionic species is a metal ion, such as Zn^+, then the metal atoms will deposit themselves on the negatuve electrode, forming a metallic cluster. Figure 11.18 shows a sketch of an electrodeposition setup and a typical electrodeposition pattern. The similarity between the electrodeposition pattern and the DLA pattern shown in Fig. 11.16 should be obvious.

The mathematical description of electrodeposition is quite similar to that of DLA. Since the flow of ions is very slow, we can describe the electric fields (or equivalently the electric potentials) that "drive" the ions as electrostatic fields. The electric potential $V(x,y,z)$ is described by Laplace's equation

$$\nabla^2 V(x, y, z) = 0 \qquad\qquad (11.6\text{-}3)$$

in the region in which the ions are moving. (We are assuming that the concentration of the mobile ions is so small that they do not significantly affect the potential "seen" by one of the ions.) For boundary conditions, we take the electric potential V as constant over the positive electrode and over the metallic cluster forming on the negative electrode. The transport relation is just Ohm's Law in the current density form:

$$\vec{j} = \sigma \vec{E} = -\sigma \, grad \, V(x, y, z) \tag{11.6-4}$$

We see that the mathematical description of electrodeposition is analogous to that for DLA.

We can also understand the physics behind the formation of the dendrites in electrodeposition. If the developing metallic cluster by chance begins to develop a sharp tip, then the electric field near that sharp tip will be larger (in magnitude) than the electric field near a smooth part of the boundary. The larger electric field will lead to an increased current density of incoming ions in that region and the tip will tend to grow. We say that the smooth surface is unstable against small, spatially localized perturbations.

Viscous Fingering

If a less-viscous fluid is injected slowly into a more viscous fluid, then the less-viscous fluid will tend to penetrate the other fluid, forming long thin "fingers," often with branches developing out from the fingers. A typical viscous fingering pattern is shown in Fig. 11.19.

The physical processes underlying viscous fingering are important in several applied areas. For example, water is often injected into oil (petroleum) fields to enhance the recovery of oil by using the water pressure to force the oil to flow toward a well site. However, the recovery attempt may be frustrated by the tendency of the less-viscous water to form viscous fingers through the oil. See WON88 for a wider discussion of applications of viscous fingering.

Let us set up the formalism for describing this viscous fingering. We assume that the fluid is so slow that we can write the fluid velocity as simply proportional to the gradient of the pressure p in a form known as **Darcy's Law**:

$$\vec{v} = -\beta \, grad \, p \tag{11.6-6}$$

where β is a parameter that depends on the fluid viscosity. Under the kinds of experimental conditions used, we can neglect the compressibility of the fluids. Hence, we may invoke the incompressibility condition $div \, \vec{v} = 0$ to write

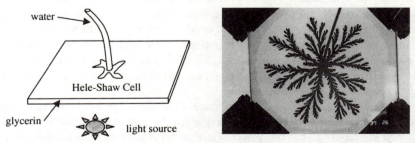

Fig. 11.19. On the left is a sketch of a Hele–Shaw cell used in the study of viscous fingering. A thin layer of viscous fluid lies between two flat plates. A less-viscous fluid is slowly injected through a small hole in the center of the cell. On the right is a typical viscous fingering pattern for water and glycerin. (From WON88.)

$$\nabla^2 p = 0 \qquad\qquad (11.6\text{-}7)$$

Thus, we see that the pressure p plays the role played by concentration in DLA and electrical potential in electrodeposition. The boundary condition on p is that the pressure is constant in the less-viscous fluid. We must also specify the pressure gradient in the viscous fluid near the boundary (BKL86).

Again, we can make a simple physics argument to understand why fingers tend to form: Imagine a flat interface between the two fluids. If a small bump occurs due to a random fluctuation, then the fluid in the bump will tend to move faster than the neighboring fluid because the gradient of the pressure will be larger near the tip of the bump. Once again, we see that a smooth interface is unstable against small, spatially localized perturbations. A small bump, once started, will tend to grow.

The mathematical description of DLA, electrodeposition, and viscous fingering may, at first sight, seem incomplete. The common equation, Laplace's Equation [Eq. (11.6-1), Eq. (11.6-3), and Eq. (11.6-7)], is a linear partial differential equation. How can a linear equation lead to bifurcation events and fractal patterns, occurrences that we now associate with nonlinear equations? The answer is that the nonlinearities, and hence the interesting physics, for the class of phenomena discussed in this section are found in the boundary conditions. With a moving boundary, the net effect at the boundary, as described for each of the systems, is nonlinear. The mathematics of this kind of somewhat unusual situation is discussed in PET89 and BLT91.

Fractal Dimensions

The physical patterns produced by DLA, electrodeposition, and viscous fingering seem to have some degree of self-similarity, the geometric feature of simple fractal objects discussed in Chapter 9. Box-counting algorithms can be applied in a fairly straightforward fashion to the two-dimensional patterns formed in most DLA, electrodeposition, and viscous fingering experiments. The box-counting dimension turns out to be close to 1.7 for many of these patterns. Further evidence for the universality of this number comes from computer simulations of DLA. A wide variety of models all lead to box-counting dimensions of about 1.7. Of course, we know that this self-similarity cannot occur on all length scales for these patterns. At the upper end, we are limited by the overall size of the system. At the lower end, we eventually arrive at individual atoms. To speak precisely, we ought to say that these systems have a fractal box-counting (scaling) dimension over such and such length range.

By now we recognize that a single fractal dimension is probably not sufficient to characterize the geometry of these patterns. The multifractal formalism discussed in Chapter 10 has been used to analyze DLA patterns. See the references at the end of the chapter for more details.

Fig. 11.20. A photograph of ice crystals formed on a smooth branch when the earlier morning temperature suddenly dropped below freezing. The humidity was particularly high just before the temperature drop. The freezing vapor formed long, sharp dendrites. (Photograph courtesy of Daniel Krause, Jr.)

Many other patterns arising in nature can be characterized by fractal and multifractal dimensions. The interested reader should consult the books on fractals listed at the end of Chapter 9 and the references at the end of this chapter.

We should point out that the fractal geometry observed for many of the systems cited here depends crucially on the nonequilibrium nature of the systems. If the systems are allowed to evolve toward equilibrium, then the fractal structures "fill in" to form relatively smooth structures. See SFM88 for a model calculation of this change from fractal geometry to smooth geometry.

Exercise 11.6-2. Figure 11.20 shows a photograph of ice crystals formed on a smooth branch on a November's morning when the air temperature suddenly dropped below freezing. Based on the kinds of reasoning introduced earlier, explain why the ice crystals formed long, sharp dendrites.

11.7 Self-Organized Criticality: The Physics of Fractals?

We mentioned at the end of the previous section that patterns formed by diffusion limited aggregation, electrodeposition, and viscous fingering could all be characterized by a fractal (box-counting) dimension, whose value turns out to be about 1.7. This result, along with the many examples cited in any book on fractal geometry, shows that the notion of fractal dimension is useful in characterizing geometric shapes that fall, in a sense, between, the traditional points, curves, surfaces, and volumes. However, from the physics point of view a crucial question remains: Why do these objects develop a pattern characterized by a fractal dimension? In other words, what is the physics behind fractals?

A mechanism called *self-organized criticality* (BAC91) has been proposed to attempt to explain the physics of fractals. The name implies that the system itself seeks out a critical state at which the spatial structure of the system has self-similarity over an extended range of length scales. Another way of characterizing the critical state is to point out that the ideal critical state has no inherent length scale: When looked at on a particular length scale, the system looks just like it does at any other length scale. The basic premise of self-organized criticality is that many spatially extended nonequilibrium systems evolve naturally toward a critical state. By way of contrast, we should note that *equilibrium* systems evolve to a critical state (such as a phase transition) only if the system parameters are carefully tuned to those values associated with the critical state. For self-organized criticality, the critical state is reached over a range of parameter values.

To date, self-organized criticality has been studied primarily in various model systems such as cellular automata, whose rules are chosen to mimic, at least crudely, various physical systems. A "sand-pile" model has been widely studied in which the cellular automata rules allow for "avalanches" if the difference in the numbers assigned to adjacent cells (corresponding to the height of the sand in a sand pile) becomes too large. It is found that these models evolve naturally to a critical state in which avalanches of all sizes occur. However, the time distribution of avalanche sizes does not seem to match the distribution of avalanches in experiments on sand piles. Therefore, at present self-organized criticality cannot claim to have explained all of the physics of fractals.

Because the systems generally do not start out in the critical state, the obvious question is, How do self-organizing systems "find" the critical state? One possible mechanism involving *singular diffusion* has been suggested (CCG90). The basic idea is that self-organized criticality could occur in a diffusive system if the diffusion coefficient itself depends on the concentration of the substance under consideration. In that case, the transport equation for the concentration C is Eq. (11.4-5), which we write here with the concentration dependence of D shown explicitly:

$$\frac{\partial C}{\partial t} = div[\,D(C)\,grad\,C\,] \tag{11.7-1}$$

If the diffusion coefficient becomes infinite for some value of the concentration, say C_*, then the system will tend to evolve toward that state in which almost all parts of its spatial domain except for a possible boundary layer, the relative size of which decreases as the size of the system increases, have this critical concentration and exhibit the scaling behavior required for self-organized criticality. CCG90 established this behavior for a simple model system and provided some numerical evidence that similar behavior occurs in more complicated systems.

Although the study of self-organized criticality is still in its infancy, it does seem to offer a possible physical explanation for the ubiquity of fractal structures in nature. It remains to be seen if self-organized criticality will be able to explain the wide variety of fractal structures found in nature, thereby giving us some understanding of the detailed mechanisms that lead to fractal structures. Other mechanisms that also lead "naturally" to scaling laws have been proposed (ABH98). Self-organized criticality has been suggested (RIZ99) as a dynamic mechanism that might explain some of the universal features of $1/f$ noise, mentioned in Section 7.3.

11.8 Summary

In this chapter, we have seen how a wide variety of model systems share the common property of "spontaneously" generating spatial patterns under nonequilibrium conditions. Although the study of such pattern formation is really just beginning to develop, we can already see that the emergence of order in nature is not a rare occurrence. In fact, we might say that nature seems to prefer order and structure to uniformity in nonequilibrium situations.

Although we have concentrated on the emergence of relatively simple patterns, we can appreciate an important consequence of the dynamical point of view of pattern formation: Complex patterns might evolve from the (nonlinear) interactions of only a few spatial modes. This realization parallels what we learned about complex temporal dynamics: At least in some cases, complex temporal dynamics, indistinguishable by eye from random behavior, can arise in systems with only a few degrees of freedom. Whether this program can be carried through to explain complex spatial patterns that emerge in strong fluid turbulence, for example, remains to be seen, though optimism is high (MES87, FRO90). Chaotic (temporal) behavior and pattern formation both arise as a result of nonlinear effects in the evolution of some systems.

The study of pattern formation really deserves a book in its own right; we can hardly do justice to this exciting area in one chapter. Several important areas have been neglected completely here. Two of these particularly deserve the attention of physicists. They are (1) chaotic scattering and (2) nonlinear wave phenomena, especially the *soliton* effect. (A soliton is a spatially localized wave disturbance that can propagate over long distances without changing its shape. In brief, many nonlinear spatial modes become synchronized to produce a stable localized disturbance.) To give the reader some sense of the extent of these and other spatiotemporal nonlinear phenomena, we have included an extensive set of suggested readings at the end of the chapter.

We should also mention several ways of characterizing spatio-temporal patterns when they do emerge. A very powerful technique called Karhunen-Loève decomposition separates the time dependence from the spatial pattern and provides a set of mathematical functions with which to analyze the spatial structure. The appropriate mathematical functions can be determined directly from the spatial pattern. AHK93 and TRS97 provide a good introduction to this technique.

Of course, we are a long way from providing a physical theory of the emergence of complex order, say, in the form of a biological cell. But at least in the case of simple kinds of spatial patterns, we now see that these patterns emerge without the (direct) intervention of a designer's hand. And we can understand why they emerge. For those who want to argue that the emergence of life was a spontaneous event, in fact an almost inevitable event given Nature's proclivity for order and structure, our meager understanding at present may offer some cause for optimism. Even if this scientific program could be carried through and if we could show that the emergence of the complex structure we call life is a consequence of the playing out of simple physical laws, there is still much room for debate about the theological implications of those results. In any case, the development of a scientific understanding of pattern formation seems to be one of the most exciting aspects of nonlinear dynamics.

11.9 Further Reading

General Introductions to Pattern Formation and Complexity

B. F. Madure and W. L. Freedman, "Self-Organizing Structures," *American Scientist* **75**, 252–59 (1987).

G. Nicolis and I. Prigogine, *Exploring Complexity* (W. H. Freeman, New York, 1989). A formulation of a rather general theory of pattern formation, at roughly the level of this book, with many examples.

J. D. Murray, *Mathematical Biology* (Springer-Verlag, Berlin, Heidelberg, New York, 1989). An extensive treatment of the mathematics of pattern formation (and many other topics) in biology.

A. V. Gaponov-Grekhov and M. I. Rabinovich, "Disorder, Dynamical Chaos, and Structures," *Physics Today* **43** (7), 30–38 (July, 1990).

H. Haken, *Synergetics, An Introduction*, 3rd ed. (Springer-Verlag, Berlin, 1983).

H. Haken, "Synergetics: An Overview," *Rep. Prog. Phys.* **52**, 515–33 (1989).

H.-T. Moon, "Approach to chaos through instabilities," *Rev. Mod. Phys.* **65**, 1535–43 (1993). Provides a general introduction to the emergence of spatio-temoral order and disorder at the upper-undergraduate, beginning graduate level.

S. C. Müller, P. Coullet, and D. Walgraef, "From oscillations to excitability: A case study in spatially extended systems," *Chaos* **4**, 439–42 (1994). The lead article in an issue devoted to spatiotemporal patterns with many examples from physics and chemistry.

Paul Meakin, *Fractals, Scaling and Growth Far from Equilibrium* (Cambridge University Press, Cambridge, 1997).

C. Bowman and A. C. Newell, "Natural patterns and wavelets," *Rev. Mod. Phys.* **70**, 289–301 (1998). A very nice introduction to pattern formation and the use of the wavelet transform to analyze patterns.

Philip Ball, *The Self-Made Tapestry, Pattern Formation in Nature* (Oxford University Press, New York, 1999).

K. Kaneko and I. Tsuda, "Constructive complexity and artificial reality: an introduction," *Physica D* **75**, 1–10 (1994). Lead article for a volume devoted to complexity.

David Ruelle, *Chance and Chaos* (Princeton University Press, Princeton, NJ, 1991). A good discussion of what "complexity theory" might be all about.

M. Mitchell Waldrop, *Complexity: The Emerging Science at the Edge of Order and Chaos* (Simon and Schuster, New York, 1992). A popular account of complexity theory and its practitioners.

Stuart A. Kaufman, *Origins of Order: Self-Organization and Selection in Evolution* (Oxford University Press, New York, 1993).

Peter Coveney and Roger Highfield, *Frontiers of Complexity: The Search for Order in a Chaotic World* (Fawcett/Columbine, New York, 1995). Another popular book on complexity.

John H. Holland, *Hidden Order: How Adaptation Builds Complexity* (Addison–Wesley, Reading, MA, 1995) A collection of essays by one of the complexity pioneers.

Stuart A. Kaufmann *At Home in the Universe, The Search for the Laws of Self-Organization and Complexity* (Oxford University Press, New York, 1995).

John Horgan, "From Complexity to Perplexity," *Scientific American* **272** (6), 104–109 (1995). A skeptical review of what complexity theory is up to.

N. Goldenfeld and L. P. Kadanoff, "Simple Lessons from Complexity," *Science* **284**, 87–89 (1999). The same issue has articles on complexity in other fields of science as well.

Coherent Structures

T. Shinbrot and J. M. Ottino, "Geometric Method to Create Coherent Structures in Chaotic Flows," *Phys. Rev. Lett.* **71**, 843–46 (1993).

M. Rabinovich, J. Torres, P. Varona, R. Huerta, and P. Weidman, "Origin of coherent structures in a discrete choatic medium," *Phys. Rev. E* **60**, R1130–33 (1999).

Fluid Mixing

The best general reference is [Ottino, 1989].

J. M. Ottino, C. Leong, H. Rising, and P. Swanson, "Morphological Structure Produced by Mixing in Chaotic Flows," *Nature* **333**, 419–25 (1988).

J. M. Ottino, "The Mixing of Fluids," *Scientific American* **260** (1), 56–67 (January, 1989).

J. Franjione, C.-W. Leong, and J. M. Ottino, "Symmetries within Chaos: A Route to Effective Mixing," *Physics of Fluids A* **1**, 1772–83 (1989).

H. A. Kusch, "Continuous Chaotic Mixing of Viscous Liquids," Ph. D. Thesis, Department of Chemical Engineering, University of Massachusetts, 1991.

H. A. Kusch and J. M. Ottino, "Experiments on Mixing in Continuous Chaotic Flows," *J. Fluid Mech.* **236**, 319–48 (1992).

J. M. Ottino, F. J. Muzzio, M. Tjahjadi, J. G. Frangione, S. C. Jano, and H. A. Kusch, "Chaos, Symmetry, and Self-Similarity: Exploiting Order and Disorder in Mixing Processes," *Science* **257**, 754–60 (1992).

G. O. Fountain, D. V. Khakhar, and J. M. Ottino, "Visualization of Three-Dimensional Chaos," *Science* **281**, 683–86 (1998). Fluid flow techniques can provide visual examples of KAM tori, island structures, and so on.

V. Rom-Kedar and G. M. Zaslavsky, "Chaotic kinetics and transport (Overview)," *Chaos* **10** (1), 1–2 (2000). The introduction to a focus issue on chaotic kinetics, mixing, and transport.

H. Aref, "Integrable, Chaotic, and Turbulent Vortex Motion in Two-dimensional Flow," *Ann. Rev. Fluid Mech.* **15**, 345–89 (1983). Notes the connection between two-dimensional fluid flow and Hamiltonian systems.

H. Aref, "Stirring by Chaotic Advection," *J. Fluid Mech.* **143**, 1–21 (1984).

A. R. Paterson, *A First Course in Fluid Dynamics* (Cambridge University Press, Cambridge, 1983). A good introduction to fluid dynamics at about the level of this book.

R. Feynman, R. Leighton, and M. Sands, Feynman, *Lectures on Physics*, Vol. II (Addison–Wesley, Reading, MA, 1963). This volume contains lots of good introductory material on fluid dynamics.

Full-scale fluid turbulence, a subject we have avoided in this chapter, remains an exciting challenge for nonlinear dynamics. Some progress in using the notions of nonlinear dynamics to understand turbulence are discussed in

C. Meneveau and K. R. Sreenivasan, "Simple Multifractal Cascade Model for Fully Developed Turbulence," *Phys. Rev. Lett.* **59**, 1424–27 (1987).

U. Frisch and S. A. Orszag, "Turbulence: Challenges for Theory and Experiment," *Physics Today* **43** (1), 24–32 (January, 1990).

Granular Flow

G. B. Lubkin, "Oscillating granular layers produce stripes, squares, hexagons, …," *Physics Today* **48** (10), 17–19 (October, 1995).

G. Metcalf, T. Shinbrot, J. J. McCarthy and J. M. Ottino, "Avalanche mixing of granular solids," *Nature* **374**, 39–41 (1995).

C. S. Daw, C. Finney, M. Vasudevan, N van Goor, K. Nguyen, D. Bruns, E. Kostelich, C. Grebogi, E. Ott, and J. A. Yorke, "Self-Organization and Chaos in a Fluidized Bed," *Phys. Rev. Lett.* **75**, 2308–11 (1995).

F. Melo, P. Umbanhowar, and H. L. Swinney, "Hexagons, Kinks, and Disorder in Oscillated Granular Layers," *Phys. Rev. Lett.* **75**, 3838–41 (1995).

H. Jaeger, S. Nagel, and R. Behringer, "The Physics of Granular Materials," *Physics Today* **49** (4), 32–38 (1996). A very nice introduction to granular materials and their unusual properties.

T. Shinbrot and F. J. Muzzio, "Reverse Buoyancy in Shaken Granular Beds," *Phys. Rev. Lett.* **81**, 4365–68 (1998).

J. S. Olafsena and J. S. Urbach, "Clustering, Order, and Collapse in a Driven Granular Monolayer," *Phys. Rev. Lett.* **81**, 4369–72 (1998).

D. V. Khakhar, J. J.McCarthy, J.F. Gilchrist, and J. M. Ottino, "Chaotic mixing of granular materials in two-dimensional tumbling mixers," *Chaos* **9**, 195–205 (1999).

Coupled Iterated Maps, Cellular Automata, and Neural Networks

Coupled Iterated Maps and Oscillators

F. Kaspar and H. G. Schuster, "Scaling at the Onset of Spatial Disorder in Coupled Piecewise Linear Maps," *Phys. Lett. A* **113**, 451–53 (1986).

K. Kaneko, "Chaotic but Regular Posi-Nega Switch among Coded Attractors by Cluster-Size Variation," *Phys. Rev. Lett.* **63**, 219–23 (1989).

K. Kaneko, "Pattern Dynamics in Spatiotemporal Chaos," *Physica D* **34**, 1–41 (1989)

K. Kaneko, "Globally Coupled Chaos Violates the Law of Large Numbers but not the Central-Limit Theorem," *Phys. Rev. Lett.* **65**, 1391–94 (1990).

M. Hassell, H. Comins, and R. May, "Spatial Structure and Chaos in Insect Population Dynamics," *Nature* **353**, 255–58 (1991).

P. Matthews, R. Mirollo, and S. Strogatz, "Dynamics of a large system of coupled nonlinear oscillators," *Physica D* **52**, 293–331 (1991).

K. Kaneko, "Overview of coupled map lattices," *Chaos* **2**, 279–82 (1992). An introduction to a focus issue on coupled iterated maps.

S. H. Strogatz and I. Stewart, "Coupled Oscillators and Biological Synchronization," *Scientific American* **269** (6), 102–109 (1993).

S. Sinha and W. L. Ditto, "Dynamics Based Computation," *Phys. Rev. Lett.* **81**, 2156–59 (1998). Coupled chaotic map functions can be used to perform simple computations.

T. Shibata, T. Chawanya, and K. Kaneko, "Noiseless Collective Motion Out of Noisy Chaos," *Phys. Rev. Lett.* **82**, 4424–27 (1999). Added noise can reduce the disorder in a coupled map model.

Y.-C. Lai and C. Grebogi, "Modeling of Coupled Oscillators," *Phys. Rev. Lett.* **82**, 4803–6 (1999). The shadowing theorem may fail for coupled chaotic oscillators.

D. A. Egolf, "Equilibrium Regained: From Nonequilibrium Chaos to Statistical Mechanics," *Science* **287**, 101–104 (2000).

Cellular Automata

M. Gardner, "The Fantastic Combinations of John Conway's New Solitaire Game of Life," *Scientific American* **223** (4), 120–23 (April, 1970).

T. Toffoli and N. Margolus, *Cellular Automata Machines* (MIT Press, Cambridge, 1987).

S. Wolfram, "Cellular Automata as Models of Complexity," *Nature* **341**, 419–24 (1984).

S. Wolfram, ed. *Theory and Applications of Cellular Automata* (World Scientific Press, Singapore, 1986).

M. Gardner, *Wheels, Life and Other Mathematical Amusements* (W. H. Freeman, New York, 1983).

S. Wolfram, "Universality and Complexity in Cellular Automata," *Physica D* **10**, 1–35 (1984).

P. Halpern, "Sticks and Stones: A Guide to Structurally Dynamic Cellular Automata," *Am. J. Phys.* **57**, 405–8 (1989).

[Jackson, 1990], Vol. II, Chapter 10. This section contains a nice discussion of cellular automata and coupled maps.

Applications of cellular automaton gas lattice models to the study of fluid dynamics:

J. Salem and S. Wolfram, "Thermodynamics and Hydrodynamics of Cellular Automata," in [S. Wolfram, ed., 1986].

N. Margolis, T. Toffoli, and G. Vichniac, "Cellular-Automata Supercomputers for Fluid Dynamics Modeling," *Phys. Rev. Lett.* **56**, 1694–96 (1986).

S. Wolfram, "Cellular Automaton Fluids 1: Basic Theory," *J. Stat. Phys.* **45**, 471-526 (1986).

P. Mammeiulle, N. Boccasa, C. Y. Vichniac, and R. Bidaux, eds. *Cellular Automata and Modeling of Complex Physical Systems* (Springer-Verlag, Berlin, 1989).

R. Zorzenon dos Santos and A. Bernardes, "Immunization and Aging: A Learning Process in the Immune Network," *Phys. Rev. Lett.* **81**, 3034–37 (1998). A cellular automata model is applied to the immune system.

Networks

C. L. Scofield and L. N. Cooper, "Development and Properties of Neural Networks," *Contemporary Physics* **26**, 125–45 (1985).

D. Tank and J. Hopfield, "Collective Computation in Neuronlike Circuits," *Scientific American* **257** (6), 104–14 (1987).

H. Sompolinksy, A. Crisanti, and H. J. Sommers, "Chaos in Random Neural Networks," *Phys. Rev. Lett.* **61**, 259–62 (1988).

Stuart A. Kaufman, "Antichaos and Adaptation," *Scientific American* **265** (2), 78–84 (August, 1991). The emergence of order from disorder in Boolean networks and its possible implications for genetics and biological adaptation.

D. Hansel and H. Sompolinsky, "Synchronization and Computation in a Chaotic Neural Network," *Phys. Rev. Lett.* **68**, 718–21 (1992).

D. Caroppo, M. Mannarelli, G. Nardulli, and S. Stramaglia, "Chaos in neural networks with a nonmonotonic transfer function," *Phys. Rev. E* **60**, 2186–92 (1999).

A.-L. Barabási and R. Albert, "Emergence of Scaling in Random Networks," *Science* **286**, 509–12 (1999). Even connections on the World Wide Web exhibit scaling behavior.

Transport Phenomena

The averaging property of the Laplacian is discussed within the context of electrostatics in two delightful books:

David J. Griffiths, *Introduction to Electrodynamics* (Prentice-Hall, Englewood Cliffs, NJ, 1981), pp.100–1.

Edward M. Purcell, *Electricity and Magnetism, 2nd ed.*, (McGraw–Hill, New York, 1985), pp. 63–65.

Reaction-Diffusion Systems

A. M. Turing, "The Chemical Basis of Morphogenesis," *Phil. Trans. Roy. Soc. London B* **237**, 37–72 (1952).

A. M. Albano, N. B. Abraham, D. E. Chyba, and M. Martelli, "Bifurcation, Propagating Solutions, and Phase Transitions in a Nonlinear Reaction with Diffusion," *Am. J. Phys.* **52**, 161–67 (1984).

A. Winfree, "Rotating Solutions to Reaction-Diffusion Equations in Simply-Connected Media," *SIAM-AMS Proceedings* **8**, 13–31 (1974). Rotating spirals of chemical reaction can be modeled with a very simple reaction-diffusion system.

Y. Kuramoto, "Rhythms and Turbulence in Populations of Chemical Oscillators," *Physica A* **106**, 128–43 (1981). If the Brusselator model of chemical reactions (Chapter 3) is used in a reaction-diffusion, then complex ("chaotic") spatial patterns emerge.

W. Y. Tam, J. A. Vastano, H. L. Swinney, and W. Horsthemke, "Regular and Chaotic Chemical Spatiotemporal Patterns," *Phys. Rev. Lett.* **61**, 2163–66 (1988). Experimental observations of complex spatiotemporal patterns in an open chemical reactor system.

V. Castets, E. Dulos, J. Boissonade, and P. DeKepper, "Experimental Evidence of a Sustained Standing Turing-Type Nonequilibrium Chemical Pattern," *Phys. Rev. Lett.* **64**, 2953–56 (1990). Experimental evidence for the existence of Turing structures and chemical turbulence in reaction-diffusion systems.

Q. Ouyang and H. L. Swinney, "Transition from a Uniform State to Hexagonal and Striped Turing Patterns," *Nature* **352**, 610–12 (1991).

Q. Ouyang and H. L. Swinney, "Transition to Chemical Turbulence," *Chaos* **1**, 411–19 (1991).

L. Tsimring, H. Levine, I. Aranson, E. Ben-Jacob, I. Cohen, O. Shochet, and W. Reynolds, "Aggregation Patterns in Stressed Bacteria," *Phys. Rev. Lett.* **75**, 1859–62 (1995).

I. B. Schwartz and I. Triandaf, "Chaos and intermittent bursting in a reaction-diffusion process," *Chaos* **6**, 229–37 (1996).

General Questions

J.-P. Eckmann and I. Procaccia, "Onset of Defect-Mediated Turbulence," *Phys. Rev. Lett.* **66**, 891–94 (1991). The use of the methods analyzing spatial modes as state space variables with nonlinear dynamics has been extended to explain "defect-mediated turbulence."

J. P. Crutchfield and K. Kaneko, "Are Attractors Relevant to Turbulence," *Phys. Rev. Lett.* **60**, 2715–18 (1988). The assumption that an underlying attractor describes all kinds of spatially complex behavior is questioned.

Fractal Growth Phenomena

T. A. Witten, Jr. and L. M. Sander, "Diffusion-Limited Aggregation, a Kinetic Critical Phenomenon," *Phys. Rev. Lett.* **47**, 1400–03 (1981).

L. Niemeyer, L. Pietronero, and H. J. Wiesmann, "Fractal Dimension of Dielectric Breakdown," *Phys. Rev. Lett.* **52**, 1033–36 (1984). The fractal dimension of dielectric breakdown is shown to be the same as that of DLA.

D. Bensimon, L. P. Kadanoff, S. Liang, B. Shraiman, and C. Tang, "Viscous Flows in Two Dimensions," *Rev. Mod. Phys.* **58**, 977–99 (1986). An excellent introduction to the physics and mathematics of viscous fingering.

L. M. Sander, "Fractal Growth," *Scientific American* **256** (1), 94–100 (January, 1987). General properties of systems that develop fractal geometric patterns.

Po-Zen Wong, "The Statistical Physics of Sedimentary Rock," *Physics Today* **41** (12), 24–32 (December, 1988). The physics and applications of diffusion limited aggregation, electrodeposition and viscous fingering.

The $f(\alpha)$ multifractal formalism is applied to DLA and other phenomena in the following two papers:

J. Lee and H. E. Stanley, "Phase Transition in the Multifractal Spectrum of Diffusion-Limited Aggregation," *Phys. Rev. Lett.* **61**, 2945–48 (1988).

H. E. Stanley and P. Meakin, "Multifractal Phenomena in Physics and Chemistry," *Nature* **335**, 405–9 (1988).

E. Sorensen, H. Fogedby, and O. Mouritsen, "Crossover from Nonequilibrium Fractal Growth to Equilibrium Compact Growth," *Phys. Rev. Lett.* **61**, 2770–73 (1988). The transition from fractal geometry to smooth geometry as a system tends to equilibrium.

T. Vicsek, *Fractal Growth Phenomena*, 2nd ed. (World Scientific Publishing, River Edge, NJ, 1991).

F. Family, B. Masters, and D. E. Platt, "Fractal Pattern Formation in Human Retinal Vessels," *Physica D* **38**, 98–103 (1989). Diffusion limited aggregation is used to understand patterns of blood vessels in the human retina.

F. Caserta, H. E. Stanley, W. D. Eldred, G. Daccord, R. E. Hausman, and J. Nittmann, "Physical Mechanisms Underlying Neurite Outgrowth: A Quantitative Analysis of Neuronal Shape," *Phys. Rev. Lett.* **64**, 95–98 (1990).

J. Walker, "Fluid Interfaces, Including Fractal Flows can be Studied in a Hele–Shaw Cell," *Scientific American* **257** (5), 134–38 (November, 1987). How to make your own viscous fingering cell.

J.-F. Gouyet, *Physics of Fractal Structures* (Springer-Verlag, New York, 1995). Covers a wide variety of fractal growth phenomena in many scientific fields at the upper-undergraduate level.

Mathematical References

Treatments of spatial modes and their associated functions are found in almost every book on mathematical physics. A treatment roughly at the level of this book is Mary L. Boas, *Mathematical Methods in the Physical Sciences*, 2nd ed. (John Wiley and Sons, New York, 1983).

M. Peterson, "Nonuniqueness in Singular Viscous Fingering," *Phys. Rev. Lett.* **62**, 284–87 (1989). Some of the mathematics of the relation between $\nabla^2 V = 0$ and viscous fingering, DLA, and electrodeposition.

E. Brener, H. Levine, and Y. Tu, "Mean-Field Theory for Diffusion-Limited Aggregation in Low Dimensions," *Phys. Rev. Lett.* **66**, 1978–81 (1991).

The use of the mathematical theory of symmetry groups to classify spatial structures that emerge in bifurcations of spatial patterns is discussed in the following:

M. Golubitsky, I. Stewart, and D. Schaeffer, *Singularities and Groups in Bifurcation Theory,* Applied Mathematical Sciences Vol. 69 (Springer-Verlag, New York, 1988), Vol. II.

C. Green, G. Mindlin, E. D'Angelo, H. Solari, and J. Tredicce, "Spontaneous Symmetry Breaking in a Laser: The Experimental Side," *Phys. Rev. Lett.* **65**, 3124–27 (1990).

Self-Organized Criticality

P. Bak, C. Tang, and K. Wiesenfeld, "Self-Organized Criticality: An Explantion of 1/*f* Noise," *Phys. Rev. Lett.* **59**, 381–84 (1987).

C. Tang, K. Wiesenfeld, P. Bak, S. Coppersmith, and P. Littlewood, "Phase Organization," *Phys. Rev. Lett.* **58**, 1161–64 (1987). Self-organized criticality is studied in a driven chain of coupled nonlinear oscillators.

P. Bak, C. Tang, and K. Wiesenfeld, "Self-Organized Criticality," *Phys. Rev. A* **38**, 364–74 (1988).

P. Bak and K. Chen, "The Physics of Fractals," *Physica D* **38**, 5–12 (1989).

J. M. Carlson and J. S. Langer, "Properties of Earthquakes Generated by Fault Dynamics," *Phys. Rev. Lett.* **62**, 2632–35 (1989).

J. Carlson, J. Chayes, E. Grannan, and G. Swindle, "Self-Organized Criticality and Singular Diffusion," *Phys. Rev. Lett.* **65**, 2547–50 (1990).

P. Bak and K. Chen, "Self-Organized Criticality," *Scientific American* **264** (1), 46–53 (January, 1991).

S. Boettcher and M. Paczuski, "Exact Results for Spatiotemporal Correlations in a Self-Organized Model of Punctuated Equilibrium," *Phys. Rev. Lett.* **76**, 348–51 (1996).

P. Bak, *How Nature Works: The Science of Self-Organized Criticality* (Springer-Verlag, New York, 1996).

Henrik Jeldtoft Hensen, *Self-Organized Criticality: Emergent Complex Behavior in Physical and Biological Systems* (Cambridge University Press, New York, 1998).

L Amaral, S. Buldyrev, S. Havlin, M. Salinger, and H. E. Stanley, "Power Law Scaling for a System of Interacting Units with Complex Internal Structure," *Phys. Rev. Lett.* **80** 1385–88 (1998). Suggests an alternative mechanism, different from Self-Organized Criticality that leads to scaling law behavior.

R. De Los Rios and Y.-C. Zhang, "Universal 1/f Noise from Dissipative Self-Organized Criticality Models," *Phys. Rev. Lett.* **82**, 472–75 (1999). Suggest that SOC models can explain the universality of 1/f noise.

Other Spatiotemporal Effects

Spatial and temporal behavior also become linked in the study of scattering problems, a subject dear to many physicists. The basic idea is that when a moving object collides with another object and is deflected by it, the scattering process may show sensitive dependence to the details of the incoming trajectory, another kind of sensitive dependence on initial conditions, resulting in "chaotic scattering." For some taste of this field see the following:

B. Eckhardt, "Irregular Scattering," *Physica D* **33**, 89–98 (1988).

E. Doron, U. Smilansky, and A. Frenkel, "Experimental Demonstration of Chaotic Scattering of Microwaves," *Phys. Rev. Lett.* **65**, 3072–75 (1990).

Y.T. Lau, J. M. Finn, and E. Ott, "Fractal Dimension in Nonhyperbolic Chaotic Scattering," *Phys. Rev. Lett.* **66**, 978–81 (1991).

E. Ott and T. Tél, "Chaotic scattering: an introduction," *Chaos* **3**, 417–426 (1993). The lead article in an issue devoted to chaotic scattering.

Another important class of nonlinear spatiotemporal effects involve (nonlinear) waves and their propagation, including the important class of soliton waves that propagate without changing their shape. An excellent introduction to nonlinear waves is:

E. Infeld and G. Rowlands, *Nonlinear Waves, Solitons and Chaos* (Cambridge University Press, Cambridge, 1990).

N. B. Trufillaro, "Nonlinear and Chaotic String Vibrations," *Am J. Phys.* **57**, 408–14 (1989). The standard introductory physics topic of elastic waves on strings has nonlinear characteristics.

Karhunen-Loève decomposition

D. Armbruster, R. Heiland, and F. Kostelich, "KLTOOL: A tool to analyze spatiotemporal complexity," *Chaos* **4**, 421–24 (1994). Describes a computer program that implements Karhunen-Loève decomposition.

I. Triandaf and I. B. Schwartz, "Karhunen-Loeve mode control of chaos in a reaction-diffusion process," *Phys. Rev. E* **56**, 204–212 (1997). Has a nice introduction to Karhunen-Loève decomposition.

11.10 Computer Exercises

CE11-1. *Chaos Demonstrations* has several examples that are useful for material in this chapter. Run the examples "Diffusion," "Deterministic Fractals," "Random Fractals," and "The Game of Life."

CE11-2. Write a program to implement a directed-diffusion version of diffusion limited aggregation as suggested in C. A. Pickover, "Markov Aggregation on a Sticky Circle," *Computers in Physics* **3**, 79–80 (July/August, 1989).

CE11-3. Some (advanced) suggestions for programming models of diffusion limited aggregation and viscous fingering are discussed in F. Family and T. Vicsek, "Simulating Fractal Aggregation," *Computers in Physics* **4**, 44–49 (January/February, 1990). Try out some of the examples suggested in this article.

CE11-4. Try the package of programs by M. Vicsek and T. Vicsek, *Fractal Growth* (instructors manual and diskette) (World Scientific Publishing, River Edge, New Jersey, 1991).

CE11-5. Try some of the programming exercises in P. Bak, "Catastrophes and Self-Organized Criticality," *Computers in Physics* **5**, 430–33 (July/August, 1991).

Information on cellular automata and related topics can be found at the following web site: http://alife.santafe.edu/alife/topics/cas/ca-faq/ca-faq.html.

12

Quantum Chaos,

The Theory of Complexity, and Other Topics

Chaos often breeds life, when order breeds habit. Henry Brooks Adams,
The Education of Henry Adams.

12.1 Introduction

In this chapter we discuss several broad issues that tie nonlinear dynamics and
chaos to fundamental questions in a variety of areas of science. The first issue is
the question of the relationship, if any, between nonlinear dynamics and physics'
most fundamental theory: quantum mechanics. As we shall see, this relationship is
somewhat controversial. A second issue raises the question of characterizing the
complexity of behavior of dynamical systems (i.e., trying to find a continuous
strand that links complete determinism and simple behavior at one end to complete
randomness at the other). Neither of these issues is likely to have a major practical
impact on the utility of nonlinear dynamics, but the intellectual questions are deep
and subtle.

   We have also included brief mention of several topics in this chapter that did
not fit in nicely elsewhere in the text: piece-wise linear models, delay-differential
equation models, stochastic resonance, controlling and synchronizing chaos, and
the possibility of chaotic behavior in computer networks. We close the chapter
with some general remarks about a vision of where the study of chaos and
nonlinear dynamics is going.

12.2 Quantum Mechanics and Chaos

When the definitive history of twentieth-century science is written, that century will
be known as the age of the quantum. During the twentieth century, physicists
invented a theory of the microscopic world: the quantum theory. This theory,
formally known as *quantum mechanics*, has been wildly successful, far beyond the
hopes and expectations of its founders, in describing and predicting phenomena in
the microscopic world of elementary particles, nuclei, atoms, and molecules. In
essence modern chemistry and molecular biology are based on quantum mechanics.
Quantum mechanics has played a crucial role in the development of condensed-
matter physics and its critical applications in solid-state electronics and computers.
This is not to say that we need all of the formal machinery of quantum mechanics

490

to do science today, but the understanding gained from that theory about the structure and stability of matter, the nature of fundamental forces, the interaction of light and matter, and so on, forms the background against which the theater of twentieth century science was played out. There is no sign that the twenty-first century will be any different.

It is widely believed, however, that quantum mechanics presents us with many conceptual and philosophical problems along with its successes. The theory tells us that the microscopic world cannot adequately be described with the words and concepts we have developed to talk about the large-scale world around us. For example, quantum mechanics tells us that it is not possible (i.e., it is not physically meaningful) to assign a well-defined position and momentum to an electron (or any other particle); therefore, it is not possible to talk about a well-defined trajectory in state space for that particle. In fact, quantum mechanics forces us to abandon, or at least to modify substantially, all of our ideas about initial conditions, forces, and trajectories.

From the point of view of quantum mechanics, all of the physics used in this book (and all other treatments not based on quantum mechanics) is wrong, at least in principle. Before we throw away all of our work in despair, however, we should point out that quantum mechanics also tells us that in the limit of everyday-size objects, the theories we have built up with the use of Newtonian classical (that is, pre-quantum) mechanics are exceedingly good in the sense that we make only numerically insignificant errors in using those theories in place of quantum mechanics. We shall explore this notion more quantitatively later.

The crucial aspect of quantum theory that forces us to abandon traditional modes of description is quantum mechanics' use of probabilities. Unlike classical Newtonian mechanics in which, in principle, we can calculate the future behavior ("trajectory") of a system if we know its force laws and its initial conditions, quantum mechanics tells us that the best we can do in general is to predict the probability of various future behaviors for the system. In fact, according to quantum mechanics, the crucial problem is that we cannot in principle (as well as in practice) prepare a system (e.g., an electron) with precisely defined initial position and velocity (or momentum). There is always some minimum "fuzziness" associated with these initial conditions. In the language of state space, all we can say is that the system's trajectories start within some fuzzy region of state space. The best we can do is predict the probabilities of the evolution of this fuzzy blob. The size of this minimum area in state space is set by *Planck's constant*, $h = 6.626... \times 10^{-34}$ Joule–sec. Note that Planck's constant has units of classical *action* (recall Chapter 8). In fact, many quantum states can be characterized by the amount of action associated with that state. In common cases, the amount must be an integer multiple of the fundamental amount h.

Given this inherent fuzziness associated with quantum state space, the most we can do is predict the probability that the electron's position will lie within some range of values and that its momentum will lie within some other corresponding range. Most quantum mechanicians argue that we should not talk about the electron's trajectory at all. Here we are close to treacherous philosophical waters in which lurk important but distracting questions: Can we talk sensibly about the position and momentum of a single electron? Does quantum mechanics describe only ensembles of particles? Fortunately, when the system is large, like a baseball or an electrical current in a standard electrical circuit or the fluid in a test tube, the fuzziness associated with quantum mechanical uncertainty is completely negligible, and we may, we believe, use classical mechanics to describe the behavior of the system.

We know, however, that there are many systems in which quantum mechanics plays a fundamental role. Given our knowledge of the new dynamical possibilities that chaos brings to classical mechanics, we ask: Do systems in which quantum mechanics is essential display (under appropriate conditions) behavior that is analogous to the chaotic behavior of classical systems? This question is surprisingly difficult to answer. The difficulties arise from the conceptual gulf separating quantum mechanics from classical mechanics. In some sense, the two theories speak different languages; therefore, we must, when talking about whether or not a quantum system displays chaotic behavior, be careful not to try to compare apples and hydrogen atoms. On the other hand, we can apply quantum mechanics and classical mechanics to the same model of a physical system. If classical mechanics predicts chaotic behavior for the system under certain conditions (that is, for certain ranges of control parameter values), we can ask what quantum mechanics predicts under those same circumstances. (Of course, we must also ask what the actual physical system does under those circumstances.)

The comparison between the predictions of classical mechanics and quantum mechanics is not easy, particularly for systems that classical mechanics says are chaotic. Chaotic systems must be nonlinear and cannot have analytic solutions in classical mechanics. If the system is a Hamiltonian system, it is nonintegrable. In quantum mechanics, then, we have an extremely difficult mathematical problem to solve and many of the techniques that apply to integrable systems fail for nonintegrable ones. Finding any prediction at all using quantum mechanics is difficult, and we must be sure that our predictions are not an artifact of the approximation scheme used to carry out the computation. More fundamentally, however, quantum mechanics does not allow us to calculate just those features of the system's behavior on which our classical notions of chaos are based; namely, the system's trajectories in state space. Thus, we need to think carefully about what it is in quantum mechanics that we are going to look at to detect chaotic behavior.

Unfortunately, as of this writing (2000), there is not yet agreement about what those features should be. Some physicists argue that there cannot be such a thing as "quantum chaos" because, as we shall discuss later, quantum mechanics, as it is usually practiced, is a linear theory. Since chaos requires nonlinearity, it might appear that quantum mechanics cannot lead to chaotic behavior. If this result is true, then we have a fundamental problem since we know that real systems do display chaotic behavior.

The general belief among quantum mechanicians is that the predictions of quantum mechanics ought to agree with the predictions of classical mechanics in an appropriate limit. This *Correspondence Principle* requires this agreement either for large values of the action or in the limit $h \to 0$. If quantum mechanics, therefore, does not describe chaotic behavior at all, it is hard to see how it would in the classical limit. Either quantum mechanics imitates chaotic behavior in some way yet to be fully understood, or, if quantum mechanics does not include such behavior, then quantum mechanics must be wrong! If such a heretical conclusion is correct, then perhaps finding out what is "wrong" with quantum mechanics will lead us on to a new and perhaps more fundamental theory that will encompass the successes of quantum mechanics, but would also include the possibility of chaos. Alternatively, quantum mechanics might be correct and what we have been calling chaos is actually an elaborate charade perpetrated by Nature. Before we come to such radical conclusions, we need to explore more fully what quantum mechanics does tell us about systems whose behavior is chaotic.

There is another reason to be concerned about chaotic behavior in quantum systems. An important extension of simple quantum mechanics is the description of systems with large numbers of particles, so-called quantum statistical mechanics. In quantum statistical mechanics, we have an added layer of statistics or randomness, above that contributed by the inherent probability distributions of quantum mechanics. A fundamental question arises: Can quantum mechanics itself account for this added randomness or must this randomness be imposed as an ad hoc feature? The quantum analog of chaos could provide a fundamental explanation of quantum statistical mechanics (KAM85). If this explanation cannot come from within quantum mechanics itself, then we must once again conclude that quantum mechanics is incomplete in its present form.

We first give a very brief synopsis of quantum mechanics to highlight those features that are important for our discussion. We then look at various approaches that have been taken to find what happens in quantum mechanics when the corresponding classical mechanics description predicts chaotic behavior. Finally we give a brief overview of some experiments that may show quantum chaos if it exists. Readers who are not familiar with quantum mechanics may wish to skip the remainder of this section.

The literature on "quantum chaos" is vast, and it would require a book in itself to do it justice. We make no pretense of covering this rapidly developing field. The interested reader is encouraged to learn more from the excellent books by Gutzwiller [Gutzwiller, 1990] and Reichl [Reichl, 1992].

A Synopsis of Quantum Mechanics

In this section we give a brief introduction to the theory of quantum mechanics to point out those features that are important in our discussion of quantum chaos. Of necessity, we must simply state results without much justification, either mathematical or physical. Fortunately, to understand the issue of quantum chaos, it is not necessary to develop the full formalism of quantum mechanics. We shall restrict the discussion to ordinary quantum mechanics as embodied in the Schrödinger equation. How the question of chaos extends to quantum field theory, which allows for the possibility of the creation and annihilation of particles and the inclusion of relativity, remains to be explored.

The Schrödinger form of quantum mechanics makes use of an important mathematical intermediary to make predictions about the behavior of a system. This mathematical intermediary is known as the **wave function** for the system because it shares many (but not all) properties of the functions used to describe waves, such as electromagnetic waves or water waves, in classical mechanics. The wave function, usually denoted as $\Psi(x, y, z, t)$ does not have a direct physical meaning, but is used to calculate properties of the system. For example, the square of the wave function (actually the absolute value squared, since in general the wave function is a complex function) gives the probability density for finding the system at a particular location. That is, for a system consisting of a single particle $|\Psi(x, y, z, t)|^2$ is the probability (per unit volume) for finding the particle at time t at the location indicated by the coordinates x,y,z.

The wave function is also used to calculate the average value ("expectation value") of other properties of the system. If we have some property of the system written as a function of x, y, and z, say, $A(x,y,z)$, then the average value of the property represented by A is found by evaluating the integral

$$< A >= \int dxdydz\, \Psi^*(x, y, z, t) A(x, y, z) \Psi(x, y, z, t) \qquad (12.2\text{-}1)$$

where the integral is over all values of the coordinates, and Ψ^* is the complex conjugate of the wave function. [In general $A(x,y,z)$ may not be an ordinary function but may instead be represented by so-called operators. This important mathematical feature is treated in all standard books on quantum mechanics, but it does not need to concern us here.] According to the usual interpretation of quantum mechanics, these probabilities contain all the information we can have about the behavior of the system.

A critical feature of quantum mechanics is the fact that the fundamental calculational quantity $\Psi(x, y, z, t)$ is a probability **amplitude**. We call $\Psi(x, y, z, t)$ an amplitude in analogy with a classical electromagnetic wave in which the electric field (the wave amplitude) $\vec{E}(x, y, z, t)$ is squared to give the intensity (the quantity measured by a power meter, for example). The use of a wave (probability) amplitude allows quantum mechanics to account for the wave properties (such as interference and diffraction) observed for electrons, protons, helium atoms, and so on.

How do we find the wave function for the system? We do this by solving the **Schrödinger equation** for the system. The Schrödinger equation makes use of the Hamiltonian for the system. (Recall from Chapter 8 that the Hamiltonian, in simple cases, is just the sum of kinetic energy and potential energy for the system.) In the so-called coordinate representation, in which everything is expressed as a function of the coordinates, the Hamiltonian for a single particle is written as $H(x,y,z,t)$. The Schrödinger equation then has the form

$$H(x, y, z, t)\Psi(x, y, z) = \frac{ih}{2\pi}\frac{\partial\Psi(x, y, z, t)}{\partial t}$$ ('12.2-2)

where $i = \sqrt{(-1)}$ and h is Planck's constant, which, as mentioned earlier, has the units of energy×time or equivalently, momentum×position. (In general, H may be an operator, and the resulting equation is a partial differential equation for the wave function.) The solutions of the Schrödinger equation give us the possible Ψs for the system. We specify the nature of the physical problem we are describing by choosing the appropriate Hamiltonian and choosing appropriate boundary conditions for Ψ. A good deal of the creativity in quantum mechanics goes into finding the "correct" Hamiltonian (e.g., for high-temperature superconductivity or for quarks inside a proton).

Once a Hamiltonian has been adopted, we can proceed to find the solutions of Schrödinger's equation. In practice, we usually look for solutions that can be written in a factored form:

$$\Psi(x, y, z, t) = g(t)\Phi(x, y, z)$$ (12.2-3)

If the Hamiltonian does not depend explicitly on time, then we find that $g(t)$ and $\Phi(x, y, z)$ are given by

$$g(t) = Ge^{-i2\pi Et/h}$$
$$H(x, y, z)\Phi(x, y, z) = E\Phi(x, y, z)$$ (12.2-4)

In Eq. (12.2-4), E can be identified as the energy of the system and the resulting $\Phi(x, y, z)$ is called the energy eigenfunction for that particular value of the energy. The crucial point for our discussion is that all of the time dependence of the wave function is contained in the simple exponential form for $g(t)$ and depends only on the energy *eigenvalue E*.

The final step in finding the wave function is to make use of the mathematical fact that the most general solution of the Schrödinger equation can be written as a linear combination of the product solution forms given in Eq. (12.2-3):

$$\Psi(x, y, z, t) = \sum_i c_i e^{-i2\pi E_i t/h}\Phi_i(x, y, z)$$ (12.2-5)

In the previous equation we have adorned the energy eigenvalue E and the energy eigenfunction $\Phi(x, y, z)$ with subscripts to label the possible eigenvalues and the eigenfunctions associated with them. The sum is over all the possible E values for the system. In the most general case we might need to include an integral over some continuous range of E values as well. The coefficients c_i are parameters (independent of position and time) that are set by the initial conditions for the system. Note that the expansion in Eq. (12.2-5) is just like the Fourier analysis

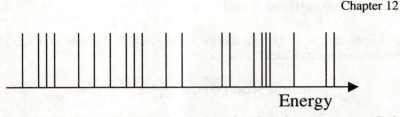

Fig. 12.1. A sketch of the energy eigenvalue spectrum for a bound quantum system. Each vertical line marks the location of one of the allowed energy eigenvalues.

sums described in Appendix A and in Chapter 11. The energy eigenfunctions $\Phi_i(x, y, z)$ play the role of the sine and cosine functions used there.

To understand the possible role of chaos in quantum mechanics, we will need to know one crucial characteristic of the solutions to the Schrödinger equation: For a bounded system (i.e., for a system that is restricted to some finite region of space), only certain discrete values of the energy eigenvalues E_i lead to "allowed" wave functions. A wave function for a bounded system is allowed if the probability density calculated from it goes to 0 for large values of the position coordinates. This is just the mathematical way of saying that the system is restricted to a finite region of space. This property holds only for discrete values of the energy E, and these values are called the energy eigenvalues for the system. (This behavior is exactly like that of spatial modes discussed in Chapter 11.) In more physical terms, we say that if we measure the energy of the bounded system, we will find not any value of the energy but only one of the discrete set of energy eigenvalues. We say that the energy of the system is "quantized." To illustrate the allowed energy values, we often use an energy level diagram, one version of which is shown in Fig. 12.1.

For a classical wave, the theory of Fourier analysis (see Appendix A) says that any vibrational pattern can be viewed as a linear superposition of waves each of which is associated with a single frequency. The superposition of energy eigenfunctions shown in Eq. (12.2-5) expresses the same result in quantum mechanics: Any wave function can be expressed as a linear superposition of energy eigenfunctions, those associated with a single energy value.

Time-dependent Hamiltonians are used to describe quantum systems that are subject to a controlled outside force, such as an oscillating electromagnetic field generated by a laser beam. In these models, the external fields are assumed to be completely under the control of the experimenter and to have a well-defined temporal behavior. Clearly, these models are idealizations because the external fields are themselves, in principle, physical systems which must obey the rules of quantum mechanics. However, the well-defined temporal behavior is a good approximation for many situations.

For a time-dependent Hamiltonian, we usually find the so-called time evolution operator $U(t, t_0)$ which acts on the initial wave function $\Psi(x, t_0)$ and gives us the system's wave function at a later time:

$$\Psi(x, t) = U(t, t_0)\Psi(x, t_0) \tag{12.2-6}$$

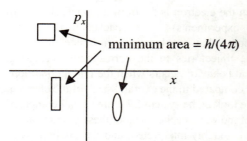

Fig. 12.2. Quantum mechanical systems must occupy an area at least as large as $h/4\pi$ in state space. The shape of the region may vary.

The time-dependence of the system in these cases can be more complicated than in the case of time-independent Hamiltonians, but as we shall see later, the evolution is still linear.

Quantum Mechanics and State Space Trajectories

In quantum mechanics, as we have mentioned, the precision with which we can specify the initial conditions of a system is limited. If we use a state space in which the axes are the x coordinate position and the x-component of the linear momentum, p_x, then quantum mechanics tells us that the initial conditions for the system must be spread over an area of this space that is larger than $h/4\pi$. See Fig. 12.2. It is impossible, according to quantum mechanics, to prepare a state that has a more restricted range of initial conditions. This result is usually stated in the form of the **Heisenberg Uncertainty Relation**:

$$\Delta x \Delta p_x \geq \frac{h}{4\pi} \qquad (12.2\text{-}7)$$

where Δx is a measure of the spread in initial x values (strictly speaking, it is the standard deviation in those values) and Δp_x is the corresponding measure of the spread in linear momentum values. The product form of the Uncertainty Relation tells us that if we try to reduce the uncertainty in the position, we must increase the uncertainty in the linear momentum. Thus, we must have a finite region within state space for our initial conditions. As the system evolves, the state space area occupied by those trajectories can only increase for an ideal quantum system.

This restriction on initial conditions is a manifestation of the wave nature of matter. For any wave, if we try to produce a wave disturbance in a very small region of space (small Δx), then, according to Fourier analysis (see Appendix A), we necessarily include a range of wavevector values. In quantum mechanics, by the famous de Broglie relation the wavevector k is related to the linear momentum ($p = hk/2\pi = h/\lambda$). So, a range of wavevectors implies a range of momentum values for the particle being described.

Thus far the search for quantum chaos has focused on systems that, under appropriate conditions, display what is called semi-classical behavior. This phrase means that the behavior of the quantum system is beginning to mimic the behavior of the corresponding classical model of the system. For example, for some highly excited electron states in atoms, when the electron is far from the nucleus, we can construct (both in theory and in experiments) "wave packets" (i.e., linear combinations of energy eigenfunctions, that are localized in space and have "trajectories" that are similar to the trajectories in the corresponding classical model). In rough terms, semi-classical behavior occurs when the initial conditions in phase space occupy a region small compared to the overall "size" of the classical trajectory in phase space and when we look at the system for a time small compared to the time for significant spreading of the wave packet. Under these circumstances the predictions of the classical model, suitably interpreted, and the predictions of quantum mechanics closely agree.

Can There Be Quantum Chaos?

With this brief review of the essentials of quantum mechanics, we are now ready to talk about the notion of quantum chaos. Since we cannot identify individual trajectories in quantum mechanics, we cannot use the notion of exponential divergence of trajectories to test for chaotic behavior. Thus, we must look at other aspects of the dynamics.

One of the arguments against the possibility of quantum chaos focuses on the time-dependence of the wave function by considering the superposition form of Eq. (12.2-5). For a bound system and a time-independent Hamiltonian, any wave function for the system can be written as the linear superposition of energy eigenfunctions where the time dependence of each part is just a combination of sine and cosine oscillations as embodied in the complex exponential factor. Since the energy eigenvalues take on only discrete values, the time dependence of any wave function can be at worst quasi-periodic in the sense defined in Chapter 4. Furthermore, since the time-dependence of the expectation values of the physical properties of the system is determined by products of wave functions, as indicated in Eq.(12.2-1), the physical properties can be quasi-periodic at worst. Thus, it seems that for a bounded system there is no possibility for chaotic time behavior if the Hamiltonian is time-independent. (Another way of stating this result is that the Schrödinger equation is linear in the wave function and hence lacks the essential nonlinearity needed for chaos.) Ford and Ilg (FOI92) have extended this argument to a wide variety of quantum properties and quantum systems.

Suppose we shift our attention to another aspect of chaotic behavior, namely, exponential divergence of nearby trajectories. Do wave functions show exponential divergence? Using the time evolution operator introduced in Eq. (12.2-6), we may write the solution to the Schrödinger equation in the following form (in which, for simplicity, we have suppressed the spatial dependence of the wave function):

$$\Psi(t) = U(t,t_o)\Psi(t_o) \qquad\qquad (12.2\text{-}8)$$

$\Psi(t_o)$ is the initial value of the wave function for the system and specifies the initial conditions for the system within the limits allowed by quantum mechanics.

If we now start the system with slightly different initial conditions, namely $\Psi(t_o) + \delta$, then we find that the solution of the Schrödinger equation is

$$\Psi(t) + U(t, t_o)\delta \qquad (12.2\text{-}9)$$

The second term of Eq. (12.2-9) is also a solution to the Schrödinger equation; hence, it cannot grow exponentially in time. Two "nearby" wave functions remain close as time goes on.

The lesson is that trying to extend the notion of exponentially diverging nearby trajectories to quantum mechanics, as a possible test for quantum chaos, will not work. There is no possibility for chaotic time evolution of the wave function.

We should point out that we have met up with an analogous situation in Chapter 8 during the discussion of Hamiltonian systems. There we saw that the time evolution of the state space (or phase space) probability distribution was given by a linear equation, the Liouville equation. We concluded that if the Hamiltonian system started with a slightly different probability distribution, then the two probability distributions would not "diverge" exponentially from each other even if the system were behaving chaotically and nearby trajectories were diverging exponentially. The situation in quantum mechanics is similar. In fact, the Arnold cat map, discussed in Chapter 8, is an exact deduction from a time-dependent model in quantum mechanics (RIS87, FMR91).

In quantum mechanics the role of the probability distribution in phase space is played by what is called the **Wigner distribution function** (see the references listed at the end of this chapter). Wishing to avoid unnecessary details, we simply state that this function is constructed from products of wave functions in such a way as to mimic the classical phase space probability distribution. One important difference is that the Wigner function can be negative while the classical probability distribution function is always positive. (The possibility of being negative is essential to capture the interference effects that are critical in wave mechanics.) The time evolution equation for the Wigner function is also linear in this function and hence cannot show sensitive dependence on initial conditions. More specifically, two "nearby" Wigner functions—ones that are only slightly different—cannot diverge exponentially with time.

We believe that this argument about the lack of exponential divergence in quantum mechanics is correct. The conclusion we must draw is that for quantum mechanical <u>models</u> with time-independent Hamiltonians for bounded systems there is no possibility for chaos in the sense of chaotic time dependence for the properties of the system. However, we want to emphasize that this conclusion applies only to a certain class of models. If we extend our models to include interactions with the system's environment (e.g., to allow for the possibility of the emission of electromagnetic radiation or for interaction with a thermal "bath"), then we find that the discrete energy eigenvalues are replaced by a continuous distribution of energy values (usually the distribution of energy values is relatively sharply peaked at or near the discrete values for the isolated system). Thus our restriction to quasi-periodic behavior is removed. Even in this case, however, the linearity of the Schrödinger equation remains an apparent obstacle to chaos.

The previous arguments have been very general. What happens if we look at actual quantum mechanics calculations? It has proven extraordinarily difficult to prove general results about quantum chaos, and we must rely on analyses of cleverly chosen systems for which some results can be computed. An example of a (model) quantum system with a time-dependent Hamiltonian is the periodically-kicked rotator. This is a model of a rigidly rotating dumbbell, which is kicked periodically. The classical model of this system, which is equivalent to the standard map of Chapter 8 (CHI79), displays chaotic behavior for some range of parameter values. The chaotic behavior is manifested by a growth of the kinetic energy (or equivalently of the classical action) of the rotating dumbbell with time.

One of the characteristics of classically chaotic behavior in Hamiltonian systems is the wandering nature of trajectories in state space: A trajectory starting in one region of state space wanders throughout the entire allowed region of state space if all the KAM tori have disintegrated. In that case we say we have fully chaotic behavior. In the analogous quantum systems, we see that wave functions that are initially localized in some region of state space (this is most obvious in the action-angle version of state space) start to diffuse much like their classical counterparts, but eventually, this diffusion ceases, and we say that the wave function remains localized in action-angle space even though the corresponding classical model continues to diffuse.

In the quantum version of this system, studied in detail by Chirikov and others (CCI79), the energy grows with time only up to a certain point called the "break time." Thereafter, the energy is essentially constant. Thus the diffusive nature of trajectories in a classically chaotic Hamiltonian system seems to be highly constrained in the corresponding quantum system.

Even worse for the prospects of quantum chaos is the reversibility of the quantum model: Chirikov found that the quantum model could be integrated backward in time from the break time to arrive exactly at its starting point. For chaotic behavior in the classical model, no such backward integration is possible because of the existence of a positive Lyapunov exponent. We believe (and there is a substantial body of information to confirm this belief) that we cannot reverse the time integration to recover the initial distribution for a truly chaotic classical system. For a classically chaotic system, the diffusion has led to a true mixing and loss of memory of the initial state. Apparently quantum mechanics, in spite of its inherent probabilities, is not random enough to satisfy the requirements of classical chaos.

We can understand why quantum mechanics, even with its inherent probabilities, is sometimes more deterministic than classical mechanics: Quantum mechanical "smearing" in phase space has important dynamical implications. This situation occurs in systems for which KAM tori begin to break up. (We saw in Chapter 8 that KAM tori confine phase space trajectories to certain regions.) In classical mechanics, the breakup of KAM tori is signaled by the spread or diffusion

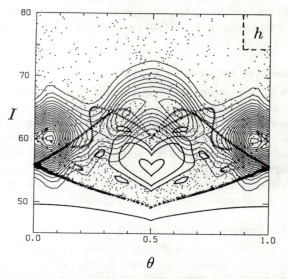

Fig. 12.3. Probability distribution (Husimi distribution) contours for a model of highly excited hydrogen atoms in a strong microwave field. The dynamics have been expressed in terms of action I and angle θ variables. The distribution shows clustering (scars) associated with the (unstable) periodic orbits of the corresponding classical model. The in-set and out-set (stable and unstable manifolds) for the unstable periodic point near the center form a homoclinic tangle in the action-angle diagram. (From JSS89.)

of trajectories through wide regions of state space, leading to stochastic (ergodic) behavior. However, in quantum mechanics, the wave function in a sense does not see openings in the KAM tori that are smaller in area than the fundamental quantum area $h/4\pi$. Hence, in quantum mechanics, probability distributions can remain confined by the remnants of KAM tori that in classical mechanics would allow trajectories to wander through phase space. In that case the quantum system is less stochastic than the equivalent classical system.

This "localization" of wave functions in phase-space is directly analogous to the spatial localization of electrons (as described by quantum mechanical waves) in a medium with random potentials (Anderson localization), an important and long-sought effect in condensed matter physics. See, for example, JEN87 and [Gutzwiller, 1990].

Another explanation of this limitation on "transport" in quantum systems is to link the quantum probability distributions and the topology of the corresponding classical dynamics. Apparently, the quantum probability distributions tend to cluster around the locations (most obvious in state space diagrams) of unstable periodic orbits in the corresponding classical system (see Fig. 12.3). (The *Husimi distribution* used in Fig. 12.3 is the overlap of the wave function with the so-called coherent states for the system. See, for example, RBW96.) These periodic orbits are said to leave "scars" in the quantum probability distributions. This analysis

(JSS89) has been used to explain the suppression (below what would be expected from a classical mechanics analysis) of ionization of highly excited hydrogen atoms by microwave fields.

Quantum Analogies to Chaotic Behavior

Because chaotic time dependence in quantum mechanics seems to be ruled out by the arguments given in the previous paragraphs and because quantum mechanics does not allow us to talk about well-defined trajectories in state space, we must ask another kind of question. Are there any significant qualitative changes in the predictions of a quantum mechanical model for a system when the corresponding classical model predicts chaotic behavior? Two quantities have been looked at in some detail over the past few years: the statistical distribution of energy eigenvalue spacings and spatial correlations of energy eigenfunctions.

Distribution of Energy Eigenvalue Spacings

One of the most widely explored possible signatures of quantum chaos has been the statistical distribution of energy eigenvalue spacings. Recall from our discussion of Hamiltonian systems that chaotic behavior may occur over significant regions of state space for a certain range of energies (with all other parameters of the system held fixed). Thus, it seems obvious to compare the distribution of energy eigenvalues in energy regions that correspond to regular behavior in the corresponding classical model with the distribution of energy eigenvalues in energy regions that correspond to widespread chaotic behavior in the classical model.

First, we should point out that the energy ranges of interest are not the lowest energy states for the system, the ones most often calculated in beginning courses in quantum mechanics. For many systems, those energy levels have very regular spacing and provide no surprises. We need to look at highly excited states with energies far above the lowest possible values for the system. In those cases the energy eigenvalues (as illustrated in Fig. 12.1) seem to have no pattern. We then ask: For a given energy range, what is the distribution of energy spacings between neighboring energy eigenvalues? It is conjectured that the distribution will be significantly different in the cases of (classical) regular motion and (classical) chaotic motion. The consensus seems to be that in the regions corresponding to regular motion, the spacing of the quantum energy values will be described by a Poisson distribution, while in the regions corresponding to chaotic behavior, the distribution will be the so-called Wigner-type distribution (JOS88, IZR90). These two distributions are shown in Fig. 12.4.

We can understand the difference between these two distributions by considering the difference between regular and chaotic behavior in the classical model of the system. When the classical system's behavior is regular (or quasi-periodic), the classical phase space behavior is dominated by the KAM surfaces as discussed in Chapter 8. In quantum mechanics each KAM surface is characterized by a set of quantum numbers associated with the quantities that remain constant

Energy-Level Spacing Distribution

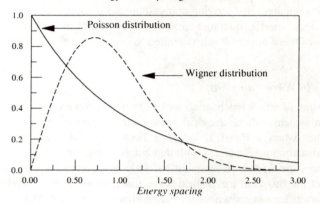

Fig. 12.4. Graphs of the Poisson distribution and Wigner distribution for energy level spacings.

(are conserved) as the system moves over the KAM surface. Hence, the energy eigenvalues are given by running through the sequence of possible quantum numbers. The resulting sequence of eigenvalues is just a "mixture" of sequences resulting from changing each of the quantum numbers while leaving the others fixed. This mixture leads to a distribution of energy eigenvalues in which the most likely spacing is very small (that is, quite frequently the energy eigenvalues will lie almost on top of each other).

On the other hand when the classical system becomes chaotic, the system is nonintegrable (in the extreme case it has no conserved quantities except the energy), and the trajectory wanders throughout state space. In that case all of the degrees of freedom start interacting, and their effects on the energy eigenvalues become strongly correlated. (By way of contrast, in the integrable case, we can say that the degrees of freedom are independent.) The net effect is that it becomes highly unlikely that two energy eigenvalues lie very close to each other. (We say we have "energy level repulsion.") (A general property of quantum states is that energy levels corresponding to interacting degrees of freedom tend to repel one another; that is, with the interaction, the energy levels are further apart than they would be if the interaction were turned off.) Thus, the distribution of energy level spacings changes to a distribution like the Wigner-type distribution.

However, the previous arguments really only apply to the extreme cases of integrability and complete nonintegrability (no constants of the motion except the energy). Thus there may be intermediate cases that are hard to distinguish according to this criterion. Moreover, studies have found cases in which the quantum counterparts of classically integrable systems do not show a Poisson distribution of energy level spacings (CCG85), cases of classically nonintegrable systems whose quantum cousins do not have a Wigner-type distribution of energy level spacings (ZDD95), and finally there are cases that are classically nonchaotic but whose energy level spacing distribution is (close to) a Wigner-type distribution. Thus, it seems apparent that looking at distributions of energy level spacings is not the way to look for the quantum analog of chaos. In any case, the practical computation of these energy levels is limited to systems with only a few degrees of freedom.

Correlations in Wave Functions

The second line of attack has been to look at spatial correlations of wave functions for quantum systems whose classical analog shows chaotic behavior. The basic notion is that when a classical system's behavior is chaotic, the state space probability distribution associated with that behavior becomes "spikey," reflecting, in many cases, the fractal geometry of the state space attractor. By analogy, we might expect the quantum mechanical wave functions to develop irregular spikes for those parameter ranges where the corresponding classical model has chaotic behavior.

The degree of "randomness" in the wave function can be expressed in terms of a spatial correlation function, defined in analogy with the temporal autocorrelation functions described in Chapter 10. Figure 12.5 shows an example of a contour plot of a wave function for a system of quantum billiards, for an energy range in which the corresponding classical system is chaotic. (Recall from Chapter 8 that a billiard model consists of a particle free to move in two dimensions inside a perfectly reflective boundary.) However, having a wave function with little

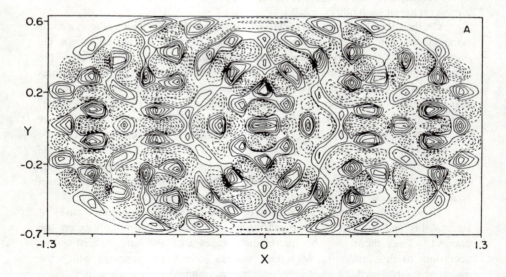

Fig. 12.5. A contour plot of the quantum mechanical wave function for a quantum billiards system whose classical counterpart has chaotic behavior. Solid lines indicate positive values of the wave function; dashed lines indicate negative values. The irregular pattern is taken as evidence for the quantum behavior analogous to chaotic trajectories. (From SHG84.)

spatial correlation is necessary, but not sufficient, for what we would want to call chaotic behavior. So again, we are left with only a partial definition.

These irregular eigenfunctions can be characterized by fractal dimensions. KKK97 show that the generalized dimension D_2 introduced in Chapter 9 can be used to characterize the energy level spectrum and the energy eigenfunctions for a quantum system whose classical counter-part has chaotic behavior. They find that the ratio of the two dimensions (one for the energy level spectrum, the other for the energy eigenfunctions) controls how rapidly wave packets spread.

Chaos and Semi-Classical Approaches to Quantum Mechanics

The exploration of the possibility of quantum chaos has led to a revival of interest in calculational techniques that make use of a mixture of quantum mechanics and classical mechanics. The idea is that for many quantum systems, the calculation of the wave function from the Schrödinger equation is impractical. Approximate wave functions, however, can be calculated by borrowing some ideas from classical mechanics for which the problem may be more tractable. Thus, the method mixes classical and quantum ideas, leading to the name semi-classical method. (We might also call this the semi-quantal method.)

To get some notion of this method, we outline one variant of the semi-classical method: In classical mechanics, one can define a function $S(q)$, the so-called Hamilton's characteristic function, such that the momentum associated with the generalized coordinate q is given by

$$p(q) = \frac{\partial S(q)}{\partial q} \qquad (12.2\text{-}10)$$

(For the sake of simplicity, we will restrict the discussion to one spatial dimension.) Equation (12.2-10) tells us that the trajectories are perpendicular to the curves of constant S (since the gradient is perpendicular to curves of constant S). If we turn the procedure around, we write $S(q)$ as an integral of the linear momentum over the path of the particle

$$S(q, q_0) = \int_{q_0}^{q} p(q')dq' \qquad (12.2\text{-}11)$$

where in general S depends on both the starting point and the end point of the trajectory. If S is calculated for a periodic orbit, then S is proportional to the *classical action* associated with that orbit.

In the early quantum theory (before the development of the Schrödinger equation), the descriptions of systems were "quantized" by setting the action equal to an integer multiple of Planck's constant h. This method, known as the Einstein–Brillouin–Keller quantization method, asserted that only those actions (and hence only certain energies) satisfying the quantization condition led to allowed orbits. It was quickly realized (EIN17) that this procedure broke down in the case of nonintegrable classical systems, which in general do not have closed orbits.

Given the characteristic function, we can also find the period T of the classical orbit by finding out how $S(q)$ varies with energy E:

$$T = \frac{\partial S(q)}{\partial E} \tag{12.2-12}$$

We shall now show how $S(q)$ can be used to get an approximate form of the quantum mechanical wave function. In quantum mechanics, the wave function for a particle with a definite value of the momentum p has the form

$$\Psi(q) = A e^{i\frac{2\pi pq}{h}} \tag{12.2-13}$$

If the particle is subject to a force, then the momentum will change. The semi-classical form of the wave function is then given by

$$\Psi(q) = A(q) e^{i\frac{2\pi}{h}S(q)} \tag{12.2-14}$$

where the action $S(q)$ is given by the integral expression in Eq. (12.2-11). Since the integral is carried out over the classical trajectory, we say we have a semi-classical approximation to the wave function. In this approximation, the curves of constant S are the "wavefronts" of the quantum mechanical waves since a wavefront is the curve of constant phase for the wave.

When the classical motion is regular, each individual quantum state is correlated with a single classical trajectory with a single value of the quantized action (e.g., the circular orbits or elliptical orbits in the well-known Bohr and Bohr–Sommerfeld models for the hydrogen atom). Nearby orbits track each other, and the wavefront represented by $S(q)$ is smooth. However, when the classical motion is chaotic, the chaotic trajectories do not provide a vector field ($grad$ $S(q)$) from which we can construct a wave function.

What do we do in the chaotic case? Gutzwiller (GUT71, BER89) has developed a method that relies on the properties of just the periodic orbits of the classical system to find the so-called Green's function for the corresponding quantum problem. (The Green's function is that mathematical function that allows us to find, at least in principle, the time evolution of the wave function.) The Gutzwiller method can be applied to systems that are classically chaotic.

Experiments on Quantum Chaos

Our discussion on quantum chaos has focused on theoretical models and issues. What do experiments have to say about the issue of quantum chaos? The short answer seems to be that all experiments that have been carried out on systems whose classical analogs show chaotic behavior have results that are in agreement with the predictions of quantum mechanics. That is, quantum mechanics apparently does not break down when the classical analog shows chaotic behavior. We briefly describe a few experiments to illustrate this point.

If an atom with a single electron is placed in a strong magnetic field, then the dynamics of the system as described by classical mechanics is nonintegrable and shows chaotic behavior for sufficiently high energies or sufficiently strong magnetic fields. Fortunately, this is a system for which quantum mechanics can provide, with a fair amount of effort, predictions for the energy level spectrum (DBG91), for example, and where precise experiments are possible (IWK91). Thus far there is good agreement between the theoretical predictions of quantum mechanics and the experiments, but the physical interpretation and understanding of the regularities observed have yet to be worked out.

In a second category of experiments, hydrogen atoms excited to high-lying energy states were ionized by being exposed to an intense electric field oscillating at microwave frequencies. The oscillating field makes the Hamiltonian for the system time-dependent and provides the extra degree of freedom needed to make chaotic behavior (in the classical description) possible. Again, the experimental results seem to be in agreement with the predictions of quantum mechanics. See JSS91 for a review of both the theoretical and experimental results.

Yet another confrontation between classical chaos and quantum mechanics occurs in the interaction of lasers and atoms. If a stream of atoms traveling through a vacuum is subject to a laser beam, the atoms can emit and absorb photons from the laser beam. When an atom absorbs a photon, it receives a small momentum "kick" as well as the energy from the photon. Similarly, when the atom emits a photon, it must recoil. With an intense laser beam, you might expect the atoms to gradually increase the spread in their momenta as they are kicked back and forth. The classical mechanics description of this process predicts a "diffusion" in momentum as a result of this chaotic behavior. Quantum mechanics, however, predicts that the momentum spread will stay limited (an effect called *dynamical localization*). Experiments (MRB94, RBM95) indicate that quantum mechanics gives the correct description even when the classical behavior is chaotic. However, LAW95 suggests that the localization can be explained solely by classical nonlinear dynamics and that the observed effects are not examples of dynamical localization. For a thorough survey of the experiments and the conclusions to be drawn from them see RAI99.

Conclusions

The story of quantum chaos, if it exists, and the connections between the predictions of quantum mechanics and the chaotic behavior of macroscopic systems now observed daily in laboratories around the world, if those connections can be made, is far from complete. Reputable scientists hold contradictory (and often strongly worded) opinions about quantum chaos. We believe that we will eventually understand how the predictions of quantum mechanics mimic classical chaos for all times of practical interest: the half-life of a graduate student or perhaps even the age of the universe. In the meantime, we will learn a lot about the subtleties of both quantum mechanics and classical mechanics.

12.3 Chaos and Algorithmic Complexity

Most of us, when pressed to think about the matter, divide dynamical behavior into two distinct categories: (1) regular, periodic, and, hence, determinable behavior and (2) random (indeterminate) behavior. As we have learned in the study of nonlinear dynamics, there is a third kind of behavior—the type we have called *chaotic*. With chaotic behavior, we have deterministic rules of time evolution leading to behavior that, at first sight, looks indistinguishable from random behavior. As we have seen, however, there are methods of analysis that allow us to determine, at least approximately, the number of active degrees of freedom, and thus to make a distinction between chaotic behavior involving a small number of degrees of freedom and "true" randomness, which in a sense can be characterized as having an infinite (or at least very large) number of degrees of freedom.

Finding this intermediate kind of dynamical behavior has led to some attempts to provide a measure of complexity of dynamical behavior that ranges continuously from simple periodic behavior at one extreme, through deterministic chaotic behavior with a few degrees of freedom as an intermediate case, to complete randomness at the other extreme. We shall describe one such scheme for quantifying complexity.

In Chapter 5 we saw how the dynamics of iterated maps (and in some cases, other dynamical systems) can be reduced to symbolic dynamics: a sequence of usually two symbols, R and L, or 1 and 0. One measure of complexity focuses attention on this string of symbols and measures the ***algorithmic complexity*** of a sequence as the length (say, in bits) of the shortest computer program (algorithm) that will reproduce the sequence. For example, the algorithmic complexity of a periodic sequence is quite small because we need to specify only the pattern of symbols that repeats and the instruction to repeat that pattern. On the other hand, a completely random pattern requires an algorithm that is as long as the sequence itself; that is, for a completely random pattern, the only algorithm that can give the sequence is the one that states the sequence in its entirety.

What is the algorithmic complexity of a sequence of symbols generated by a chaotic system (that is, one with at least one positive Lyapunov exponent)? We might expect that the sequence of symbols generated by a chaotic system, such as the logistic map function with $A = A_\infty$, is of intermediate algorithmic complexity: Only a modest computer program is needed to iterate the logistic map function. For the Bernoulli shift map, we need only a very short program to shift the sequence of bits one place to the left.

We need to think through this procedure more carefully, however, because to produce a <u>particular</u> sequence, we also need to specify the initial value, say, x_0. To be concrete, suppose we want to reproduce the bit sequence (0s and 1s) resulting from 10 iterations of the logistic map function with $A = 4$. As we saw in Chapter 5, the logistic map with $A = 4$ is equivalent to a Bernoulli shift map; therefore, each iteration of the logistic map is equivalent to shifting the sequence of bits one place to the left. Suppose we record the first bit to the right of the binary point for each iteration. Thus to produce a particular sequence of 100 bits, we would need to specify x_0 to 100 bits. The exponential divergence of nearby trajectories would otherwise cause the final x to differ from our desired x. Thus, we need to specify essentially the entire sequence, and we have high algorithmic complexity. In this sense, a sequence of numbers produced by a chaotic system is equivalent to a random sequence.

The straightforward notion of algorithmic complexity, however, seems to miss something. In the case of a sequence of numbers generated by a chaotic system, say the logistic map with $A = 4$, we can expose the underlying deterministic algorithm by plotting the $(n+1)$th iterate as a function of the nth to display the functional relationship for a one-dimensional map system. In some cases this procedure can be extended to systems with higher dimensionality.

We can also test for determinism by looking at the sequence of numbers directly. Since the numbers are necessarily specified to some finite accuracy, the sequence will necessarily eventually return to any one of the specified numbers. Once we return to a specified number, we can ask if the next number in the sequence is the same as the one that followed the first occurrence of our specified number. If it is not, we know that we do not have a one-dimensional iterated map sequence. However, the sequence might be the result of a two- (or higher-) dimensional map function, and we would need to look for the recurrence of two (or more) numbers in sequence to test for determinism. As the number of possible dimensions rises, the difficulty of testing for determinism obviously increases greatly. Alternatively, the embedding schemes discussed in Chapter 10 can give some indication, at least for low-dimensional systems, of the number of dimensions and hence can separate "true" randomness from deterministic chaos. In practice, there may not be much difference between "true" randomness and high dimensionality deterministic chaos.

Thus, it would be premature to claim that all randomness is due to deterministic chaos or to equate chaos with all forms of randomness. (But, see FOR89 for forceful arguments along these lines.) There would seem to be heuristic, if not practical, reasons for retaining a distinction between deterministic chaos (at least for low-dimensionality systems) and "pure" randomness, as defined by algorithmic complexity or some other scheme.

J. Ford (FOR89) has argued that algorithmic complexity provides a definition of randomness (which he equates with chaotic behavior), which could provide a definitive test for the existence or nonexistence of chaos in quantum mechanics. If quantum mechanics fails this test, then, Ford argues, we must be prepared to modify (or perhaps replace) quantum mechanics because we know that there exist systems in nature that exhibit chaos as defined by algorithmic complexity. Of course, we must point out that there are yet other possibilities: We might decide that algorithmic complexity does not provide the appropriate definition of chaotic (random) behavior or quantum mechanics might know how to mimic behavior of high algorithmic complexity in a way not currently understood.

12.4 Miscellaneous Topics: Piece-Wise Linear Models, Time-Delay Models, Information Theory, Stochastic Resonance, Computer Networks, Controlling Chaos, and Synchronizing Chaos

In this section we discuss several topics that did not fit in nicely elsewhere in this book. The topics, however, illuminate several interesting aspects of nonlinear dynamics and thus deserve some attention.

The first two topics are two types of models that are widely used in nonlinear dynamics. However, their mathematical characteristics are sufficiently different from the models discussed elsewhere in this book that they need some special comments.

Piece-Wise Linear Systems

Throughout this book, we have emphasized the necessity of having nonlinear time evolution equations to see the effects of frequency-locking, chaotic behavior, and a host of other intriguing effects. However, there is a class of models that make use of linear differential equations to model nonlinear systems. How do nonlinear effects emerge from these apparently linear models? These models are called piece-wise linear models because they use a linear equation to describe the time evolution of the system for some time interval. At the end of the time interval, the model switches to a different linear equation. In a sense, it is the switching process that contains the nonlinear elements.

The advantage of using a piece-wise linear model lies in our ability to find a closed-form solution for the system's behavior in the time between switches. Thus, part of the time evolution can be solved exactly. At the switching time, we match values of the relevant physical variables (since we do not want most physical observables to change discontinuously). After the switch, with the new linear equation, we can again solve the time evolution exactly up to the next switch. Piece-wise linear models are widely used in engineering for the modeling of nonlinear systems.

As an example of a piece-wise linear model, let us examine a model (ROH82) used to describe the behavior of the semiconductor diode circuit introduced in Chapter 1. In that model, when the forward-bias potential across the diode reaches a set value V_f, the semiconductor diode is treated as a fixed voltage (emf) source V_f. When the forward-bias current passing through the diode drops to 0, the diode continues to conduct for a "recovery time" τ that depends on the magnitude of the most recent maximum value of the forward current. When the recovery time is past, the semiconductor diode is treated as a fixed capacitor. The two circuit models are shown in Fig. 12.6.

Using standard ac circuit analysis, we can solve the equations describing the time behavior of these models exactly for the time periods between switching. The

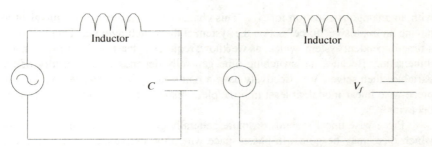

Fig. 12.6. The two circuit models used for a piece-wise linear description of the semiconductor diode circuit introduced in Chapter 1. The diode is treated as a fixed-value capacitor C unless its forward-bias potential difference reaches V_f. The model then switches to a fixed potential difference equal to V_f for the diode. The circle represents the sinusoidal emf signal generator.

nonlinearity is introduced via the recovery time because the exact time of switching depends on the history of the current flowing through the diode.

With a slight modification of the model for the reverse-recovery time, the piece-wise linear model of the semiconductor diode circuit can account for higher-dimensional behavior observed under some conditions (HUR84). A different piece-wise linear model (with different fixed capacitance values for the diode when forward- or reverse-biased) also gives rise to chaotic effects (MCT84).

Let us now look at piece-wise linear models in more general terms. For the sake of concreteness, we consider a model with two state-space dimensions and two state-space variables x_1 and x_2. For a linear system, the state space evolution is given by our now familiar equations

$$\begin{aligned}
\dot{x}_1 &= f_{11}x_1 + f_{12}x_2 \\
\dot{x}_2 &= f_{21}x_1 + f_{22}x_2
\end{aligned} \tag{12.4-1}$$

where the fs are constant parameters. A piece-wise linear model follows this evolution for some time and then switches to another linear set of equations

$$\begin{aligned}
\dot{x}_1 &= f'_{11}x_1 + f'_{12}x_2 \\
\dot{x}_2 &= f'_{21}x_1 + f'_{22}x_2
\end{aligned} \tag{12.4-2}$$

with new values of the fs. It is these discontinuous changes in the fs that may induce nonlinear effects. But between switches, we can solve the time evolution equations exactly. In other words, we can find the mapping function that takes the state space variables just after one switch up to their values just before the next switch.

To see how this works out, let us assume that the switch occurs when the x_1 and x_2 variables take on certain values X_1 and X_2. The time derivatives then show discontinuous changes at the switch. For example for $\dot{x}_1$, we have

$$\dot{x}_{1\,after} - \dot{x}_{1\,before} = (f'_{11} - f_{11})X_1 + (f'_{12} - f_{12})X_2 \tag{12.4-3}$$

with an analogous equation for $\dot{x}_2$. This change in the derivative is equivalent to having an impulsive "force" act on the system at the switching time. Thus, we have a time-dependent system, which, as we know, requires at least one more state space dimension. Because the switching time generally depends on the history of the variables themselves, we effectively have a nonlinear system. Hence, this kind of piece-wise linear model, at least in principle, satisfies our requirements for chaotic behavior.

Piece-wise linear systems or, more generally piece-wise smooth systems (for which there may be points in phase space where the derivatives of the functions governing the dynamics may not be defined) can exhibit novel kinds of bifurcations. In these bifurcations several attractors can be created simultaneously as a parameter of the system is varied. The creation of multiple attractors means the system becomes very sensitive to noise and we lose ability to predict even the attractor to which the system will evolve. The theory of such a bifurcation is discussed in DNO99.

Time-Delay Differential Equation Models

In an interesting class of models, the time evolution equations give the derivative of the state space variables in terms of functions that depend not only on the current state of the system at time t but also on the state at an earlier time $t - \tau$, where τ is called the delay time for the system. In formal terms, the time evolution equation for one state space variable would look like

$$\frac{dx}{dt} = f(x(t), x(t - \tau)) \qquad (12.4\text{-}4)$$

A well-known example of a delay-differential model is the Glass–Mackey model [Glass and Mackey, 1988]:

$$\frac{dx}{dt} = \frac{ax(t - \tau)}{1 + x^c(t - \tau)} - bx \qquad (12.4\text{-}5)$$

where a, b, and c are numerical parameters. Such a model can be used to describe time delay effects in biological systems, for example.

In order to determine uniquely the solution of a delay-differential model, we must specify the initial condition of x over a time interval of length τ. In practice, we break up that interval into N discrete time units as an approximation. In that case, the delay-differential equation is equivalent to an N-dimensional state space evolution. In the limit $N \to \infty$, we have an infinite dimensional state space. Thus, a delay-differential equation model is equivalent to a system with an infinite number of degrees of freedom. The immediate question that arises is: How do the dynamics of a system with an infinite number of degrees of freedom differ from the dynamics when we have only a finite (and usually small) number of degrees of freedom? Are the conceptual and quantitative tools developed for low-dimensionality systems applicable to a system with an infinite number of degrees of freedom? Farmer (FAR82a), in a quite readable paper, has explored the behavior of the Glass–Mackey model, its Lyapunov exponents and the fractal dimension of the attractors for the system. He has found that for some range of parameter values, the system's behavior is well-accounted for by a finite dimensional attractor. The connections between the infinite number of degrees of freedom in a delay-differential model and the behavior of spatial systems with an infinite number of degrees of freedom (see Chapter 11) remain to be explored.

Information and Chaotic Behavior

Ever since the development of a formal theory of information and information transmission by Claude Shannon in the 1940s, physicists have applied information concepts to illuminate the physics of statistical systems. (See, for example, [Baierlein, 1971].) Thus, it should come as no surprise that the notions of information have been applied to the behavior of chaotic systems. However, two apparently contradictory statements can be found in the literature: (1) Chaotic systems create information; (2) chaotic systems destroy information. As we shall see, both statements are correct.

First, we need to specify what we mean by *information*. The technical definition is akin to the definition of entropy found in Chapter 9. We focus our attention on the symbolic dynamics coding of the behavior of some system by using a string of symbols: 1s and 0s or Rs and Ls. The information content of that string of symbols is defined as the logarithm (usually to the base 2) of the number of symbols needed to define the sequence:

$$I = \log_2 N \qquad\qquad (12.4\text{-}6)$$

For a repetitive sequence, for example all 0s, we need specify only one symbol, and the information content is 0. For a purely random sequence, the number of symbols needed is the length of the sequence, and the information content is high. (We see that this technical definition equates the notion of unpredictability or surprise with high information. [Baierlien, 1971] urges us to call this quantity "missing information.")

For a chaotic system, we can say that for short times, the chaotic behavior generates information. To see this, think of starting several trajectories within a very small region of state space. If the initial conditions are close enough, we cannot tell that we have distinct trajectories. Because of the exponential divergence of nearby trajectories, however, the trajectories eventually separate sufficiently for us to see that there are in fact distinct trajectories. We have gained information about the system. This behavior is in agreement with the notion that a chaotic system has positive Kolmogorov entropy as discussed in Chapter 9.

Chaotic behavior, however, destroys information in the long run. To see this aspect, consider again starting several distinct trajectories, in this case, in different parts of state space. After letting the system run for a long time, we find that the trajectories will settle onto an attractor and be stretched and folded many times for dissipative systems. Eventually, we will no longer be able to tell which trajectory came from which set of initial conditions. We have lost information about the system.

If we focus our attention on the probability that a trajectory visits various cells (or sequences of cells) in state space, as we did in Chapters 9 and 10, then the (missing) information can be expressed in terms of those probabilities:

$$I = -\sum_i p_i \ln p_i \qquad (12.4\text{-}7)$$

(If the information is defined without the minus sign, it is sometimes called *negentropy*, because it is then the negative of the usual entropy function given in terms of the probabilities.)

Our conclusion from this brief discussion is that using information terminology to talk about chaotic dynamics helps us understand some of the implications of information theory, but it does not appear to be useful for understanding dynamics.

Stochastic Resonance

Nonlinear dynamics challenges our intuition on many fronts. One of the most interesting of those challenges arises in what is called *stochastic resonance*. Quite contrary to our ordinary notions, there are circumstances under which adding noise to a system makes it easier to recover weak signals. This effect has been observed in laser systems, electrical systems, chemical reactions, and even in human perception. MOW95 and GHJ98 provide excellent introductions to stochastic resonance and its applications.

To see stochastic resonance, we need one essential feature: the system must have some sort of "activation barrier" or "threshold." (That is, the system shows a response only if the stimulus exceeds a certain level.) Then if both a weak periodic signal and some noise are applied as a stimulus to the system, the response of the system at the frequency associated with the weak periodic signal will show a maximum amplitude when the noise amplitude is different from zero (a kind of resonance effect). In other words, for zero noise, the response will be weak; for large noise amplitudes, the response is lost in the noise. For some intermediate value of the noise amplitude, however, the response at the desired frequency is a maximum. In some sense, the noise helps the system get over the activation barrier or threshold. The references at the end of the chapter give some taste of the many applications of stochastic resonance.

Chaos in Computer Networks

As we have mentioned at several points in this book, computers and computer graphics in particular have been indispensable in the modern development of the study of nonlinear dynamics and chaos; however, computers themselves are complex systems. Could they exhibit chaotic behavior? Of course, most single computers run programs that are in essence deterministic and carefully controlled: They give definite responses to definite inputs. The interesting possibility for complex behavior seems to arise in computer networks—systems in which many computers are interconnected. In some sense, computers connected together are like coupled oscillators: Each one by itself might act quite predictably, but the interconnections, like the coupling between oscillators, may make the system as a whole behave chaotically. However, it turns out, such behavior is not necessarily deleterious. The ramifications of this possibility have been explored in several papers listed in the references at the end of the chapter.

Taming and Controlling Chaos

In thinking about the occurrence of chaotic behavior in real systems such as lasers, electronic circuits, mechanical oscillators, and so on, we are tempted to view such behavior as an (intriguing) annoyance: We want the laser or electronic circuit or mechanical oscillator to behave predictably, and thus we tend to design the system to avoid those parameter ranges for which chaotic behavior might occur.

Some recent work, however, has demonstrated that there may be some advantage to having a system operate in a range of parameter values for which chaos is possible. The essential idea is that when a system is behaving chaotically, it explores a relatively large region of state space. Embedded in that region of state space are many (unstable) periodic orbits. By providing a weak control signal, we can induce the system to follow an orbit that is close to almost any one of these periodic orbits. Thus, in a sense, chaotic behavior provides a varied landscape for the dynamical behavior of the system, and by changing the weak control signal, we can cause the system to take on a wide variety of different types of periodic behavior.

This type of adaptive control does not require prior knowledge of the system's dynamics. In fact, we can use the system's behavior itself to learn what kind of small perturbation is necessary to induce it to follow a particular periodic orbit (OGY90). Such control schemes have been demonstrated in several experimental systems (BRG91, HUN91, RMM92 and other references at the end of the chapter). With some clever adaptations, the technique can be extended to systems with high-dimensional attractors (AGO92) and can operate over extended parameter ranges (GIR92). [Kapitaniak, 1996], [Schuster, 1999], and DIS97 give good introductions to the various control techniques.

Given the flexible type of response possible for a system exhibiting chaotic behavior, it is tempting to speculate about the biological implications of such behavior. Might it be advantageous in some sense for biological systems to arrange themselves to "live" in a chaotic regime with a wide variety of possible behaviors, allowing them to adapt more easily to changing conditions? By way of contrast, if their behavior were strictly periodic, they might have difficulty responding to a changed environment. On a more specific level, we might speculate (and indeed this speculation is far beyond any experimental evidence) about the possibility that the brain's behavior might be modeled as that of a chaotic system. With the help of weak external perturbations (some sort of perceptual clue, for example) the pattern of neural firings may settle onto some periodic orbit constituting a particular memory. While we are trying to remember, for example, someone's name, the brain's "orbit" wanders chaotically through a relatively large region of "state space" until, induced by the clue that got us thinking about that particular person, it locks onto the "periodic orbit," which constitutes the memory of the person's name.

Clearly, this is very (if not overly) speculative. The important point, however, is that chaotic behavior and its possible control gives us a new model to think about the behavior of complex systems.

Synchronizing Chaos

As we have emphasized throughout this book, chaotic behavior is characterized by sensitive dependence on initial conditions, which leads to apparently random, disorderly behavior. We might conclude based on that notion that it would be nearly impossible to get two chaotic systems to behave exactly in the same way— that is, it would not be possible to get the two systems synchronized so that their chaotic trajectories would be the same. Once again, nature challenges our intuition. It indeed is possible to synchronize two (or more) chaotic systems so that their trajectories (in state space) track each other quite closely over extended periods of time. In essence we make use of the sensitive dependence on initial conditions to gently nudge one system so that is stays synchronized with the other.

Why would we want to have two chaotic systems synchronized? There are many reasons. For example, in many technological applications, it is important to have two "oscillators" be synchronized. Often we think of getting two motors synchronized or synchronizing two power generators that produce ac power. We can extend that notion to nonperiodic (chaotic) behavior. The healthy human heart and the human brain both apparently exhibit slightly chaotic behavior. In the heart it is important to have the various parts of the heart, which in principle can beat independently, remain synchronized even though there are slight variations in the heart rate from moment to moment. As another example, consider synchronization in communications systems. Chaotic signals can be used to hide the "true" message signal from eavesdroppers. In order to extract the message at the receiving end, in many cases there must be synchronization between the chaotic behavior of the sender and the chaotic behavior of the receiver.

The references at the end of the chapter provide a sample of the many applications of synchronized chaos.

Using Chaos to Predict the Future

Using time-series data to develop models for the dynamics of systems and to predict the future has a long and venerable history. In recent years there has been some success in using notions based on chaotic behavior in nonlinear systems to improve prediction and forecasting schemes for systems displaying complex behavior. The same ideas can be used to reduce the effects of noise in the time-series data: that is, we can use these methods to say what the time-series would have been in the absence of noise. A detailed discussion of forecasting and noise reduction methods would take us far afield; therefore, the interested reader is directed to the references at the end of the chapter.

12.5 Roll Your Own: Some Simple Chaos Experiments

Throughout this book we have mentioned the application of nonlinear dynamics to various physical, chemical, and biological systems. In recent years, many articles have been written describing simple (mostly physical) experimental systems that can (and should) be used to demonstrate bifurcations, routes to chaos, and so on. We strongly encourage you to try some of these systems. Most can be set up with equipment readily available in every college or university physics department.

The diode circuit described in Chapter 1 is one of the most versatile and easily controlled nonlinear systems. An extensive description of nonlinear behavior for such a system is given in BUJ85. Using this system permits precise quantitative measurements of many aspects of nonlinear dynamics. But it does require some knowledge of electronics and the use of an oscilloscope.

Many mechanical systems can be used to see (in a literal sense) the effects of nonlinear dynamics, although quantitative measurements are usually more difficult. Among those described in the literature are experiments on a ball bouncing from a rigid vibrating platform (TUA86, MET87), various magnets driven by oscillating currents (BSS90, MES86, OMH91), and a dripping faucet (DRH91). Several other electronic and mechanical nonlinear systems are discussed in BRI87. In addition, Appendix C of [Moon, 1992] shows various "chaotic toys." Commercial versions of the driven damped pendulum are available from Daedalon, Inc. (Danvers, MA) and Tel-Atomic, (Jackson, MI) (complete with computer interface and software). Several readily available "executive toys" found in gift shops and science museum shops, for example, illustrate chaotic motion in coupled pendulums and magnetic pendulum systems.

12.6 General Comments and Overview: The Future of Chaos

Two revolutions occurred in twentieth-century physics: The development of Einstein's theory of relativity and the development of quantum mechanics. These two revolutions have forced us to acknowledge limits on physical reality and hence on our knowledge of the world: Relativity has taught us that there is an upper speed limit to the transmission of energy and information (the speed of light c) and forced a complete rethinking of our notions of space and time. At the quantum level, quantum mechanics, in its usual interpretation, tells us that there is a lower limit to the size of the interaction between systems, a limit determined by Planck's constant h. Thus, any physical interaction, and hence any measurement necessarily produces an effect on the system being observed, an effect that cannot be made negligibly small.

The birth and flowering of nonlinear dynamics and chaos may also constitute a revolution in our understanding of nature. The existence of chaos provides us with yet another limitation on our knowledge of the natural world: For chaotic systems, we are limited in our ability to predict the long-term behavior of systems, even ones that are deterministic in principle. In some ways this limitation is just as fundamental and just as important as the limitations imposed by relativity and quantum mechanics.

Whether nonlinear dynamics and chaos constitute a revolution in physics as fundamental as those of relativity and quantum mechanics is open to debate. We would argue that the recent developments in nonlinear dynamics and chaos are the (beginnings of the) completion of the "program" of dynamics, the study of how the fundamental forces of nature act together to give us the complex world around us. This program got bogged down when Poincaré, and others realized that, in general, nonlinear dynamics leads to problems not solvable by closed formula expressions, a signal that what we now call chaotic behavior, with its inherent lack of predictability, is possible. Moreover, Poincaré, realized that nonlinearity, not linearity, is the paradigm for most of nature's behavior. Poincaré, however, lacked a computer. More importantly, he lacked computer graphics to explore the geometric aspects of nonlinear dynamics. Although he realized that these geometric aspects are crucial for unraveling complex behavior, he had no way to generate the required pictures.

To reinforce the importance of nonlinear dynamics in the overall scheme of science, we quote Nobel-Prize-winner and quintessential physicist, Richard Feynman:

> [T]here is a physical problem that is common to many fields, that is very old, and that has not been solved. It is not the problem of finding new fundamental particles, but something left over from a long time ago—over a hundred years. Nobody in physics has really been able to analyze it mathematically satisfactorily in spite of its importance to the sister sciences. It is the analysis of circulating or turbulent fluids .... We cannot analyze the weather. We do not know the patterns of motion that there should be inside the earth. [Feynman, Leighton, Sands, 1963], Vol. I, p. 3–9.

It is clear from the context that Feynman had the entire class of nonlinear problems in mind, including turbulence, pattern formation, and what we now call chaos.

In a sense, the study of nonlinear dynamics and chaos needed to wait for the development of quantum mechanics and the consequent understanding of solid-state physics, which permitted the development of fast and readily-available computers. In the meantime, the efforts of mathematicians, particularly in the Russian school, and the progress in understanding statistical mechanics laid the conceptual groundwork for rapid development of nonlinear dynamics in the last two decades or so.

What does constitute at least a minor uprising in physics is the discovery of the universal classes of qualitative and quantitative behavior in nonlinear systems. We have found no hints in the early literature on nonlinear dynamics that this universality was anticipated at all. Even if it was anticipated, that anticipation was highly limited and localized. The scientific community at large was taken by surprise.

Another surprising feature of the "revolution" of chaos has been its applicability to experiments in almost every field of science and engineering. Chaos and other "typical" nonlinear behavior have been found, as we have tried to emphasize throughout this book, in now hundreds of experiments in practically every field of science, engineering, and technology. The theory of nonlinear dynamics and chaos, as rudimentary as it is, has helped us describe, organize, and even quantify much complex behavior. It is this contact with experiment that leads us to believe that the current developments in nonlinear dynamics will make a lasting contribution to our scientific world view.

Of course, there has been much speculation about the applicability of these concepts to phenomena outside the natural sciences and engineering. Here, we are on much softer ground, and the case has yet to be made convincingly that nonlinear dynamics will help us understand complex phenomena such as perception, economics, and sociology. Even analogous developments in literary theory have been pointed out [Hayles, 1991]. In any case, nonlinear dynamics and chaos have enriched our vocabularies, both verbal and conceptual, and have given us new models to help us think about the world around us. The adventure has just begun.

12.7 Further Reading

Chaos and Quantum Mechanics

General References

P. V. Elyatin, "The Quantum Chaos Problem," *Sov. Phys. Usp.* **31**, 597-622 (1988).

B. Eckhardt, "Quantum Mechanics of Classically Non-Integrable Systems," *Phys. Rep.* **163**, 205–97 (1988).

J. Ford, "What Is Chaos That We Should Be Mindful of It?" in P. W. Davies, ed. *The New Physics* (Cambridge University Press, Cambridge, 1989). A thoughtful and provocative essay that reviews many of the issues of quantum chaos (as well as many other issues in classical chaos).

M. C. Gutzwiller, *Chaos in Classical and Quantum Mechanics* (Springer-Verlag, New York, 1990).

J. Ford and M. Ilg, "Eigenfunctions, Eigenvalues, and Time Evolution of Finite Bounded, Undriven, Quantum Systems are not Chaotic," *Phys. Rev. A* **45**, 6165–73 (1992).

J. Ford and G. Mantica, "Does Quantum Mechanics Obey the Correspondence Principle? Is it Complete?" *Am. J. Phys.* **60**, 1086–98 (1992).

M. C. Gutzwiller, "Quantum Chaos," *Scientific American* **266** (1), 78–84 (January, 1992). An introductory essay on the questions raised by quantum chaos.

L. Reichl, *The Transition to Chaos in Conservative Classical Systems: Quantum Manifestations* (Springer-Verlag, New York, 1992).

[Schuster, 1995] Chapter 8.

H.-J. Stockman, *Quantum Chaos, An Introduction* (Cambridge University Press, Cambridge, 1999). Assumes some background in quantum mechanics at the advanced undergraduate or beginning graduate level.

Semiclassical Quantum Mechanics

A. Einstein, "Zum Quantensatz vom Sommerfeld und Epstein," *Verh. Dtsch. Phys. Ges.* **19**, 82–92 (1917). Einstein pointed out the difficulties with "early" quantum theory for nonintegrable systems.

M. Gutzwiller, "Phase-Integral Approximation in Momentum Space and the Bound State of Atoms," *J. Math. Phys.* **8**, 1979–2000 (1967). The connection between semi-classical wave functions and classical periodic orbits.

M. C. Gutzwiller, "Periodic Orbits and Classical Quantization Conditions," *J. Math. Phys.* **12**, 343–58 (1971).

M. V. Berry, "Quantum Scars of Classical Closed Orbits in Phase Space," *Proc. Roy. Soc. London A* **423**, 219–31 (1989).

R. Aurich and F. Steiner, "Periodic-Orbit Sum Rules for the Hadamard-Gutzwiller Model," *Physica D* **39**, 169–93 (1989).

S. A. Meyer, M. Morgenstern, S. Knudson, and D. Noid, "Novel Method for WKB Analysis of Multidimensional Systems," *Am. J. Phys.* **59**, 145–51 (1991). This paper illustrates a simple semiclassical method for finding energy eigenvalues of periodic and quasi-periodic systems.

G. Tanner, P. Scherer, E. B. Bogomolny, B. Eckhardt, and D. Wintgen, "Quantum Eigenvalues from Classical Periodic Orbits," *Phys. Rev. Lett.* **67**, 2410–13 (1991).

Chaos **2** (1) (1992). This issue focuses on periodic orbit theory in semiclassical quantum mechanics.

E. J. Heller and S. Tomsovic, "Postmodern Quantum Mechanics," *Physics Today* **46** (7), 38–46 (1993).

Map Models in Quantum Mechanics

G. Casati, B. V. Chirikov, F. M. Izrailev, and J. Ford, in *Stochastic Behavior in Classical and Hamiltonian Systems*, G. Casati and J. Ford, eds. Lecture Notes in Physics, Vol. 93 (Springer-Verlag, New York, 1979). Kicked rotator model (Chirikov model).

J. H. Hannay and M. V. Berry, "Quantization of Linear Maps on a Torus—Fresnel Diffraction by a Periodic Grating," *Physica D* **1**, 267–90 (1980). The application of the Arnold cat map to a problem in quantum mechanics.

G. H. Ristow, "A Quantum Mechanical Investigation of the Arnold Cat Map," Master's Thesis, School of Physics, Georgia Institute of Technology, 1987.

S. Adachi, M. Toda, and K. Ikeda, "Quantum-Classical Correspondence in Many-Dimensional Quantum Chaos," *Phys. Rev. Lett.* **61**, 659–61 (1988). Coupled kicked-rotors show effects closer to those of classical chaos.

R. E. Prange and S. Fishman, "Experimental Realizations of Kicked Quantum Chaotic Systems," *Phys. Rev. Lett.* **63**, 704–7 (1989). An experiment (modes in optical fibers) described by the kicked rotator model.

J. Ford, G. Mantica, and G. H. Ristow, "The Arnol'd Cat: Failure of the Correspondence Principle," *Physica D* **50**, 493–520 (1991).

Wigner Phase Space Distribution

E. Wigner, "On the Quantum Correction for Thermodynamic Equilibrium," *Phys. Rev.* **40**, 749–59 (1932). The Wigner distribution function plays for quantum mechanics the role played by a phase space probability distribution in classical mechanics.

R. F. O'Connell, "The Wigner Distribution Function—50th Birthday," *Found. Phys.* **13**, 83–93 (1983).

Y. S. Kim and E. P. Wigner, "Canonical Transformations in Quantum Mechanics," *Am. J. Phys.* **58**, 439–48 (1990). A nice introduction to the Wigner distribution.

W. A. Lin and L. E. Ballentine, "Quantum Tunneling and Chaos in a Driven Anharmonic Oscillator," *Phys. Rev. Lett.* **65**, 2927–30 (1990). Another phase space distribution called the Husimi function is also used to compare classical and quantal behavior.

Energy Level Distributions

G. Casati, B. V. Chirikov, and I. Guarneri, "Energy-Level Statistics of Integrable Quantum Systems," *Phys. Rev. Lett.* **54**, 1350–53 (1985). Exactly solvable quantum systems do not always have a Poisson distribution of energy levels.

J. Ford, "Quantum Chaos, Is There Any?" in [Hao, 1988], pp. 128–47.

J. V. José, "Quantum Manifestations of Classical Chaos: Statistics of Spectra," in [Hao, 1988]. A good review of the issues concerning energy level statistics as a symptom of classical chaos in quantum mechanics.

F. M. Izrailev, "Simple Models of Quantum Chaos: Spectrum and Eigenfunctions," *Phys. Rep.* **196**, 299–392 (1990).

M. Shapiro and G. Goelman, "Onset of Chaos in an Isolated Energy Eigenstate," *Phys. Rev. Lett.* **53**, 1714–7 (1984). Assesses chaotic behavior of a quantum system in terms of the correlation function of the wave function.

J. Zakrzewski, K. Dupret, and D. Delande, "Statistical Properties of Energy Levels of Chaotic Systems: Wigner or Non-Wigner," *Phys. Rev. Lett.* **74**, 522–25 (1995).

R. Ketzmerick, K. Kruse, S. Kraut, and T. Geisel, "What Determines the Spreading of a Wave Packet?," *Phys. Rev. Lett.* **79**, 1959–63 (1997). The generalized dimension D_2 is applied to the energy spectrum and eigenfunctions. The ratio determines how rapidly wave packets spread.

Atomic Physics

Atomic physics provides an ideal testing ground for the notions of quantum chaos because calculations can be done in many cases to high accuracy and high precision experiments on carefully controlled systems are its forte.

Two very readable general reviews are given in

R. V. Jensen, "Chaos in Atomic Physics," in *Atomic Physics 10*, H. Narumi and I. Shimamura, eds. (Elsevier North-Holland, Amsterdam, 1987).

R. V. Jensen, "The Bohr Atom Revisited: A Test Case for Quantum Chaos," *Comments At. Mol. Phys.* **25**, 119–31 (1990).

Comments At. Mol. Phys. **25** (1-6), 1–362 (1991). An entire volume dedicated to the question of "Irregular Atomic Systems and Quantum Chaos."

K. A. H. van Leeuwen, et al. "Microwave Ionization of Hydrogen Atoms: Experiment versus Classical Dynamics," *Phys. Rev. Lett.* **55**, 2231–34 (1985). Highly excited hydrogen atoms ionized by microwave oscillating fields have been a test bed for both theory and experiment dealing with quantum chaos.

E. J. Galvex, B. E. Sauer, L. Moorman, P. M. Koch, and D. Richards, "Microwave Ionization of H Atoms: Breakdown of Classical Dynamics for High Frequencies," *Phys. Rev. Lett.* **61**, 2011–14 (1988).

J. E. Bayfield, G. Casati, I. Guarneri, and D. W. Sokol, "Localization of Classically Chaotic Diffusion for Hydrogen Atoms in Microwave Fields," *Phys. Rev. Lett.* **63**, 364–67 (1989).

R. V. Jensen, M. M. Sanders, M. Saraceno, and B. Sundaram, "Inhibition of Quantum Transport Due to 'Scars' of Unstable Periodic Orbits," *Phys. Rev. Lett.* **63**, 2771–15 (1989).

R. V. Jensen, S. M. Susskind, and M. M. Sanders, "Chaotic Ionization of Highly Excited Hydrogen Atoms: Comparison of Classical and Quantum Theory with Experiment," *Physics Reports* **201**, 1–56 (1991).

Single electron atoms in strong magnetic fields have provided an impressive test of quantum predictions in a regime for which the corresponding classical system is nonintegrable and shows chaotic behavior. The quantum mechanical calculations are described are in the following papers:

M. L. Du and J. B. Delos, "Effect of Closed Classical Orbits on Quantum Spectra: Ionization of Atoms in a Magnetic Field," *Phys. Rev. Lett.* **58**, 1731– 33 (1987).

P. Leboeuf, J. Kurchan, M. Feingold, and D. P. Arovas, "Phase-Space Localization: Topological Aspects of Quantum Chaos," *Phys. Rev. Lett.* **65**, 3076– 79 (1990). Some further insight into the reasons for phase-space localization.

D. Delande, A. Bommier, and J. C. Gay, "Positive-Energy Spectrum of the Hydrogen Atom in a Magnetic Field," *Phys. Rev. Lett.* **66**, 141–44 (1991).

C.-H. Iu, G. R. Welch, M. M. Kasch, D. Kleppner, D. Delande, and J. C. Gay, "Diamagnetic Rydberg Atom: Confrontation of Calculated and Observed," *Phys. Rev. Lett.* **66**, 145–48 (1991). The corresponding experiment.

F. L. Moore, J. C. Robinson, C. Bharucha, P. E. Williams, and M. G. Raizen, "Observation of Dynamical Localization in Atomic Momentum Transfer: A New Testing Ground for Quantum Chaos," *Phys. Rev. Lett.* **73**, 2974–77 (1994).

J. C. Robinson, C. Bharucha, F. L. Moore, R. Jahnke, G. A. Georgakis, Q. Niu, and M. G. Raizen, and B. Sundaram, "Study of Quantum Dynamics in the Transition from Classical Stability to Chaos," *Phys. Rev. Lett.* **74**, 3963–66 (1995).

M. Latka and B. J. West, "Nature of Quantum Localization in Atomic Momentum Transfer Experiments," *Phys. Rev. Lett.* **75**, 4202–5 (1995).

R. Blümel and W. P. Reinhardt, *Chaos in Atomic Physics* (Cambridge University Press, New York, 1997).

M. Raizen, "Quantum Chaos and Cold Atoms," in *Advances in Atomic, Molecular, and Optical Physics* **41**, 43–81. B. Bederson and H. Walther, eds. (Academic Press, San Diego, 1999).

Other Areas of Quantum Physics

J. P. Pique, Y. Chen, R. W. Field, and J. L. Kinsey, "Chaos and Dynamics on 0.5–300 ps Time Scales in Vibrationally Exited Acetylene: Fourier Transform of Stimulated-Emission Pumping Spectrum," *Phys. Rev. Lett.* **58**, 475– 78 (1987). Chaotic effects in molecular energy levels.

M. Schreiber and H. Grussbach, "Multifractal Wave Functions at the Anderson Transition," *Phys. Rev. Lett.* **67**, 607–10 (1991). A nice example of the application of the $\tau(q)$ and D_q formalism of Chapter 10 to describe quantum probability distributions.

Chaotic effects may show up in nuclear physics. For example:

G. E. Mitchell, E. G. Bipluch, P. M. Endt, and J. F. Shringer, Jr., "Broken Symmetries and Chaotic Behavior in ^{26}Al," *Phys. Rev. Lett.* **61**, 1473– 76 (1988).

O. Bohigas and H. A. Weidenmller, "Aspects of Chaos in Nuclear Physics," *Ann. Rev. Nucl. Part. Science* **38**, 421–53 (1988).

S. Berg, "Onset of Chaos in Rapidly Rotating Nuclei," *Phys. Rev. Lett.* **64**, 3119–22 (1990).

So-called spin systems provide another useful testing ground for the notions of quantum chaos, in particular, the search for thermodynamic behavior in quantum systems. For a readable introduction to these issues, see

G. Müller, "Nature of Quantum Chaos in Spin Systems," *Phys. Rev. A* **34**, 3345–55 (1986).

N. G. van Kampen, in *Chaotic Behavior in Quantum Systems. Theory and Applications.* G. Casati, ed. (Plenum, NY, 1985). The general issue of finding an explanation for quantum statistical mechanics.

S. Weinberg, "Testing Quantum Mechanics," *Annals of Physics* **194**, 336–86 (1989). A proposed nonlinear generalization of quantum mechanics.

J. Wilkie and P. Brumer, "Time-Dependent Manifestations of Quantum Chaos," *Phys. Rev. Lett.* **67**, 1185–88 (1991). Chaotic effects may have time-dependent manifestations in quantum mechanics.

The following two papers provide a nice introduction to quantum chaos and various ways of visualizing how chaos shows up in quantum systems.

N. Srivastava, C. Kaufman, and G. Müller, "Hamiltonian Chaos III," *Computers in Physics* **6**, 84–88 (1992).

N. Regez, W. Breyman, S. Weigert, C. Kaufman, and G. Müller, "Hamiltonian Chaos IV," *Computers in Physics* **10**, 39–45 (1996).

Quantum optics has been a fertile field for nonlinear dynamics. For a good introduction see

J. R. Ackerhalt, P. W. Milonni, and M. L. Shih, "Chaos in Quantum Optics," *Phys. Rep.* **128**, 205–300 (1985).

Two issues of the Journal of the Optical Society of America have been devoted to nonlinear dynamics in optics:

"Instabilities in Active Optical Media," *J. Opt. Soc. Am. B* **2** (1) (January, 1985).

N. B. Abraham, L. A. Lugiato, and L. M. Narducci, "Overview of Instabilities in Laser Systems," *J. Opt. Soc. Am. B* **2**, 7–13 (1985). This article is particularly worth reading.

"Nonlinear Dynamics of Lasers," *J. Opt. Soc. Am. B* **5** (5) (May, 1988).

Piece-Wise Linear and Delay-Differential Models

R. W. Rollins and E. R. Hunt, "Exactly Solvable Model of a Physical System Exhibiting Universal Chaotic Behavior," *Phys. Rev. Lett.* **49**, 1295–98 (1982).

J. N. Schulman, "Chaos in Piecewise-Linear Systems," *Phys. Rev. A* **28**, 477–79 (1983).

E. R. Hunt and R. W. Rollins, "Exactly Solvable Model of a Physical System Exhibiting Multidimensional Chaotic Behavior," *Phys. Rev. A* **29**, 1000–2 (1984).

T. Matsumoto, L. O. Chua, and S. Tanaka, "Simplest Chaotic Nonautonomous Circuit," *Phys. Rev.* **30**, 1155–57 (1984).

S. Banerjee, J. A. Yorke, and C. Grebogi, "Robust Chaos," *Phys. Rev. Lett.* **80** 3049–52 (1998). A piecewise smooth system can have parameter intervals that have no periodic orbits, only chaotic orbits.

M. Dutta, H. E. Nusse, E. Ott, J. A. Yorke, and G. Yuan, "Multiple Attractor Bifurcations: A Source of Unpredictability in Piecewise Smooth Systems," *Phys. Rev. Lett.* **83**, 4281–4284 (1999).

J. D. Farmer, "Chaotic Attractors of an Infinite-Dimensional Dynamical System," *Physica D* **4**, 366–93 (1982). Discusses delay-differential models and their Lyapunov exponents and fractal dimensions.

[Kaplan and Glass, 1995] pp. 183–188 has a nice introduction to time-delay models.

G. Giacomelli and A. Politi, "Relationship between Delayed and Spatially Extended Dynamical Systems," *Phys. Rev. Lett.* **76** 2686–89 (1996).

Stochastic Resonance

F. Moss and K. Wiesenfeld, "The Benefits of Background Noise," *Scientific American* **273** (3) 66–69 (1995).

A. R. Bulsara and L. Gammaitoni, "Tuning in to Noise," *Physics Today* **49** (3), 39–45 (1996).

E. Simonott, M. Riani, C. Seife, M. Roberts, J. Twitty, and F. Moss, "Visual Perception of Stochastic Resonance," *Phys. Rev. Lett.* **78**, 1186–89 (1997).

L. Gammaitoni, P. Hänggi, P. Jung, F. Marchesoni, "Stochastic Resonance," *Rev. Mod. Phys.* **70**, 223–87 (1998).

K. Richardson, T. Imhoff, P. Grigg, and J. J. Collins, "Using electrical noise to enhance the ability of humans to detect subthreshold mechanical cutaneous stimuli," *Chaos* **8**, 599–603 (1998).

Algorithmic Complexity and Information Theory

R. Baierlein, *Atoms and Information Theory* (W. H. Freeman, San Francisco,1971). A nice introduction to information theory in the context of physics.

G. J. Chaitin, "Randomness and Mathematical Proof," *Scientific American* **232** (5), 47–52 (1975).

V. M. Alekseev and M. V. Yakobson "Symbolic Dynamics and Hyperbolic Dynamic Systems," *Phys. Rep.* **75**, 287–325 (1981). The first few pages of this paper provide a good introduction to algorithmic complexity in chaotic systems. The rest of the paper is quite sophisticated mathematically.

J. Ford, "How Random is a Coin Toss," *Physics Today* **36** (4), 40–47 (1983). See also, J. Ford in [Hao, 1988], pp. 139–47.

G. J. Chaitin, *Algorithmic Information Theory* (Cambridge University Press, Cambridge, 1987).

A. Shudo, "Algorithmic Complexity of the Eigenvalue Sequence of a Nonintegrable Hamiltonian System," *Phys. Rev. Lett.* **63**, 1897–901 (1989). Argues that algorithmic complexity is not the appropriate measure for quantum chaos.

Several interesting papers on the use of concepts from nonlinear dynamics and the possibility of chaos in computer networks:

B. A. Huberman and T. Hogg, "Adaptation and Self-Repair in Parallel Computing Structures," *Phys. Rev. Lett.* **52**, 1048–51 (1984).

B. A. Huberman, "Computing with Attractors: From Self-repairing Computers, to Ultradiffusion, and the Application of Dynamical Systems to Human Behavior," in *Emerging Syntheses in Science*, D. Pines, ed. Vol. I, (Addison–Wesley, Redwood City, CA, 1988).

B. A. Huberman, "An Ecology of Machines, How Chaos Arises in Computer Networks," *The Sciences* (New York Academy of Sciences), 38–44 (July/August, 1989).

Controlling Chaos

E. Ott, C. Grebogi, and J. A. Yorke, "Controlling Chaos," *Phys. Rev. Lett.* **64**, 1196–99 (1990).

· W. L. Ditto, S. N. Rauseo, and M. L. Spano, "Experimental Control of Chaos," *Phys. Rev. Lett.* **65**, 3211–14 (1990).

T. Shinbrot, E. Ott, C. Grebogi, and J. A. Yorke, "Using Chaos to Direct Trajectories to Targets," *Phys. Rev. Lett.* **65**, 3215–18 (1990).

Y. Braiman and I. Goldhirsch, "Taming Chaotic Dynamics with Weak Periodic Perturbations," *Phys. Rev. Lett.* **66**, 2545–48 (1991).

L. Pecora and T. Carroll, "Pseudoperiodic Driving: Eliminating Multiple Domains of Attraction Using Chaos," *Phys. Rev. Lett.* **67**, 945–48 (1991).

E. R. Hunt, "Stabilizing High-Period Orbits in a Chaotic System: The Diode Resonator," *Phys. Rev. Lett.* **67**, 1953–55 (1991).

U. Dressler and G. Nitsche, "Controlling Chaos Using Time Delay Coordinates," *Phys. Rev. Lett.* **68**, 1–4 (1992).

R. Roy, T. Murphy, T. Maier, Z. Gills, and E. R. Hunt, "Dynamical Control of a Chaotic Laser: Experimental Stabilization of a Globally Coupled System," *Phys. Rev. Lett.* **68**, 1259–62 (1992).

T. Shinbrot, W. Ditto, C. Grebogi, E. Ott, M. Spano, and J. A. Yorke, "Using the Sensitive Dependence of Chaos (the "Butterfly Effect") to Direct Trajectories in an Experimental Chaotic System," *Phys. Rev. Lett.* **68**, 2863–66 (1992).

Z. Gills, C. Iwata, R. Roy, I. Schwartz, and I. Triandof, "Tracking Unstable Steady States: Extending the Stability Regime of a Multimode Laser System," *Phys. Rev. Lett.* **69**, 3169–72 (1992).

D. Auerback, C. Grebogi, E. Ott, and J. A. Yorke, "Controlling Chaos in High Dimensional Systems," *Phys. Rev. Lett.* **69**, 3479–82 (1992).

W. L. Ditto and L. Pecora, "Mastering Chaos," *Scientific American* **269** (2), 78–84 (1993).

T. Shinbrot, C. Grebogi, E. Ott, and J. A. Yorke, "Using small perturbations to control chaos," *Nature* **363**, 411–17 (1993).

D. J. Gauthier, D. W. Sukow, H. M. Concannon, and J. E. S. Socolar, "Stablizing unstable periodic orbits in a fast diode resonator using continuous time-delay autosynchronization," *Phys. Rev. E* **50**, 2343–46 (1994).

E. Ott and M. Spano, "Controlling Chaos," *Physics Today* **48** (5), 34–40 (1995).

T. Kapitaniak, *Controlling Chaos* (Academic Press, San Diego, 1996). An introduction to controlling chaos with a collection of thirteen reprints.

K. Hall, D. J. Christini, M. Tremblay, J. J. Collins, L. Glass, and J. Billette, "Dynamic Control of Cardiac Alternans," *Phys. Rev. Lett.* **78**, 4518–21 (1997).

W. L. Ditto and K. Showalter, "Introduction: Control and synchronization of chaos," *Chaos* **7**, 509–11 (1997). Lead article in an issue devoted to control and synchronization of chaos.

D. J. Christini and J. J. Collins, "Control of chaos in excitable physiological systems: A geometric analysis," *Chaos* **7**, 544–49 (1997).

K. Myneni , T. A. Barr, N. J. Corron, S. d. Pethel, "New Method for the Control of Fast Chaotic Oscillations," *Phys. Rev. Lett.* **83**, 2175–78 (1999).

H. G. Schuster, *Handbook of Chaos Control* (Wiley, New York, 1999). Gives an introduction to chaos theory and discusses a wide range of chaos control techniques.

Synchronizing Chaos

L. Pecora and T. Carroll, "Driving Systems with Chaotic Signals," *Phys. Rev. A* **44**, 2374–83 (1991). Chaotic systems can be synchronized with suitable interconnections.

K. M. Cuomo and A. V. Oppenheim, "Circuit Implementation of Synchronized Chaos with Applications to Communications," *Phys. Rev. Lett.* **71**, 65–68 (1993).

S. H. Strogatz and I. Stewart, "Coupled Oscillators and Biological Synchronization," *Scientific American* **269** (6), 102–109 (1993).

T. C. Newell, P. M. Alsing, A. Gavrielides, and V. Kovanis, "Synchronization of Chaotic Diode Resonators by Occasional Proportional Feedback," *Phys. Rev. Lett.* **72**, 1647–50 (1994).

L. Kocarev and U. Parlitz, "General Approach for Chaotic Synchronization with Applications to Communication," *Phys. Rev. Lett.* **74**, 5028–31 (1995).

Y. Braiman, J. F. Lindner, annd W. L. Ditto, "Taming spatiotemporal chaos with disorder," *Nature* **378**, 465–67 (1995). Adding noise can help synchronize oscillators.

See DIS97 for a review article and an issue devoted to synchronization and control.

R. Fitzgerald, "Phase Synchronization May Reveal Communication Pathways in Brain Activity," *Physics Today* **52** (3), 17–19 (1999).

A. Neiman, L. Schimanksy-Geier, F. Moss, B. Shulgin, and J. J. Collins, "Synchronization of noisy systems by stochastic signals," *Phys. Rev. E* **60**, 284–92 (1999).

A. Neiman, L. Schimanksy-Geier, A. Cornell-Bell, and F. Moss, "Noise-Enhanced Phase Synchronization in Excitable Media," *Phys. Rev. Lett.* **83**, 4896–99 (1999).

Forecasting with Nonlinear Dynamics and Chaos

J. D. Farmer and J. J. Sidorowich, "Predicting Chaotic Time Series," *Phys. Rev. Lett.* **59**, 845–48 (1987).

M. Casdagli, "Nonlinear Predictions of Chaotic Time Series," *Physica D* **35**, 335–56 (1989).

J. D. Farmer and J. J. Sidorowich, "Optimal Shadowing and Noise Reduction," *Physica D* **47**, 373–92 (1991).

C. G. Schroer, T. Sauer, E. Ott, and J. A. Yorke, "Predicting Chaos Most of the Time from Embeddings with Self-Intersections," *Phys. Rev. Lett.* **80**, 1410–13 (1998).

Applications of Nonlinear Dynamics and Chaos in Other Fields

Acoustics

W. Lauterborn and U. Parlitz, "Methods of chaos physics and their application to acoustics," *J. Acoust. Soc. Am.* **84**, 1975–93 (1988).

D. Crighton and L. A. Ostrovsky, "Introduction to acoustical chaos," *Chaos* **5** (3), 495 (1995). Introduction to a focus issue on acoustical chaos.

D. M. Campbell, "Nonlinear dynamics of musical reed and brass wind instruments," *Contemporary Physics* **40**, 415–431 (1999).

Astrophysics

G. J. Sussman and J. Wisdom, "Chaotic Evolution of the Solar System," *Science* **257**, 56–62 (1992). A special computer was used to track the trajectories of the planets for the equivalent of 100 million years. The solar system has a positive Lyapunov exponent with $1/\lambda = 4$ million years.

J. R. Buchler, T. Serre, and Z. Kolláth, "A Chaotic Pulsing Star: The Case of R Scuti," *Phys. Rev. Lett.* **73**, 842–45 (1995).

W. Vieira and P. Letelier, "Chaos around a Hénon-Heiles-Inspired Exact Perturbation of a Black Hole," *Phys. Rev. Lett.* **76**, 1409–12 (1996).

M. Paluš and D. Novotná, "Sunspot Cycle: A Driven Nonlinear Oscillator?" *Phys. Rev. Lett.* **83**, 3406–9 (1999).

Biomedical

B. J. West, *Fractal Physiology and Chaos in Medicine* (World Scientific Publishing, River Edge, NJ, 1990).

L. Glass, "Nonlinear dynamics of physiological function and control," *Chaos* **1**, 247–50 (1991). Lead article for an issue devoted to nonlinear dynamics of physiological control.

F. Witkowski, K. Kavanagh, P. Penkoske, R. Plonsey, M. Spano, W. Ditto, and D. Kaplan, "Evidence for Determinism in Ventricular Fibrillation," *Phys. Rev. Lett.* **75**, 1230–33 (1995).

J. Bélair, L. Glass, U. an der Heiden, and J. Milton, "Dynamical disease: Identification, temporal aspects and treatment strategies," *Chaos* **5** (1), 1–7 (1995). Lead-off article for focus issue on dynamical disease: mathematical analysis of human illness.

L. Glass, "Dynamics of Cardiac Arrhythmias," *Physics Today* **49** (8), 40–45 (1996).

A. Mandell and K. Selz, "Entropy conservation as $h_{T\mu} \approx \bar{\lambda}_\mu^+ d_\mu$ in neurobiological dynamical systems," *Chaos* **7**, 67–81 (1997). Application of generalized entropy and Lyapunov exponents can be used to characterize brain activity.

P. Muruganandam and M. Lakshmanan, "Bifurcation analysis of the travelling waveform of FitzHugh-Nagumo nerve conduction model equation," *Chaos* **7**, 476–87 (1997).

R. Larter, B. Speelman, and R. W. Worth, "A coupled ordinary differential equation lattice model for the simulation of epileptic seizures," *Chaos* **9**, 795–804 (1999).

Chemical Engineering

[Ottino, 1989]
Chaos **9** (1) (1999). An issue devoted to chaos in chemical engineering.

Chemistry

S. K. Scott, *Oscillations, Waves, and Chaos in Chemical Kinetics* (Oxford University Press, New York, 1994).

W. G. Rothschild, *Fractals in Chemistry* (Wiley, New York, 1998).

Irving R. Epstein and John A. Pojman, *An Introduction to Nonlinear Chemical Dynamics: Oscillations, Waves, Patterns, and Chaos* (Oxford University Press, New York, 1998).

M. Dolnik, T. Gardner, I. Epstein, and J. Collins, "Frequency Control of an Oscillatory Reaction by Reversible Binding of an Autocatalyst," *Phys. Rev. Lett.* **82**, 1582–85 (1999).

I. R. Epstein and J. A. Pojman, "Overview: Nonlinear dynamics related to polymeric systems," *Chaos* **9**, 255–59 (1999). The lead article for an issue devoted to nonlinear dynamics of polymer systems.

Condensed Matter Physics

Philip E. Wigen, ed., *Nonlinear Phenomena and Chaos in Magnetic Materials* (World Scientific, Singapore, New Jersey, London, Hong Kong, 1994).

Cryptography and Communications

E. Bollt, Y.-C. Lai, and C. Grebogi, "Coding, Channel Capacity, and Noise Resistance in Communicating with Chaos," *Phys. Rev. Lett.* **79**, 3787–90 (1997).

M. S. Baptista, "Cryptography with chaos," *Phys. Lett. A* **240**, 50–54 (1998).

G. VanWiggeren and R. Roy, "Optical Communication with Chaotic Waveforms," *Phys. Rev. Lett.* **81**, 3547–50 (1998).

G. VanWiggeren and R. Roy, "Communication with Chaotic Lasers," *Science* **279**, 1198–1200 (1998). Perhaps high–dimensional chaotic systems can be used to encrypt messages.

J.-P. Goedgebuer, L. Larger, and H. Porte, "Optical Cryptosystem Based on Synchronization of Hyperchaos Generated by a Delayed Feedback Tunable Laser Diode," *Phys. Rev. Lett.* **80**, 2249–52 (1998).

J. Geddes, K. Short, and K. Black, "Extraction of Signals from Chaotic Laser Data," *Phys. Rev. Lett.* **83**, 5389–92 (1999). At least for the laser system of VAR98a,b the encryption can be circumvented.

Ecology

S. Ellner and P. Turchin, "Chaos in a Noisy World: New Methods and Evidence from Time-Series Analysis," *The American Naturalist* **145**, 343–75 (1995). Applications to population dynamics in ecology.

D. Zanette and S. Manrubia, "Role of Intermittency in Urban Development: A Model of Large-Scale City Formation," *Phys. Rev. Lett.* **79**, 523–26 (1997).

J. Maron and S. Harrison, "Spatial Pattern Formation in an Insect Host-Parasitoid System," *Science* **278**, 1619–21 (1997).

P. Rohani, T. Lewis, D. Grünbaum, and G. Ruxton, "Spatial self-organization in ecology: pretty patterns or robust reality," *TREE* **12** (2), 70–74 (1997).

Economics

W. A. Brooks and C. L. Sayers, "Is the Business Cycle Characterized by Deterministic Chaos?" *J. of Monetary Economics* **22**, 71–90 (1988).

W. J. Baumol and J. Benhabib, "Chaos: Significance, Mechanism, and Economic Applications," *J. Econ. Perspectives* **3**, 77–105 (1989).

C. L. Sayers, "Statistical Inference Based Upon Nonlinear Science," *European Economic Review* **35**, 306–12 (1991).

E. E. Peters, *Chaos and Order in the Capital Markets* (Wiley, New York, 1991).

R. Palmer, W. B. Arthur, J. Holland, B. LeBaron, and P. Tayler, "Artificial economic life: a simple model of a stockmarket," *Physica D* **75**, 264–74 (1994).

B. B. Mandelbrot, *Fractals and Scaling in Finance: Discontinuity, Concentration, Risk* (Springer-Verlag, New York, 1997).

Y. Lee, L. Amaral, D. Canning, M. Meyer and H. E. Stanley, "Universal Features in the Growth Dynamics of Complex Organizations," *Phys. Rev. Lett.* **81**, 3275–78 (1998).

Rosario N. Mantegna and H. Eugene Stanley, *An Introduction to Econophysics* (Cambridge University Press, New York, 1999).

Electrical Engineering

Proc. IEEE, August (1987). A special issue on Chaotic Systems with an emphasis on electrical systems.

Many electrical circuits show chaotic behavior. See the list of references for simple experiments at the end of this section.

Geosciences

R. Pool, "Ecologists Flirt with Chaos," *Science* **243**, 310–13 (1989).

C. H. Scholz and B. B. Mandelbrot, eds. *Fractals in Geophysics* (Birkhauser Verlag, Basel and Boston, 1989).

M. Fleischmann and D. J. Teldesley, eds. *Fractals in the Natural Sciences* (Princeton University Press, Princeton, NJ, 1990).

Geophys. Res. Lett. 18 (8) (1991) has a special section: "Chaos and Stochasticity in Space Plasmas."

N. Lam and L. De Cola, *Fractals in Geography* (PTR Prentice-Hall, 1993).

A. Sharma, "Assessing the magnetoshpere's nonlinear behavior: Its dimension is low, its predictability, high," *Rev. Geophysics Suppl.* 645–50 (July, 1995).

D. L. Turcotte, *Fractals and Chaos in Geology and Geophysics*, 2nd ed. (Cambridge University Press, 1997).

Literary Theory

Chaos and Order: Complex Dynamics in Literature and Science. N. Katherine Hayles, ed. (University of Chicago Press, Chicago, 1991).

Mechanical Engineering

T. Burns and M. Davies, "Nonlinear Dynamics Model for Chip Segmentation in Machining," *Phys. Rev. Lett.* **79**, 447–50 (1997).

C. Caravati, F. Delogu, G. Cocco, and M. Rustici, "Hyperchaotic qualities of the ball motion in a ball milling device," *Chaos* **9**, 219–26 (1999).

Simple Nonlinear Dynamics and Chaos Experiments

E. V. Mielczarek, J. S. Turner, D. Leiter, and L. Davis, "Chemical Clocks: Experimental and Theoretical Models of Nonlinear Behavior," *Am. J. Phys.* **51**, 32–42 (1983). Chemical reactions show chaotic behavior.

R. Shaw, *The Dripping Faucet as a Model Chaotic System* (Aerial Press, Santa Cruz, CA, 1984). A dripping faucet may exhibit chaotic behavior.

R. Van Buskirk and C. Jeffries, "Observation of Chaotic Dynamics of Coupled Oscillators," *Phys. Rev.* **31**, 3332–57 (1985). Semiconductor diodes show many aspects of nonlinear behavior.

P. Martien, S. C. Pope, P. L. Scott, and R. S. Shaw, "The Chaotic Behavior of the Leaky Faucet," *Phys. Lett. A* **110**, 399–404 (1985).

H. Meissner and G. Schmidt, "A Simple Experiment for Studying the Transition from Order to Chaos," *Am. J. Phys.* **54**, 800–4 (1986). Various experiments with magnets driven by oscillating currents.

N. B. Tufillaro and A. M. Albano, "Chaotic Dynamics of a Bouncing Ball," *Am. J. Phys.* **54**, 939–44 (1986).

T. M. Mello and N. B. Tufillaro, "Strange Attractors of a Bouncing Ball," *Am. J. Phys.* **55**, 316–20 (1987).

K. Briggs, "Simple Experiments in Chaotic Dynamics," *Am. J. Phys.* **55**, 1083–89 (1987). A survey of various electronic and mechanical experiments.

R. Landauer, "Nonlinearity: Historical and Technological Review," in *Nonlinearity in Condensed Matter*, A. R. Bishop, D. K. Campbell, P. Kumar, and S. E. Trullinger, eds. (Springer-Verlag, Berlin, Heidelberg, New York, London, Paris, Tokyo, 1987). Describes a simple chaotic neon-bulb experiment. Lots of interesting historical details of nonlinearity in technology.

M. J. Ballico, M. L. Sawley, and F. Skiff, "The Bipolar Motor: A Simple Demonstration of Deterministic Chaos," *Am. J. Phys.* **58**, 58–61 (1990).

R. F. Cahalan, H. Leidecker, and G. D. Cahalan, "Chaotic Rhythms of a Dripping Faucet," *Computers in Physics* **4**, 368–83 (1990).

A. Ojha, S. Moon, B. Hoeling, and P. B. Siegel, "Measurements of the Transient Motion of a Simple Nonlinear System," *Am. J. Phys.* **59**, 614–19 (1991).

K. Dreyer and F. R. Hickey, "The Route to Chaos in a Dripping Faucet," *Am. J. Phys.* **59**, 619–27 (1991).

M. T. Levinson, "The Chaotic Oscilloscope," *Am. J. Phys.* **61**, 155–165 (1993). An oscilloscope coupled with a photodiode shows chaotic behavior.

B. K. Clark, R. F. Martin, Jr., R. J. Moore, and K. E. Jesse, "Fractal dimension of the strange attractor of the bouncing ball circuit," *Am. J. Phys.* **63**, 157–63 (1995).

Thomas L. Carroll, "A simple circuit for demonstrating regular and synchronized chaos," *Am. J. Phys.* **63**, 377–379 (1995).

Th. Pierre, G. Bonhomme, and A. Atipo, "Controlling the Chaotic Regime of Nonlinear Ionization Waves using the Time-Delay Autosynchronization Method," *Phys. Rev. Lett.* **76**, 2290–93 (1996).

Stephen J. Van Hook and Michael F. Schatz, "Simple Demonstrations of Pattern Formation," *The Physics Teacher* **35**, 391–95 (1997).

A. Siahmakoun, V. A. French, and J. Patterson, "Nonlinear dynamics of a sinusoidally driven pendulum in a repulsive magnetic field," *Am. J. Phys.* **65** (5), 393–400 (1997).

E. Lanzara, R. N. Mangegna, B. Spagnolo, and R. Zangara, "Experimental study of a nonlinear system in the presence of noise: The stochastic resonance," *Am. J. Phys.* **65**, 341–49 (1997). A tunnel diode circuit is used to study stochastic resonance: adding noise can help increase the signal-to-noise ratio in certain nonlinear systems.

J. E. Berger and G. Nunes, Jr., "A mechanical Duffing oscillator for the undergraduate laboratory," *Am. J. Phys.* **65**, 841–846 (1997).

[Epstein and Pojman, 1998]. The two appendices describes some nice demonstrations and the design of experiments for undergraduate laboratories.

W. L. Shew, H. A. Coy, and J. F. Lindner, "Taming chaos with disorder in a pendulum array," *Am. J. Phys.* **67**, 703–708 (1999). How to construct an array of ten pendulums and show that disorder can lead to synchronization.

Appendix A

Fourier Power Spectra

A.1 Introduction and Basic Definitions

In this appendix we give a brief introduction to the methods of Fourier analysis and synthesis and the resulting Fourier power spectra. For more mathematical details, a good reference at this level of treatment is Mary L. Boas, *Mathematical Methods for the Physical Sciences*, 2nd ed. (Wiley, New York, 1983).

The crucial notion of Fourier analysis is contained in Fourier's Theorem, which allows us to decompose any periodic function into a series of sine and cosine functions.

Definition: A periodic function with period T satisfies $f(t+T) = f(t)$ for all t.

Fourier's Theorem: Any periodic function with period T can be written as

$$f(t) = \frac{a_0}{2} + \sum_{n=1}^{\infty} a_n \cos(n\omega_0 t) + \sum_{n=1}^{\infty} b_n \sin(n\omega_0 t) \tag{A.1-1}$$

where $\omega_0 = 2\pi/T$.

Fourier's Theorem says that we can express a periodic function as a sum of a constant term and a series of cosine and sine terms, where the frequencies associated with the sines and cosines are integer multiples ("harmonics") of the "fundamental frequency." (We shall see later why the 2 is included in the denominator of the a_0 term.)

Given a periodic function $f(t)$, how do we find the "Fourier amplitudes" (or "Fourier coefficients") a_n and b_n? We find the coefficients by using what is sometimes called "Fourier's trick": To determine the coefficient a_m, we multiply both sides of Eq. (A.1-1) by $\cos(m\omega_0 t)$, where m is a positive integer, and then integrate with respect to t from $t = 0$ to $t = T$:

$$\int_0^T f(t) \cos(m\omega_0 t) dt = \int_0^T dt \frac{a_0}{2} \cos(m\omega_0 t) + \sum_{n=1}^{\infty} a_n \int_0^T dt \cos(m\omega_0 t) \cos(n\omega_0 t)$$

$$+ \sum_{n=1}^{\infty} b_n \int_0^T dt \cos(m\omega_0 t) \sin(n\omega_0 t) \tag{A.1-2}$$

All of the integrals on the right-hand side of Eq.(A.1-2) are 0 except the second when $m = n$. (We say that the sines and cosines with different m and n are "orthogonal with respect to integration over the interval 0 to T.") After carrying out a similar procedure by multiplying by $\sin(m\omega_0 t)$, we have

$$a_n = \frac{2}{T}\int_0^T dt\, f(t)\cos(m\omega_0 t)$$

(A.1-3)

$$b_n = \frac{2}{T}\int_0^T dt\, f(t)\sin(m\omega_0 t)$$

Note that $a_0/2$ is just the average of $f(t)$ over one period.

Exercise A.1. Show that in evaluating the integrals in Eqs. (A.1-2) and (A.1-3), as long as we integrate over a time interval whose length is the period T, we can position that interval anywhere along the t axis.

Equations (A.1-3) give us a recipe for finding the Fourier amplitudes for any specified $f(t)$. We call this process **Fourier analysis** because we are analyzing the frequency content of the "signal" $f(t)$. The reverse process, constructing an $f(t)$ from a series of appropriately weighted sines and cosines is called *Fourier synthesis*. In rough terms, each coefficient or amplitude tells how much of the original function $f(t)$ is associated with each harmonic. If $f(t)$ "looks" a lot like $\cos(m\omega_0 t)$ then as we integrate from 0 to T, $f(t)$ and $\cos(m\omega_0 t)$ will stay "in phase" producing an integrand proportional to $\cos^2$, and we get a large contribution to the integral. The resulting coefficient a_m will be large. On the other hand, if $f(t)$ does not look at all like $\cos(m\omega_0 t)$, then the integrand will oscillate between positive and negative values as we integrate from 0 to T, and the resulting integral will be small. We will refine the notion of what the coefficients physically represent later.

Mathematical Conditions

A Fourier series representation of $f(t)$ exists if $f(t)$ satisfies the so-called Dirichlet conditions:

1. $f(t)$ has a finite number of discontinuities in the period T (it is piece-wise differentiable).
2. $f(t)$ has a finite average value.
3. $f(t)$ has a finite number of relative maxima and minima in the period T.

If these conditions are met (which we would expect for any data from an experiment and for any reasonably realistic model), then the series converges to $f(t)$ at the values of t where $f(t)$ is continuous and converges to the mean of $f(t_+)$ and $f(t_-)$ at a finite discontinuity.

A.2 Exponential Series

A great deal of algebraic simplification occurs if we rewrite the sine and cosine series in terms of complex exponential functions. To see how this goes, let us write out the first few terms of the sine and cosine series and then regroup those terms using the Euler formula:

$$
\begin{aligned}
f(t) &= \frac{a_0}{2} + a_1\cos(\omega_0 t) + b_1\sin(\omega_0 t) + \ldots \\
&= \frac{a_0}{2} + \frac{a_1}{2}\left[e^{i\omega_0 t} + e^{-i\omega_0 t}\right] + \frac{b_1}{2i}\left[e^{i\omega_0 t} - e^{-i\omega_0 t}\right] + \ldots \\
&= \frac{a_0}{2} + e^{i\omega_0 t}\left[\frac{a_1}{2} - \frac{ib_1}{2}\right] + e^{-i\omega_0 t}\left[\frac{a_1}{2} + \frac{ib_1}{2}\right] + \ldots
\end{aligned}
\tag{A.2-1}
$$

Thus, we see that we can write $f(t)$ as a sum of complex exponentials

$$
f(t) = \sum_{n=-\infty}^{\infty} c_n e^{in\omega_0 t}
\tag{A.2-2}
$$

with $c_n = \frac{1}{2}(a_n - ib_n)$ and $c_{-n} = \frac{1}{2}(a_n + ib_n)$.

The coefficients c_n can be found directly by multiplying each side of Eq. (A.2-2) by $e^{-im\omega_0 t}$ and then integrating from 0 to T. The resulting integral on the right-hand side is equal to 0 unless $m = n$ and in that case, the integral is equal to T. We thus find that

$$
c_m = \frac{1}{T}\int_0^T dt\, f(t)e^{-im\omega_0 t}
\tag{A.2-3}
$$

Note: If $f(t)$ is a real function, then we must have $c_n = (c_{-n})^*$.

Exercise A.2. For the function shown below, prove that for m not equal to 0

$$
c_m = \frac{iA}{2\pi m}\left[e^{-im2\pi/T} - 1\right]
$$

and that $c_0 = A\tau/T$. The function $f(t)$ is a periodic "rectangular wave" defined by $f(t) = A$ for $0 < t < \tau$, $f(t) = 0$ for $\tau < t < T$ and $f(t) = f(t+T)$.

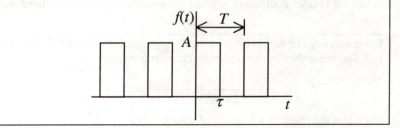

A.3 Power Spectrum

As mentioned earlier, the Fourier coefficients tell us how much (in a rough sense) the various frequency components contribute to the signal. Here, we will be more precise about this relationship. First, we compute the *time-average* of the *square* of the signal. We do this because in many (but not all) situations instruments respond to this average. For example, consider a resistor in which the average power dissipated is proportional to the time-average of the square of the current flowing through the resistor. As a second example, we recall that the average power transported by an electromagnetic wave is proportional to the time-average of the square of the electric field amplitude. This time-average is defined as

$$\left\langle f(t)^2 \right\rangle \equiv \frac{1}{T} \int_0^T dt \left[f(t) \right]^2 \tag{A.3-1}$$

The notation on the left of Eq. (A.3-1) reminds us that we are calculating a time average. The t in parentheses does not mean that the average depends on time. In fact, for a perfectly periodic function, the average so defined is independent of time.

Now, we write $[f(t)]^2$ in terms of Fourier exponential sums.

$$\begin{aligned} \left[f(t) \right]^2 &= \left(\sum_m c_m e^{im\omega_0 t} \right) \left(\sum_n c_n e^{in\omega_0 t} \right) \\ &= \sum_{m,n} c_m c_n e^{i(n+m)\omega_0 t} \end{aligned} \tag{A.3-2}$$

It should be fairly obvious that when we integrate the time-dependent part of the right-hand side of the previous equation from 0 to T to calculate the time average, we will get 0 unless $m = -n$, and in that case we get the value T. Thus, we see that the average of the square of the signal is given by

$$\left\langle f(t)^2 \right\rangle = \sum_n c_n c_{-n} = \sum_n \left| c_n \right|^2 \tag{A.3-3}$$

where we have assumed that $f(t)$ is real; so that $c_{-n} = c_n{}^*$.

A plot of $|c_n|^2$ as a function of harmonic number n is called the "power spectrum" for the signal. The previous result tells us that the absolute-value-squared of the Fourier coefficient c_n gives the amount of power associated with the nth harmonic.

Exercise A.3. For the rectangular wave train signal illustrated in Exercise A.2, show that the Fourier power spectrum is given by

$$\left| c_n \right|^2 = \frac{A^2}{2\pi^2 n^2} \left[1 - \cos(2\pi n\tau / T) \right]$$

Then plot the result as a function of n for $\tau = T/2$.

A.4 Nonperiodic Functions

Fourier analysis can be generalized to deal with nonperiodic functions. (This is clearly the situation of interest in discussing possibly chaotic signals.) The basic idea is to start with the Fourier description of a periodic waveform and then to let the period become infinite. For example, we could analyze a single rectangular pulse signal by using the result for the periodic series of rectangular pulses given earlier and letting the period T become infinite as illustrated in Fig. A.1.

To see how this works out, let us begin with the Fourier exponential series with the explicit form for the coefficient inserted in the sum

$$f(t) = \sum_{m=-\infty}^{\infty} \left[\frac{1}{T} \int_{-T/2}^{+T/2} dt' f(t') e^{-im\omega_0 t'} \right] e^{im\omega_0 t} \qquad (A.4-1)$$

For convenience, the integration range has been shifted to be symmetric around $t = 0$.

We now introduce a new variable $\omega_m = m\omega_0$. Note that the difference between adjacent ω_m values is given by

$$\Delta\omega_m = \omega_0 \qquad (A.4-2)$$

Thus, we can write Eq. (A.4-1) as

$$f(t) = \sum_{m=-\infty}^{\infty} \left[\frac{\Delta\omega_m}{2\pi} \int_{-T/2}^{+T/2} dt' f(t') e^{-i\omega_m t'} \right] e^{i\omega_m t} \qquad (A.4-3)$$

In the limit $T \to \infty$, we let $\Delta\omega_m \to d\omega$. We may then replace the sum with an integral and replace ω_m with ω, a continuous variable. In that limit, we write

$$f(t) = \frac{1}{\sqrt{2\pi}} \int_{-\infty}^{+\infty} d\omega\, c(\omega) e^{i\omega t} \qquad (A.4-4)$$

where

$$c(\omega) = \frac{1}{\sqrt{2\pi}} \int_{-\infty}^{+\infty} dt\, f(t) e^{-i\omega t} \qquad (A.4-5)$$

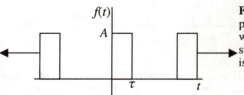

Fig. A.1. If the period of a rectangular pulse train becomes infinite (while the width of each pulse stays fixed), the signal becomes equivalent to a single isolated rectangular pulse.

$c(\omega)$ is called the **Fourier transform** of $f(t)$. $f(t)$ and $c(\omega)$ are said to form a **Fourier transform pair.** (We have split up the $1/2\pi$ symmetrically; therefore, there is a factor of $1/\sqrt{2\pi}$ associated with both $f(t)$ and its Fourier transform.)

As an aside, we should note that the Fourier transform integral in Eq. (A.4-5) includes an integration over negative values of the variable ω. These negative values are a consequence of the "repackaging" of the sine and cosine sums into complex exponentials, which gave rise to a sum over positive and negative values of the integer n in Eq. (A.2-2), and the subsequent switch to a continuous variable proportional to the integer index of those complex exponentials. There is no mystical significance to negative frequencies.

In the case of a nonperiodic function, the power spectrum is given by the absolute-value-squared of the Fourier transform function and is now a <u>continuous</u> function of the transform variable. We interpret $|c(\omega)|^2\, d\omega$ as proportional to the amount of power contained in the frequency range between ω and $\omega + d\omega$. As an example, for the rectangular pulse example given earlier, the power spectrum is given by

$$|c(\omega)|^2 = \frac{A^2}{\pi\omega^2}[1 - \cos\omega\tau] \tag{A.4-6}$$

This result is plotted in the Fig. A.2.

Note that the power spectrum falls off rapidly with ω. As a rough measure of

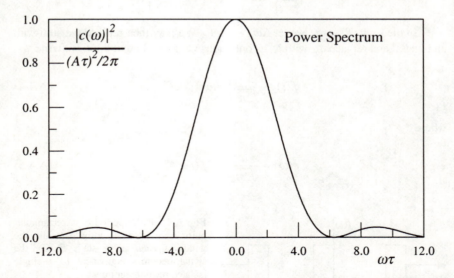

Fig. A.2. A plot of the Fourier power spectrum for a nonperiodic signal consisting of a single rectangular pulse of length τ. The power spectrum is now a continuous function of the frequency variable ω. The maximum value is $(A\tau)^2/2\pi$.

the "width" of the power spectrum (as a function of ω), we can use the range between the first two zeroes of the spectrum, which occur at $\omega = \pm 2\pi / \tau$. Thus, we define the width to be $\Delta\omega = 4\pi / \tau$. Note that the product of $\Delta\omega$ and τ is a constant:

$$(\Delta\omega)\tau = 4\pi \qquad (A.4\text{-}7)$$

This result is, in fact, an example of a rather general relationship between the time duration of the nonperiodic signal and the "width" of the corresponding Fourier power spectrum, or, equivalently, of the range of frequencies contained in the signal. (This product is sometimes called the **_pulse-duration-bandwidth product_**.) If we introduce the "ordinary frequency" $f = \omega/2\pi$, the product takes the form

$$(\Delta f)\tau = 2 \qquad (A.4\text{-}8)$$

A more complete (and more difficult proof) shows that in general $(\Delta f)\tau \geq 1$. [For example, see M. Born and E. Wolf, *Principles of Optics*, 5th ed. (Pergamon Press, New York, 1975), pp. 540–44]. (In the literature other forms of the product appear depending on the precise definitions of Δf and τ. The important point is that there is some lower bound on the product. If τ is made smaller, then Δf, the spread of frequencies in the signal, is necessarily made larger.)

Exercise A.4. Compute the Fourier transform and the corresponding power spectrum and pulse-duration-bandwidth product for the following nonperiodic waveform:

$$f(t) = 0 \text{ for } t < 0 \text{ and } t > \tau$$
$$f(t) = A\cos(\omega t) \text{ for } 0 \leq t \leq \tau$$

Compare your results to those stated in Eq. (A.4-6) and Eq. (A.4-7).

A.5 Fourier Analysis and Nonlinear Dynamics

What is the utility of Fourier analysis for nonlinear dynamics? As we have seen, Fourier analysis lets us determine the frequency content of some signal. If the signal is periodic or quasi-periodic, then the Fourier power spectrum will consist of a sequence of "spikes" at the fundamental frequencies, their harmonics, and the frequencies that are the sums and differences of the various frequencies. The crucial point is that the spectrum will consist of a *discrete* set of frequencies. However, if the signal is neither periodic or quasi-periodic (for example, if it is chaotic), then the Fourier power spectrum will be continuous. See Appendix H, Fig. H.2. Thus, the sudden appearance of a continuous power spectrum from a discrete spectrum, as some parameter of the system is changed, is viewed as an indicator of the onset of chaotic behavior.

However, a continuous Fourier power spectrum can also arise if external noise is present. In addition, if many degrees of freedom are active, then there may be so many fundamental frequencies and harmonics present that the Fourier power spectrum *appears* to be continuous for a given experimental resolution, even though the distinct frequencies might be resolved with higher resolution. Thus, the presence of a continuous power spectrum cannot necessarily be taken as conclusive evidence for the existence of (low-dimensional) chaos unless you can show that external noise is absent and the experimental resolution is sufficient to see all the frequencies that might be present for the expected number of degrees of freedom.

A.6 Spatial Fourier Transforms and Comments on Numerical Methods

We have calculated the Fourier transform (a function of a frequency variable) for a time-dependent function. We can play the same game for a function of position. In that case the Fourier transform variable is usually called k, the **wavevector**. (k is called a wave<u>vector</u> because we can generalize the concept to apply to functions of more than one spatial dimension. There is then a "component" of k for each spatial dimension.) For the spatial Fourier transform, the Fourier expressions are (in one dimension)

$$f(x) = \frac{1}{\sqrt{2\pi}} \int_{-\infty}^{+\infty} dk\, F(k) e^{-ikx}$$

$$F(k) = \frac{1}{\sqrt{2\pi}} \int_{-\infty}^{+\infty} dx\, f(x) e^{+ikx}$$

(A.6-1)

The plus and minus signs in the exponentials have been chosen so that if we include both space and time dependence for the wave, the exponential function in the first part of Eq. (A.6-1) will read $e^{i(\omega t - kx)}$, which describes a wave traveling in the positive x direction.

Numerics and Software

Many software packages are available to compute Fourier transforms and power spectra via an algorithm called the Fast Fourier Transform (FFT). See, for example, W. H. Press, et al. *Numerical Recipes, The Art of Scientific Computing* (Cambridge University Press, Cambridge, 1986).

There are several variations on the Fourier transform technique, such as the Hartley transform, wavelet transform, and so on, that have advantages, both computational and conceptual, for various applications. A nice survey of these techniques is given in R. N. Bracewell, "The Fourier Transform," *Scientific American* **260** (6), 86–95 (June, 1989).

Appendix B

Bifurcation Theory

B.1 Introduction

In this appendix, we give a brief introduction to *bifurcation theory*, the systematic treatment of sudden changes that occur for nonlinear dynamical systems. Bifurcation theory is a vast field, but is still incomplete today. We shall provide only a sketchy outline with just enough information to give you an overview of the theory. In studying bifurcation theory it is easy to lose sight of the forest because of the profusion of trees of many shapes and sizes. In this introduction, therefore, we shall step back to take a broad overview of this lush and complex landscape. For more mathematical and pictorial details, we have provided a guide to several other treatments at the end of the appendix.

Bifurcation theory attempts to provide a systematic classification of the sudden changes in the qualitative behavior of dynamical systems. The effort is divided into two parts. The first part of the theory focuses attention on bifurcations that can be linked to the change in stability of either fixed points or limit cycles (which can be treated as fixed points in Poincaré sections). We call these bifurcations *local* because they can be analyzed in terms of the local behavior of the system near the relevant fixed point or limit cycle. The other part of the theory, the part which is much less well-developed, deals with bifurcation events that involve larger scale behavior in state space and hence are called *global* bifurcations. These global events involve larger scale structures such as basins of attraction and homoclinic and heteroclinic orbits for saddle points.

For both types of bifurcations, bifurcation theory attempts to classify the kinds of bifurcations that can occur for dynamical systems as a function of the dimensionality of the state space (or, more importantly, of the <u>effective</u> dimensionality associated with trajectories near the relevant fixed points or limit cycles). Bifurcation theory is also concerned with the parameter dependence of the bifurcation. In particular the number of parameters that must change to "cause" a bifurcation is called the *co-dimension* of the bifurcation. For the most part, we will focus our attention on bifurcations with co-dimension equal to 1 (i.e., only one parameter varies).

Co-dimension can also be defined from a more geometric point of view. If we have a "surface" (or manifold, in more mathematical language) of dimension m that lives in a space of dimension n, then the co-dimension of the manifold is $n - m$. For example, a surface area in a three-dimensional space has a co-dimension of 1, while a (one-dimensional) curve in that three-dimensional space has a co-dimension of 2. The connection between the parameter co-dimension and the geometric co-dimension for a bifurcation is explored in [Guckenheimer and Holmes, 1990], but the essential idea is that the number of parameters must equal the geometric co-dimension in order to satisfy certain mathematical requirements to be discussed shortly.

In most discussions of bifurcations, we focus our attention on bifurcations that satisfy a so-called transversality condition; bifurcations that satisfy the conditions are called *generic*. The idea is that the bifurcation event can be considered to be a crossing of curves, surfaces, and so on, in a *parameter space* (creating the now familiar bifurcation diagram). If the crossings are transverse, then small perturbations will not affect the nature of the crossings. However, nontransverse crossings will change character, in general, under the effects of perturbations. For example, in a three-dimensional space, two (two-dimensional) surfaces cross transversely to form a curve (a one-dimensional manifold). If the surfaces are slightly perturbed, the crossing moves but remains a one-dimensional manifold. On the other hand, two curves crossing at a point in three-dimensional space constitute a nontransverse intersection since a slight perturbation of the curves will cause the intersection to disappear.

This transversality condition can be expressed in terms of the co-dimensions of the intersecting manifolds. If the sum of the co-dimensions of the intersecting manifolds is equal to the co-dimension of the manifold that constitutes the intersection, then the intersection is transverse. For example, with two (two-dimensional) surfaces in a three-dimensional space, the sum of the co-dimensions is 2, which is the co-dimension of the curve generated by the intersection. Hence, the intersection is transverse. For two curves, however, the sum of the co-dimensions is 4, and the intersection, resulting in a point, is not transverse. As yet another example, we note that the intersection of a curve and a surface in a three-dimensional space is in general transverse.

By imposing transversality conditions on the bifurcations, we restrict our attention to those that are likely to survive small perturbation effects [Guckenheimer and Holmes, 1990]. Non-transverse bifurcations are occasionally important if there are other constraints on the system, such as the imposition of certain symmetries.

Both the co-dimension and the effective dimensionality play an important role in determining the so-called structural stability of a particular type of bifurcation event. (Note that the notion of structural stability is usually applied to the nature of the solutions of the time evolution equations. At a bifurcation point, the solutions are structurally unstable. Here, we are extending this idea to another level: the stability of the bifurcation event itself.) The essential question is whether the type of bifurcation stays the same if the nature of the dynamical system is perturbed by adding small terms to the dynamical equations for the system. If the bifurcation stays the same, the bifurcation is said to be structurally stable. If it changes, then it is structurally unstable. The notion is that structurally unstable bifurcations are not likely to occur in actual systems because many effects such as "noise" or unaccounted for degrees of freedom effectively play the role of perturbations of the system. It would take an unlikely conspiracy of these perturbations to allow a structurally unstable bifurcation to be seen.

B.2 Local Bifurcations

As we mentioned earlier, local bifurcations are those in which fixed points or limit cycles appear, disappear, or change their stability. Since we can treat limit cycles as fixed points of Poincaré sections, we shall use the term *fixed point bifurcation* for both types. This change in stability is signaled by a change in the real part of one (or more) of the characteristic exponents associated with that fixed point: At a local bifurcation, the real part becomes equal to 0 as some parameter (or parameters) of the system is changed. (Recall that in general the characteristic exponents can be complex numbers.) As the real part of the characteristic exponent changes from negative to positive, for example, the motion associated with that characteristic direction goes from being stable (attracted toward the fixed point) to being unstable (being repelled by the fixed point). For a Poincaré map fixed point, this criterion is equivalent to having the absolute value of the characteristic multiplier equal to unity.

An important theorem, the ***Center-Manifold Theorem***, tells us that at a local bifurcation, we can focus our attention on just those degrees of freedom associated with the characteristic exponents whose real parts go to 0. It is the number of characteristic exponents with real parts equal to 0 that determines the number of effective dimensions for the bifurcation. The ***Center Manifold*** is that "subspace" associated with the characteristic exponents whose real parts are 0. (It is to be contrasted with the stable manifold, where the real parts are negative, and the unstable manifold, where the real parts are positive.)

In order to classify the types of bifurcations, it is traditional to reduce the dynamical equations to a standard form, the ***normal form*** [See, for example, Kahn and Zarmi, 1997], in which the bifurcation event occurs when a parameter value μ reaches 0 and a fixed point, located at $x = 0$, has a characteristic exponent with real part equal to 0. More general situations can be reduced to these normal forms by appropriate coordinate and parameter transformations; therefore, there is no loss in generality in using the simplified normal forms.

For a system whose center manifold dynamics are one-dimensional and described by a differential equation, we write the time evolution equation as

$$
\begin{aligned}
\frac{dx}{dt} = {} & A_0 + B_0 x + C_0 x^2 + \ldots \\
& + \mu \left(A_1 + B_1 x + C_1 x^2 + \ldots \right) \\
& + \mu^2 \left(A_2 + B_2 x + C_2 x^2 + \ldots \right) \\
& + \ldots
\end{aligned}
\tag{B.2-1}
$$

where the subscripts on the constants A, B, C, and so on, tell us with which power of the control parameter μ they are to be associated. By choosing the values of the constants, we can develop a systematic classification of the bifurcations. For systems (such as Poincaré mappings of limit cycles) described by iterated map functions, we can write a similar expression for the iterated map dynamics near the bifurcation point:

$$x_{n+1} = A_0 + B_0 x_n + C_0 x_n^2 + \ldots$$
$$+ \mu \left(A_1 + B_1 x_n + C_1 x_n^2 + \ldots \right) \tag{B.2-2}$$
$$+ \ldots$$

To see how the classification scheme works, let us try out a few examples. For a system described by differential equations, if we set $A_1 = 1$, $C_0 = -1$ and all the other constants are 0, we arrive at the equation

$$\frac{dx}{dt} = \mu - x^2 \tag{B.2-3}$$

which we recognize from Section 3.17 as the equation describing the behavior at a repellor-node (or, in higher dimensions, saddle-node) bifurcation.

If we use the iterated map form with $B_0 = -1$, $B_1 = -1$ and $D_0 = 1$, the equation

$$x_{n+1} = -(1+\mu)x_n + x_n^3 \tag{B.2-4}$$

describes a period-doubling bifurcation such as those in the Feigenbaum cascade. By adding a second state space dimension with its own normal form equation, we can model Hopf bifurcations as described in Chapter 3.

Bifurcations are also classified as **subtle** (or equivalently, **supercritical**) and **catastrophic** (or equivalently, **subcritical**). In a subtle bifurcation, the location of the (stable) fixed point changes smoothly with parameter value near the bifurcation point. The Hopf bifurcation of Chapter 3 is an example of a subtle bifurcation. For a catastrophic bifurcation, the (stable) fixed point suddenly appears (as in a saddle-node bifurcation) or disappears or jumps discontinuously to a new location.

As an example of a subcritical bifurcation, consider the following normal form equation:

$$\dot{x} = \mu x + x^3 - x^5 \tag{B.2-5}$$

where μ is the control parameter. For $\mu < 0$, the steady-state solution $x = 0$ is stable. For $\mu > 0$, there are two stable states given by

$$x_\pm = \sqrt{\tfrac{1}{2} \pm \tfrac{1}{2}\sqrt{1+4\mu}} \tag{B.2-6}$$

A bifurcation occurs at $\mu = 0$ and the stable state locations suddenly jump to a new location.

Listings of various normal forms and pictorial representations of the corresponding bifurcations can be found in [Thompson and Stewart, 1986] and [Abraham and Marsden, 1978].

B.3 Global Bifurcations

Global bifurcations are bifurcation events that involve changes in basins of attraction, homoclinic or heteroclinic orbits, or other structures that extend over significant regions of state space. Such bifurcations include intermittency and crises as described in Chapter 7. Since we need to take into account behavior over a wide range of state space, a different means of classifying and studying such bifurcations is obviously needed. Unfortunately, the theory of global bifurcations is both more difficult and less articulated than is the theory of local bifurcations. Specific cases, such as homoclinic tangencies and crises, have been studied in some detail, but a general classification scheme is yet to be devised. A schematic classification of bifurcations involving chaotic attractors is given in [Thompson and Stewart, 1986, Chapter 13] and [Wiggins, 1988].

B.4 Further Reading

A Guided Tour of Bifurcation Theory

We suggest the following "reading course" to learn more about the theory of bifurcations. A treatment of bifurcations roughly at the level of this book, with some nice illustrations of the accompanying bifurcation diagrams is given by
[Thompson and Stewart, 1986], Chapters 7, 8, and 13.

An illustrated guide to both local and global bifurcations but without any mathematical support is given by [Abraham and Shaw, 1992] and [Abraham, Abraham, and Shaw, 1996]. With the basic picture in mind you are ready for more sophisticated and general mathematical treatments such as those given in
[Guckenheimer and Holmes, 1990] Chapters 3, 6, and 7.

J. D. Crawford, "Introduction to Bifurcation Theory," *Rev. Mod. Phys.* **63**, 991–1037 (1991).

Other General Introductions to Bifurcations in Dynamical Systems

R. Abraham and J. E. Marsden, *Foundations of Mechanics*, 2nd ed. (Benjamin/Cummings, Reading, MA, 1978).

G. Iooss and D. D. Joseph, *Elementary Stability and Bifurcation Theory* (Springer-Verlag, New York, 1980).

V. I. Arnol'd, *Geometrical Methods in the Theory of Ordinary Differential Equations* (Springer, New York, 1983).

P. Kahn and Y. Zarmi, *Nonlinear Dynamics: Exploration through Normal Forms* (Wiley, New York, 1997).

Y. A. Kuznetsov, *Elements of Applied Bifurcation Theory*, 2nd ed. (Springer-Verlag, New York, 1998). Aimed at upper-level undergraduates and beginning graduate students. Nice diagrams, examples, and explanations.

Examples of Bifurcation Analysis

C. Robert, K. T. Alligood, E. Ott, and J. A. Yorke, "Outer Tangency Bifurcations of Chaotic Sets," *Phys. Rev. Lett.* **80**, 4867–70 (1998).

M. C. Eguia and G. B. Mindlin, "Semiconductor laser with optical feedback: From excitable to deterministic low-frequency fluctuations," *Phys. Rev. E* **60**, 1551–57 (1999).

Global Bifurcations

The following book contains an excellent first chapter, which provides an introduction to the more mathematically inclined treatments of dynamical systems. The later chapters contain a detailed discussion of what is known about global bifurcations.

S. Wiggins, *Global Bifurcations and Chaos, Analytical Methods* (Springer-Verlag, New York, 1988).

Gluing Bifurcations

As our experience with nonlinear dynamics increases, new types of bifurcations are being recognized. For example, gluing bifurcations occur when two periodic orbits approach a saddle point and then combine to form a new periodic orbit with a period equal to the sum of the two original periods. For example, see

E. Meron and I.Proccacia, "Gluing Bifurcations in Critical Flows: The Route to Chaos in Parametrically Excited Surface Waves," *Phys. Rev. A* **35**, 4008-11 (1987).

Appendix C

The Lorenz Model

C.1 Introduction

In this appendix we show how the Lorenz model equations introduced in Chapter 1 are developed (derived is too strong a word) from the Navier–Stokes equation for fluid flow and the equation describing thermal energy diffusion. This development provides a prototype for the common process of finding approximate, but useful, model equations when we cannot solve the fundamental equations describing some physical situation.

The Lorenz model has become almost totemistic in the field of nonlinear dynamics. Unfortunately, most derivations of the Lorenz model equations leave so much to the reader that they are essentially useless for all but specialists in fluid dynamics. In this appendix, we hope to give a sufficiently complete account that readers of this text come away with a good understanding of both the physics content and the mathematical approximations that go into this widely cited model.

The Lorenz model describes the motion of a fluid under conditions of Rayleigh–Bénard flow: an incompressible fluid is contained in a cell which has a higher temperature T_w at the bottom and a lower temperature T_c at the top. The temperature difference $\delta T = T_w - T_c$ is taken to be the control parameter for the system. The geometry is shown in Fig. C.1.

Before launching into the formal treatment of Rayleigh–Bénard flow, we should develop some intuition about the conditions that cause convective flow to begin. In rough terms, when the temperature gradient between the top and bottom plates becomes sufficiently large, a small packet of fluid that happens to move up a bit will experience a net upward buoyant force because it has moved into a region of lower temperature and hence higher density: It is now less dense than its surroundings. If the upward force is sufficiently strong, the packet will move upward more quickly than its temperature can drop. (Since the packet is initially warmer than its surroundings, it will tend to loose thermal energy to its

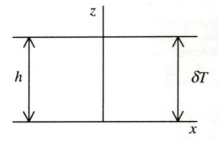

Fig. C.1. A diagram of the geometry for the Lorenz model. The system is infinite in extent in the horizontal direction and in the direction in and out of the page.

environment.) Then convective currents will begin to flow. On the other hand if the buoyant force is relatively weak, the temperature of the packet will drop before it can move a significant distance, and it remains stable in position.

We can be slightly more quantitative about this behavior by using our knowledge (gained in Chapter 11) about thermal energy diffusion and viscous forces in fluids. Imagine that the fluid is originally at rest. We want to see if this condition is stable. We begin by considering a small packet of fluid that finds itself displaced upward by a small amount Δz. The temperature in this new region is lower by the amount $\Delta T = (\delta T / h)\Delta z$. According to the thermal energy diffusion equation (Chapter 11), the rate of change of temperature is equal to the thermal diffusion coefficient D_T multiplied by the Laplacian of the temperature function. For this small displacement, we may approximate the Laplacian by

$$\nabla^2 T \approx \frac{\delta T}{h^2}\frac{\Delta z}{h} \tag{C.1-1}$$

We then define a thermal relaxation time δt_T such that

$$\delta t_T \frac{dT}{dt} = \Delta T = \delta t_T D_T \nabla^2 T \tag{C.1-2}$$

where the second equality follows from the thermal diffusion equation. Using our approximation for the Laplacian, we find that

$$\delta t_T = \frac{h^2}{D_T} \tag{C.1-3}$$

Let us now consider the effect of the buoyant force on the packet of fluid. This buoyant force is proportional to the difference in density between the packet and its surroundings. This difference itself is proportional to the thermal expansion coefficient α (which gives the relative change in density per unit temperature change) and the temperature difference ΔT. Thus, we find for the buoyant force

$$F = \alpha \rho_o g \Delta T = \alpha \rho_o g \frac{\delta T}{h}\Delta z \tag{C.1-4}$$

where ρ_o is the original density of the fluid and g is the strength of the local gravitational field. (Near the surface of Earth, $g = 9.8$ N/kg $= 9.8$ m/s^2.)

We assume that this buoyant force just balances the fluid viscous force; therefore, the packet moves with a constant velocity v_z. It then takes a time $\tau_d = \Delta z / v_z$ for the packet to be displaced through the distance Δz. As we learned in Chapter 11, the viscous force is equal to the viscosity of the fluid multiplied by the Laplacian of the velocity. Thus, we approximate the viscous force as

$$F_v = \mu \nabla^2 v_z \approx \mu \frac{v_z}{h^2} \tag{C.1-5}$$

where the right-most equality states our approximation for the Laplacian of v_z.

If we now require that the buoyant force be equal in magnitude to the viscous force, we find that v_z can be expressed as

$$v_z = \frac{\alpha \rho_o g h \delta T}{\mu} \Delta z \qquad \text{(C.1-6)}$$

The displacement time is then given by

$$\tau_d = \frac{\mu}{\alpha \rho_o g h \delta T} \qquad \text{(C.1-7)}$$

The original nonconvecting state is stable if the thermal diffusion time is less than the corresponding displacement time. If the thermal diffusion time is longer, then the fluid packet will continue to feel an upward force, and convection will continue. The important factor is the ratio of the thermal diffusion time to the displacement time. This ratio is called the *Rayleigh number R* and takes the form

$$R = \frac{\alpha \rho_o g h^3 \delta T}{D_T \mu} \qquad \text{(C.1-8)}$$

As we shall see, the Rayleigh number is indeed the critical parameter for Rayleigh–Bénard convection, but we need a more detailed calculation to tell us the actual value of the Rayleigh number at which convection begins.

C.2 The Navier–Stokes Equations

Because of the geometry assumed, the fluid flow can be taken to be two-dimensional. Thus, we need consider only the x (horizontal) and z (vertical) components of the fluid velocity. The Navier–Stokes equations (see Chapter 11) for the x and z components of the fluid velocity are

$$\rho \frac{\partial v_z}{\partial t} + \rho \vec{v} \cdot grad \ \ v_z = -\rho g - \frac{\partial p}{\partial z} + \mu \nabla^2 v_z$$
$$\rho \frac{\partial v_x}{\partial t} + \rho \vec{v} \cdot grad \ \ v_x = -\frac{\partial p}{\partial x} + \mu \nabla^2 v_x \qquad \text{(C.2-1)}$$

In Eq. (C.2-1), ρ is the mass density of the fluid; g is again the strength of the local gravitational field; p is the fluid pressure, and μ is the fluid viscosity. Note that gravity appears explicitly only in the z component equation.

The temperature T of the fluid is described by the thermal diffusion equation (see Chapter 11), which takes the form

$$\frac{\partial T}{\partial t} + \vec{v} \cdot grad \ \ T = D_T \nabla^2 T \qquad \text{(C.2-2)}$$

where, as before, D_T is the thermal diffusion coefficient.

In the steady nonconvecting state (when the fluid is motionless) the temperature varies linearly from bottom to top:

$$T(x,z,t) = T_w - \frac{z}{h}\delta T \qquad (C.2\text{-}3)$$

For the purposes of our calculation, we will focus our attention on a function $\tau(x,z,t)$ that tells us how the temperature deviates from this linear behavior:

$$\tau(x,z,t) = T(x,z,t) - T_w + \frac{z}{h}\delta T \qquad (C.2\text{-}4)$$

If we use Eq. (C.2-4) in Eq. (C.2-2), we find that τ satisfies

$$\frac{\partial \tau}{\partial t} + \vec{v} \cdot grad\,\tau - v_z \frac{\delta T}{h} = D_T \nabla^2 \tau \qquad (C.2\text{-}5)$$

We now need to take into account the variation of the fluid density with temperature. (It is this decrease of density with temperature that leads to a buoyant force, which initiates fluid convection.) We do this by writing the fluid density in terms of a power series expansion:

$$\rho(T) = \rho_o + \frac{\partial \rho}{\partial T}(T - T_w) + \ldots \qquad (C.2\text{-}6)$$

where ρ_o is the fluid density evaluated at T_w.

Introducing the thermal expansion coefficient α, which is defined as

$$\alpha = -\frac{1}{\rho_o}\frac{\partial \rho}{\partial T} \qquad (C.2\text{-}7)$$

and using $T - T_w$ from Eq. (C.2-4), we may write the temperature variation of the density as

$$\rho(T) = \rho_o - \alpha \rho_o \left[-\frac{z}{h}\delta T + \tau(x,z,t) \right] \qquad (C.2\text{-}8)$$

The fluid density ρ appears in several terms in the Navier–Stokes equations. The **Boussinesq approximation**, widely used in fluid dynamics, says that we may ignore the density variation in all the terms except the one that involves the force due to gravity. This approximation reduces the v_z equation in Eq. (C.2-1) to

$$\rho_o \frac{\partial v_z}{\partial t} + \rho_o \vec{v} \cdot grad\ v_z = -\rho_o g - \alpha g \rho_o \frac{z}{h}\delta T - \frac{\partial p}{\partial z}$$
$$+ \alpha g \rho_o \tau(x,z,t) + \mu \nabla^2 v_z \qquad (C.2\text{-}9)$$

We then recognize that when the fluid is not convecting, the first three terms on the right-hand side of the previous equation must add to 0. Hence, we introduce an effective pressure gradient, which has the property of being equal to 0 when no fluid motion is present:

$$p' = p + \rho_o gz + \alpha g \rho_o \frac{z^2}{2} \frac{\delta T}{h}$$
$$\frac{\partial p'}{\partial z} = \frac{\partial p}{\partial z} + \rho_o g + \alpha g \rho_o \frac{z}{h} \delta T \tag{C.2-10}$$

Finally, we use this effective pressure gradient in the Navier–Stokes equations and divide through by ρ_o to obtain

$$\frac{\partial v_z}{\partial t} + \vec{v} \cdot grad \ v_z = -\frac{1}{\rho_o} \frac{\partial p'}{\partial z} + \alpha \tau g + v \nabla^2 v_z$$
$$\frac{\partial v_x}{\partial t} + \vec{v} \cdot grad \ v_x = -\frac{1}{\rho_o} \frac{\partial p'}{\partial x} + v \nabla^2 v_x \tag{C.2-10}$$

where $v = \mu/\rho_o$ is the so-called kinematic viscosity.

C.3 Dimensionless Variables

Our next step in the development of the Lorenz model is to express the Navier–Stokes equations Eq. (C.2-11) in terms of dimensionless variables. By using dimensionless variables, we can see which combinations of parameters are important in determining the behavior of the system. In addition, we generally remove the dependence on specific numerical values of the height h and temperature difference δT, and so on, thereby simplifying the eventual numerical solution of the equations.

First, we introduce a dimensionless time variable t'

$$t' = \frac{D_T}{h^2} t \tag{C.3-1}$$

[You should recall from Eq. (C.1-3) (and from Chapter 11) that h^2/D_T is a typical time for thermal diffusion over the distance h.] In a similar fashion, we introduce dimensionless distance variables and a dimensionless temperature variable:

$$x' = \frac{x}{h} \qquad z' = \frac{z}{h} \qquad \tau' = \frac{\tau}{\delta T} \tag{C.3-2}$$

We can also define a dimensionless velocity using the dimensionless distance and dimensionless time variables. For example, the x component of the dimensionless velocity is

$$v'_x = \frac{dx'}{dt'} = \frac{D_T}{h} v_x \tag{C.3-3}$$

Finally, the Laplacian operator can also be expressed in terms of the new variables with the replacement

$$\nabla'^2 = h^2 \nabla^2 \tag{C.3-4}$$

If we use these new variables in the Navier–Stokes equations (C.2-11) and multiply through by $h^3/(v D_T)$, we arrive at

$$\frac{D_T}{v}\left[\frac{\partial v'_z}{\partial t'} + \vec{v}' \cdot grad' v'_z\right] = -\frac{h^2}{v D_T \rho_o}\frac{\partial p'}{\partial z'} \tag{C.3-5}$$
$$+\frac{\alpha \delta T g h^3}{v D_T}\tau' + \nabla'^2 v'_z$$

$$\frac{D_T}{v}\left[\frac{\partial v'_x}{\partial t'} + \vec{v}' \cdot grad' v'_x\right] = -\frac{h^2}{v D_T \rho_o}\frac{\partial p'}{\partial x'} + \nabla'^2 v'_x \tag{C.3-6}$$

We recognize that certain dimensionless ratios of parameters appear in the equations. First, the **Prandtl number** σ gives the ratio of kinematic viscosity to the thermal diffusion coefficient:

$$\sigma = \frac{v}{D_T} \tag{C.3-7}$$

The Prandtl number measures the relative importance of viscosity (dissipation of mechanical energy due to the shearing of the fluid flow) compared to thermal diffusion, the dissipation of energy by thermal energy (heat) flow. The Prandtl number is about equal to 7 for water at room temperature.

The **Rayleigh number** R tells us the balance between the tendency for a packet of fluid to rise due to the buoyant force associated with thermal expansion relative to the dissipation of energy due to viscosity and thermal diffusion. R is defined as the combination

$$R = \frac{\alpha g h^3}{v D_T}\delta T \tag{C.3-8}$$

The Rayleigh number is a dimensionless measure of the temperature difference between the bottom and top of the cell. In most Rayleigh–Bénard experiments, the Rayleigh number is the control parameter, which we adjust by changing that temperature difference.

Finally, we introduce a dimensionless pressure variable Π defined as

$$\Pi = \frac{p'h^2}{\nu\rho_o D_T} \tag{C.3-9}$$

We now use all these dimensionless quantities to write the Navier–Stokes equations and the thermal diffusion equation in the following form, in which, for the sake of simpler typesetting, we have dropped the primes (but we remember that all the variables are dimensionless):

$$\frac{1}{\sigma}\left[\frac{\partial v_z}{\partial t} + \vec{v}\cdot grad\, v_z\right] = -\frac{\partial \Pi}{\partial z} + R\tau + \nabla^2 v_z$$

$$\frac{1}{\sigma}\left[\frac{\partial v_x}{\partial t} + \vec{v}\cdot grad\, v_x\right] = -\frac{\partial \Pi}{\partial x} + \nabla^2 v_x \tag{C.3-10}$$

$$\frac{\partial \tau}{\partial t} + \vec{v}\cdot grad\, \tau - v_z = \nabla^2 \tau$$

We should point out that in introducing the dimensionless variables and dimensionless parameters, we have not changed the physics content of the equations, nor have we introduced any mathematical approximations.

C.4 The Streamfunction

As we discussed in Chapter 11, for two-dimensional fluid flows, we may introduce a streamfunction $\Psi(x,z,t)$, which carries all the information about the fluid flow. The actual fluid velocity components are obtained by taking partial derivatives of the streamfunction:

$$v_x = -\frac{\partial \Psi(x,z,t)}{\partial z} \qquad v_z = \frac{\partial \Psi(x,z,t)}{\partial x} \tag{C.4-1}$$

(We are free to place the minus sign on either of the velocity components. The sign choice made here gives us the conventional signs in the Lorenz model equations.) We now use the streamfunction in the thermal diffusion equation:

$$\frac{\partial \tau}{\partial t} - \frac{\partial \Psi}{\partial z}\frac{\partial \tau}{\partial x} + \frac{\partial \Psi}{\partial x}\frac{\partial \tau}{\partial z} - \frac{\partial \Psi}{\partial x} = \nabla^2 \tau \tag{C.4-2}$$

in which we have expanded the *grad* term explicitly in terms of components. (Mathematically experienced readers may recognize the middle two terms on the left-hand side of the previous equation as the Jacobian determinant of the functions Ψ and τ with respect to the variables x and z.)

The fluid flow equations can also be written in terms of the streamfunction. Unfortunately, the equations become algebraically messy before some order emerges. The v_z equation becomes

$$\frac{1}{\sigma}\left[\frac{\partial^2\Psi}{\partial t\partial x}-\frac{\partial\Psi}{\partial z}\frac{\partial^2\Psi}{\partial x^2}+\frac{\partial\Psi}{\partial x}\frac{\partial^2\Psi}{\partial z\partial x}\right]=-\frac{\partial\Pi}{\partial z}+R\tau+\nabla^2\frac{\partial\Psi}{\partial x} \qquad (C.4\text{-}3)$$

The υ_x equation becomes

$$\frac{1}{\sigma}\left[-\frac{\partial^2\Psi}{\partial t\partial z}+\frac{\partial\Psi}{\partial z}\frac{\partial^2\Psi}{\partial x\partial z}-\frac{\partial\Psi}{\partial x}\frac{\partial^2\Psi}{\partial z\partial x}\right]=-\frac{\partial\Pi}{\partial x}-\nabla^2\frac{\partial\Psi}{\partial z} \qquad (C.4\text{-}4)$$

If we now take $\partial/\partial x$ of Eq. (C.4-3) and subtract from it $\partial/\partial z$ of Eq. (C.4-4), the pressure terms drop out, and we have

$$\frac{1}{\sigma}\left[+\frac{\partial}{\partial t}(\nabla^2\Psi)-\frac{\partial}{\partial z}\left\{\frac{\partial\Psi}{\partial z}\frac{\partial^2\Psi}{\partial x\partial z}-\frac{\partial\Psi}{\partial x}\frac{\partial^2\Psi}{\partial z^2}\right\}-\frac{\partial}{\partial x}\left\{\frac{\partial\Psi}{\partial z}\frac{\partial^2\Psi}{\partial x^2}-\frac{\partial\Psi}{\partial x}\frac{\partial^2\Psi}{\partial z\partial x}\right\}\right]$$
$$= R\frac{\partial\tau}{\partial x}+\nabla^4\Psi \qquad (C.4\text{-}5)$$

Eq. (C.4-2) and the rather formidable looking Eq. (C.4-5) contain all the information on the fluid flow.

C.5 Fourier Expansion, Galerkin Truncation, and Boundary Conditions

Obviously, we face a very difficult task in trying to solve the partial differential equations that describe our model system. The usual practice in solving partial differential equations is to seek solutions that can be written as products of functions, each of which depends on only one of the independent variables x, z, t. Since we have a rectangular geometry, we expect to be able to find a solution of the form

$$\Psi(x,z,t)=\sum_{m,n}e^{\omega_{m,n}t}\{A_m\cos\lambda_m z+B_m\sin\lambda_m z\}$$
$$\times\{C_n\cos\lambda_n x+D_n\sin\lambda_n x\} \qquad (C.5\text{-}1)$$

where the λs are the wavelengths of the various Fourier spatial modes and ω_{mn} are the corresponding frequencies. We would, of course, have a similar equation for τ, the temperature variable. (Appendix A contains a concise introduction to Fourier analysis.)

As we saw in Chapter 11, the standard procedure consists of using this sine and cosine expansion in the original partial differential equations to develop a corresponding set of (coupled) ordinary differential equations. This procedure will lead to an infinite set of ordinary differential equations. To make progress, we must somehow reduce this infinite set to a finite set of equations. This truncation process is known as the *Galerkin procedure*.

For the Lorenz model, we look at the boundary conditions that must be satisfied by the streamfunction and the temperature deviation function and choose a very limited set of sine and cosine terms that will satisfy these boundary conditions.

It is hard to justify this truncation a priori, but numerical solutions of a larger set of equations seem to indicate (SAL62) that the truncated form captures most of the dynamics over at least a limited range of parameter values.

The boundary conditions for the temperature deviation function are simple. Since τ represents the deviation from the linear temperature gradient and since the temperatures at the upper and lower surfaces are fixed, we must have

$$\tau = 0 \quad \text{at} \quad z = 0,1 \tag{C.5-2}$$

For the streamfunction, we look first at the boundary conditions on the velocity components. We assume that at the top and bottom surfaces the vertical component of the velocity v_z must be 0. We also assume that we can neglect the shear forces at the top and bottom surfaces. As we saw in Chapter 11, these forces are proportional to the gradient of the tangential velocity component; therefore, this condition translates into having $\partial v_x / \partial z = 0$ at $z = 0$ and $z = 1$. For the Lorenz model, these conditions are satisfied by the following *ansatz* for the streamfunction and temperature deviation function:

$$\Psi(x,z,t) = \psi(t)\sin(\pi z)\sin(ax)$$
$$\tau(x,z,t) = T_1(t)\sin(\pi z)\cos(ax) - T_2(t)\sin(2\pi z) \tag{C.5-3}$$

where the parameter a is to be determined. As we shall see, this choice of functions not only satisfies the boundary conditions, but it also greatly simplifies the resulting equations.

The particular form of the spatial part of the streamfunction Ψ models the convective rolls observed when the fluid begins to convect. You may easily check this by calculating the velocity components from Eq. (C.4-1). The form for the temperature deviation function has two parts. The first, T_1, gives the temperature difference between the upward and downward moving parts of a convective cell. The second, T_2, gives the deviation from the linear temperature variation in the center of a convective cell as a function of vertical position z. (The minus sign in front of the T_2 term is chosen so that T_2 is positive: The temperature in the fluid must lie between T_w and T_c.)

C.6 Final Form of the Lorenz Equations

We now substitute the assumed forms for the streamfunction and the temperature deviation function into Eqs. (C.4-2) and (C.4-5). As we do so, we find that most terms simplify. For example, we have

$$\nabla^2 \Psi = -(a^2 + \pi^2)\Psi$$
$$\nabla^4 \Psi = +(a^2 + \pi^2)^2 \Psi \tag{C.6-1}$$

The net result is that some of the complicated expressions that arise from $\vec{v} \cdot grad \, v$ terms disappear, and we are left with

$$-\frac{d\psi(t)}{dt}(a^2+\pi^2)\sin\pi z\sin ax = -\sigma RT_1(t)\sin\pi z\sin ax$$

$$+\sigma(a^2+\pi^2)^2\psi(t)\sin\pi z\sin ax \tag{C.6-2}$$

The only way the previous equation can hold for all values of x and z is for the coefficients of the sine terms to satisfy

$$\frac{d\psi(t)}{dt}=\frac{\sigma R}{\pi^2+a^2}T_1(t)-\sigma(\pi^2+a^2)\psi(t) \tag{C.6-3}$$

The temperature deviation equation is a bit more complicated. It takes the form

$$\dot{T}_1\sin\pi z\cos ax-\dot{T}_2\sin 2\pi z+(\pi^2+a^2)T_1\sin\pi z\cos ax$$

$$-4\pi^2 T_2\sin 2\pi z-a\psi\sin\pi z\cos ax$$

$$=-[\pi\psi\cos\pi z\sin ax][aT_1\sin\pi z\sin ax] \tag{C.6-4}$$

$$-[a\psi\sin\pi z\cos ax][\pi T_1\cos\pi z\cos ax]$$

$$+[\psi\sin\pi z\cos ax][2\pi T_2\cos 2\pi z]$$

We first collect all those terms which involve $\sin\pi z\cos ax$. We note that the last of these terms in Eq. (C.6-4) is $2a\pi\psi T_2\sin\pi z\cos 2\pi z$. Using standard trigonometric identities, this term can be written as the following combination of sines and cosines: $(-\frac{1}{2}\sin\pi z + \frac{1}{2}\sin 3\pi z)$. The $\sin 3\pi z$ term has a spatial dependence more rapid than allowed by our *ansatz*; so, we drop that term. We may then equate the coefficients of the terms in Eq. (C.6-4) involving $\sin\pi z\cos ax$ to obtain

$$\dot{T}_1=a\psi-(\pi^2+a^2)T_1-\pi a\psi T_2 \tag{C.6-5}$$

All the other terms in the temperature deviation equation are multiplied by $\sin 2\pi z$ factors. Again, equating the coefficients, we find

$$\dot{T}_2=\frac{\pi a}{2}\psi T_1-4\pi^2 T_2 \tag{C.6-6}$$

To arrive at the standard form of the Lorenz equations, we now make a few straightforward changes of variables. First, we once again change the time variable by introducing a new variable $t''=(\pi^2+a^2)t'$. We then make the following substitutions:

$$X(t)=\frac{a\pi}{(\pi^2+a^2)\sqrt{2}}\psi(t)$$

$$Y(t)=\frac{r\pi}{\sqrt{2}}T_1(t) \tag{C.6-7}$$

$$Z(t)=\pi rT_2(t)$$

where r is the so-called reduced Rayleigh number:

$$r = \frac{a^2}{(a^2 + \pi^2)^3} R \tag{C.6-8}$$

We also introduce a new parameter b defined as

$$b = \frac{4\pi^2}{a^2 + \pi^2} \tag{C.6-9}$$

With all these substitutions and with the replacement of σ with p for the Prandtl number, we finally arrive at the standard form of the Lorenz equations:

$$
\begin{aligned}
\dot{X} &= p(Y - X) \\
\dot{Y} &= rX - XZ - Y \\
\dot{Z} &= XY - bZ
\end{aligned}
\tag{C.6-10}
$$

At this point we should pause to note one important aspect of the relationship between the Lorenz model and the reality of fluid flow. The truncation of the sine–cosine expansion means that the Lorenz model allows for only one spatial mode in the x direction with "wavelength" $2\pi/a$. If the actual fluid motion takes on more a complex spatial structure, as it will if the temperature difference between top and bottom plates becomes too large, then the Lorenz equations no longer provide a useful model of the dynamics.

Let us also take note of where nonlinearity enters the Lorenz model. We see from Eq. (C.6-10) that the product terms XZ and XY are the only nonlinear terms. These express a coupling between the fluid motion (represented by X, proportional to the streamfunction) and the temperature deviation (represented by Y and Z, proportional to T_1 and T_2, respectively. The Lorenz model does not include, because of the choice of spatial mode functions, the usual $\vec{v} \cdot grad\,v$ nonlinearity from the Navier–Stokes equation.

C.7 Stability Analysis of the Nonconvective State

The parameter a is determined by examining the conditions on the stability of the nonconvective state. The nonconvective state has $\psi = 0$ and $\tau = 0$ and hence corresponds to $X, X, Z = 0$. If we let x, y, and z represent the values of X, Y, and Z near this fixed point, and drop all nonlinear terms from the Lorenz equations, the dynamics near the fixed point is modeled by the following *linear* differential equations:

$$
\begin{aligned}
\dot{x} &= p(x - y) \\
\dot{y} &= rx - y \\
\dot{z} &= -bz
\end{aligned}
\tag{C.7-1}
$$

Note that $z(t)$ is exponentially damped since the parameter b is positive. Thus, we need consider only the x and y equations. Using our now familiar results from Section 3.11, we see that the nonconvective fixed point becomes unstable when $r >$ 1. Returning to the original Rayleigh number, we see that the condition is

$$R \geq \frac{(\pi^2 + a^2)^3}{a^2} \qquad \text{(C.7-2)}$$

We choose the parameter a to be the value that gives the lowest Rayleigh number for the beginning of convection. In a sense, the system selects the wavelength $2\pi/a$ by setting up a convection pattern with the wavelength $2\pi/a$ at the lowest possible Rayleigh number. This condition yields $a = \pi/\sqrt{2}$. Hence, the Rayleigh number at which convection begins is $R = 27\pi^4/4$. The parameter b is then equal to 8/3, the value used in most analyses of the Lorenz model.

C.8 Further Reading

E. N. Lorenz, "Deterministic Nonperiodic Flow," *J. Atmos. Sci.* **20**, 130–41 (1963). Reprinted in [Cvitanovic, 1984]. The Lorenz model first appeared in this pioneering and quite readable paper.

B. Saltzman, "Finite Amplitude Free convection as an Initial Value Problem-I." *J. Atmos. Sci.* **19**, 329–41 (1962). The Lorenz model was an outgrowth of an earlier model of atmospheric convection introduced by Saltzman.

[Berg, Pomeau, Vidal, 1984]. Appendix D, contains a slightly different development of the Lorenz model equations, and in addition, provides more details on the how the dynamics evolve as the reduced Rayleigh number r changes.

[Sparrow, 1982]. A detailed treatment of the Lorenz model and its behavior.

S. Chandrasekhar, *Hydrodynamic and Hydromagnetic Stability* (Dover, New York, 1984). Chapter II. A wide-ranging discussion of the physics and mathematics of Rayleigh–Bénard convection along with many historical references.

H. Haken, "Analogy between Higher Instabilities in Fluids and Lasers," *Phys. Lett. A* **53**, 77–78 (1975). Certain laser systems are modeled by equations that are identical in form to the Lorenz model equations.

R. Graham, "Onset of Self-Pulsing in Lasers and the Lorenz Model," *Phys. Lett. A* **58**, 440–41 (1976).

C. O. Weiss and J. Brock, "Evidence for Lorenz-Type Chaos in a Laser," *Phys. Rev. Lett.* **57**, 2804–6 (1986).

C. O. Weiss, N. B. Abraham, and U. Hübmer. "Homoclinic and Heteroclinic Chaos in a Single-Mode Laser," *Phys. Rev. Lett.* **61**, 1587–90 (1988).

S. H. Strogatz, *Nonlinear Dynamics and Chaos* (Addison–Wesley, Reading, MA, 1994). Chapter 9 includes a nice discussion of the Lorenz model, including a detailed description of a waterwheel that can be modeled with the same equations.

Appendix D

The Research Literature on Chaos

A Survey

The research literature on nonlinear dynamics and chaos has spread through journals in almost every field of science and engineering. Here we list the titles of journals in which we have found a large number of articles dealing both with specific nonlinear systems and their analysis and with fundamental problems in nonlinear science. These journals are available in almost all college and university science libraries.

- *American Journal of Physics* (American Association of Physics Teachers)
- *Chaos* (American Institute of Physics)
- *Complexity International* (an electronic journal available at http://www.csu.edu.au/ci/ci.html
- *International Journal of Bifurcations and Chaos in Applied Sciences and Engineering* (World Scientific)
- *Nonlinear Dynamics* (Kluwer)
- *Physica D* (North-Holland)
- *Physical Review Letters* (American Physical Society)
- *Physical Review A* (American Physical Society)
- *Physical Review E* (American Physical Society)
- *Physics Letters A* (North-Holland)

Appendix E

Computer Programs

E.1 General Comments

This appendix contains the listings of three simple computer programs. The first calculates trajectory points for the logistic map and then plots them as a function of iteration number. The second plots a bifurcation diagram for the logistic map. You may select a range of parameter values and x values. The third displays the graphic iteration method. Sample screen outputs from the programs are included.

The programs have been written in the language QuickBasic, but they are easily adaptable to other languages. As you can see, most of the program provides the user with information, requests information from the user and sets up the screen display. The actual computations are quite simple.

Comments are preceded by the ' symbol and are ignored by the computer.

E.2 Program to Calculate and Plot Trajectory Points for the Logistic Map

```
'The map function is  X = A X (1 - X).
CLS:' clear the screen

'Set up screen mode and colors
        SCREEN 11:  VGA 640x480 graphics, text has 80 columns and 30 rows
        CNUMW = 65536*63+256*63+63: 'gives bright white
        CNUMB = 0: 'gives black
        PALETTE 0, CNUMW: 'this gives a white background
        PALETTE 1, CNUMW: 'this gives  black text and drawing lines
'Alternatively SCREEN 12 allows for more colors if desired.
'print information for the user
        PRINT "This program calculates and plots trajectory points for the logistic
map."
        PRINT
        PRINT "The map function is X = A * X * (1 - X)"
        PRINT
        PRINT "You are asked for the value of the parameter A   (0 < A < 4)"
        PRINT "and for the initial value of x   (0 < x < 1)"
        PRINT "and for the number of points to be calculated and plotted."
        PRINT
        PRINT "Hit any key to proceed."
' this is a wait loop
```

```
cbase:   A$ = INKEY$: IF A$ = "" THEN GOTO cbase
CLS:' clear the screen


again: ' program returns here for a repeat performance
CLS 0: 'clear the screen of all text and graphics
LOCATE 1, 1: ' put the cursor back at the top left of the screen
AIN:     INPUT " VALUE OF A (0 < A < 4)"; A
IF A < 0 OR A > 4 THEN GOTO AIN ' check range of A
XIN:     INPUT "Initial x value (0 < x < 1)"; xi
IF xi < 0 OR xi > 1 THEN GOTO XIN: ' check range of xi
INPUT "Number of points to calculate and plot"; NP
x = xi: ' use this as the initial x value
INPUT "Allow transients to go away [y or n]"; T$
IF T$ = "y" OR T$ = "Y" THEN
FOR I = 1 TO 200: ' get rid of transients
x = A * x * (1 - x): 'logistic map function
NEXT I: ' END OF TRANSIENT LOOP
END IF
FOR J = 1 TO NP: 'print next NP points to the screen
x = A * x * (1 - x): ' logistic map function
PRINT J, x
NEXT J: ' end of print loop
```

'COMMENTS: This next section uses QuickBasic graphics commands VIEW, WINDOW, LINE and PSET. These may need to be changed for other versions of Basic and for other languages. The LOCATE command assumes we are using 30 rows and 80 columns of characters on the screen.

'Ask the user if the points should be plotted:

```
        INPUT "Plot these [y or n]"; B$


        IF B$ = "y" OR B$ = "Y" THEN CLS 0:
' clear all text and graphics
        VIEW (1, 1)-(639, 439): ' set viewport for graphics
        WINDOW (0, 0)-(NP, 1): 'set coordinates for plotting
        LINE (0, 0)-(NP, 1), , B: 'draw box
'Draw hash marks
        FOR J = 1 TO NP
        LINE (J, 0)-(J, .05)
        NEXT J
        LOCATE 30, 40: PRINT "Iteration number";
        LOCATE 2, 2: PRINT "X        A = "; A;
        IF T$ = "N" OR T$ = "n" THEN x = xi:
'start at beginning to see transient
```

```
PSET (0, x): 'plot first point
FOR J = 1 TO NP
x = A * x * (1 - x)
LINE -(J, x): ' draw line to next point
NEXT J
END IF ' Now print messages for further action by the user
LOCATE 30, 1: PRINT "Hit r to repeat.  Hit q to quit";
'Wait for response
Here:  A$ = INKEY$: IF A$ = "" THEN GOTO Here
IF A$ = "q" OR A$ = "Q" THEN GOTO qbase
IF A$ = "r" OR A$ = "R" THEN GOTO again
qbase: 'quit program
END
```

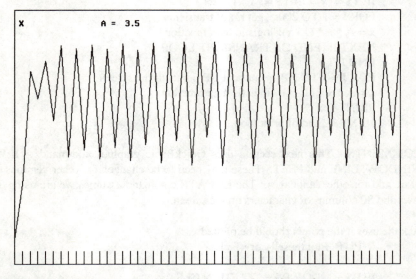

Fig. E.1. Typical output from the logistic map program. Here $A = 3.5$ and $x_0 = 0.1$. After an initial transient, the behavior is period-4.

E.3 BIFUR, A Program to Plot a Bifurcation Diagram for the Logistic Map

```
'The map function is X = A * X * (1-X)
' clear the screen and print messages
CLS 0

PRINT "This program plots a bifurcation diagram for the logistic map."
PRINT
```

```
            PRINT "You are asked for the range of parameter values"
            PRINT "and for the range of x values (vertical section)."
            PRINT
            PRINT "Hit any key to proceed."
      abase: A$ = INKEY$: IF A$ = "" THEN GOTO abase: 'a wait loop

'Set up screen mode and colors
            SCREEN 11:  VGA 640x480 graphics, text has 80 columns and 30 rows
            CNUMW = 65536*63+256*63+63: 'gives bright white
            CNUMB = 0: 'gives black
            PALETTE 0, CNUMW: 'this gives a white background
            PALETTE 1, CNUMW: 'this gives  black text and drawing lines
'Alternatively SCREEN 12 allows for more colors if desired.


            rbase: 'back to here to repeat program
            CLS 0: ' clear the screen
' Ask for range of A and range of x
            INPUT "Initial value of A ( 0 < A < 4) "; AI
            INPUT "Final value of A (0 < A < 4) "; AF
            PRINT
            INPUT "Lower value of x to be plotted (0 < x < 1)"; XL
            INPUT "Lower upper of x to be plotted (0 < x < 1)";XU
            PRINT
            PRINT "Suggestion: plot NP = 50/(XU -XL) points."
            PRINT
            INPUT "Number of points to plot for each value of A"; NP
            CLS 0: ' clear the screen


'set view port with room for labels


'COMMENTS:  This next section uses QuickBasic graphics commands VIEW,
WINDOW, LINE and PSET.  These may need to be changed for other versions of
Basic and for other languages.  The LOCATE command assumes we are using 30
rows and 80 columns of characters on the screen.


'set up view area for graphics
            VIEW (50, 0)-(639, 479)
'set up Window with axes between XL and XU
            WINDOW (AI, XL)-(AF, XU)
'draw box
            LINE (AI, XL)-(AF, XU), , B
'Print Labels for the graph
            LOCATE 5, 1: PRINT "X";
            LOCATE 28, 1: PRINT XL;
```

```
          LOCATE 1, 1: PRINT XU;
          LOCATE 29, 40: PRINT "A";
          LOCATE 29, 3: PRINT AI;
          LOCATE 29, 80 - LEN(STR$(AF)): PRINT AF;
' print information on action keys
          LOCATE 30, 1: PRINT "Hit r to repeat. Hit q to quit.";
' Now calculate the trajectories and plot the diagram
          FOR K = 0 TO 639: ' use 640 steps across the screen
          A = AI + (AF - AI) * K / 639: 'ACTUAL PARAMETER VALUE
          X = .2: 'initial value
          FOR I = 1 TO 200: ' get rid of transients
          X = A * X * (1 - X)
          NEXT I: ' END OF TRANSIENT LOOP
          FOR J = 1 TO NP: 'plot next NP points
          X = A * X * (1 - X)
          PY = X
          PSET (A, PY)
          NEXT J: ' END OF PLOT LOOP
'check for key stroke
          A$ = INKEY$
          IF A$ = "q" OR A$ = "Q" OR A$ = "r" OR A$ = "R" THEN EXIT FOR
' exit the loop if an action key has been hit
          NEXT K: ' END OF PARAMETER VALUES LOOP
' decide on which action to take
          IF A$ = "q" OR A$ = "Q" THEN END
          IF A$ = "r" OR A$ = "R" THEN GOTO rbase
          HERE:  A$ = INKEY$: IF A$ = "" THEN GOTO HERE
'wait loop for screen capture if desired
          IF A$ = "r" OR A$ = "R" THEN GOTO rbase
END
```

E.4 A Program to Display Graphic Iterations of Map Functions

This program puts the iterated map function in a QuickBasic FUNCTION statement so that you can easily substitute other map functions. The control parameter is A and the iterated variable is x.

```
          DECLARE FUNCTION MF! (A!, x!)
          DECLARE SUB Waitsub (A$)
' Plots map function and steps through graphical version of iteration scheme
' The mapping function is defined in the Function MF(A,x)

'Set up graphics screen
'Set up screen mode and colors
          SCREEN 11:  VGA 640x480 graphics, text has 80 columns and 30 rows
```

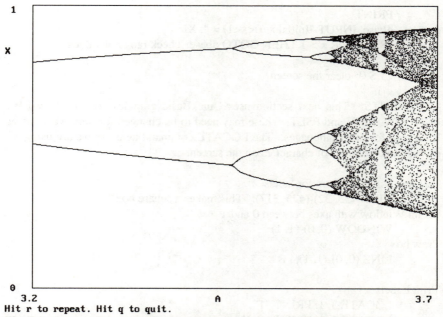

Fig. E.2. Typical bifurcation diagram generated by the program BIFUR.

```
        CNUMW = 65536*63+256*63+63: 'gives bright white
        CNUMB = 0: 'gives black
        PALETTE 0, CNUMW: 'this gives a white background
        PALETTE 1, CNUMW: 'this gives black text and drawing lines
'Alternatively SCREEN 12 allows for more colors if desired.


        CLS 0: 'clear the screen
' Give first message
        PRINT "This program iterates the logistic map graphically"
        PRINT "step by step."
        PRINT
        PRINT "After each step, the program pauses.  Hit any key for the next
step."
        PRINT
        PRINT "Hit any key to proceed."
cbase: A$ = INKEY$: IF A$ = "" THEN GOTO cbase: 'wait loop


        rbase:' return here for repetition of the program
        CLS 0: ' clear the screen
'Input parameter value and initial x value
        INPUT "Parameter A  (0 < A < 4) ="; A
        IF A < 0 OR A > 4 THEN GOTO rbase: 'check range of values
```

```
        PRINT
dbase: INPUT "Initial x (0<x<1) = "; XI
        IF x < 0 OR x > 1 THEN GOTO dbase: 'check range of values

        CLS 0:' clear the screen
```

'COMMENTS: This next section uses QuickBasic graphics commands VIEW, WINDOW, LINE and PSET. These may need to be changed for other versions of Basic and for other languages. The LOCATE command assumes we are using 25 rows and 80 columns of characters on the screen.

```
'set view port with room for labels
        VIEW (35, 32)-(425, 317): ' This makes a square box
'set up Window with axes between 0 and 1
        WINDOW (0, 0)-(1, 1)
'draw box
        LINE (0, 0)-(1, 1), , B

'Print Labels
        LOCATE 3, 1: PRINT "1";
        LOCATE 7, 1: PRINT "f(x)";
        LOCATE 29, 28: PRINT "x";
        LOCATE 29, 2: PRINT "0";
        LOCATE 29, 55: PRINT "1";
        LOCATE 5, 10: PRINT "A="; A;
        LOCATE 5, 60: PRINT "Hit a key:";
        LOCATE 10, 70: PRINT "q = quit";
        LOCATE 15, 70: PRINT "r = restart";
        LOCATE 20, 70: PRINT "n = next step";
'Plot x = y line
        LINE (0, 0)-(1, 1)
'Plot function
        FOR j = 0 TO 639
        x = j / 639
        y = MF(A, x)
        PSET (x, y)
        NEXT j
'Draw line from XI to f(x) then to x = y line
        y = MF(A, XI)
        LINE (XI, 0)-(XI, y)
        LINE -(y, y)
'go to Waitsub to pause
        CALL Waitsub(A$)
```

```
'Now set up loop to repeat this process
        xx = y:  'transfer the y value to xx


        AGAIN:
        y = MF(A, xx)
        LINE -(xx, y)
        LINE -(y, y)
        CALL Waitsub(A$)
        IF A$ = "r" OR A$ = "R" THEN GOTO rbase
        IF A$ = "q" OR A$ = "Q" THEN GOTO qbase
        xx = y
        GOTO AGAIN
        qbase: ' quit the program
        END
' The FUNCTION statement defines the iterated map function being used.
        FUNCTION MF (A, x)
        MF = A * x * (1 - x)
        END FUNCTION

        SUB Waitsub (A$)
'This routine waits for the user to hit a key.
'Screen capture can be invoked while waiting.
        DO
        FOR j = 1 TO 1000: NEXT j
        A$ = INKEY$
        LOOP WHILE A$ = ""
        END SUB
```

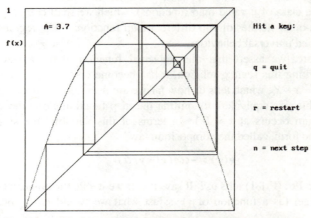

Fig. E.3. Typical graphical iteration output for the logistic map.

Appendix F

Theory of the Universal Feigenbaum Numbers

F.1 The Feigenbaum α

In Chapter 2, we saw that many one-dimensional iterated maps share common quantitative, as well as qualitative, features. In particular, we saw that there is a common convergence ratio (the Feigenbaum number δ) and scaling ratio (the Feigenbaum number α) as the period-doubling sequence proceeds toward chaos. It was the discovery of these universal quantitative features that, at least in part, alerted scientists and engineers to the potential importance and usefulness of the new approaches to nonlinear dynamics.

In this appendix we will give a heuristic theory of the universality of the Feigenbaum numbers α and δ. This theory will lead to actual (but approximate) numerical values for these important numbers. The ideas involved here are at once both simple and subtle. Although the mathematics used in this theory does not require more sophistication than we have introduced before, the argument is sufficiently complicated that some readers may wish to skip this appendix. Before they do, however, we would like to state briefly, using mostly words and simple symbols, the main ideas involved.

The essential idea is that to find α and δ, we need concentrate our attention only on the behavior of the iterated function f near the value x_c for which it has a maximum. We are going to concentrate on trajectories that involve x_c (the supercycles). As the period-doubling sequence proceeds, we shift our attention to $f^{(n)}$. It turns out that if we rescale our graphs by the factor $\alpha = 2.502...$ for each period-doubling, then $f^{(n)}$ near x_c approaches a *universal* function (that is, it is the same for a wide class of iterated map functions), which we shall call $g(x)$. Figure F.1 shows high-order iterates of two different map functions. The regions near x_c show the expected universal behavior.

Furthermore, we assert that this universal function obeys a size scaling relation. In writing this scaling relation, it has become customary to introduce a new variable $y = x - x_c$, which tells us how far we are from the value x_c. (Note that using the variable y is equivalent to shifting the map function along the x axis so that its maximum occurs at $x = 0$.) In terms of this variable, the size-scaling relation takes the form, called the composition law:

$$g(y) = -\alpha g(g(-y/\alpha)) \tag{F.1-1}$$

What does Eq. (F.1-1) tells us? It says that if we iterate the universal function $g(y)$, what we get (as a function of y) is just what we would have obtained by

568

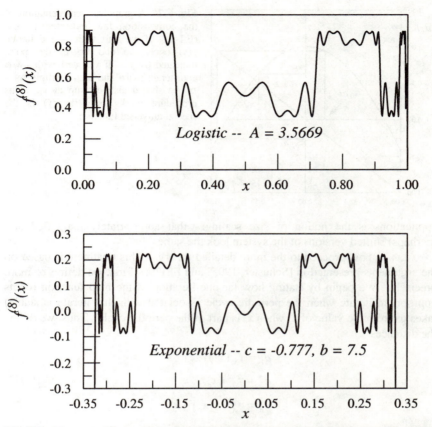

Fig. F.1. The upper diagram is a plot of the eighth iterate of the logistic map. The lower diagram is the eighth iterate of the exponential map function. Parameters are close to those of the period-doubling accumulation point. Note that near x_c (at the center of each plot) both functions have a similar appearance. $x_c = 0.5$ for the logistic map and $x_c = 0$ for the exponential map. Differences show up near for x values further from x_c. With appropriate scaling of the ordinate and abscissa, the central parts of the diagrams match almost exactly.

starting with $-y/\alpha$, iterating the function twice and then multiplying the result by $-\alpha$. Figure F.2 shows the graphic representation of this equivalence.

As we shall see, the existence of this universal function, the size scaling near x_c, and the shape of f near x_c are all that we need to find the values of α and δ. This type of reasoning is widely used in contemporary theoretical physics, where it goes under the name of "renormalization theory." The term *renormalization* means essentially the same as the size scaling in our discussion. Renormalization theory has been very important in the theory of phase transitions in statistical mechanics and in quantum field theory. In all cases the common feature linking these diverse

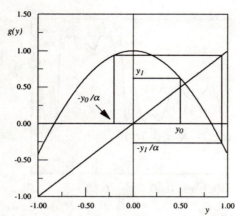

Fig. F.2. A graphical representation of the composition law indicated in Eq. (F.1-1). If we start with y_0 and iterate $g(y)$ once, we end up at the point indicated by y_1. If we start with $-y_0/\alpha$ and iterate twice and then multiply by $-\alpha$, we should also end up at y_1. This procedure works only if $g(y)$ is the special universal function.

applications is the notion of size scaling: that appropriately magnified and (perhaps) shifted versions of the system look the same.

Let us now proceed to the more detailed theory. (This treatment is based on the arguments presented in [Schuster, 1995] and FEI79.) First, we define α more precisely. We begin by stating how far one iteration of the map function (or its appropriate iterate when the periodic cycle associated with that iterate is stable) takes us from the value of x_c when x_c is part of the periodic cycle. Thus, we define the distances

$$d_1 = f_{A_1}(x_c) - x_c$$
$$d_2 = f_{A_2}^{(2)}(x_c) - x_c \qquad \text{(F.1-2)}$$
$$d_n = f_{A_n}^{(2^{n-1})}(x_c) - x_c$$

In the previous equations, A_n is now the supercycle parameter value, for which x_c is part of the stable cycle with period-2^n (that is, the period is 2, 4, 8, 16, . . .). (In Chapter 2, we called these A_n^s, but here we will drop the superscript for the sake of simplicity of typesetting). d_n is the difference between x_c and the value of x obtained by repeated application of the appropriate iterate of the map function f to take us <u>half-way</u> through the cycle. Figure F.3 illustrates these definitions. From that figure, it should be "obvious" that by going half-way through the periodic cycle, we arrive at the trajectory value closest to x_c. Thus, the d_ns measure the distance from x_c to its nearest neighbor in the cycle.

It will be convenient to use the variable y defined earlier, in terms of which the definition of d_n can be written

$$d_n = f_{A_n}^{(2^{n-1})}(0) \qquad \text{(F.1-3)}$$

The basic observation of size scaling for the period-doubling sequence is embodied in the statement that

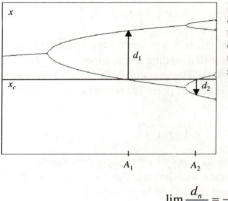

Fig. F.3. The definition of the distances d_1, d_2, and so on, used in the definition of the Feigenbaum α. Note that d_n is the distance from x_c to its nearest neighbor in the periodic cycle for that particular supercycle parameter value.

$$\lim_{n \to \infty} \frac{d_n}{d_{n+1}} = -\alpha \qquad \text{(F.1-4)}$$

The minus sign is included in the previous equation to account for the fact that the nearest neighbor of x_c is alternately above and below x_c for successive supercycles as seen in Fig. F.3.

If we assume that Eq. (F.1-4) holds all the way down to $n = 1$ (we might call this the "exact scaling" assumption), then we can write

$$(-\alpha)^n d_{n+1} = d_1 \qquad \text{(F.1-5)}$$

(We saw in Chapter 2 that although this scaling is not exact for low values of n, for the logistic map, the scaling is a very good approximate description even for $n = 2$, 3, and so on.) Using Eq. (F.1-3), we can write this last relation in terms of the iterated function:

$$(-\alpha)^n f_{A_n}^{(2^n)}(0) = d_1 \qquad \text{(F.1-6)}$$

We now take the following bold step, based partly on the behavior seen in Fig. F.1, in which we see evidence for a universal function. We assume that at least in the limit of large n, Eq. (F.1-6) also holds for values of y around $y = 0$ in the sense that

$$\lim_{n \to \infty} (-\alpha)^n f_{A_n}^{(2^n)} \left(y / (-\alpha)^n \right) = g_1(y) \qquad \text{(F.1-7)}$$

defines a universal function of y. This important equation needs some explanation. Basically, it tells us that if we scale the size of the function by the factor $(-\alpha)^n$ (embodying the size scaling of the distance d_n) and simultaneously change the range of the y axis by the factor $1/(-\alpha)^n$, then these appropriately "renormalized" or rescaled functions approach a universal function, which we call $g_1(y)$. (The reason for the subscript on g will become apparent in the next paragraph.)

Our next step may seem to be rather mysterious at first, but it is at the heart of the renormalization argument. We introduce a whole series of functions $g_i(y)$:

$$g_i(y) = \lim_{n \to \infty} (-\alpha)^n f_{A_{n+i-1}}^{(2^n)} \left(y/(-\alpha)^n \right) \quad i = 1, \ 2, \ 3, \ldots \quad \text{(F.1-8)}$$

We see from the definition of $g_i(y)$ that this series of functions evaluates high-order iterates of f at the parameter values A_n corresponding to the supercycles. [It is easy to see that Eq. (F.1-8) is satisfied for $y = 0$. The importance of Eq. (F.1-8) lies in its extension to nonzero values of y.] The crucial point for our argument is that these $g_i(y)$s obey a *composition law*:

$$g_{i-1} = -\alpha g_i \left[g_i (-y/\alpha) \right] \quad \text{(F.1-9)}$$

The proof of this last result proceeds simply by writing out the composition of functions indicated on the right-hand side and showing that they reduce to the left-hand side. As we work through the details, we will temporarily leave off the "lim." First, we use Eq. (F.1-8) to write out the right-hand side of Eq. (F.1-9):

$$-\alpha g_i[g_i(-y/\alpha)] = (-\alpha)(-\alpha)^n f_{A_{n+i-1}}^{(2^n)} \left[\frac{(-\alpha)^n}{(-\alpha)^n} f_{A_{n+i-1}}^{(2^n)} (-y/\alpha(-\alpha)^n) \right] \quad \text{(F.1-10)}$$

$$= (-\alpha)^{n+1} f_{A_{n+i-1}}^{(2^n)} \left[f_{A_{n+i-1}}^{(2^n)} (y/(-\alpha)^{n+1}) \right]$$

We now recognize (from Exercise 5.4-3) that

$$f^{(2^n)} \left[f^{(2^n)}(y) \right] = f^{(2 \cdot 2^n)}(y) = f^{(2^{n+1})}(y) \quad \text{(F.1-11)}$$

We set $n + 1 = m$ and find that the last entry of Eq. (F.1-10) reduces to

$$(-\alpha)^m f_{A_{m+i-2}}^{(2^m)} \left(\frac{y}{(-\alpha)^m} \right) \quad \text{(F.1-12)}$$

With the restoration of the lim ($m \to \infty$), we see that (F.1-12) is just what we mean by $g_{i-1}(y)$. This completes our proof of the composition law.

The final and critical step of the analysis is to assert (see FEI79 and [Collet and Eckmann, 1980] for proofs) that in the limit $i \to \infty$ (in which case there is no difference between i and $i - 1$), the functions $g_i(y)$ approach a universal function, which we shall call $g(y)$. Note that taking the limit $i \to \infty$ is equivalent to taking the parameter A to the period-doubling accumulation value A_∞. The universal function satisfies the composition law

$$\boxed{\boxed{g(y) = -\alpha g \left[g(-y/\alpha) \right] \quad \text{(F.1-13)}}}$$

We put a double box around the previous equation because it embodies the entire renormalization argument. Asserting (or better yet, proving) that such a function exists, is equivalent to asserting (or proving) the universality of the Feigenbaum numbers.

In order to find the numerical value of α, we need to specify how the function $g(y)$ behaves near $y = 0$. We now see what determines the various "universality classes" for iterated maps: It is just the behavior of the function near $y = 0$ (or near $x = x_c$ for our original variable). We expect that many functions with a maximum at $y = 0$ will be described near $y = 0$ by the following quadratic form

$$g(y) = b - cy^2 \tag{F.1-14}$$

where b and c are fixed parameters, whose values we shall find.

Exercise F.1-1. Why do we expect Eq. (F.1-14) to be very common? A partial answer: Think of expanding $g(y)$ in a Taylor series about $y = 0$. The first derivative term vanishes. (Why?) What would we expect for the first nonzero term?

Exercise F.1-2. Show that the logistic map and the sine map of Chapter 2 can be put in the form of Eq. (F.1-14) by evaluating a Taylor series expansion near x_c and finding b and c.

If we now accept Eq. (F.1-14) as an approximate representation of $g(y)$, we can find a value for α by using Eq. (F.1-14) in Eq. (F.1-13). This substitution leads to the following algebraic steps:

$$
\begin{aligned}
b - cy^2 &= -\alpha \left[b - c \left\{ b - c \frac{y^2}{\alpha^2} \right\}^2 \right] \\
&= -\alpha \left[b - c \left\{ b^2 - 2bc \frac{y^2}{\alpha^2} + c^2 \frac{y^4}{\alpha^4} \right\} \right] \\
&= -\alpha \left[b - cb^2 \right] - \frac{2bc^2}{\alpha} y^2 + \frac{c^3 y^4}{\alpha^3}
\end{aligned}
\tag{F.1-15}
$$

We now need to make an additional approximation. If we let y be very small, then the last term in the previous equation, which involves y^4, will be very small compared to the first two terms. We assume that it can be ignored compared to the other terms. Then for the equation to hold, the coefficients of the terms that are independent of y on the two sides of the equation must be equal and the coefficients multiplying y^2 must be equal. Thus, we require

$$b = -\alpha[b - cb^2] \tag{F.1-16}$$

$$-c = -\frac{2bc^2}{\alpha} \tag{F.1-17}$$

Using Eq. (F.1-17) in (F.1-16) yields the following quadratic equation for α:

$$\alpha^2 - 2\alpha - 2 = 0 \tag{F.1-18}$$

which has the solutions

$$\alpha = 1 \pm \frac{1}{2}\sqrt{12} \tag{F.1-19}$$

If we take α positive, then we find $\alpha = 2.73...$, which is within 10% of Feigenbaum's 2.502... [We could improve our approximation by using a more elaborate power series expansion for $g(y)$ in Eq. (F.1-14) and following essentially the same procedures.]

Question: Does the negative root solution have any significance?

Note that in Eq. (F.1-14), the parameter b is the value of $g(y)$ at $y = 0$. We can set $b = 1$ without loss of generality since b just sets the scale of our vertical axis. Using this value, along with our approximate value for α in Eq. (F.1-17), we can find a value for c and hence get an approximate representation of the universal function $g(y)$:

$$g(y) \approx 1 - 1.36 y^2 \tag{F.1-20}$$

Eq. (F.1-20) looks exactly like the equation for any run-of-the-mill quadratic function with a maximum at $y = 0$, but it is important to realize that the composition law Eq. (F.1-13) holds only for one specific value of α and one specific value of c. You might try the graphic representation of the composition law as illustrated in Fig. F.2 to convince yourself of the uniqueness of $g(y)$.

Using a procedure similar to the one outlined here, Lanford (LAN82) has calculated a power series expansion of $g(y)$:

$$\begin{aligned} g(y) \approx {}& 1 - 1.52763 y^2 \\ & + 0.104815 y^4 \\ & + 0.026705 y^6 \dots \end{aligned} \tag{F.1-21}$$

which then provides a more precise representation of this universal function.

F.2 *Derivation of the Feigenbaum Number δ

Now we want to use a similar analysis to find a value for the Feigenbaum δ. To carry out this analysis, we will introduce a slightly different notation for the composition law Eq. (F.1-9):

$$g_{i-1}(y) = -\alpha g_i \left[g_i \left(-y/\alpha \right) \right] \equiv T g_i(y) \tag{F.2-1}$$

The last equality defines an "operator" T, which is called the **doubling transformation** since it corresponds to applying $g_i(y)$ twice. We say that the doubling transformation operator T operates on the function $g_i(y)$ to generate another function $g_{i-1}(y)$. The universal function $g_i(y)$ is then identified as that function which is a "fixed point" of this transformation

$$g(y) = Tg(y) \qquad\qquad (F.2-2)$$

If you recall the definition of fixed point for some iterated map function $x^* = f(x^*)$), you can see why $g(y)$ is called a fixed point of the transformation. In the present case, we are dealing with functions rather than numbers, but the ideas are the same.

Next we note that we are concerned with parameter values near A_∞, the parameter value at which the period-doubling bifurcations accumulate and beyond which chaotic behavior occurs. In this region repeated iterations of f are supposed to converge, with the appropriate rescalings as described in the previous section, to the universal function $g(y)$. So, we express f as a Taylor series expansion in terms of the <u>control parameter</u> (not in terms of the variable y) about the point A_∞, that is, we focus on how the function depends on the parameter A:

$$f_A(y) = f_{A_\infty}(y) + (A - A_\infty)\frac{\partial f_A(y)}{\partial A}\bigg|_{A_\infty} + \dots \qquad (F.2-3)$$

Note that we use a <u>partial</u> derivative with respect to the parameter A since the function f depends on both y and A.

The next step is to observe that repeated application of the doubling operator to $f_{A_\infty}(y)$ should bring us closer and closer to $g(y)$. Specifically, if we apply T n-times, we write

$$T^{(n)} f_{A_\infty}(y) = (-\alpha)^n f_{A_\infty}^{(2^n)}\left(y/(-\alpha)^n\right) \approx g(y) \qquad (F.2-4)$$

for $n \gg 1$. Thus, we rewrite Eq. (F.2-3) in the following form

$$T^{(n)} f_A(y) \approx g(y) + (A - A_\infty)T^{(n)}\frac{\partial f_A(y)}{\partial A}\bigg|_{A_\infty} + \dots \qquad (F.2-5)$$

If we specialize to the case of $A = A_n$ (a supercycle value), we have (by the definition of the supercycle)

$$T^{(n)} f_{A_n}(0) = (-\alpha)^n f_{A_n}^{(2^n)}(0) = 0 \qquad (F.2-6)$$

Thus, we must have

$$-g(0) = (A_n - A_\infty)T^{(n)}\frac{\partial f_{A_n}}{\partial A}\bigg|_{A_\infty} \qquad (F.2-7)$$

where we have kept only the first-derivative term on the right-hand side of Eq. (F.2-5). If Eq. (F.2-7) is to hold for all n, then we must have (you may need to convince yourself of this point)

$$T^{(n)} \frac{\partial f_{A_n}}{\partial A}\bigg|_{A_\infty} = \delta^n \times \text{constant} \qquad \text{(F.2-8)}$$

where the Feigenbaum δ is defined as

$$\delta = \lim_{n \to \infty} \frac{A_n - A_{n-1}}{A_{n+1} - A_n} \qquad \text{(F.2-9)}$$

in terms of the control parameter values at which supercycles occur. Eq. (F.2-8) asserts that the parameter scaling holds for all values of n.

To see that Eq. (F.2-8) is reasonable, recall that in Chapter 2 we established that

$$(A_\infty - A_n) \times \delta^n = (A_2 - A_1) \frac{\delta^2}{\delta - 1} = \text{constant} \qquad \text{(F.2-10)}$$

In essence what Eq. (5.7-8) requires is that when T acts on $g(y)$, we must have, to first-order in $A - A_\infty$

$$Tg(y) = \delta g(y) \qquad \text{(F.2-11)}$$

We can now employ Eq. (F.2-11) to find a value of δ by writing a "linearized" form of Eq. (F.2-1) by using a Taylor series expansion about $g(0)$:

$$Tg(y) = -\alpha g\left[g\left(-y/\alpha\right)\right]$$
$$= -\alpha \left\{ g\left[g(0)\right] + \frac{\partial g}{\partial y}\bigg|_{g(0)} [g(y) - g(0)] + \dots \right\} \qquad \text{(F.2-12)}$$

We invoke our choice of vertical scale to set $g(0) = 1$. We also note that $g(1) = -1/\alpha$ and that

$$\frac{\partial g}{\partial y}\bigg|_{y=1} = -\alpha \qquad \text{(F.2-13)}$$

Using these results in Eq. (F.2-12) and requiring that Eq. (F.2-11) also be satisfied leads to the following relationship between the two Feigenbaum numbers:

$$-\alpha(-\alpha + 1) = \delta \qquad \text{(F.2-14)}$$

If we use our previously calculated value of α, we find that $\delta \sim 4.72$, which is remarkably close to the more precise value 4.669... determined by a more refined version of the present calculation.

Exercise F.2-1. Prove the result stated in Eqs. (F.2-13) and (F.2-14) and then verify the numerical result stated earlier for δ. Note that our remarkable agreement is partly accidental: If we use the more precise value $\alpha = 2.502...$ in Eq. (F.2-14) our agreement with the more precise value of δ is worse.

In Chapter 2, we introduced the Feigenbaum δ as the ratio of parameter value differences at which bifurcations occur because it is these bifurcations (rather than the supercycles) that are detectable experimentally. If we denote the bifurcation parameter values by a_n and the supercycle values by A_n, then we see that we must have $a_n < A_n < a_{n+1} < A_{n+1}$. Hence in the limit $n \to \infty$, we expect the ratio of the parameter differences for successive bifurcations and the ratio of parameter differences for successive supercycles to be the same.

Let us summarize the arguments used to calculate the Feigenbaum numbers. (It is tempting to call these important numbers "*Feigenvalues*.") We have essentially implemented three (related) ideas: (1) functional convergence to a universal function depending only on the nature of the iterated map function near its maximum value, (2) universal size scaling of that function as we approach the period-doubling accumulation point, and (3) parameter difference scaling in the approach to the period-doubling accumulation value. We have admittedly handled these notions in a nonrigorous mathematical fashion, but we want to convey the essence of the renormalization arguments without getting involved in the formalism needed to make the derivations into proofs. The net result of implementing these ideas, together with the specification of the nature of the iterated map function near its maximum value, is that the Feigenbaum α and δ values are determined "automatically." These values are universal for all iterated map functions that have the same mathematical character near their maxima.

We can now see why there are ***universality classes*** for iterated map functions. The universality classes are determined by the nature of the map function near its maximum value. As we have seen, map functions that have a quadratic maximum give rise to the universal function defined earlier and lead to the α and δ values 2.502 and 4.669. If the map function behaves as y^4, then the equivalent convergence ratio and scaling ratio will be different. Applying techniques similar to those developed here, Hu and Satija (HUS83) have determined α and δ values for iterated map functions of the form $f(x) = 1 - a|x|^z$, where the exponent z determines the universality class. They found that the Feigenbaum numbers vary smoothly with z. In particular, α seems to decrease smoothly as z increases and approaches the value 1.27 for large z. On the other hand, δ seems to increase linearly for large values of z.

Other Periodic Windows

As we have seen, many iterated map functions give rise to periodic windows within the bands of chaos (see Fig. 5.9, for example). Each of these windows contains a period-doubling sequence leading to yet more chaotic bands. We might expect that there are Feigenbaum numbers analogous to the Feigenbaum δ and α for each of these windows. These numbers have been computed for two different map functions in DHK85. They find that the numbers increase exponentially with the period of the periodic window. For example, for a period-3 window, they find that $\delta_3 = 55.26$ and $\alpha_3 = 9.277$. To a good approximation, for all periodic windows the relation $3\delta_N = 2\alpha_N^2$ holds for iterated map functions with a quadratic extremum, where N is the period of the periodic window.

F.3 Further Reading

M. Feigenbaum, "The Universal Metric Properties of Nonlinear Transformations," *J. Stat. Phys.* **21**, 669–706 (1979) (reprinted in [Hao, 1984]). Provides a proof of the universality of α and δ.

M. J. Feigenbaum, "Universal Behavior in Nonlinear Systems," *Los Alamos Science* **1**, 4–27 (1980) (reprinted in [Cvitanovic, 1984]). A quite readable introduction to the universal features of one-dimensional iterated maps.

O. E. Lanford III, "A Computer-Assisted Proof of the Feigenbaum Conjectures," *Bull. Am. Math. Soc.* **6**, 427–34 (1982) (reprinted in [Cvitanovic, 1984]). A power series representation of the universal $g(y)$ function.

Appendix G

The Duffing Double-Well Oscillator

G.1. The Model

Many systems in nature have several stable states separated by energy barriers. When the system can move among the stable states, the dynamics can become quite complex. A simple model that illustrates some of these features is the Duffing double-well oscillator. This model was first introduced to understand forced vibrations of industrial machinery [Duffing, 1918]. In this model, a particle is constrained to move in one spatial dimension. An external force acts on the particle. The force is described by

$$F = +kx - bx^3 \qquad (G\text{-}1)$$

The name "double-well" enters because the corresponding potential energy function has a double well structure. Formally, the potential energy function is written as

$$U(x) = -\tfrac{1}{2}kx^2 + \tfrac{1}{4}bx^4 \qquad (G\text{-}2)$$

Figure G.1 shows a plot of the potential energy function. We see that there are two stable equilibrium states at $x = \pm\sqrt{k/b}$. There is an unstable equilibrium point at $x = 0$.

To build a simple mechanical model of the Duffing oscillator, mount a flexible metal strip vertically with the base rigidly clamped. Then place a movable mass on the flexible strip. If the mass is mounted low enough, the stable position will occur with the strip directly vertical. Small deviations from this position will result in oscillations around the vertical position. If the mass is moved further up the strip, eventually the vertical position becomes unstable, and the mass will "flop" to one side or the other. There are now two stable positions—one on each side of the vertical—with a "barrier" in between. See BEN97 for the details of setting up such a system.

Exercise G.1. There are several variations on the Duffing oscillator. If we take $k < 0$, we retrieve for $b = 0$ the usual simple harmonic oscillator. If $b \neq 0$, we say that the oscillator is **anharmonic**. For $b > 0$, the force gets weaker for larger displacements from $x = 0$, and we say we have a *softening* spring situation. For $b < 0$, the force gets stronger for large displacements, and we say we have a *hardening* spring. For each of those cases, plot $U(x)$ as a function of x and identify equilibrium points.

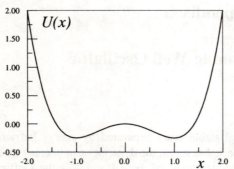

Fig. G.1. A plot of the Duffing double-well potential energy function with $k = 1$ and $b = 1$.

The equations describing the dynamics in state space are usually written with $k = 1$ and $b = 1$ (with no loss in generality) as

$$\dot{x} = y$$
$$\dot{y} = x - x^3 - \gamma y \qquad \text{(G-3)}$$

where the γ term represents damping proportional to the velocity of the particle. The motion of the particle in this situation is relatively simple. If started off with a certain amount of kinetic energy, the particle oscillates back and forth, gradually losing energy via damping and finally comes to rest at the bottom of one of the wells. What makes the oscillator interesting is that the period of the oscillations depends on the amplitude. A typical trajectory is illustrated in Fig. G.2. Note the difference in oscillation period for the initial large oscillation compared to the smaller amplitude oscillations. For the small oscillations, the oscillation period is $\pi\sqrt{2}$; that is, the natural oscillation frequency ω_0 (for small oscillation amplitudes) is equal to $\sqrt{2}$.

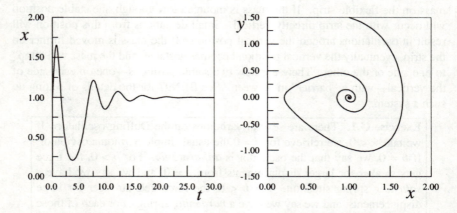

Fig. G.2. On the left, a plot of x as a function of time. On the right, a state space plot of a Duffing model trajectory from Eq. (G-3) starting with $x = 1$ and $y = 1.5$. The damping coefficient is $\gamma = 0.5$.

Exercise G.2. Verify that the period of oscillation for small amplitude oscillations is $\pi\sqrt{2}$. Hint: Set $\delta = 1 + x$ and use that in Eq. (G-3). Assume $\delta \ll 1$, use a binomial expansion on the cubic term, and reduce the equations to those for a simple harmonic oscillator.

The behavior becomes much more interesting if we "jiggle" the particle with another external force that varies, say, periodically in time. In that case, the state space equations become

$$\dot{x} = y$$
$$\dot{y} = x - x^3 - \gamma\, y + F\cos(\omega t) \qquad\qquad (G\text{-}4)$$

We might expect, based on experience with the simple harmonic oscillator, that the particle will respond with relatively large amplitude motion when the frequency of the external force matches the natural oscillation frequency of the particle. The complication is that the natural oscillation frequency depends on the amplitude of the motion. So as the particle begins to respond to the external oscillating force, its amplitude changes and hence its natural oscillation frequency changes. Several novel features can appear:

1. The response curve of the system changes shape as the amplitude of the external oscillating force increases.
2. The response curve of the system shows hysteresis: the response amplitude depends on whether we increase the frequency through the resonance region or decrease the frequency through the resonance region.
3. The system can display chaotic behavior.

The first two features are discussed in some detail, supported by analytical calculations, in [Strogatz, 1994], pp. 226–7 and 238–40 and [Jackson, 1991], Vol. 1, pp. 308-314. Here, we will focus our attention on the third item, chaotic behavior.

Let's begin with a relatively small value of the amplitude of the external oscillating force. In that case the system behaves, after initial transients die out, much like a simple harmonic oscillator with the oscillations confined around $x = \pm 1$. Fig. G.3 illustrates some possibilities.

We see that there are two attractors: a limit cycle centered on $x = 1$ and a limit cycle centered on $x = -1$. Which initial conditions (in state space) lead to which attractor? The answer turns out to be rather complicated because the two basins of attractions are thoroughly intertwined and their boundaries form a fractal structure in state space (MOL85). The frontispiece of this book illustrates the complex basins of attractions for $F = 0.25$ and $\gamma = 0.25$.

For values of F between 0.38 and 0.84 (for $\gamma = 1$ and $\omega = 1.0$), we get a complex mix of chaotic behavior interspersed with periodic windows. Two

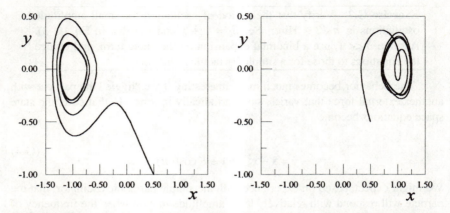

Fig. G.3. State space trajectories for the Duffing model with $F = 0.25$, $\gamma = 0.5$, and $\omega = 1.0$. On the left, $y_0 = -1.0$, $x_0 = 0.5$. On the right, $y_0 = -0.5$, $x_0 = 0.5$. The two trajectories lead to the two different limit cycles.

examples are shown in Fig. G.4. Figure G.5 shows a Poincaré section (stroboscopic portrait) of the state space taken at the phase when the driving force has its largest value, that is, when $\cos(\omega t) = 1$. The complex structure of the chaotic attractor is apparent. For larger values of F, the behavior is periodic with the period of the driving force.

   The Duffing oscillator model, though relatively simple mathematically, yields surprisingly rich behavior. References for further reading are given in the next section.

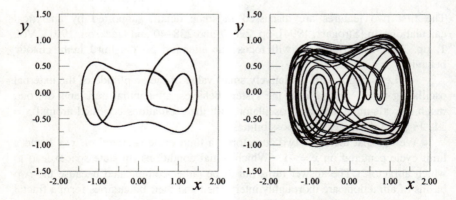

Fig. G.4. On the left, a period-4 attractor with $F = 0.5$. On the right, a chaotic attractor with $F = 0.7$. In both cases $\gamma = 0.5$ and $\omega = 1.0$.

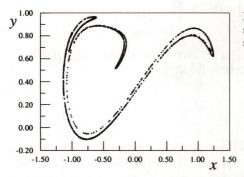

Fig. G.5. The Poincaré section of the state space for the chaotic attractor shown on the right in Fig. G.4.

G-2. Further Reading

G. Duffing, *Erzwungene Schwingungen bei veränderlicher Eigenfrequenz and ihre technische Bedeutung* (Friedr. Vieweg & Sohn, Braunschweig, 1918).

F. C. Moon and G.-X. Li, "Fractal Basin Boundaries and Homoclinic Orbits for Periodic Motion in a Two-Well Potential," *Phys. Rev. Lett.* **55**, 1439–42 (1985).

[Moon, 1992]. Contains a good discussion of the Duffing model.

C. L. Olson and M. G. Olsson, "Dynamical symmetry breaking and chaos in Duffing's equation," *Am. J. Phys.* **59**, 907–11 (1991). This paper gives a detailed analysis of the single well, hardening spring version of the Duffing model.

J. E. Berger and G. Nunes, Jr., "A mechanical Duffing oscillator for the undergraduate laboratory," *Am. J. Phys.* **65**, 841–846 (1997).

G-3. Computer Exercises

CEG-1. Use *Chaotic Dynamics Workbench* to explore the dynamics of the Duffing oscillator. For a fixed value of c and ω find the range of F that leads to chaotic dynamics. Locate a period-doubling sequence if you can.

CEG-2. Use *Dynamics: Numerical Explorations* [Nusse and Yorke, 1998] (or, more challenging, write your own program) to explore the basin of attraction for the two types of limit cycles for the Duffing model as illustrated in Fig. G.3. Note: this computation may take a long time if you want a high resolution picture of the basins.

CEG-3. Use *Dynamics: Numerical Explorations* [Nusse and Yorke, 1998] (or, more challenging, write your own program) to generate a bifurcation diagram for the Duffing oscillator with F the variable parameter. How are you sure that you have got all the attractors in the bifurcation diagram?

Appendix H

Other Universal Features for One-Dimensional Iterated Maps

H.1 Introduction

In this appendix we describe, without any derivations or proofs, some additional universal quantitative features that appear for one-dimensional iterated maps, beyond those introduced in Chapter 5. The numerical values for these features can be derived by methods similar to those used in Chapter 5 and Appendix F. However, the details of the derivations would take us too far afield; therefore, the treatment will be purely descriptive.

H.2 Power Spectrum

We have seen that as the period-doubling bifurcations proceed, new subharmonics of the fundamental period of the system appear with each new bifurcation. Thus, if we calculate a power spectrum of the trajectories of the iterated map function, we would expect to see new components appear at the 2^n subharmonic when period 2^n is "born" at the nth bifurcation. However, because the system is nonlinear, we also expect frequency components at the frequencies corresponding to all the possible sum and difference frequencies for all the harmonics present. These notions are perhaps best understood through a simple example.

For a parameter value below the first period-doubling bifurcation for an iterated map function, the trajectory has a period that we shall call $T = 1$. After transients die away, every iteration value is the same. Thus, the power spectrum would have just a single frequency component, namely at the frequency $v = 1/T = 1$. After the first period-doubling bifurcation occurs, we now have period $T = 2$ behavior. Thus, the power spectrum has a component at $v = 1/2$. Because the system is nonlinear, however, there will also be a second harmonic component at $v = 2 \times (1/2) = 1$. (We will ignore any higher frequency harmonics.) After the second period-doubling bifurcation, the system exhibits period $T = 4$ behavior, and hence has a component of the power spectrum at $v = 1/4$. Again, however, the nonlinearities produce power spectrum components at $v = 1/2$, 3/4, and 1. This evolution of the power spectrum is shown in Fig. H.1.

Feigenbaum (FEI80) showed that the total "intensity" (the sum of the Fourier transform amplitudes) associated with the new frequency components (e.g. 1/4 and 3/4 after the second period-doubling bifurcation at the supercycle parameter value after a period-doubling bifurcation) are smaller than those associated with the previous bifurcation by a universal constant factor, whose value is approximately

584

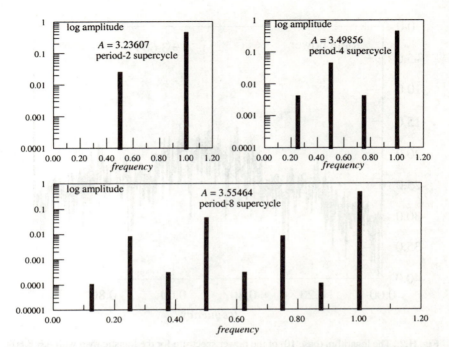

Fig. H.1. The evolution of the power spectrum for several period-doubling bifurcations. The logarithm of the Fourier amplitude is plotted as a function of frequency. The frequency scale is chosen such that period 1 behavior has a frequency of 1 associated with it. All frequencies higher than $v = 1$ are ignored here. Top left: period-2 supercycle. Top right: period-4 supercycle. Note the new components at $v = 1/4$ and $v = 3/4$. Bottom: period-8 supercycle. New components appear at $v = 1/8$, $3/8$, $5/8$, and $7/8$. According to theory the sum of the amplitudes of the new frequency components at each period-doubling should be 8.17 dB smaller than the components associated with the previously existing frequency. On these diagrams, 8.17 dB corresponds to about 0.8 of a vertical division.

0.1525. [Since power spectra are often plotted on logarithmic scales, this factor corresponds to a logarithmic difference of $10 \log 10 \, (0.1525) = - 8.17$ dB (dB = decibels).] This power spectrum ratio has been observed in a few experiments (GMP81, TPJ82). The observations seem to agree reasonably well with this universality prediction, though it is often difficult to determine these average values from the experimental data because the different odd-number harmonics have different strengths. (See the references cited for a discussion of various averaging techniques.)

   If we find the Fourier power spectrum for a chaotic signal, we find that the power is distributed continuously as a function of frequency. Figure H.2 shows the power spectrum for a signal from the logistic map function with $A = 3.609$. This parameter value results in chaotic trajectories that alternate between two bands as shown in Fig. 5.9. The power spectrum is continuous (like that of a "noisy" signal)

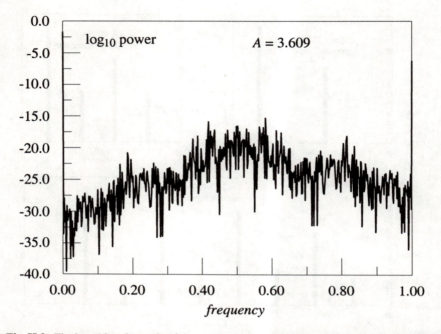

Fig. H.2. The logarithm (base 10) of the power spectrum for the logistic map with $A = 3.609$ corresponding to two-band chaotic behavior. Note the continuous range of frequencies present for chaotic behavior. The alternation between the two bands results in a broad maximum in the spectrum near $v = 0.5$.

but shows a broad maximum near $v = 0.5$ corresponding to the alternation between the two chaotic bands.

H.3 Effects Due to Noise

In our discussion of nonlinear systems, we have so far completely avoided the question of noise, that is, of uncontrollable outside influences, usually of a random nature, that limit the level of precision possible in any real scientific measurement. For nonlinear systems, this noise has some obvious effects. For example, if the experimental quantity corresponding to the control parameter for a system is "noisy," then any effect that occurs (theoretically) at some well-defined value of the control parameter will be smeared out by this noise. To be concrete, let us think about the electrical voltage used as a control parameter in the diode circuit of Chapter 1. Even though we try to control this voltage rather precisely, there is always some small amount of electrical "noise" present. In this case, the noise manifests itself as small fluctuations in the control parameter voltage. Thus, when we say that the first period-doubling bifurcation occurs at $V = 1.3345$ volts, we really mean that the bifurcation occurs at $V = 1.3345 \pm 0.0002$ volts if the noise level is about 0.0002 volts. Clearly, when we proceed to higher-order bifurcations, the voltage difference between successive bifurcations will eventually become smaller than the fluctuating noise level and the bifurcations, including the transition to chaos, will be smeared together.

Noise can also be present in the dependent variable being monitored. (This would correspond to noise in the value of x used in a one-dimensional iterated map function.) This kind of noise can be studied numerically for iterated maps by adding to the value of x_{n+1} calculated from the map function a (usually) small amount of noise. This is done in practice by using a "random number generator" available in most computer languages. Formally, we write

$$x_{n+1} = f(x_n) + \sigma_n \qquad \text{(H.3-1)}$$

where σ_n is a random number, usually chosen so that the average value of the σ_n s is 0 (we choose the average value to be 0 so that positive and negative values are equally likely). The average of the squares of the random numbers is some fixed value, whose square root is denoted by σ. Crutchfield, Farmer and Huberman (CFH82) studied the effects of such "additive noise" on the iterates of the logistic map. They found, as we might anticipate, that in the presence of noise, chaotic behavior begins apparently at lower values of the control parameter A. In fact, they showed that the difference between the parameter value at which chaos begins in the absence of noise A_∞ and the value at which it begins in the presence of noise A^* obeys a universal power law expression

$$A_\infty - A^* \approx k\sigma^\gamma \qquad \text{(H.3-2)}$$

where $\gamma = \log \delta / \log \mu$ and $\mu = 1/0.1525$, the reciprocal of the power spectrum scaling number. (δ is, of course, the Feigenvalue $4.669...$.) This result can be derived by recognizing that if noise of average size σ_m is large enough to "hide" all subharmonics whose index m is greater than some value, then chaos is apparently present for $A = A_m$. To hide the next lower subharmonic, that is, to push chaotic behavior to lower values of A, we need to increase σ by an amount proportional to μ, since it is μ that gives the relative size of successive components in the power spectrum as discussed in the previous section. Thus, we can write

$$\frac{\sigma_m}{\sigma_1} \sim \mu^{-m} \qquad \text{(H.3-3)}$$

where σ_1 is the amount of noise needed to push chaos all the way to the parameter value A_1. We can take the logarithm of Eq. (H.3-3) to solve for m and then use that value in the result derived in Exercise 2.4-1 to obtain

$$(A_\infty - A_m) = (A_2 - A_1) \frac{\delta^2}{\delta - 1} \left(\frac{\sigma_m}{\sigma_1} \right)^\gamma \qquad \text{(H.3-4)}$$

which is the same as Eq. (H.3-2). This behavior has been verified numerically for the logistic map (CFH82) and in a few experiments on electronic oscillators (TPJ82 and YEK82).

Exercise H-1. Work through the algebraic details leading from Eq. (H.3-3) to Eq. (H.3-4)

In practice, the situation is a bit more complicated. In some cases, noise can simply mask the period-doubling cascade without actually inducing chaotic behavior (characterized, as usual, by a positive average Lyapunov exponent) while in other cases, the noise can indeed induced chaotic behavior. See GHL99 for a nice study of these two cases in the context of the logistic map model.

Since systems with chaotic behavior are sensitive to small changes in initial conditions, those systems can serve as "noise amplifiers" (FOE93). That is, the sensitive dependence can amplify small, microscope noise up to macroscopic levels.

H.4 Further Reading

M. J. Feigenbaum, "Universal Behavior in Nonlinear Systems," *Los Alamos Science* **1**, 4–27 (1980) (reprinted in [Cvitanovic, 1984]). Provides a quite readable introduction to the universal features of one-dimensional iterated maps.

M. Giglio, S. Musazzi, and U. Perini, "Transition to Chaotic Behavior Via a Reproducible Sequence of Period-Doubling Bifurcations," *Phys. Rev. Lett.* **47**, 243–46 (1981).

J. Testa, J. Perez, and C. Jeffries, "Evidence for Universal Chaotic Behavior of a Driven Nonlinear Oscillator," *Phys. Rev. Lett.* **48**, 714–17 (1982).

J. P. Crutchfield, J. D. Farmer, and B. A. Huberman, "Fluctuations and Simple Chaotic Dynamics," *Phys. Reports* **92**, 45–82 (1982).

The following three papers are reprinted in both [Hao, 1984] and [Cvitanovic, 1984]:

J. P. Crutchfield and B. A. Huberman, "Fluctuations and the onset of chaos," *Phys. Lett. A* **77**, 407–10 (1980).

J. Crutchfield, M. Nauenberg, and J. Rudnick, "Scaling for external noise at the onset of chaos," *Phys. Rev. Lett.* **46**, 933–35 (1981).

B. Shraiman, C. E. Wayne, and P. C. Martin, "Scaling theory for noisy period-doubling transitions to chaos," *Phys. Rev. Lett.* **46**, 935–9 (1981).

R. F. Fox and T. C. Elston, "Amplification of intrinsic fluctuations by the Lorenz equations," *Chaos* **3**, 313–23 (1993).

J. B. Gao, S. K. Hwang, and J. M. Liu, "When Can Noise Induce Chaos?" *Phys. Rev. Lett.* **82**, 1132–35 (1999).

Appendix I

The van der Pol Oscillator

I.1 The van der Pol model

Limit cycles describe the spontaneous occurrence of periodic time-dependent behavior in some models. Since this behavior may be at odds with our linearly trained intuition, we examine a two-dimensional model in some detail to understand at a more physical, intuitive level how the various parts of the system interact to produce limit cycle behavior. The less-mathematically inclined reader should feel free to skim this appendix.

The model we shall describe has a venerable history in nonlinear dynamics. It was originally developed by van der Pol in the 1920s (VDP26) to describe the dynamics of a triode electronic oscillator. (A *triode* is an electronic vacuum tube with three elements.) We will not describe the details of van der Pol's derivation. Instead, we will try to make some plausibility arguments and then see how we can understand the appearance of limit cycle behavior. The rest of this appendix will be devoted to working through some of the analytic methods that can give us an approximate description of the van der Pol oscillator.

In the van der Pol model, the electrical charge [denoted by $q(t)$] passing through the triode tube is assumed to be described by an equation that is similar to that for a linear, damped, simple harmonic oscillator:

$$\frac{d^2q}{dt^2} + \gamma\frac{dq}{dt} + \omega^2 q = 0 \tag{I-1}$$

where γ is the so-called damping rate (representing a damping or energy loss mechanism) and ω is the frequency with which the charge would oscillate in the absence of damping. Van der Pol's insight was to model the behavior of the triode tube by allowing the damping parameter to depend on the amount of charge q. For small q the tube would tend to increase the amplitude of the oscillation due to the circuit's behavior as an amplifier. For large q, however, the amount of oscillating charge q would be limited by so-called saturation effects in the tube and the associated circuitry. (In rough terms, the tube elements and associated circuitry can supply only so much charge in a given period of time.) This behavior is modeled by making γ depend on q in such a way that for small q, γ is less than 0. A negative value for "dissipation" means that the amplitude of the oscillations grows, rather than decays. For large q, however, γ becomes positive, representing a dissipation of energy from the oscillating charge. Van der Pol chose the simple function

589

$$\gamma = -\gamma_o \left(1 - \frac{q^2}{q^2_o} \right)$$ (I-2)

We take $\gamma_o > 0$. Then for $q < q_o$, the damping parameter γ is negative, and for $q > q_o$, we have γ positive. If we now use Eq. (I-2) in Eq. (I-1) and introduce the variables

$$Q = \frac{q}{q_o} \sqrt{\gamma_o / \omega}$$

$$\tau = t\omega$$ (I-3)

$$R = \gamma_o / \omega$$

the van der Pol equation becomes

$$\frac{d^2Q}{d\tau^2} - (R - Q^2) \frac{dQ}{dt} + Q = 0$$ (I-4)

We now put Eq. (I-4) into our standard first-order form by introducing the variable $U = \dot{Q} = dQ / d\tau$ to obtain

$$\dot{Q} = U \equiv f_1(Q,U)$$ (I-5)

$$\dot{U} = (R - Q^2)U - Q \equiv f_2(Q,U)$$ (I-6)

Next we find the fixed points for the system by setting $f_1(Q,U) = 0$ and $f_2(Q,U) = 0$. It is obvious that the only fixed point is $U = 0$, $Q = 0$, which corresponds to the no oscillation condition. Is this point stable or unstable? To answer this question, we evaluate the Jacobian matrix for the system of Eqs. (I-5) and (I-6).

$$J = \begin{pmatrix} \dfrac{\partial f_1}{\partial Q} & \dfrac{\partial f_1}{\partial U} \\ \dfrac{\partial f_2}{\partial Q} & \dfrac{\partial f_2}{\partial U} \end{pmatrix} = \begin{pmatrix} 0 & 1 \\ -2QU - 1 & R - Q^2 \end{pmatrix}$$ (I-7)

Thus we see that the determinant $\Delta = 1$ at $Q = 0$, $U = 0$ and that $TrJ = R$. Since R is positive by definition, we see that the no-oscillation fixed point is unstable. The characteristic values for this fixed point are

$$\lambda_\pm = \frac{1}{2} \left(R \pm \sqrt{R^2 - 4} \right)$$ (I-8)

Hence, for $R < 2$, the fixed point is a spiral repellor. For $R > 2$, the fixed point is a simple repellor. The behavior for $R = 0.3$ is shown in Fig. I.1.

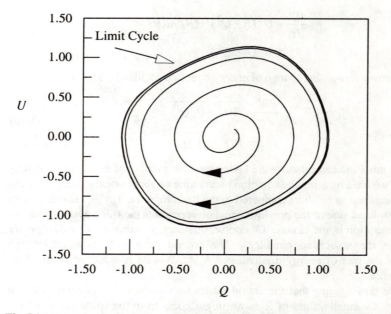

Fig. I.1. A state space trajectory starting near $U = 0.1$, $Q = 0.1$ for the van der Pol oscillator with $R = 0.3$. The fixed point at $U = 0$, $Q = 0$ is clearly a spiral repellor. Trajectories approach the limit cycle as $t \to \infty$.

We should remind ourselves that the analysis of the fixed point's stability does not tell us what happens to those trajectories repelled by the fixed point. However, our intuition tells us that the trajectories cannot get too far from (0,0) because eventually the damping term becomes positive and the corresponding dissipation of energy will limit the size of the trajectory.

We gain some insight into what happens by considering the time dependence of the energy associated with the charge oscillations. We can write this energy as a sum of terms that are analogous to the kinetic energy and potential energy for a mechanical oscillator ([Berg, Pomeau and Vidal, 1986], pp. 28-29):

$$W(U,Q) = \frac{1}{2}LU^2 + \frac{1}{2}\frac{Q^2}{C} \tag{I-9}$$

Here L represents the inductance of the oscillator circuit, and C is its capacitance. We can choose units such that $L = 1$ and $C = 1$. We then find that the rate of change of this energy is given by

$$\frac{dW}{d\tau} = U\frac{dU}{d\tau} + Q\frac{dQ}{d\tau} \tag{I-10}$$

Using Eqs. (I-5) and (I-6) for the time derivatives of U and Q, we obtain

$$\frac{dW}{d\tau} = U^2(R - Q^2) - QU + QU \tag{I-11}$$

$$= U^2(R - Q^2)$$

We now average this change of energy over one oscillation period

$$\overline{\frac{dW}{d\tau}} = \overline{RU^2} - \overline{U^2Q^2} \tag{I-12}$$

$$= \overline{R\dot{Q}^2} - \overline{\dot{Q}^2Q^2}$$

The bar symbol indicates that we are taking the time average of each term. (Strictly speaking, we are averaging the equation over a time of about one oscillation period, which is assumed to be short compared to the damping time $1/\gamma_o$). The first term on the right-hand side of the previous equation represents the (time average) rate of energy generation in the circuit. Of course, this energy is provided by the "power supply" in the associated circuitry. The second term with its negative sign represents the rate of energy dissipation. A steady-state is achieved when these terms balance.

If we now <u>assume</u> that the circuit settles into sinusoidal oscillations (this in fact occurs for small values of R as we might guess from the spiral nature of the repellor), we may write

$$Q(t) = Q_o \sin \omega t = Q_o \sin \tau \tag{I-13}$$

where Q_o is the amplitude of the charge oscillations and ω is the oscillation frequency. We can use this *ansatz* to determine how the amplitude of the oscillations depend on R. First we evaluate the time-averaged quantities:

$$\overline{\dot{Q}^2} = \overline{(\omega Q_o)^2 \cos^2 \omega t} = \frac{1}{2}\omega^2 Q_o^2 \tag{I-14}$$

$$\overline{Q^2 \dot{Q}^2} = \frac{1}{8}\omega^2 Q_o^4 \tag{I-15}$$

We set $dW/dt = 0$ in Eq. (I-12) for a steady-state oscillation and then use Eqs. (I-14) and (I-15) to find

$$Q_o = 2\sqrt{R} \tag{I-16}$$

Exercise I-1. Work through the details of the calculations leading to Eq. (I-16).

We see that in the limit of small R, we expect to have sinusoidal oscillations, represented in state space by a circle whose radius is given by $2\sqrt{R}$. For larger values of R, the oscillations become non-sinusoidal. Typical behavior is shown in

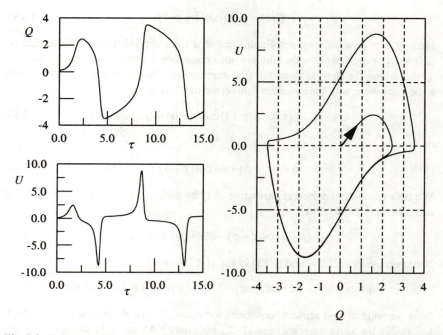

Fig. I-2. Behavior of the van der Pol oscillator for $R = 3.0$. On the left are plotted $Q(\tau)$ and $U(\tau)$. On the right is the Q–U state space behavior. Notice the repelling nature of the fixed point at the origin.

Fig. I-2 for $R = 3.0$. These oscillations, which switch rapidly from one extreme value to another, were called "relaxation oscillations" by van der Pol. The state space trajectory is shown on the right-hand side of Fig. I-2.

We now show how to analyze the stability of the limit cycle. That is, we want to know if trajectories near the limit cycle are attracted toward it or are repelled from it. The procedure we shall use is called "the method of slowly varying amplitude and phase." It finds many applications in the study of nonlinear dynamics (see, for example, [Sanders and Verhulst, 1984]). The method is also called the KBM averaging method after the mathematicians Krylov, Bogoliubov, and Mitropusky, who developed the general formalism.

Let us begin by rewriting the differential equation (I-4):

$$\frac{d^2Q}{d\tau^2} + Q = (R - Q^2)\frac{dQ}{d\tau} \tag{I-17}$$

If the right-hand side of the previous equation were 0, then Q would oscillate sinusoidally in time. This observation leads us to introduce the following expressions for Q and its time derivative (keeping in mind the small R limitation):

$$Q(\tau) = a(\tau)\sin(\tau + \phi(\tau)) \tag{I-18a}$$

$$\dot{Q}(\tau) = a(\tau)\cos(\tau + \phi(\tau)) \tag{I-18b}$$

Here a is a time-varying amplitude, and ϕ a time-varying phase. Note that we define $a(\tau)$ and $\phi(\tau)$ so that the previous equations are true. We do <u>not</u> calculate $\dot{Q}$ by taking the derivative of the first expression. However, for these definitions to be consistent with the usual derivative, we must have

$$\dot{a}\sin(\tau + \phi(\tau)) + a(\dot{\phi} + 1)\cos(\tau + \phi) = a\cos(\tau + \phi) \tag{I-19}$$

or

$$\dot{a}\sin(\tau + \phi) + a\dot{\phi}\cos(\tau + \phi) = 0 \tag{I-20}$$

We now compute the second derivative of Q by taking the derivative of the second equation in (I-18) to obtain

$$\ddot{Q} = \dot{a}\cos(\tau + \phi) - a(1 + \dot{\phi})\sin(\tau + \phi) \tag{I-21}$$

Next, we use Eqs. (I-21) and (I-18) in Eq. (I-17) to find

$$\dot{a}\cos(\tau + \phi) - a\dot{\phi}\sin(\tau + \phi) = [R - a^2\sin^2(\tau + \phi)]a\cos(\tau + \phi) \tag{I-22}$$

Now we want to find separate equations for $\dot{a}$ and $\dot{\phi}$. To do that, we first multiply Eq. (I-20) by $\sin(\tau + \phi)$ and Eq. (I-22) by $\cos(\tau + \phi)$ and add the resulting two equations. We then multiply Eq. (I-20) by $\cos(\tau + \phi)$ and Eq. (I-22) by $-\sin(\tau + \phi)$ and add those. After all these algebraic manipulations we finally arrive at the desired results:

$$\dot{a} = \{R - a^2\sin^2(\tau + \phi)\}a\cos^2(\tau + \phi)$$
$$\dot{\phi} = -[R - a^2\sin^2(\tau + \phi)]\sin(\tau + \phi)\cos(\tau + \phi) \tag{I-23}$$

We should point out that Eqs. (I-23) are exactly equivalent to Eq. (I-17); we have just implemented a change of variables. Up to this point, no approximations have been made. Now we want to invoke the following crucial notion: When a trajectory gets near the limit cycle, its amplitude a and its phase ϕ vary slowly over the time scale of the period of oscillation. Hence, the time derivatives of these quantities are nearly constant over one period of oscillation. If these arguments indeed apply to trajectories near the limit cycle, then we can get approximate equations for the amplitude and phase by integrating the right-hand sides of Eqs. (I-23) over one period and treating in those integrations the amplitude and phase as constants. In carrying out those integrations, we make use of the following integrals:

$$\frac{1}{2\pi}\int_0^{2\pi} d\tau \cos^2(\tau) = \frac{1}{2} \tag{I-24a}$$

$$\frac{1}{2\pi} \int_0^{2\pi} d\tau \sin^2(\tau) \cos^2(\tau) = \frac{1}{8} \tag{I-24b}$$

$$\int_0^{2\pi} d\tau \sin^2(\tau) \sin\tau \cos\tau = 0 \tag{I-24c}$$

After evaluating those integrals, we arrive at the approximate equations

$$\dot{a} = \frac{a}{2}\left[R - \frac{a^2}{4} \right] \tag{I-25}$$

$$\dot{\phi} = 0$$

Exercise I-2. Verify the calculations leading to Eqs. (I-25).

Note that the limit cycle is reached when $\dot{a} = 0$; that is, when $a = 2\sqrt{R}$ the same value we found before. The present method, however, allows us to find the rate at which a nearby trajectory approaches the limit cycle. To find this rate of approach, we expand the right-hand side of the first of Eqs. (I-25) in a Taylor series about the limit cycle value $a^* = 2\sqrt{R}$:

$$\dot{a} = f(a)$$
$$= f(a^*) + \frac{df}{da}(a - a^*) + \dots. \tag{I-26}$$
$$= -R(a - a^*) + \dots$$

where the last equality follows because the derivative of f evaluated at a^* is equal to $-R$. Equation (I-26) tells us that the trajectory approaches the limit cycle exponentially. If we let d be the difference between the trajectory amplitude $a(t)$ and the limit cycle amplitude a^*, we see that

$$d(t) = d_o e^{-R\tau} \tag{I-27}$$

where d_o is the value of that difference at time $\tau = 0$. Thus, we conclude that the limit cycle is stable because trajectories on either side of the limit cycle approach it as time goes on.

We can use what we have just learned to construct an approximate Poincaré map function for the van der Pol oscillator. If we choose the Poincaré section to be the positive Q axis in state space, then the amplitude $a(\tau)$ gives us the location of the point at which the trajectory crosses that section. Since the time between crossings is 2π in our units for time, Eq. (I-27) tells us that the Poincaré map function, expressed in terms of the distance from the limit cycle amplitude, must be

$$d_{n+1} = d_n e^{-2\pi R} \tag{I-28}$$

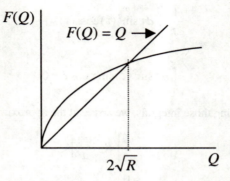

Fig. I-3. Sketch of the Poincaré map function for the van der Pol oscillator. We have assumed that $R < 1$.

and we see that the limit cycle crossing point, for which $d_n = 0$, is a stable fixed point of the Poincaré map function.

Exercise I-3. Graphical analysis of the Poincaré map for the van der Pol oscillator with $R < 1$. For values of the parameter for which the origin ($U = 0$, $Q = 0$) is a repellor and for which $Q = 2\sqrt{R}$ is an attractor, we know that the Poincaré map function $F(Q)$ must look roughly like that shown in Fig. I-3. Use the graphic technique introduced in Section 1.4 to show how values of Q approach the fixed point value $2\sqrt{R}$ both from above and from below.

Exercise I-4. In the limit of $R < 1$ and for small Q oscillations (for which the behavior is sinusoidal), it is possible to find the Poincaré map function for <u>any</u> section in the Q–U state space plane. Find that map function.

I.2 Further Reading

B. van der Pol, "On Relaxation Oscillations," *Phil. Mag.*(7) **2**, 978–92 (1926). This paper describes the original van der Pol oscillator.

I.3 Computer Exercises

CEI-1. Use *Chaos Demonstrations* to study the van der Pol equation limit cycles in state space. Vary the parameter h (equivalent to the parameter R used in the text) to see how the oscillations change from simple harmonic (for small values) to relaxation oscillations for larger values.

CEI-2. Use *Chaotic Dynamics Workbench* to study the Shaw-Van der Pol Oscillator with the force term set to 0 (to make the state space two-dimensional). Observe the time dependence of the dynamical variables and the state space diagrams as the coefficient A (corresponding to R in the text) increases.

Appendix J

Simple Laser Dynamics Models

J.1 A Simple Model

In this appendix we apply the ideas developed in Chapter 3 to a model of the dynamics of a laser. We shall see that the notion of bifurcations helps us understand some of the physics of the laser's operation. Let us consider a very simple model of the light amplification processes that occur in a laser. Here we will focus on the time evolution of the number of photons ("particles" of light energy) in the laser. The intensity of the light beam emitted by the laser is proportional to the number of photons in the laser system. Let that number be denoted by N.

In a laser, the light beam is amplified by its interaction with an "active medium," which we will take to be a collection of atoms. When light interacts with the atoms, the atoms can absorb light by making a transition from a lower energy state to a higher energy state. Conversely, the atoms can emit light by making the reverse transition, from a higher energy state to a lower energy state. The change in energy of the atom in making this transition is equal to the energy of the photon emitted or absorbed in that transition. The rate at which transitions occur is proportional to the number of atoms. Hence, the net amplification for the light beam is proportional to the population difference, that is the difference between the number of atoms in the higher energy level and the number in the lower energy level involved in the laser transition. Let us call that difference N_a.

A phenomenological model of the time behavior of N, the number of photons is given by

$$\dot{N} = GNN_a - \gamma N \qquad (J\text{-}1)$$

where both N and N_a are given per unit volume. γ is the rate at which photons "leak" out of the laser. G is called the "gain" or "amplification" coefficient for the laser. The previous equation tells us that the number of photons in the laser increases due to the amplification process of *stimulated emission* (the gain part) and decreases due to losses of photons from the laser (the loss part). The type of light emission important for lasers is called stimulated emission because the presence of light (photons) stimulates the emission of yet more light.

The population difference for the atoms depends on the number of photons since stimulated emission (the probability of which is proportional to the number of photons) brings an atom from its excited-state to some lower energy state, while absorption brings the atom from the lower state to the higher state. Thus we write that

$$N_a = N_{ap} - BN \qquad (J\text{-}2)$$

where N_{ap} is the atom population difference produced by some "pumping" process (usually collisions due to an electrical current or sometimes the absorption of light from a flashlamp or another laser) and B is the coefficient for stimulated emission, which process, as we mentioned, is proportional to the number of photons present. (You might reasonably question the validity of Eq. (J-2). We shall give a more "realistic" description of the change in the population difference shortly. For now let us view Eq. (J-2) simply as an assumption of this particular model.) Using the previous equation in Eq. (J-1), we find that the time evolution of N, the number of photons, is described by

$$\dot{N} = k_1 N - k_2 N^2 \equiv f(N) \qquad (J\text{-}3)$$

where $k_1 = GN_{ap} - \gamma$ and $k_2 = GB$.

This model is a one-dimensional state space model, and the steady-state behavior of the laser corresponds to the fixed points of Eq. (J-3). It is easy to see that there are two fixed points, which we shall label as N_o and N_*, given by

$$N_o = 0$$
$$N_* = \frac{k_1}{k_2} \qquad (J\text{-}4)$$

If we view N_{ap}, the population difference produced by the pumping mechanism, as the control parameter, then we see from the expressions for k_1 and k_2 that for $N_{ap} < \gamma / G$, N_* is negative, which is not a physically relevant fixed point since the number of photons must be either 0 or positive. For $N_{ap} > \gamma / G$, we have two relevant fixed points N_o and N_*. $N_{ap} = \gamma / G$ is called the "threshold" value, since laser amplification begins above that value.

Let us look at the stability of these fixed points by calculating the appropriate derivatives and evaluating those derivatives at the two fixed points:

$$\left. \frac{df}{dN} \right|_o = +k_1$$
$$\left. \frac{df}{dN} \right|_* = -k_1 \qquad (J\text{-}5)$$

We see that for N_{ap} below the threshold value γ/G (in the language of nonlinear dynamics, below the bifurcation value), N_o is a stable fixed point while N_* is unstable. Thus, below threshold, the number of photons tends to 0, and the laser does not emit its characteristic beam of light. (There still will be some light due to "incoherent" spontaneous emission, but there will be no amplified beam.) Above this threshold value, N_* becomes the stable fixed point, and the number of photons rises linearly with the control parameter N_{ap}. Mathematically, N_* is an unstable fixed point below threshold (even though it is not relevant for the physical problem), and we say that there is an ***exchange of stability*** at the bifurcation point.

The time evolution equation for this laser model can actually be solved exactly because it can be transformed into a linear differential equation by changing to a new variable $u = k_1/(k_2 N)$. The time evolution equation for u is

$$\dot{u} = k_1(1 - u) \tag{J-6}$$

which has the solution

$$1 - u(t) = \{1 - u(0)\}e^{-k_1 t} \tag{J-7}$$

Converting back to our original variable $N(t)$ yields

$$N(t) = \frac{k_1/k_2}{1 - \left(1 - \dfrac{k_1}{k_2 N(0)}\right)e^{-k_1 t}} \tag{J-8}$$

Thus, we see that below threshold, when $k_1 < 0$, the exponential function in the denominator of the previous equation dominates, and $N(t) \rightarrow 0$ as time goes on. Above threshold, when $k_1 > 0$, $N(t)$ approaches the value

$$N(t) \rightarrow \frac{k_1}{k_2} = \frac{N_{ap}}{B} - \frac{\gamma}{GB} \tag{J-9}$$

which we see increases linearly with the control parameter N_{ap}, that is, the more energy we pump in, the more photons we get out.

> **Exercise J-1.** Verify the calculations leading to Eq. (J-8).

There are several important lessons to be learned from this example. First, not all nonlinear differential equations are insoluble. In this example, Eq. (J-3) is nonlinear in the variable N, but it can be transformed into a linear differential equation (J-6) by a change of variable. The second lesson is that solutions to linear differential equations can also display bifurcations. Eq. (J-6) has a fixed point at $u = 1$. For $k_1 < 0$, this is an unstable fixed point, and the solution $u(t) \rightarrow \infty$ (which corresponds to the number of photons approaching 0) as time goes on. For $k_1 > 0$, this fixed point is stable. Note that in the linear version, Eq. (J-6), the fixed point at $N = 0$ is no longer apparent. The third lesson is that this model is very simplified and does not capture the more complex dynamics of actual lasers. In particular, the simple connection between N_a and N embodied in Eq. (J-2) is unjustified for most lasers.

J.2 An Improved Laser Model

A better model for a laser asserts that the <u>time rate of change</u> of the population difference N_a (rather than the number itself) is proportional to a pumping *rate* minus a transition *rate* due to the stimulated emission process. Thus, we write

$$\dot{N}_a = -BN_aN + \gamma_r(N_{ap} - N_a) \tag{J-10}$$

This equation tells us that in the absence of amplification ($B = 0$), the number of atoms approaches a value N_{ap}, which is determined by the external pumping process. The parameter γ_r tells us the rate at which the number of excited state atoms "relaxes" to that value. The first term in the equation gives us the rate of change (a decrease) in the number of excited atoms due to the amplification process, which takes an atom from the excited state to some lower state. Eq. (J-1) and Eq. (J-10) together constitute a two-dimensional model of laser action. The state space variables are N_a and N. Now let us use the techniques described in Chapter 3 to study the fixed points and bifurcations for this system.

First, we set $G = B$ in Eq. (J-1) since the first term refers to the same stimulated emission process indicated in Eq. (J-10). Next, it is useful to put the time evolution equations into a dimensionless form by introducing the new variables

$$P = \frac{BN_{ap}}{\gamma} \qquad R = \frac{\gamma}{\gamma_r} \qquad n_a = \frac{BN_{ap}}{\gamma}$$
$$\tau = t\gamma_r \qquad n = \frac{BN}{\gamma_r} \tag{J-11}$$

P is essentially the amplification rate of photons due to the pumped excited atoms relative to the leakage rate for the photons. R is the photon leakage rate relative to the excited state relaxation rate. τ is a dimensionless time variable giving time in units of the inverse of the excited state relaxation rate. n and n_a are "normalized" photon numbers and the atom population difference, respectively. To get an idea of the sizes of these numbers, we quote the following results [Tarasov, 1983]: For a ruby laser, $P = 30$, $R = 10^5$. For a Nd-YAG laser, $P = 2$ and $R = 10^4$. Using these variables, we write the time evolution equations as

$$\frac{dn}{d\tau} = Rn(n_a - 1)$$
$$\frac{dn_a}{d\tau} = P - n_a(n+1) \tag{J-12}$$

Exercise J-2. Use the definitions of the dimensionless variables to verify that Eq. (J-12) follows from Eqs. (J-1) and (J-10).

Next, we need to identify the fixed points for this two-dimensional system, whose variables are n and n_a. By setting the time derivatives equal to 0 in Eq. (J-12), we find that there are two fixed points:

$$\text{fixed point I:} \quad n = 0, \quad n_a = P$$
$$\text{fixed point II:} \quad n = P - 1, \quad n_a = 1 \tag{J-13}$$

The fixed point II, the lasing condition, occurs only for $P > 1$. The other fixed point corresponds to the number of photons being equal to 0, and as we shall see is the stable fixed point for $P < 1$. Thus, we call $P = 1$ the threshold (or bifurcation) value.

Now we need to examine the stability of the fixed points to determine the character of the solutions in the neighborhood of the fixed points. We calculate the Jacobian matrix of the partial derivatives:

$$J = \begin{pmatrix} R(n_a - 1) & Rn \\ -n_a & -(n+1) \end{pmatrix} \tag{J-14}$$

Proceeding as before, we evaluate the matrix at each fixed point in turn, and find the corresponding eigenvalues. For fixed point I, we have two eigenvalues

$$\lambda_+ = R(P - 1)$$
$$\lambda_- = -1 \tag{J-15}$$

Thus, we see that for $P < 1$, both eigenvalues are negative, and fixed point I is stable and is, in fact, a node. For $P < 1$ (below the threshold value), the system goes to the no-photon ($n = 0$) state for (almost) all initial conditions. For $P > 1$, fixed point I is unstable. In fact, it is a saddle point and all state space trajectories (except for those that start with $n = 0$ exactly) move away from that fixed point.

For fixed point II (with $n_a = 1$ and $n = P - 1$), the two eigenvalues can be written as

$$\lambda_\pm = -\frac{P}{2} \pm \frac{1}{2}\sqrt{P^2 - 4R(P - 1)} \tag{J-16}$$

We have seen that for ruby and Nd-YAG lasers, $4R(P\text{-}1) \gg P^2$. In that case, the argument of the square root function is negative for $P > 1$ and we write

$$\lambda_\pm = -\frac{P}{2} \pm i\sqrt{R(P - 1)} \tag{J-17}$$

We see that for $P > 1$, fixed point II is a stable spiral node since the eigenvalues have negative real parts and nonzero imaginary parts. A detailed analysis of state space trajectories for this system can be found in [Tarasov, 1983]. Figure J.1 shows a typical state space trajectory for this model. The time dependence of the photon number is also shown. The number of photons shows a series of sharp "spikes" as

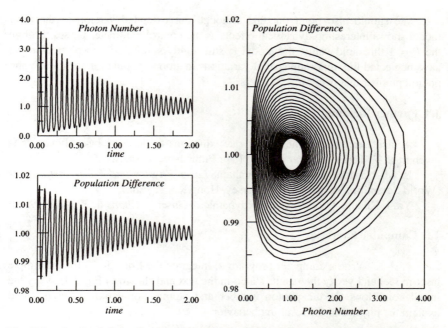

Fig. J.1. On the left is the time dependence of the photon number n and the atomic population difference n_a for our laser model with $R = 10^4$ and $P = 2.0$. On the right is the corresponding state space trajectory. The initial conditions are $n = 0.1$ and $n_a = 1$. The state space fixed point II is clearly a spiral node.

the laser approaches its stable operating point (our fixed point II). This kind of behavior is actually observed in ruby and Nd-YAG lasers.

> **Exercise J-3.** Work through the calculation of the eigenvalues for both fixed points and verify the results given earlier. Examine the stability of fixed point II for $P < 1$. What kind of fixed point is it? Are there any initial conditions for $P < 1$ that do not give rise to trajectories that end up on fixed point I? What kind of bifurcation occurs at $P = 1$?

These oscillations in photon number and in atomic population difference are called **relaxation oscillations** since the oscillations occur as the system "relaxes" to its steady-state conditions. The oscillations also occur whenever some disturbance, such as a fluctuation in the pumping rate or an additional photon from spontaneous emission, causes the system to move away from its steady-state behavior. Although these relaxation oscillations are important in many solid-state and semiconductor lasers, they do not occur in most gas lasers because there the ratio R of photon loss rate compared to population relaxation rate is small. In the small R case, the P^2 term under the square root in Eq. (J-16) dominates and the eigenvalues are purely real.

We should also point out that although our extended version of the laser model shows interesting dynamical effects as the control parameter passes through the $P = 1$ threshold value, the model is still highly simplified. In particular, we have neglected the important spatial variations in atomic population difference and photon number that occur in real lasers.

J.4 Further Reading

A nice treatment of the use of rate equations to describe laser dynamics is given in L. V. Tarasov, *Laser Physics* (Mir Publishers, Moscow, 1983).

L. M. Narducci and N. B. Abraham, *Laser Physics and Laser Instabilities* (World Scientific, Singapore, New Jersey, Hong Kong, 1988).

Ya. I. Khanin, "Low-frequency dynamics of lasers," *Chaos* **6**, 373–80 (1996).

J.5 Computer Exercise

CEJ-1. Write a computer program to integrate the Eqs. (J-12) for the simple laser model and verify the results stated in the text and shown in Fig. J.1. Vary the initial conditions for the photon number and atomic population difference and explain the physics of the resulting behavior.

References

Books

[Abarbanel, 1996]. H. Abarbanel, *Analysis of Observed Chaotic Data* (Springer-Verlag, New York, 1996).

[Abraham and Marsden, 1978]. R. Abraham and J. Marsden, *Foundations of Classical Mechanics* (Benjamin/Cummings, Reading, MA, 1978).

[Abraham and Shaw, 1992]. R. H. Abraham and C. D. Shaw, *Dynamics—The Geometry of Behavior* (Addison–Wesley, Reading, MA, 1992).

[Abraham, Abraham, and Shaw, 1996]. F. D. Abraham, R. H. Abraham, and C. D. Shaw, *Dynamical Systems*: *A Visual Introduction* (Science Frontier Express, 1996].

[Alligood, Sauer, and Yorke, 1997]. Kathleen Alligood, Timothy Sauer and James A. Yorke, *Chaos, An Introduction to Dynamic Systems* (Springer-Verlag, New York, 1997).

[Arnold, 1978]. V. I. Arnold, *Mathematical Methods in Classical Mechanics* (Springer-Verlag, New York, 1978).

[Arnold, 1983]. V. I. Arnold, *Geometric Methods in the Theory of Ordinary Differential Equations* (Springer-Verlag, New York, 1983).

[Arnold and Avez, 1968]. V. I. Arnold and A. Avez, *Ergodic Problems of Classical Mechanics* (Benjamin, New York, 1968).

[Baierlein, 1971]. R. Baierlein, *Atoms and Information Theory* (W. H. Freeman, San Francisco, 1971).

[Bak, 1996]. P. Bak, *How Nature Works*: *The Science of Self-Organized Criticality* (Springer-Verlag, New York, 1996).

[Baker and Gollub, 1996]. Gregory Baker and Jerry Gollub, *Chaotic Dynamics, An Introduction*, 2nd ed. (Cambridge University Press, New York, 1996).

[Ball, 1999]. Philip Ball, *The Self-Made Tapestry*, *Pattern Formation in Nature* (Oxford University Press, New York, 1999).

[Barnsley, 1988]. M. Barnsley, *Fractals Everywhere* (Academic Press, San Diego, 1988).

[Bergé, Pomeau, and Vidal, 1986]. P. Bergé, Y. Pomeau, and C. Vidal, *Order within Chaos* (Wiley, New York, 1986).

[Blümel and Reinhardt, 1997]. R. Blümel and W. P. Reinhardt, *Chaos in Atomic Physics* (Cambridge University Press, New York, 1997).

[Boas, 1983]. Mary L. Boas, *Mathematical Methods in the Physical Sciences*, 2nd ed. (John Wiley and Sons, New York, 1983).

[Born and Wolf, 1975]. M. Born and E. Wolf, *Principles of Optics*, 5th ed. (Pergamon Press, New York, 1975).

[Briggs and Peat, 1989]. John Briggs and F. David Peat, *Turbulent Mirror* (Harper & Row, New York, 1989).

[Burrus, Gopinath, and Guo, 1997]. C. S. Burrus, R. A. Gopinath, and H. Guo, *Introduction to Wavelets and Transforms: A Primer* (Prentice–Hall, 1997).

[Cartwright, 1983]. Nancy Cartwright, *How the Laws of Physics Lie* (Oxford University Press, Oxford and New York, 1983).

[Chaitin, 1987]. G. J. Chaitin, *Algorithmic Information Theory* (Cambridge University Press, Cambridge, 1987).

[Chandler, 1987]. D. Chandler, *Introduction to Modern Statistical Mechanics* (Oxford University Press, New York, 1987).

[Chandrasekhar, 1981]. S. Chandrasekhar, *Hydrodynamic and Hydromagnetic Stability* (Dover, New York, 1984).

[Collet and Eckman, 1980]. P. Collet and J. P. Eckmann, *Iterated Maps on the Interval as Dynamical Systems* (Birkhauser, Cambridge, MA, 1980).

[Coveney and Highfield, 1995]. Peter Coveney and Roger Highfield, *Frontiers of Complexity: The Search for Order in a Chaotic World* (Fawcett/Columbine, New York, 1995).

[Crichton, 1990]. Michael Crichton, *Jurassic Park* (Ballentine Books, New York, 1990).

[Cvitanovic, 1989]. Pedrag Cvitanovic, *Universality in Chaos*, 2nd ed. (Adam Hilger, Bristol, 1989).

[Davies, 1999]. Brian Davies, *Exploring Chaos: Theory and Experiment* (Perseus Books, Reading, MA, 1999).

[Devaney, 1986]. R. L. Devaney, *An Introduction to Chaotic Dynamical Systems* (Benjamin–Cummings, Menlo Park, CA, 1986).

[Devaney, 1990]. R. L. Devaney, *Chaos, Fractals, and Dynamics, Computer Experiments in Mathematics* (Addison–Wesley, Reading, MA, 1990).

[Devaney, 1992]. R. L. Devaney, *A First Course in Chaotic Dynamical Systems* (Addison–Wesley, Reading, MA, 1992).

[Duffing, 1918]. G. Duffing, *Erzwungene Schwingungen bei veränderlicher Eigenfrequenz and ihre technische Bedeutung* (Friedr. Vieweg & Sohn, Braunschweig, 1918).

[Enns and McGuire, 1997]. R. H. Enns and G. C. McGuire, *Nonlinear Physics with Maple for Scientists and Engineers* (Birkhäuser, Boston, 1997) and *Nonlinear Physics with Maple for Scientists and Engineers: A Laboratory Manual* (Birkhäuser, Boston, 1997).

[Epstein and Pojman, 1998]. Irving R. Epstein and John A. Pojman, *An Introduction to Nonlinear Chemical Dynamics: Oscillations, Waves, Patterns, and Chaos* (Oxford University Press, New York, 1998).

[Feynman, Leighton, and Sands, 1963]. R. Feynman, R. Leighton, and M. Sands, *Feynman Lectures on Physics* (Addison–Wesley, Reading, MA, 1963).

[Fleischmann and Teldesley, 1990]. M. Fleischmann and D. J. Teldesley, eds. *Fractals in the Natural Sciences* (Princeton University Press, Princeton, NJ, 1990).

[Frank, 1957]. Philipp Frank, *Philosophy of Science* (Prentice–Hall, Englewood Cliffs, NJ, 1957).

[Gardner, 1983]. M. Gardner, *Wheels, Life and Other Mathematical Amusement* (W. H. Freeman, New York, 1983).

[Gibbs, 1902]. J. W. Gibbs, *Elementary Principles in Statistical Mechanics* (C. Scribner's Sons, New York, 1902).

[Gibbs, 1948]. J. W. Gibbs, *The Collected Works of J. W. Gibbs* (Yale University Press, New Haven, 1948.)

[Glass and Mackey, 1988]. L. Glass and M. C. Mackey, *From Clocks to Chaos, The Rhythms of Life* (Princeton University Press, Princeton, NJ, 1988).

[Gleick, 1987]. James Gleick, *Chaos, Making a New Science* (Viking, New York, 1987).

[Goldstein, 1980]. H. Goldstein, *Classical Mechanics*, 2nd ed. (Addison–Wesley, Reading, MA, 1980).

[Golubitsky, Stewart, and Schaeffer, 1988]. M. Golubitsky, I. Stewart, and D. Schaeffer, *Singularities and Groups in Bifurcation Theory*, Applied Mathematical Sciences Vol. 69 (Springer-Verlag, New York, 1988), Vol. II.

[Gould and Tobochnik, 1996]. Harvey Gould and Jan Tobochnik, *An Introduction to Computer Simulation Methods*, 2nd ed. (Addison–Wesley, Reading, Mass., 1996).

[Gouyet, 1995]. J.-F. Gouyet, *Physics of Fractal Structures* (Springer-Verlag, New York, 1995).

[Griffiths, 1981]. David J. Griffiths, *Introduction to Electrodynamics* (Prentice–Hall, Englewood Cliffs, NJ, 1981).

[Guckenheimer and Holmes, 1990]. J. Guckenheimer and P. Holmes, *Nonlinear Oscillations, Dynamical Systems, and Bifurcations of Vector Fields*, 3rd ed. (Springer-Verlag, New York, 1990).

[Gulick, 1992]. D. Gulick, *Encounters with Chaos* (McGraw–Hill, New York, 1992).

[Gutzwiller, 1990]. M. C. Gutzwiller, *Chaos in Classical and Quantum Mechanics* (Springer-Verlag, New York, 1990).

[Guyon and Stanley, 1991]. E. Guyon and H. E. Stanley, *Fractal Forms* (Elsevier, New York, 1991).

[Hagedorn, 1981]. P. Hagedorn, *Nonlinear Oscillations* (Clarendon Press, Oxford, 1981).

[Haken, 1983]. H. Haken, *Synergetics, An Introduction*, 3rd ed. (Springer-Verlag, Berlin, 1983).

[Hale and Koçak, 1991]. J. Hale and H. Koçak, *Dynamics and Bifurcations* (Springer-Verlag, New York, 1991).

[Hao, 1984]. Hao Bai-Lin, ed. *Chaos,* Vol. 1 (World Scientific, Singapore, 1984).

[Hao, 1988]. Hao Bai-Lin, ed. *Directions in Chaos*, Vol. 2 (World Scientific, Singapore, 1988).

[Hao, 1989]. Hao Bai-Lin, *Chaos, Vol II* (World Scientific, Singapore, 1989).

[Hao, 1990]. Hao Bai-Lin, ed. *Directions in Chaos*, Vol. 3. (World Scientific, Singapore, 1990).

[Hassani, 1991]. S. Hassani, *Foundations of Mathematical Physics* (Allyn and Bacon, Boston, 1991).

[Hayashi, 1964]. C. Hayashi, *Nonlinear Oscillations in Physical Systems* (McGraw–Hill, New York, 1964; reprinted by Princeton University Press, Princeton, NJ, 1985)

[Hilborn and Tufillaro]. Robert C. Hilborn and Nicholas B. Tufillaro, *Chaos and Nonlinear Dynamics* (American Association of Physics Teachers, College Park, MD, 1999).

[Hayles, 1991]. N. Katherine Hayles, ed. *Chaos and Order: Complex Dynamics in Literature and Science* (University of Chicago Press, Chicago, 1991).

[Hille, 1969]. E. Hille, *Lectures on Ordinary Differential Equations* (Addison–Wesley, Reading, MA, 1969).

[Hirsch and Smale, 1974]. M. W. Hirsch and S. Smale, *Differential Equations, Dynamical Systems, and Linear Algebra* (Academic Press, New York, 1974).

[Holden, 1986]. A. V. Holden, ed. *Chaos* (Princeton University Press, Princeton, NJ, 1986).

[Holland, 1995]. John H. Holland, *Hidden Order: How Adaptation Builds Complexity* (Addison–Wesley, Reading, MA, 1995)

[Huntley, 1970]. H. E. Huntley, *The Divine Proportion* (Dover, New York, 1970).

[Infeld and Rowlands, 1990]. E. Infeld and G. Rowlands, *Nonlinear Waves, Solitons and Chaos* (Cambridge University Press, Cambridge, 1990).

[Iooss and Joseph, 1980]. G. Iooss and D. D. Joseph, *Elementary Stability and Bifurcation Theory* (Springer-Verlag, New York, 1980).

[Jackson, 1989, 1991]. E. Atlee Jackson, *Perspectives of Nonlinear Dynamics*, Vol. 1 and Vol. 2 (Cambridge University Press, New York, 1989, 1991).

[Jensen, 1998]. Henrik Jeldtoft Hensen, *Self-Organized Criticality: Emergent Complex Behavior in Physical and Biological Systems* (Cambridge University Press, New York, 1998).

[Kahn and Zarmi, 1997]. P. Kahn and Y. Zarmi, *Nonlinear Dynamics: Exploration through Normal Forms* (Wiley, New York, 1997).

[Kaiser, 1994]. G. Kaiser, *A Friendly Guide to Wavelets* (Springer-Verlag, New York, 1994).

[Kapitaniak, 1996]. T. Kapitaniak, *Controlling Chaos* (Academic Press, San Diego, 1996).

[Kapitaniak and Bishop, 1999]. T. Kapitaniak and S. R. Bishop, *The Illustrated Dictionary of Nonlinear Dynamics and Chaos* (Wiley, Chicester and New York, 1999).

[Kaplan and Glass, 1995]. Daniel Kaplan and Leon Glass, *Understanding Nonlinear Dynamics* (Springer-Verlag, New York, Berlin, Heidelberg, 1995).

[Kaufman, 1993]. Stuart A. Kaufman, *Origins of Order: Self-Organization and Selection in Evolution* (Oxford University Press, New York, 1993).

[Kaufman, 1995]. Stuart A. Kaufman, *At Home in the Universe, The Search for the Laws of Self-Organization and Complexity* (Oxford University Press, New York, 1995).

[Khinchin, 1992]. A. Ya Khinchin, *Continued Fractions* (The University of Chicago Press, Chicago, 1992).

[Lam and De Cola, 1993]. N. Lam and L. De Cola, *Fractals in Geography* (PTR Prentice–Hall, Upper Saddle Rive, NJ, 1993).

[Landau and Lifshitz, 1959]. L. D. Landau and E. M. Lifshitz, *Fluid Mechanics* (Pergamon, London, 1959).

[Laplace, 1812]. Pierre Simon de Laplace, *A Philosophical Essay on Probabilities* (Dover, New York, 1951).

[Liebovitch, 1998]. Larry S. Liebovitch, *Fractals and Chaos Simplified for the Life Sciences* (Oxford University Press, New York, 1998).

[Lichtenberg and Liebermann, 1992]. A. J. Lichtenberg and M. A. Liebermann, *Regular and Chaotic Dynamics*, 2nd ed. (Springer-Verlag, New York, Heidelberg, Berlin, 1992).

[Lipschutz, 1968]. Seymour Lipschutz, *Linear Algebra* (Schaum's Outline Series) (McGraw–Hill, New York, 1968).

[Mackay and Meiss, 1987]. R. S. Mackay and J. D. Meiss, *Hamiltonian Dynamical Systems* (Adam Hilger, Bristol, 1987).

[Mackey and Glass, 1988]. L. Glass and M. C. Mackey, *From Clocks to Chaos, the Rhythms of Life* (Princeton University Press, Princeton, NJ, 1988).

[Mammeiulle, Boccasa, Vichniac, and Bidaux, 1989]. P. Mammeiulle, N. Boccasa, C. Y. Vichniac, and R. Bidaux, eds. *Cellular Automata and Modeling of Complex Physical Systems* (Springer-Verlag, Berlin, 1989).

[Mandelbrot, 1982]. B. B. Mandelbrot, *The Fractal Geometry of Nature* (W. H. Freeman, San Francisco, 1982).

[Mandelbrot, 1997]. B. B. Mandelbrot, *Fractals and Scaling in Finance: Discontinuity, Concentration, Risk* (Springer-Verlag, New York, 1997).

Mandelbrot, 1999]. B. B. Mandelbrot, *Multifractals and 1/f Noise: Wild Self-Affinity in Physics* (Springer-Verlag, New York, 1999).

[Mantegna and Stanley, 1999]. Rosario N. Mantegna and H. Eugene Stanley, *An Introduction to Econophysics* (Cambridge University Press, New York, 1999).

[Marion and Thornton, 1988]. J. B. Marion and S. T. Thornton, *Classical Dynamics of Particles and Systems* (Harcourt Brace Jovanovic, San Diego, 1988).

[Meakin, 1997]. Paul Meakin, *Fractals, Scaling and Growth Far from Equilibrium* (Cambridge University Press, Cambridge, 1997).

[Meyer and Hall, 1992]. K. R. Meyer and G. R. Hall, *Introduction to Hamiltonian Dynamical Systems and the N-body Problem* (Springer-Verlag, New York, 1992).

[Moon, 1992]. Francis C. Moon, *Chaotic and Fractal Dynamics, An Introduction for Applied Scientists and Engineers* (Wiley, New York, 1992).

[Morse, 1964]. P. M. Morse, *Thermal Physics* (W. A. Benjamin, New York, 1964).

[Moser, 1973]. J. Moser, *Stable and Random Motions in Dynamical Systems* (Princeton University Press, Princeton, NJ, 1973).

[Murray, 1989]. J. D. Murray, *Mathematical Biology* (Springer-Verlag, Berlin, Heidelberg, New York, 1989).

[Narducci and Abraham, 1988]. L. M. Narducci and N. B. Abraham, *Laser Physics and Laser Instabilities* (World Scientific, Singapore, New Jersey, Hong Kong, 1988).

[Nicolis and Prigogine, 1989]. G. Nicolis and I. Prigogine, *Exploring Complexity* (W. H. Freeman, San Francisco, 1989).

[Nusse and Yorke, 1998]. Helena Nusse and James A. Yorke, *Dynamics: Numerical Explorations*, 2nd ed. (Springer-Verlag, New York, 1998).

[Ott, 1993]. E. Ott, *Chaos in Dynamical Systems* (Cambridge University Press, Cambridge, 1993).

[Ott, Sauer, and Yorke, 1997]. E. Ott., T. Sauer, and J. A. Yorke, *Coping with Chaos* (Wiley, New York, 1994).

[Ottino, 1989]. J. M. Ottino, *The Kinematics of Mixing: Stretching, Chaos, and Transport* (Cambridge University Press, Cambridge, 1989).

[Paterson, 1983]. A. R. Paterson, *A First Course in Fluid Dynamics* (Cambridge University Press, Cambridge, 1983).

[Peitgen and Richter, 1986]. H.-O. Peitgen and P. H. Richter, *The Beauty of Fractals* (Springer-Verlag, Berlin, Heidelberg, New York, Tokyo, 1986).

[Peters, 1991]. E. E. Peters, *Chaos and Order in the Capital Markets* (Wiley, New York, 1991).

[Pippard, 1985]. A. P. Pippard, *Response and Stability* (Cambridge University Press, Cambridge, 1985).

[Poincaré, 1892]. H. Poincaré, *Les Méthods Nouvelles de la Méchanique Céleste*, Vol. I, Chap. III, Art. 36. (Gauthier–Villars, Paris, 1892).

[Poincaré, 1993]. H. Poincaré, *New Methods of Celestial Mechanics* (American Institute of Physics,Woodbury, NY, 1993)

[Press, Flannery, Teukolsky, and Vetterling, 1986]. W. H. Press, B. P. Flannery, S. A. Teukolsky, and W. T. Vetterling, *Numerical Recipes, The Art of Scientific Computing* (Cambridge University Press, Cambridge, 1986).

[Purcell, 1985]. Edward M. Purcell, *Electricity and Magnetism*, 2nd ed. (McGraw–Hill, New York, 1985).

[Rasband, 1990]. S. Neil Rasband, *Chaotic Dynamics of Nonlinear Systems* (Wiley, New York, 1990).

[Reichl, 1992]. L. Reichl, *The Transition to Chaos in Conservative Classical Systems: Quantum Manifestations* (Springer-Verlag, New York, 1992).

[Renyi, 1970]. A. Renyi, *Probability Theory* (North-Holland, Amsterdam, 1970).

[Robinson, 1995]. C. Robinson, *Dynamical Systems: Stability, Symbolic Dynamics, and Chaos* (CRC Press, Boca Raton, 1995).

[Rothschild, 1998]. W. G. Rothschild, *Fractals in Chemistry* (Wiley, New York, 1998).

[Ruelle, 1989]. D. Ruelle, *Chaotic Evolution and Strange Attractors* (Cambridge University Press, New York, 1989).

[Ruelle, 1991]. David Ruelle, *Chance and Chaos* (Princeton University Press, Princeton, NJ, 1991).

[Sanders and Verhulst, 1984]. J. A. Sanders and F. Verhulst, *Averaging Methods in Nonlinear Dynamical Systems*, Applied Mathematical Sciences (Springer-Verlag, New York, Berlin, Heidelberg, 1984).

[Scholz and Mandelbrot, 1989]. C. H. Scholz and B. B. Mandelbrot, eds. *Fractals in Geophysics* (Birkhauser Verlag, Basel and Boston, 1989).

[Scott, 1994]. S. K. Scott, *Oscillations, Waves, and Chaos in Chemical Kinetics* (Oxford University Press, New York, 1994).

[Schuster, 1995]. H. G. Schuster, *Deterministic Chaos, An Introduction*, 3rd rev. ed. (Wiley, New York, 1995).

[Schuster, 1999]. H. G. Schuster, *Handbook of Chaos Control* (Wiley, New York, 1999).

[Shaw, 1984]. R. Shaw, *The Dripping Faucet as a Model Chaotic System* (Aerial Press, Santa Cruz, CA, 1984).

[Smith, 1998]. Peter Smith, *Explaining Chaos* (Cambridge University Press, Cambridge, 1998).

[Sparrow, 1982]. C. T. Sparrow, *The Lorenz Equations: Bifurcations, Chaos, and Strange Attractors* (Springer-Verlag, New York, Heidelberg, Berlin, 1982).

[Stewart, 1989]. Ian Stewart, *Does God Play Dice? The Mathematics of Chaos* (Blackwell, New York, 1989).

[Stockman, 1999]. H.-J. Stockman, *Quantum Chaos, An Introduction* (Cambridge University Press, Cambridge, 1999).

[Stoppard, 1993]. Tom Stoppard, *Arcadia* (Faber and Faber, London and Boston, 1993).

[Strogatz, 1994]. Steven H. Strogatz, *Nonlinear Dynamics and Chaos: With Applications in Physics, Biology, Chemistry and Engineering* (Addison–Wesley, Reading, MA, 1994).

[Tarasov, 1983]. L. V. Tarasov, *Laser Physics* (Mir Publishers, Moscow, 1983).

[Thompson and Stewart, 1986]. J. M. T. Thompson and H. B. Stewart, *Nonlinear Dynamics and Chaos* (Wiley, New York, 1986).

[Toffoli and Margolus, 1987]. T. Toffoli and N. Margolus, *Cellular Automata Machines* (MIT Press, Cambridge, 1987).

[Tufillaro, Abbott, and Reilly, 1992]. N. Tufillaro, T. Abbott, and J. Reilly, *An Experimental Approach to Nonlinear Dynamics and Chaos* (Addison–Wesley, Reading, MA, 1992).

[Turcotte, 1997]. D. L. Turcotte, *Fractals and Chaos in Geology and Geophysics*, 2nd ed. (Cambridge University Press, 1997).

[Vicsek, 1991]. T. Vicsek, *Fractal Growth Phenomena*, 2nd ed. (World Scientific Publishing, River Edge, NJ, 1991).

[Waldrop, 1992]. M. Mitchell Waldrop, *Complexity: The Emerging Science at the Edge of Order and Chaos* (Simon and Schuster, New York, 1992).

[West, 1990]. B. J. West, *Fractal Physiology and Chaos in Medicine* (World Scientific Publishing, River Edge, NJ, 1990).

[Wigen, 1994]. Philip E. Wigen, ed., *Nonlinear Phenomena and Chaos in Magnetic Materials* (World Scientific, Singapore, New Jersey, London, Hong Kong, 1994).

[Wiggins, 1988]. S. Wiggins, *Global Bifurcations and Chaos, Analytical Methods* (Springer-Verlag, New York, 1988).

[Wilhelm, 1991]. Kate Wilhelm, *Death Qualified, A Mystery of Chaos* (Fawcett Crest, New York, 1991).

[Williams, 1997]. Garnett P. Williams, *Chaos Theory Tamed* (National Academy Press, Washington, DC, 1997).

[Winfree, 1980]. A. T. Winfree, *The Geometry of Biological Time* (Springer-Verlag, New York, 1980).

[Wright and Hardy, 1980]. E. M. Wright and G. H. Hardy, *An Introduction* to the *Theory of Numbers* (Oxford University Press, Oxford, New York, 1980).

[Wolfram, 1986]. S. Wolfram, ed. *Theory and Applications of Cellular Automata* (World Scientific Press, Singapore, 1986).

[Yuznetsov, 1998]. Y. A. Kuznetsov, *Elements of Applied Bifurcation Theory*, 2nd ed. (Springer-Verlag, New York, 1998).

[Zaslavsky, Sagdeev, Usikov, and Chernikov, 1991] G. M. Zaslavsky, R. Z. Sagdeev, D. A. Usikov, and A. A. Chernikov, *Weak Chaos and Quasi-Regular Patterns* (Cambridge University Press, Cambridge, 1991).

[Zaslavsky, 1998]. G. M. Zaslavsky, *Physics of Chaos in Hamiltonian Systems* (Imperial College Press, London, 1998).

[Zuckerman, Montgomery, Niven, and Niven, 1991]. H. S. Zuckerman, H. L. Montgomery, I. M. Niven, and A. Niven, *An Introduction to the Theory of Numbers*, 5th ed. (John Wiley, New York, 1991).

Articles

AAC84. A. M. Albano, N. B. Abraham, D. E. Chyba, and M. Martelli, "Bifurcation, Propagating Solutions, and Phase Transitions in a Nonlinear Reaction with Diffusion," *Am. J. Phys.* **52**, 161–67 (1984).

AAD86. N. B. Abraham, A. M. Albano, B. Das, G. De Guzman, S. Yong, R. S. Gioggia, G. P. Puccioni, and J. R. Tredicce, "Calculating the Dimension of Attractors from Small Data Sets," *Phys. Lett. A* **114**, 217–21 (1986).

AAG88. F. Argoul, A. Arneodo, G. Grasseau, and H. L. Swinney, "Self-Similarity of Diffusion-Limited Aggregates and Electrodeposition Clusters," *Phys. Rev. Lett.* **61**, 2558–61 (1988).

ABB86. A. K. Agarwal, K. Banerjee, and J. K. Bhattacharjee, "Universality of Fractal Dimension at the Onset of Period-Doubling Chaos," *Phys. Lett. A* **119**, 280–83 (1986).

ABE90. S. Åberg, "Onset of Chaos in Rapidly Rotating Nuclei," *Phys. Rev. Lett.* **64**, 3119–22 (1990).

ABG95. A. Arneodo, E. Bacry, P. V. Graves, and J. F. Muzy, "Characterizing Long-Range Correlations in DNA Sequences from Wavelet Analysis," *Phys. Rev. Lett.* **74**, 3293–96 (1995).

ABH98. L Amaral, S. Buldyrev, S. Havlin, M. Salinger, and H. E. Stanley, "Power Law Scaling for a System of Interacting Units with Complex Internal Structure," *Phys. Rev. Lett.* **80** 1385–88 (1998).

ABM95. A. Arneodo, E. Bacry, and J. F. Muzy, "Oscillating Singularities in Locally Self-Similar Functions," *Phys. Rev. Lett.* **74**, 4823–26 (1995).

ABS86. R. H. Abraham and C. Simo, "Bifurcations and Chaos in Forced van der Pol Systems," in *Dynamical Systems and Singularities* (S. Pnevmatikos, ed.) (North-Holland, Amsterdam, 1986), pp. 313–23.

ABS93. H. D. I. Abarbanel, R. Brown, J. J. Sidorowich, and L. Sh. Tsimring, "The analysis of observed chaotic data in physical systems," *Rev. Mod. Phys.* **65**, 1331–92 (1993).

ACE87. D. Auerbach, P. Cvitanovic, J.-P. Eckmann, G. Gunaratne, and I. Procaccia, "Exploring Chaotic Motion Through Periodic Orbits," *Phys. Rev. Lett.* **58**, 2387–89 (1987).

ACL88. P. Alstrom, B. Christiansen, and M. T. Levinsen, "Nonchaotic Transition from Quasi-periodicity to Complete Phase Locking," *Phys. Rev. Lett.* **61**, 1679–82 (1988).

AGO92. D. Auerbach, C. Grebogi, E. Ott, and J. A. Yorke, "Controlling Chaos in High Dimensional Systems," *Phys. Rev. Lett.* **69**, 3479–82 (1992).

AGS84. N. B. Abraham, J. P. Gollub, and H. L. Swinney, "Testing Nonlinear Dynamics," *Physica D* **11**, 252–64 (1984).

AHK94. D. Armbruster, R. Heiland, and F. Kostelich, "KLTOOL: A tool to analyze spatiotemporal complexity," *Chaos* **4**, 421–24 (1994).

ALN85. N. B. Abraham, L. A. Lugiato, and L. M. Narducci, "Overview of Instabilities in Laser Systems," *J. Opt. Soc. Am. B* **2**, 7–13 (1985).

ALY81. V. M. Alekseev and M. V. Yakobson "Symbolic Dynamics and Hyperbolic Dynamic Systems," *Phys. Rep.* **75**, 287–325 (1981).

AMS88. A. M. Albano, J. Muench, C. Schwartz, A. I. Mees, and P. E. Rapp, "Singular-Value Decomposition and the Grassberger–Procaccia Algorithm," *Phys. Rev. A* **38**, 3017–26 (1988).

ANT98. C. Anteneodo and C. Tsallis, "Breakdown of Exponential Sensitivity to Initial Conditions: Role of the Range of Interactions," *Phys. Rev. Lett.* **80**, 5313–16 (1998).

ARE83. H. Aref, "Integrable, Chaotic, and Turbulent Vortex Motion in Two-Dimensional Flow," *Ann. Rev. Fluid Mech.* **15**, 345–89 (1983).

ARE84. H. Aref, "Stirring by Chaotic Advection," *J. Fluid Mech.* **143**, 1–21 (1984).

ASV88. H. Atmanspacker, H. Scheingraber, and W. Voges, "Global Scaling Properties of a Chaotic Attractor Reconstructed from Experimental Data," *Phys. Rev. A* **37**, 1314–22 (1988).

ATI88. S. Adachi, M. Toda, and K. Ikeda, "Quantum-Classical Correspondence in

Many-Dimensional Quantum Chaos," *Phys. Rev. Lett.* **61**, 659–61 (1988).

ATY91.   K. T. Alligood, L. Tedeschi-Lalli, and J. A. Yorke, "Metamorphoses: Sudden Jumps in Basin Boundaries," *Commun. Math. Phys.* **141**, 1–8 (1991).

AUS89.   R. Aurich and F. Steiner, "Periodic-Orbit Sum Rules for the Hadamard-Gutzwiller Model," *Physica D* **39**, 169–93 (1989).

AYY87.   K. T. Alligood, E. D. Yorke, and J. A. Yorke, "Why Period-Doubling Cascades Occur: Periodic Orbit Creation Followed by Stability Shedding," *Physica D* **28**, 197–203 (1987).

AYY92.   J. C. Alexander, J. A. Yorke, Z. You, and I. Kan, "Riddled Basins," *Int. J. Bifur. and Chaos* **2**, 795–80 (1992).

BAA99.   A.-L. Barabási and R. Albert, "Emergence of Scaling in Random Networks," *Science* **286**, 509–12 (1999).

BAB89.   W. J. Baumol and J. Benhabib, "Chaos: Significance, Mechanism, and Economic Applications," *J. Econ. Perspectives* **3**, 77–105 (1989).

BAC89.   P. Bak and K. Chen, "The Physics of Fractals," *Physica D* **38**, 5–12 (1989).

BAC91.   P. Bak and K. Chen, "Self-Organized Criticality," *Scientific American* **264** (1), 46–53 (January, 1991).

BAD97.   R. Badii, "Generalized entropies of chaotic maps and flows: A unified approach," *Chaos* **7**, 694–700 (1997).

BAP98.   M. S. Baptista, "Cryptography with chaos," *Phys. Lett. A* **240**, 50–54 (1998).

BAT88.   P. M. Battelino, "Persistence of Three–Frequency Quasiperiodicity under Large Perturbation," *Phys. Rev. A* **38**, 1495–502 (1988).

BBD88.   R. Badii, G. Broggi, B. Derighetti, M. Ravani, S. Ciliberto, A. Politi, and M. A. Rubio, "Dimension Increase in Filtered Chaotic Signals," *Phys. Rev. Lett.* **60**, 979–82 (1988).

BCG80.   G. Benettin, C. Cercignanni, L. Galgani, and A. Giorgilli, "Universal Properties in Conservative Dynamical Systems," *Lett. Nouvo. Cim.* **28**, 1–4 (1980).

BCG89.   J. E. Bayfield, G. Casati, I. Guarneri, and D. W. Sokol, "Localization of Classically Chaotic Diffusion for Hydrogen Atoms in Microwave Fields," *Phys. Rev. Lett.* **63**, 364–67 (1989).

BEN97.   J. E. Berger and G. Nunes, Jr., "A mechanical Duffing oscillator for the undergraduate laboratory," *Am. J. Phys.* **65**, 841–846 (1997).

BER89.   M. V. Berry, "Quantum Scars of Classical Closed Orbits in Phase Space," *Proc. Roy. Soc. London A* **423**, 219–31 (1989).

BES78.   G. Benettin and J.-M. Strelcyn, "Numerical Experiments on the Free Motion of a Point Mass Moving in a Plane Convex Region: Stochastic Transition and Entropy," *Phys. Rev. A* **17**, 773–85 (1978).

BGH95.   J. Bélair, L. Glass, U. an der Heiden, and J. Milton, "Dynamical disease: Identification, temporal aspects and treatment strategies," *Chaos* **5**, 1–7 (1995).

BGS85.   G. Benettin, L. Galgani, and J.-M. Strelcyn, "Kolmogorov Entropy and Numerical Experiments," *Phys. Rev. A* **14**, 2338–45 (1976). Reprinted in [Hao, 1985].

BHG97. E. Barreto, B. R. Hunt, C. Grebogi, and J. A. Yorke, "From High Dimensional Chaos to Stable Periodic Orbits: The Structure of Parameter Space," *Phys. Rev. Lett.* **78**, 4561–64 (1997).

BHM93. M. Bauer, H. Heng, and W. Martienssen, "Characterizing of Spatiotemporal Chaos from Time Series," *Phys. Rev. Lett.* **71**, 521–24 (1993).

BIB84. M. Bier and T. C. Bountis, "Re-merging Feigenbaum Trees in Dynamical Systems," *Phys. Lett. A* **104**, 239–44 (1984).

BIR35. G. D. Birkhoff, "Nouvelle recherches sur les systèmes dynamique," *Pont. Acad. Sci. Novi Lyncaei*, **1**, 85 (1935).

BIW89. O. Biham and W. Wentzel, "Characterization of Unstable Periodic Orbits in Chaotic Attractors and Repellers," *Phys. Rev. Lett.* **63**, 819–22 (1989).

BKL86. D. Bensimon, L. P. Kadanoff, S. Liang, B. Shraiman, and C. Tang, "Viscous Flows in Two Dimensions," *Rev. Mod. Phys.* **58**, 977–99 (1986).

BLD95. Y. Braiman, J. F. Lindner, annd W. L. Ditto, "Taming spatiotemporal chaos with disorder," *Nature* **378**, 465–67 (1995).

BLG97. E. Bollt, Y.-C. Lai, and C. Grebogi, "Coding, Channel Capacity, and Noise Resistance in Communicating with Chaos," *Phys. Rev. Lett.* **79**, 3787–90 (1997).

BLR95. A. Bonasera, V. Latora, and A. Rapisarda, "Universal Behavior of Lyapunov Exponents in Unstable Systems," *Phys. Rev. Lett.* **75**, 3434–37 (1995).

BLT91. E. Brener, H. Levine, and Y. Tu, "Mean-Field Theory for Diffusion-Limited Aggregation in Low Dimensions," *Phys. Rev. Lett.* **66**, 1978–81 (1991).

BLW93. D. Beigie, A. Leonard, and S. Wiggins, "Statistical Relaxation under Nonturbulent Chaotic Flows: Non-Gaussian High-Stretch Tails of Finite-Time Lyapunov Exponent Distributions," *Phys. Rev. Lett.* **70**, 275–78 (1993).

BOA96. Th. Pierre, G. Bonhomme, and A. Atipo, "Controlling the Chaotic Regime of Nonlinear Ionization Waves using the Time-Delay Autosynchronization Method," *Phys. Rev. Lett.* **76**, 2290–93 (1996).

BON98. C. Bowman and A. C. Newell, "Natural patterns and wavelets," *Rev. Mod. Phys.* **70**, 289–301 (1998).

BOP96. S. Boettcher and M. Paczuski, "Exact Results for Spatiotemporal Correlations in a Self-Organized Model of Punctuated Equilibrium," *Phys. Rev. Lett.* **76**, 348–51 (1996).

BOT88. T. Bohr and T. Tél, "The Thermodynamics of Fractals." In B. Hao (ed.) *Chaos*, Vol. 2 (World Scientific, Singapore, 1988).

BOU81. T. C. Bountis, "Period Doubling Bifurcations and Universality in Conservative Systems," *Physica D* **3**, 577–89 (1981).

BOW88. O. Bohigas and H. A. Weidenmüller, "Aspects of Chaos in Nuclear Physics," *Ann. Rev. Nucl. Part. Science* **38**, 421–53 (1988).

BRA89. R. N. Bracewell, "The Fourier Transform," *Scientific American* **260** (6), 86–95 (June, 1989).

BRG91. Y. Braiman and I. Goldhirsch, "Taming Chaotic Dynamics with Weak Periodic Perturbations," *Phys. Rev. Lett.* **66**, 2545–48 (1991).

BRI87. K. Briggs, "Simple Experiments in Chaotic Dynamics," *Am. J. Phys.* **55**, 1083–89 (1987).

BRJ87. P. Bryant and C. Jeffries, "The Dynamics of Phase Locking and Points of Resonance in a Forced Magnetic Oscillator," *Physica D* **25**, 196–232 (1987).

BRS87. A. Brandstater and H. L. Swinney, "Strange Attractors in Weakly Turbulent Couette–Taylor Flow," *Phys. Rev. A* **35**, 2207–20 (1987).

BRS88. W. A. Brooks and C. L. Sayers, "Is the Business Cycle Characterized by Deterministic Chaos?" *J. of Monetary Economics* **22**, 71–90 (1988).

BSK95. J. R. Buchler, T. Serre, and Z. Kolláth, "A Chaotic Pulsating Star: The Case of R Scuti," *Phys. Rev. Lett.* **73**, 842–45 (1995).

BSS90. M. J. Ballico, M. L. Sawley, and F. Skiff, "The Bipolar Motor: A Simple Demonstration of Deterministic Chaos," *Am. J. Phys.* **58**, 58–61 (1990).

BTW87. P. Bak, C. Tang, and K. Wiesenfeld, "Self-Organized Criticality: An Explanation of 1/f Noise," *Phys. Rev. Lett.* **59**, 381–84 (1987).

BTW88. P. Bak, C. Tang, and K. Wiesenfeld, "Self-Organized Criticality," *Phys. Rev. A* **38**, 364–74 (1988).

BUD97. T. Burns and M. Davies, "Nonlinear Dynamics Model for Chip Segmentation in Machining," *Phys. Rev. Lett.* **79**, 447–50 (1997).

BUG96. A. R. Bulsara and L. Gammaitoni, "Tuning in to Noise," *Physics Today* **49** (3), 39–45 (1996).

BUJ85. R. Van Buskirk and C. Jeffries, "Observation of Chaotic Dynamics of Coupled Nonlinear Oscillators," *Phys. Rev. A* **31** 3332–57 (1985).

BYG98. S. Banerjee, J. A. Yorke, and C. Grebogi, "Robust Chaos," *Phys. Rev. Lett.* **80** 3049–52 (1998).

CAG88. J. K. Cannizzo and D. A. Goodings, "Chaos in SS Cygni?" *Astrophys. J.* **334**, L31–34 (1988).

CAL45. M. L. Cartwright and J. E. Littlewood, "On Nonlinear Differential Equations of the Second Order. I. The Equation $\ddot{y} - k(1 - y^2)\dot{y} + y = b\lambda k \cos(\lambda t + \alpha)$, k Large," *J. London Math. Soc.* **20**, 180–89 (1945).

CAL89. J. M. Carlson and J. S. Langer, "Properties of Earthquakes Generated by Fault Dynamics," *Phys. Rev. Lett.* **62**, 2632–35 (1989).

CAM99. D. M. Campbell, "Nonlinear dynamics of musical reed and brass wind instruments," *Contemporary Physics* **40**, 415–431 (1999).

CAR48. M. L. Cartwright, "Forced Oscillations in Nearly Sinusoidal Systems," *J. Inst. Electr. Eng.* **95**, 88–96 (1948).

CAR95. Thomas L. Carroll, "A simple circuit for demonstrating regular and synchronized chaos," *Am. J. Phys.* **63**, 377–379 (1995).

CAR97. C. Cellucci, A. Albano, R. Rapp, R. Pittenger, and R. Josiassen, "Detecting noise in a time series," *Chaos* **7**, 414–22 (1997).

CAS89. M. Casdagli, "Nonlinear Predictions of Chaotic Time Series," *Physica D* **35**, 335–56 (1989).

CCG85. G. Casati, B. V. Chirikov, and I. Guarneri, "Energy-Level Statistics of Integrable Quantum Systems" *Phys. Rev. Lett.* **54**, 1350–53 (1985).

CCG90. J. Carlson, J. Chayes, E. Grannan, and G. Swindle, "Self-Organized Criticality and Singular Diffusion," *Phys. Rev. Lett.* **65**, 2547–50 (1990).

CCI79. G. Casati, B. V. Chirikov, F. M. Izrailev, and J. Ford, in *Stochastic Behavior in Classical and Hamiltonian Systems*, G. Casati and J. Ford, eds. Lecture Notes in Physics, Vol. 93 (Springer-Verlag, New York, 1979).

CDB90. V. Castets, E. Dulos, J. Boissonade, and P. DeKepper, "Experimental Evidence of a Sustained Standing Turing-Type Nonequilibrium Chemical Pattern," *Phys. Rev. Lett.* **64**, 2953–56 (1990).

CDC99. C. Caravati, F. Delogu, G. Cocco, and M. Rustici, "Hyperchaotic qualities of the ball motion in a ball milling device," *Chaos* **9**, 219–26 (1999).

CEF91. M. Casdagli, S. Eubank, J. D. Farmer, and J. Gibson, "State Space Reconstruction in the Presence of Noise," *Physica D* **51**, 52–98 (1991).

CFH82. J. P Crutchfield, J. D. Farmer, and B. A. Huberman, "Fluctuations and Simple Chaotic Dynamics," *Phys. Reports* **92**, 45–82 (1982).

CFH92. J. J. Collins, M Fanciulli, R. G. Hohlfeld, D. C. Finch, G. V. H. Sandri, and E. S. Shtatland, "A random number generator based on the logit transform of the logistic variable," *Computers in Physics* **6**, 630–2 (1992).

CFP86. J. P. Crutchfield, J. D. Farmer, N. H. Packard, and R. S. Shaw, "Chaos," *Scientific American* **255** (6), 46–57 (1986).

CGB89. M. Courtemancho, L. Glass, J. Belari, D. Scagliotti, and D. Gordon, "A Circle Map in a Human Heart," *Physica D* **40**, 299–310 (1989).

CGM90. J. K. Cannizzo, D. A. Goodings, and J. A. Mattei, "A Search for Chaotic Behavior in the Light Curves of Three Long-Period Variables," *Astrophys. J.* **357**, 235–42 (1990).

CGP88. P. Cvitanovic, G. H. Gunaratne, and I. Procaccia, "Topological and metric properties of Hénon-type strange attractors," *Phys. Rev. A* **38**, 1503–1520 (1988).

CHA75. G. J. Chaitin, "Randomness and Mathematical Proof," *Scientific American* **232** (5), 47–52 (1975).

CHA88. A. Chao et al., "Experimental Investigation of Nonlinear Dynamics in the Fermilab Tevatron," *Phys. Rev. Lett.* **61**, 2752–55 (1988).

CHC97. D. J. Christini and J. J. Collins, "Control of chaos in excitable physiological systems: A geometric analysis," *Chaos* **7**, 544–49 (1997).

CHI79. B. V. Chirikov, "A Universal Instability of Many Dimensional Oscillator Systems," *Physics Reports* **52**, 263–379 (1979).

CHJ89. A. Chhabra and R. V. Jensen, "Direct Determination of the $f(\alpha)$ Singularity Spectrum," *Phys. Rev. Lett.* **62**, 1327–30 (1989).

CJK85. P. Cvitanovic, M. H. Jensen, L. P. Kadanoff, and I. Procaccia, "Renormalization, Unstable Manifolds, and the Fractal Structure of Mode Locking," *Phys. Rev. Lett.* **55**, 343–46 (1985).

CLC90. R. F. Cahalan, H. Leidecker, and G. D. Cahalan, "Chaotic Rhythms of a Dripping Faucet," *Computers in Physics* **4**, 368–83 (1990).

CLP97. U. M. S. Costa, M. L. Lyra, A. R. Plastino, and C. Tsallis, "Power-law sensitivity to initial conditions within a logistic-like family of maps: Fractality and nonextensivity," *Phys. Rev. E.* **56**, 245–50 (1997).

CMM95. B. K. Clark, R. F. Martin, Jr., R. J. Moore, and K. E. Jesse, "Fractal dimension of the strange attractor of the bouncing ball circuit," *Am. J. Phys.* **63**, 157–63 (1995).

CMN99. D. Caroppo, M. Mannarelli, G. Nardulli, and S. Stramaglia, "Chaos in neural networks with a nonmonotonic transfer function," *Phys. Rev. E* **60**, 2186–92 (1999).

CMS86. K. Coffman, W. D. McCormick, and H. L. Swinney, "Multiplicity in a Chemical Reaction with One-Dimensional Dynamics," *Phys. Rev. Lett.* **56**, 999–1002 (1986).

CNR81. J. Crutchfield, M. Nauenberg, and J. Rudnick, "Scaling for External Noise at the Onset of Chaos," *Phys. Rev. Lett.* **46**, 933–35 (1981).

COF85. P. Constantin and C. Foias, *Commun. Pure Appl. Math.* **38**, 1 (1985).

COP99. S. N. Coppersmith, "A simpler derivation of Feigenbaum's renormalization group equation for the period-doubling bifurcation sequence," *Am. J. Phys.* **67**, 52–54 (1999).

COR00. E. Cohen and L. Rondoni, "Comment on 'Universal Relation between the Kolmogorov–Sinai Entropy and the Thermodynamic Entropy in Simple Liquids,' " *Phys. Rev. Lett.* **84**, 394 (2000).

CRA91. J. D. Crawford, "Introduction to Bifurcation Theory," *Rev. Mod. Phys.* **63**, 991–1037 (1991).

CRH80. J. P. Crutchfield and B. A. Huberman, "Fluctuations and the Onset of Chaos," *Phys. Lett. A* **77**, 407–10 (1980).

CRK88. J. P. Crutchfield and K. Kaneko, "Are Attractors Relevant to Turbulence," *Phys. Rev. Lett.* **60**, 2715–18 (1988).

CRO95. D. Crighton and L. A. Ostrovsky, "Introduction to acoustical chaos," *Chaos* **5** (3), 495 (1995).

CSE90. F. Caserta, H. E. Stanley, W. D. Eldred, G. Daccord, R. E. Hausman, and J. Nittmann, "Physical Mechanisms Underlying Neurite Outgrowth: A Quantitative Analysis of Neuronal Shape," *Phys. Rev. Lett.* **64**, 95–98 (1990).

CSZ88. A. Chernikov, R. Sagdeev, and G. Zaslavsky, "Chaos: How Regular Can It Be?" *Physics Today* **41** (11), 27–35 (November, 1988).

CUL87. A. Cumming and P. S. Linsay, "Deviations from Universality in the Transition from Quasi-Periodicity to Chaos," *Phys. Rev. Lett.* **59**, 1633–36 (1987).

CUL88. A. Cumming and P. S. Linsay, "Quasiperiodicity and Chaos in a System with Three Competing Frequencies," *Phys. Rev. Lett.* **60**, 2719–22 (1988).

CUO93. K. M. Cuomo and A. V. Oppenheim, "Circuit Implementation of Synchronized Chaos with Applications to Communications," *Phys. Rev. Lett.* **71**, 65–68 (1993).

CVI88a. P. Cvitanovic, "Invariant Measurement of Strange Sets in Terms of Cycles," *Phys. Rev. Lett.* **61**, 2729–32 (1988).

CVI88b. P. Cvitanovic, "Topological and Metric Properties of Hénon-Type Strange Attractors," *Phys. Rev. A* **38**, 1503–20 (1988).

DAV98. M. Dzugutov, E. Aurell, and A. Vulpiani, "Universal Relation between the Kolmogorov–Sinai Entropy and the Thermodynamic Entropy in Simple Liquids," *Phys. Rev. Lett.* **81**, 1762–65 (1998).

DBG91. D. Delande, A. Bommier, and J. C. Gay, "Positive-Energy Spectrum of the Hydrogen Atom in a Magnetic Field," *Phys. Rev. Lett.* **66**, 141–44 (1991).

DEN90. P. J. Denning, "Modeling Reality," *American Scientist* **78**, 495–98 (1990).

DFV95. C. S. Daw, C. Finney, M. Vasudevan, N van Goor, K. Nguyen, D. Bruns, E. Kostelich, C. Grebogi, E. Ott, and J. A. Yorke, "Self-Organization and Chaos in a Fluidized Bed," *Phys. Rev. Lett.* **75**, 2308–11 (1995).

DGE99. M. Dolnik, T. Gardner, I. Epstein, and J. Collins, "Frequency Control of an Oscillatory Reaction by Reversible Binding of an Autocatalyst," *Phys. Rev. Lett.* **82**, 1582–85 (1999).

DGS94. S. Dawson, C. Grebogi, T. Sauer, and J. A. Yorke, "Obstructions to Shadowing When a Lyapunov Exponent Fluctuates about Zero," *Phys. Rev. Lett.* **73**, 1927–30 (1994).

DGH86. D. Dangoisse, P. Glorieux, and D. Hennequin, "Laser Chaotic Attractors in Crisis," *Phys. Rev. Lett.* **57**, 2657–60 (1986).

DGK96. T. W. Dixon, T. Gherghetta, and B. G. Kenny, "Universality in the quasi-periodic route to chaos," *Chaos* **6**, 32–42 (1996).

DGO93a. M. Ding, C. Grebogi, E. Ott, T. Sauer, and J. A. Yorke, "Plateau Onset for Correlation Dimension: When Does It Occur?" *Phys. Rev. Lett.* **70**, 3872–75 (1993).

DGO93b. M. Ding, C. Grebogi, E. Ott, T. Sauer, and J. A. Yorke, "Estimating correlation dimension from a chaotic time series: when does plateau onset occur?" *Physica D* **69**, 404–24 (1993).

DHB92. K. Davies, T. Huston, and M. Baranger, "Calculations of Periodic Trajectories for the Hénon –Heiles Hamiltonian Using the Monodromy Method," *Chaos* **2**, 215–24 (1992).

DHK85. R. Delbourgo, W. Hart, and B. G. Kenny, "Dependence of Universal Constants upon Multiplication Period in Nonlinear Maps," *Phys. Rev. A* **31**, 514–6 (1985).

DIH96. M. Ding and R. C. Hilborn, "Optimal Reconstruction Space for Estimating Correlation Dimension," *Int. J. Bifur. Chaos* **6**, 377–381 (1996).

DIP93. W. L. Ditto and L. Pecora, "Mastering Chaos," *Scientific American* **269** (2), 78–84 (1993).

DIS97. W. L. Ditto and K. Showalter, "Introduction: Control and synchronization of chaos," *Chaos* **7**, 509–11 (1997).

DNO99. M. Dutta, H. E. Nusse, E. Ott, J. A. Yorke, and G. Yuan, "Multiple Attractor Bifurcations: A Source of Unpredictability in Piecewise Smooth Systems," *Phys. Rev. Lett.* **83**, 4281–4284 (1999).

DOO88. M. F. Doherty and J. M Ottino, "Chaos in Deterministic Systems: Strange Attractors, Turbulence, and Applications in Chemical Engineering," *Chemical Engineering Science* **43**, 139–83 (1988).

DRB83.   M. Dubois, M. A. Rubio, and P. Berg,, "Experimental Evidence of Intermittency Associated with a Subharmonic Bifurcation," *Phys. Rev. Lett.* **51**, 1446–49 (1983).

DRC89.   W. L. Ditto, S. Rauseo, R. Cawley, C. Grebogi, G.-H. Hsu, E. Kostelich, E. Ott, H. T. Savage, R. Segnan, M. L. Spano, and J. A. Yorke, "Experimental Observation of Crisis-Induced Intermittency and Its Critical Exponent," *Phys. Rev. Lett.* **63**, 923–26 (1989).

DRE92.   Max Dresden, "Chaos: A New Scientific Paradigm—or Science by Public Relations," *The Physics Teacher* **30**, 10–14 and 74–80 (1992).

DRH91.   K. Dreyer and F. R. Hickey, "The Route to Chaos in a Dripping Faucet," *Am. J. Phys.* **59**, 619–27 (1991).

DRJ90.   S. DeSouza-Machado, R. W. Rollins, D. T. Jacobs, and J.L. Hartman, "Studying Chaotic Systems Using Microcomputer Simulations and Lyapunov Exponents," *Am. J. Phys.* **58**, 321–29 (1990).

DRN92.   U. Dressler and G. Nitsche, "Controlling Chaos Using Time Delay Coordinates," *Phys. Rev. Lett.* **68**, 1–4 (1992).

DRS90.   W. L. Ditto, S. N. Rauseo, and M. L. Spano, "Experimental Control of Chaos," *Phys. Rev. Lett.* **65**, 3211–14 (1990).

DSB98.   F. K. Diakonos, P. Schmelcher, and O. Biham, "Systematic Computation of the Least Unstable Periodic Orbits in Chaotic Attractors," *Phys. Rev. Lett.* **81**, 4349–52 (1998).

DSF90.   E. Doron, U. Smilansky, and A. Frenkel, "Experimental Demonstration of Chaotic Scattering of Microwaves," *Phys. Rev. Lett.* **65**, 3072–75 (1990).

DUD87.   M. L. Du and J. B. Delos, "Effect of Closed Classical Orbits on Quantum Spectra: Ionization of Atoms in a Magnetic Field," *Phys. Rev. Lett.* **58**, 1731–33 (1987).

DUZ00.   M. Dzugutov, "Dzugtov Replies," *Phys. Rev. Lett.* **84**, 395 (2000).

ECK81.   J.-P. Eckmann, "Roads to Turbulence in Dissipative Dynamical Systems," *Rev. Mod. Phys.* **53**, 643–54 (1981).

ECK88a.   B. Eckhardt, "Quantum Mechanics of Classically Non-Integrable Systems," *Phys. Rep.* **163**, 205–97 (1988).

ECK88b.   B. Eckhardt, "Irregular Scattering," *Physica D* **33**, 89–98 (1988).

ECP86.   J.-P. Eckmann and I. Procaccia, "Fluctuations of Dynamical Scaling Indices in Nonlinear Systems," *Phys. Rev. A* **34**, 659–61 (1986).

ECP91.   J.-P. Eckmann and I. Procaccia, "Onset of Defect-Mediated Turbulence," *Phys. Rev. Lett.* **66**, 891–94 (1991).

ECR85.   J.-P. Eckmann and D. Ruelle, "Ergodic Theory of Chaos and Strange Attractors," *Rev. Mod Phys.* **57**, 617–56 (1985).

EGM99.   M. C. Eguia and G. B. Mindlin, "Semiconductor laser with optical feedback: From excitable to deterministic low-frequency fluctuations," *Phys. Rev. E* **60**, 1551–57 (1999).

EGO00.   D. A. Egolf, "Equilibrium Regained: From Nonequilibrium Chaos to Statistical Mechanics," *Science* **287**, 101–104 (2000).

EIN17. A. Einstein, "Zum Quantensatz vom Sommerfeld und Epstein," *Verh. Dtsch. Phys. Ges.* **19**, 82–92 (1917).

EKR86. J.-P. Eckmann, S. O. Kamphorst, D. Ruelle, and S. Ciliberto, "Liapunov Exponents from Time Series," *Phys. Rev. A* **34**, 4971–79 (1986).

ELT95. S. Ellner and P. Turchin, "Chaos in a Noisy World: New Methods and Evidence from Time-Series Analysis," *The American Naturalist* **145**, 343–75 (1995).

ELY88. P. V. Elyatin, "The Quantum Chaos Problem," *Sov. Phys. Usp.* **31**, 597–622 (1988).

ERE97. A. Erzan and J.-P. Eckmann, "q-analysis of Fractal Sets," *Phys. Rev. Lett.* **78**, 3245–48 (1997).

EPP99. I. R. Epstein and J. A. Pojman, "Overview: Nonlinear dynamics related to polymeric systems," *Chaos* **9**, 255–59 (1999).

ESN90. C. Essex and M. Nerenberg, "Fractal Dimension: Limit Capacity or Hausdorff Dimension?" *Am. J. Phys.* **58**, 986–88 (1990).

ESN91. C. Essex and M. A. H. Nerenberg, "Comments on 'Deterministic Chaos: the Science and the Fiction' by D. Ruelle," *Proc. Roy. Soc. London A* **435**, 287–92 (1991).

EYU86. R. Eykholt and D. K. Umberger, "Characterization of Fat Fractals in Nonlinear Dynamical Systems," *Phys. Rev. Lett.* **57**, 2333–36 (1986).

FAR82a. J. D. Farmer, "Chaotic Attractors of an Infinite-Dimensional Dynamical System," *Physica D* **4**, 366–93 (1982).

FAR82b. J. D. Farmer, "Information Dimension and the Probabilistic Structure of Chaos," *Z. Naturforsch.* **37a**, 1304–25 (1982).

FAS87. J. D. Farmer and J. J. Sidorowich, "Predicting Chaotic Time Series," *Phys. Rev. Lett.* **59**, 845–48 (1987).

FAS91. J. D. Farmer and J. J. Sidorowich, "Optimal Shadowing and Noise Reduction," *Physica D* **47**, 373–92 (1991).

FEI79. M. Feigenbaum, "The Universal Metric Properties of Nonlinear Transformations," *J. Stat. Phys.* **21**, 669–706 (1979).

FEI80. M. J. Feigenbaum, "Universal Behavior in Nonlinear Systems," *Los Alamos Science* **1**, 4–27 (1980).

FFP92. M. Finardi, L. Flepp, J. Parisi, R. Holzner, R. Badii, and E. Brun, "Topological and Metric Analysis of Heteroclinic Crisis in Laser Chaos," *Phys. Rev. Lett.* **68**, 2989–2991 (1992).

FIT99. R. Fitzgerald, "Phase Synchronization May Reveal Communication Pathways in Brain Activity," *Physics Today* **52** (3), 17–19 (1999).

FKO98. G. O. Fountain, D. V. Khakhar, and J. M. Ottino, "Visualization of Three-Dimensional Chaos," *Science* **281**, 683–86 (1998).

FKS82. M. J. Feigenbaum, L. P. Kadanoff, and S. J. Shenker, "Quasiperiodicity in Dissipative Systems: A Renormalization Group Analysis," *Physica D* **5**, 370–86 (1982). (Reprinted in [Hao, 1984].)

FLO89. J. Franjione, C.-W. Leong, and J. M. Ottino, "Symmetries within Chaos: A Route to Effective Mixing," *Physics of Fluids A* **1**, 1772–83 (1989).

FMP89. F. Family, B. Masters, and D. E. Platt, "Fractal Pattern Formation in Human Retinal Vessels," *Physica D* **38**, 98–103 (1989).

FMR91. J. Ford, G. Mantica, and G. H. Ristow, "The Arnol'd Cat: Failure of the Correspondence Principle," *Physica D* **50**, 493–520 (1991).

FOE93. R. F. Fox and T. C. Elston, "Amplification of intrinsic fluctuations by the Lorenz equations," *Chaos* **3**, 313–23 (1993).

FOI92. J. Ford and M. Ilg, "Eigenfunctions, Eigenvalues, and Time Evolution of Finite Bounded, Undriven, Quantum Systems are not Chaotic," *Phys. Rev. A* **45**, 6165–73 (1992).

FOM92. J. Ford and G. Mantica, "Does Quantum Mechanics Obey the Correspondence Principle? Is it Complete?" *Am. J. Phys.* **60**, 1086–98 (1992).

FOR83. J. Ford, "How Random is a Coin Toss," *Physics Today* **36** (4), 40–47 (1983).

FOR88. J. Ford, "Quantum Chaos, Is There Any?" in [Hao, 1989], pp. 128–47.

FOR89. J. Ford, "What Is Chaos That We Should Be Mindful of It?" In P. W. Davies, ed. *The New Physics* (Cambridge University Press, Cambridge, 1989).

FOY83. J. D. Farmer, E. Ott, and J. A. Yorke, "The Dimension of Chaotic Attractors," *Physica D* **7**, 153–80 (1983).

FRO90. U. Frisch and S. A. Orszag, "Turbulence: Challenges for Theory and Experiment," *Physics Today* **43** (1), 24–32 (January, 1990).

FVO98. G. O. Fountain, D. V. Khakhar, and J. M. Ottino, "Visualization of Three-Dimensional Chaos," *Science* **281**, 683–86 (1998).

GAO99. J. B. Gao, "Recurrence Time Statistics for Chaotic Systems and Their Applications," *Phys. Rev. Lett.* **83**, 3178–81 (1999).

GAR70. M. Gardner, "The Fantastic Combinations of John Conway's New Solitaire Game of Life," *Scientific American* **223**, 120–23 (April, 1970).

GER90. N. Gershenfeld, "An Experimentalist's Introduction to the Observation of Dynamical Systems." In [Hao, 1990], pp. 310–84.

GGB84. L. Glass, M. R. Guevar, J. Belair, and A. Shrier, "Global Bifurcations and Chaos in a Periodically Forced Biological Oscillator," *Phys. Rev. A* **29**, 1348–57 (1984).

GGR90. A. V. Gaponov-Grekhov and M. I. Rabinovich, "Disorder, Dynamical Chaos, and Structures," *Physics Today* **43** (7), 30–38 (July, 1990).

GGY93. J. A. C. Gallas, C. Grebogi, and J. A. Yorke, "Vertices in Parameter Space: Double Crises Which Destroy Chaotic Attractors," *Phys. Rev. Lett.* **71**, 1359–1362 (1993).

GHJ98. L. Gammaitoni, P. Hänggi, P. Jung, F. Marchesoni, "Stochastic Resonance," *Rev. Mod. Phys.* **70**, 223–87 (1998).

GHL99. J. B. Gao, S. K. Hwang, and J. M. Liu, "When Can Noise Induce Chaos?" *Phys. Rev. Lett.* **82**, 1132–35 (1999).

GHY90. C. Grebogi, S. M. Hammel, J. A. Yorke, and T. Sauer, "Shadowing of Physical Trajectories in Chaotic Dynamics: Containment and Refinement," *Phys. Rev. Lett.* **65**, 1527–30 (1990).

GIP96. G. Giacomelli and A. Politi, "Relationship between Delayed and Spatially Extended Dynamical Systems," *Phys. Rev. Lett.* **76**, 2686–89 (1996).

GIR92. Z. Gills, C. Iwata, R. Roy, I. Schwartz, and I. Triandof, "Tracking Unstable Steady States: Extending the Stability Regime of a Multimode Laser System," *Phys. Rev. Lett.* **69**, 3169–72 (1992).

GLA75. L. Glass, "Combinatorial and topological methods in nonlinear chemical kinetics," *J. Chem. Phys.* **63**, 1325–35 (1975).

GLA91. L. Glass, "Nonlinear dynamics of physiological function and control," *Chaos* **1**, 247–50 (1991).

GLA96. L. Glass, "Dynamics of Cardiac Arrhythmias," *Physics Today* **49** (8), 40–45 (1996).

GLP98. J.-P. Goedgebuer, L. Larger, and H. Porte, "Optical Cryptosystem Based on Synchronization of Hyperchaos Generated by a Delayed Feedback Tunable Laser Diode," *Phys. Rev. Lett.* **80**, 2249–52 (1998).

GLV89. G. H. Gunaratne, P. S. Linsay, and M. J. Vinson, "Chaos beyond Onset: A Comparison of Theory and Experiment," *Phys. Rev. Lett.* **63**, 1–4 (1989).

GMD90. C. Green, G. Mindlin, E. D'Angelo, H. Solari, and J. Tredicce, "Spontaneous Symmetry Breaking in a Laser: The Experimental Side," *Phys. Rev. Lett.* **65**, 3124–27 (1990).

GMO83. C. Grebogi, S. W. McDonald, E. Ott, and J. A. Yorke, "Final State Sensitivity: An obstruction to predictability," *Phys. Lett. A* **99**, 415–418 (1983).

GMP81. M. Giglio, S. Musazzi, and U. Perini, "Transition to Chaotic Behavior Via a Reproducible Sequence of Period-Doubling Bifurcations," *Phys. Rev. Lett.* **47**, 243–46 (1981).

GMV81. J. M. Greene, R. S. MacKay, F. Vivaldi, and M. J. Feigenbaum, "Universal Behavior in Families of Area-Preserving Maps," *Physica D* **3**, 468–86 (1981).

GOK99. N. Goldenfeld and L. P. Kadanoff, "Simple Lessons from Complexity," *Science* **284**, 87–89 (1999).

GOP84. C. Grebogi, E. Ott, S. Pelikan, and J. A. Yorke, "Strange Attractors that are not Chaotic," *Physica D* **13**, 261–68 (1984).

GOR87. C. Grebogi, E. Ott, F. Romeiras, and J. A. Yorke, "Critical Exponents for Crisis-Induced Intermittency," *Phys. Rev. A* **36**, 5365–80 (1987).

GOS75. J. P. Gollub and H. L. Swinney, "Onset of Turbulence in a Rotating Fluid," *Phys. Rev. Lett.* **35**, 927–30 (1975).

GOT90. H. Gould and J. Tobochnik, "More on Fractals and Chaos: Multifractals," *Computers in Physics* **4** (2), 202–7 (March/April 1990).

GOY82. C. Grebogi, E. Ott, and J. A. Yorke, "Chaotic Attractors in Crisis," *Phys. Rev. Lett.* **48**, 1507–10 (1982).

GOY83. C. Grebogi, and E. Ott, and J. A. Yorke, "Crises, Sudden Changes in Chaotic Attractors and Transient Chaos," *Physica D* **7**, 181–200 (1983).

GOY87. C. Grebogi, E. Ott, and J. A. Yorke, "Basin Boundary Metamorphoses: Changes in Accessible Boundary Orbits," *Physica D* **24**, 243–62 (1987).

GRA76. R. Graham, "Onset of Self-Pulsing in Lasers and the Lorenz Model," *Phys. Lett. A* **58**, 440–41 (1976).

GRA81. P. Grassberger, "On the Hausdorff Dimension of Fractal Attractors," *J. Stat. Phys.* **26**, 173–79 (1981).

GRA83. P. Grassberger, "Generalized Dimensions of Strange Attractors," *Phys. Lett. A* **97**, 227–30 (1983).

GRP83a. P. Grassberger and I. Procaccia, "Characterization of Strange Attractors," *Phys. Rev. Lett.* **50**, 346–49 (1983).

GRP83b. P. Grassberger and I. Procaccia, "Estimation of the Kolmogorov Entropy from a Chaotic Signal," *Phys. Rev. A* **28**, 2591–93 (1983).

GRP84. P. Grassberger and I. Procacci, "Dimensions and Entropies of Strange Attractors from a Fluctuating Dynamics Approach," *Physica D* **13**, 34–54 (1984).

GSB86. L. Glass, A. Shrier, and J. Belair, "Chaotic Cardiac Rhythms," in *Chaos* (A. V. Holden, ed.) (Princeton University Press, Princeton, NJ, 1986).

GSB99. J. Geddes, K. Short, and K. Black, "Extraction of Signals from Chaotic Laser Data," *Phys. Rev. Lett.* **83**, 5389–92 (1999).

GSC94. D. J. Gauthier, D. W. Sukow, H. M. Concannon, and J. E. S. Socolar, "Stabilizing unstable periodic orbits in a fast diode resonator using continuous time-delay autosynchronization," *Phys. Rev. E* **50**, 2343–46 (1994).

GSM88. E. J. Galvex, B. E. Sauer, L. Moorman, P. M. Koch, and D. Richards, "Microwave Ionization of H Atoms: Breakdown of Classical Dynamics for High Frequencies," *Phys. Rev. Lett.* **61**, 2011–14 (1988).

GUC81. J. Guckenheimer, "One-Dimensional Dynamics," *Ann. N.Y. Acad. Sci.* **357**, 343–47 (1981).

GUT67. M. Gutzwiller, "Phase-Integral Approximation in Momentum Space and the Bound State of Atoms," *J. Math. Phys.* **8**, 1979–2000 (1967).

GUT71. M. C. Gutzwiller, "Periodic Orbits and Classical Quantization Conditions," *J. Math. Phys.* **12**, 343–58 (1971).

GUT92. M. C. Gutzwiller, "Quantum Chaos," *Scientific American* **266** (1), 78–84 (January, 1992).

GWW85. E. G. Gwinn and R. M. Westervelt, "Intermittent Chaos and Low-Frequency Noise in the Driven Damped Pendulum," *Phys. Rev. Lett.* **54**, 1613–16 (1985).

GWW86a. E. G. Gwinn and R. M. Westervelt, "Horseshoes in the Driven, Damped Pendulum," *Physica D* **23**, 396–401 (1986).

GWW86b. E. G. Gwinn and R. M. Westervelt, "Fractal Basin Boundaries and Intermittency in the Driven Damped Pendulum," *Phys. Rev. A* **33**, 4143–55 (1986).

HAB80. J. H. Hannay and M. V. Berry, "Quantization of Linear Maps on a Torus—Fresnel Diffraction by a Periodic Grating," *Physica D* **1**, 267–90 (1980).

HAE89. J. W. Havstad and C. L. Ehlers, "Attractor Dimension of Nonstationary Dynamical Systems from Small Data Sets," *Phys. Rev. A* **39**, 845–53 (1989).

HAK75. H. Haken, "Analogy between Higher Instabilities in Fluids and Lasers," *Phys. Lett. A* **53**, 77–78 (1975).

HAK83. H. Haken, "At Least One Lyapunov Exponent Vanishes if the Trajectory of an Attractor does not Contain a Fixed Point," *Phys. Lett. A* **94**, 71–74 (1983).

HAK89. H. Haken, "Synergetics: An Overview," *Rep. Prog. Phys.* **52**, 515–33 (1989).

HAL89. P. Halpern, "Sticks and Stones: A Guide to Structurally Dynamic Cellular Automata," *Am. J. Phys.* **57**, 405–8 (1989).

HAL99. M. A. Harrison and Y.-C. Lai, "Route to high-dimensional chaos," *Phys. Rev. A* **59**, R3799–R3802 (1999).

HAS92. D. Hansel and H. Sompolinsky, "Synchronization and Computation in a Chaotic Neural Network," *Phys. Rev. Lett.* **68**, 718–21 (1992).

HCM91. M. Hassell, H. Comins, and R. May, "Spatial Structure and Chaos in Insect Population Dynamics," *Nature* **353**, 255–58 (1991).

HCP94. J. F. Hagey, T. L. Carroll, and L. M. Pecora, "Experimental and Numerical Evidence for Riddled Basins in Coupled Chaotic Systems," *Phys. Rev. Lett.* **73**, 3528–31 (1994).

HCT97. K. Hall, D. J. Christini, M. Tremblay, J. J. Collins, L. Glass, and J. Billette, "Dynamic Control of Cardiac Alternans," *Phys. Rev. Lett.* **78**, 4518–21 (1997).

HEG99. R. Hegger, "Estimating the Lyapunov spectrum of time delay feedback systems from scalar time series," *Phys. Rev. E* **60**, 1563–66 (1999).

HEH64. M. Hénon and C. Heiles, "The Applicability of the Third Integral of Motion: Some Numerical Experiments," *Astrophys. J.* **69**, 73–79 (1964).

HEI90. J. Heidel, "The Existence of Periodic Orbits of the Tent Map," *Phys. Lett. A* **43**, 195–201 (1990).

HEL80. R. H. G. Helleman, "Self-Generated Chaotic Behavior in Nonlinear Mechanics," in *Fundamental Problems in Statistical Mechanics*, Vol. 5. (E. G. D. Cohen, ed.) (North-Holland, Amsterdam, 1980), pp. 165–233.

HEN76. M. Hénon "A Two-Dimensional Mapping with a Strange Attractor," *Comm. Math. Phys.* **50**, 69–77 (1976).

HEP83. H. Hentschel and I. Procaccia, "The Infinite Number of Generalized Dimensions of Fractals and Strange Attractors," *Physica D* **8**, 435–44 (1983).

HET93. E. J. Heller and S. Tomsovic, "Postmodern Quantum Mechanics," *Physics Today* **46** (7), 38–46 (1993).

HIH83. H, Hayashi, S. Ishizuka, and K. Hirakawa, "Transition to Chaos Via Intermittency in the Onchidium Pacemaker Neuron," *Phys. Lett. A* **98**, 474–76 (1983).

HIL85. R. C. Hilborn, "Quantitative Measurement of the Parameter Dependence of the Onset of a Crisis in a Driven Nonlinear Oscillator," *Phys. Rev. A* **31**, 378–82 (1985).

HIT97. R. C. Hilborn and N. B. Tufillaro, "Resource Letter: ND-1: Nonlinear Dynamics," *Am. J. Phys.* **65**, 822–834 (1997).

HJK86. T. C. Halsey, M. H. Jensen, L. P. Kadanoff, I. Procaccia, and B. I. Schraiman, "Fractal Measures and Their Singularities: The Characterization of Strange Sets," *Phys. Rev. A* **33**, 1141–51 (1986).

HKG82. F. A. Hopf, D. L. Kaplan, H. M. Gibbs, and R. L. Shoemaker, "Bifurcations to Chaos in Optical Bistability," *Phys. Rev. A* **25**, 2172–82 (1982).

HKS99. R. Hegger, H. Kantz, and T. Schreiber, "Practical implementation of nonlinear time series methods: The TISEAN package," *Chaos* **9**, 413–35 (1999).

HOF81. D. R. Hofstadter, "Metamagical Themas," *Scientific American* **245**, (5) 22–43 (1981).

HOL90. P. Holmes, "Poincaré, Celestial Mechanics, Dynamical-Systems Theory and 'Chaos'," *Phys. Rep.* **193**, 137–63 (1990).

HOR95. John Horgan, "From Complexity to Perplexity," *Scientific American* **272** (6), 104–109 (1995).

HPH94. P. W. Hammer, N. Platt, S. M. Hammel, J. F. Heagy, and B. D. Lee, "Experimental Observation of On-Off Intermittency," *Phys. Rev. Lett.* **73**, 1095–98 (1994).

HUB88. B. A. Huberman, "Computing with Attractors: From Self-repairing Computers, to Ultradiffusion, and the Application of Dynamical Systems to Human Behavior," in *Emerging Syntheses in Science*, D. Pines, ed. Vol. I, (Addison–Wesley, Redwood City, CA, 1988).

HUB89. B. A. Huberman, "An Ecology of Machines, How Chaos Arises in Computer Networks," *The Sciences* (New York Academy of Sciences), 38–44 (July/August, 1989).

HUH84. B. A. Huberman and T. Hogg, "Adaptation and Self-Repair in Parallel Computing Structures," *Phys. Rev. Lett.* **52**, 1048–51 (1984).

HUK87. J.-Y. Huang and J.-J. Kim, "Type-II Intermittency in a Coupled Nonlinear Oscillator: Experimental Observation," *Phys. Rev. A* **36**, 1495–97 (1987).

HUN91. E. R. Hunt, "Stabilizing High-Period Orbits in a Chaotic System: The Diode Resonator," *Phys. Rev. Lett.* **67**, 1953–55 (1991).

HUO96. B. R. Hunt and E. Ott, "Optimal Periodic Orbits of Chaotic Attractors," *Phys. Rev. Lett.* **76**, 2254–57 (1996).

HUR80. B. A. Huberman and J. Rudnick, "Scaling Behavior of Chaotic Flows," *Phys. Rev. Lett.* **45**, 154–56 (1980).

HUR82. B. Hu and J. Rudnick, "Exact Solutions to the Feigenbaum Renormalization Equations for Intermittency," *Phys. Rev. Lett.* **48**, 1645–48 (1982).

HUR84. E. R. Hunt and R. W. Rollins, "Exactly Solvable Model of a Physical System Exhibiting Multidimensional Chaotic Behavior," *Phys. Rev. A* **29**, 1000–2 (1984).

HUS83. B. Hu and I. Satija, "A spectrum of universality classes in period doubling and period tripling," *Phys. Lett. A* **98**, 143-146 (1983).

HVP90. B. Hu, A. Valinai, and O. Piro, "Universality and Asymptotic Limits of the Scaling Exponents in Circle Maps," *Phys. Lett. A* **144**, 7–10 (1990).

IPS96. D. K. Ivanov, H. A. Posch, and Ch. Stumpf, "Statistical measures derived from the correlation integrals of physiological time series," *Chaos*, **6**, 243–53 (1996).

IWK91. C.-H. Iu, G. R. Welch, M. M. Kasch, D. Kleppner, D. Delande, and J. C. Gay, "Diamagnetic Rydberg Atom: Confrontation of Calculated and Observed," *Phys. Rev. Lett.* **66**, 145–48 (1991).

IZR90. F. M. Izrailev, "Simple Models of Quantum Chaos: Spectrum and Eigenfunctions," *Phys. Rep.* **196**, 299–392 (1990).

JAT94. I. M. Jánosi and T. Tél, "Time-Series Analysis of Transient Chaos," *Phys. Rev. E* **49**, 2756–63 (1994).

JBB83. M. H. Jensen, P. Bak, and T. Bohr, "Complete Devil's Staircase, Fractal Dimension and Universality of Mode-Locking Structure in the Circle Map," *Phys. Rev. Lett.* **50**, 1637–39 (1983).

JBB84. M. H. Jensen, P. Bak, and T. Bohr, "Transition to Chaos by Interaction of Resonances in Dissipative Systems I, II," *Phys. Rev. A* **30**, 1960–69 and 1970–81 (1984).

JBT94. A. Jedynak, M. Bach, and J. Timmer, "Failure of dimension analysis in a simple five-dimensional system," *Phys. Rev. E* **50**, 1170–80 (1994).

JEM85. R. V. Jensen and C. R. Myers, "Images of the Critical Points of Nonlinear Maps," *Phys. Rev. A* **32**, 1222–4 (1985).

JEN87a. R. V. Jensen, "Classical Chaos" *American Scientist* **75**, 168–81 (1987).

JEN87b. R. V. Jensen, "Chaos in Atomic Physics," in *Atomic Physics* **10**, H. Narumi and I. Shimamura, eds. (Elsevier North-Holland, Amsterdam, 1987).

JEN90. R. V. Jensen, "The Bohr Atom Revisited: A Test Case for Quantum Chaos," *Comments At. Mol. Phys.* **25**, 119–31 (1990).

JEP82. C. Jeffries and J. Perez, "Observation of a Pomeau–Manneville Intermittent Route to Chaos in a Nonlinear Oscillator," *Phys. Rev. A* **26**, 2117–22 (1982).

JKL85. M. H. Jensen, L. Kadanoff, A. Libchaber, I. Proccacia, and J. Stavans, "Global Universality at the Onset of Chaos: Results of a Forced Rayleigh–Bénard Experiment," *Phys. Rev. Lett.* **55**, 2798–801 (1985).

JNB96. H. Jaeger, S. Nagel, and R. Behringer, "The Physics of Granular Materials," *Physics Today* **49** (4), 32–38 (1996).

JOH88. S. C. Johnston and R. C. Hilborn, "Experimental Verification of a Universal Scaling Law for the Lyapunov Exponent of a Chaotic System," *Phys. Rev. A* **37**, 2680–82 (1988).

JOS88. J. V. José, "Quantum Manifestations of Classical Chaos: Statistics of Spectra," in [Hao, 1988].

JPS90. H. Jurgens, H.-O. Peitgen, and D. Saupe, "The Language of Fractals," *Scientific American* **263** (2), 60–67 (1990).

JSS89. R. V. Jensen, M. M. Sanders, M. Saraceno, and B. Sundaram, "Inhibition of Quantum Transport Due to 'Scars' of Unstable Periodic Orbits," *Phys. Rev. Lett.* **63**, 2771–15 (1989).

JSS91. R. V. Jensen, S. M. Susskind, and M. M. Sanders, "Chaotic Ionization of Highly Excited Hydrogen Atoms: Comparison of Classical and Quantum Theory with Experiment," *Phys. Rep.* **201**, 1–56 (1991).

KAM85. N. G. van Kampen in *Chaotic Behavior in Quantum Systems. Theory and Applications.* G. Casati, ed. (Plenum, NY, 1985).

KAN89a.   K. Kaneko, "Chaotic but Regular Posi-Nega Switch among Coded Attractors by Cluster-Size Variation," *Phys. Rev. Lett.* **63**, 219–23 (1989).

KAN89b. K. Kaneko, "Pattern Dynamics in Spatiotemporal Chaos," *Physica D* **34**, 1–41 (1989).

KAN90.   K. Kaneko, "Globally Coupled Chaos Violates the Law of Large Numbers but not the Central-Limit Theorem," *Phys. Rev. Lett.* **65**, 1391–94 (1990).

KAN92.   K. Kaneko, "Overview of coupled map lattices," *Chaos* **2**, 279–82 (1992).

KAP87.   H. Kaplan, "A Cartoon-Assisted Proof of Sarkowskii's Theorem," *Am. J. Phys.* **55,** 1023–32 (1987).

KAS86.   F. Kaspar and H. G. Schuster, "Scaling at the Onset of Spatial Disorder in Coupled Piecewise Linear Maps," *Phys. Lett. A* **113**, 451–53 (1986).

KAT94.   K. Kaneko and I. Tsuda, "Constructive complexity and artificial reality: an introduction," *Physica D* **75**, 1–10 (1994).

KAU91.   Stuart A. Kaufman, "Antichaos and Adaptation," *Scientific American* **265** (2), 78–84 (August, 1991).

KAY79.   J. Kaplan and J. A. Yorke, *Springer Lecture Notes in Mathematics* **730**, 204 (1979).

KER88.   R. A. Kerr, "Pluto's Orbital Motion Looks Chaotic," *Science* **240**, 986–87 (1988).

KES93.   Th.-M. Kruel, M. Eiswirth, and F. W. Schneider, "Computation of Lyapunov spectra: Effect of interactive noise and application to a chemical oscillator," *Physica D* **63**, 117–37 (1993).

KHA96.   Ya. I. Khanin, "Low-frequency dynamics of lasers," *Chaos* **6**, 373–80 (1996).

KIO89.   S. Kim and S. Ostlund, "Universal Scaling in Circle Maps," *Physica D* **39**, 365–92 (1989).

KIW90.   Y. S. Kim and E. P. Wigner, "Canonical Transformations in Quantum Mechanics," *Am. J. Phys.* **58**, 439–48 (1990).

KKK97.   R. Ketzmerick, K. Kruse, S. Kraut, and T. Geisel, "What Determines the Spreading of a Wave Packet?" *Phys. Rev. Lett.* **79**, 1959–63 (1997).

KMG99.   D. V. Khakhar, J. J.McCarthy, J.F. Gilchrist, and J. M. Ottino, "Chaotic mixing of granular materials in two-dimensional tumbling mixers," *Chaos* **9**, 195–205 (1999).

KOL58.   A. N. Kolmogorov, "A New Invariant for Transitive Dynamical Systems," *Dokl. Akad. Nauk. SSSR* **119**, 861–64 (1958).

KOP95.   L. Kocarev and U. Parlitz, "General Approach for Chaotic Synchronization with Applications to Communication," *Phys. Rev. Lett.* **74**, 5028–31 (1995).

KRG85.   H. Krantz and P. Grassberger, "Repellers, Semi-attractors, and Long-lived Chaotic Transients," *Physica D* **17**, 75–86 (1985).

KUH87.   J. Kurths and H. Herzel, "An Attractor in a Solar Time Series," *Physica D* **25**, 165–72 (1987).

KUO92. H. A. Kusch and J. M. Ottino, "Experiments on Mixing in Continuous Chaotic Flows," *J. Fluid Mech.* **236**, 319–48 (1992).

KUR81. Y. Kuramoto, "Rhythms and Turbulence in Populations of Chemical Oscillators," *Physica A* **106**, 128–43 (1981).

KUS91. H. A. Kusch, "Continuous Chaotic Mixing of Viscous Liquids," Ph.D. Thesis (Department of Chemical Engineering, University of Massachusetts, 1991).

KYR98. C.-M. Kim, G.-S. Yim, J.-W. Ryu, and Y.-J. Park, "Characteristic Relations of Type-III Intermittency in an Electronic Circuit," *Phys. Rev. Lett.* **80**, 5317–20 (1998).

LAB99. V. Latora and M. Baranger, "Kolmogorov–Sinai Entropy Rate versus Physical Entropy," *Phys. Rev. Lett.* **82**, 520–23 (1999).

LAC98. Y. Lee, L. Amaral, D. Canning, M. Meyer and H. E. Stanley, "Universal Features in the Growth Dynamics of Complex Organizations," *Phys. Rev. Lett.* **81**, 3275–78 (1998).

LAG99. Y.-C. Lai and C. Grebogi, "Modeling of Coupled Chaotic Oscillators," *Phys. Rev. Lett.* **82**, 4803–06 (1999).

LAI99. Y.-C. Lai and C. Grebogi, "Riddling of Chaotic Sets in Periodic Windows," *Phys. Rev. Lett.* **83**, 2926–29 (1999).

LAN44. L. D. Landau, "On the Problem of Turbulence," *Akad. Nauk. Doklady* **44**, 339 (1944).

LAN82. O. E. Lanford III, "A Computer-Assisted Proof of the Feigenbaum Conjectures," *Bull. Am. Math. Soc.* **6**, 427–34 (1982).

LAN87. R. Landauer, "Nonlinearity: Historical and Technological Review," in *Nonlinearity in Condensed Matter*, A. R. Bishop, D. K. Campbell, P. Kumar, and S. E. Trullinger, eds. (Springer-Verlag, Berlin, Heidelberg, New York, London, Paris, Tokyo, 1987).

LAP88. W. Lauterborn and U. Parlitz, "Methods of chaos physics and their application to acoustics," *J. Acoust. Soc. Am.* **84**, 1975–93 (1988).

LAW94. Y.-C. Lai and R. L. Winslow, "Riddled Parameter Space in Spatio-temporal Chaotic Dynamical Systems," *Phys. Rev. Lett.* **72**, 1640–43 (1994).

LAW95. M. Latka and B. J. West, "Nature of Quantum Localization in Atomic Momentum Transfer Experiments," *Phys. Rev. Lett.* **75**, 4202–5 (1995).

LEE85. K. A. H. van Leeuwen, et al. "Microwave Ionization of Hydrogen Atoms: Experiment versus Classical Dynamics," *Phys. Rev. Lett.* **55**, 2231–34 (1985).

LEE98. K. Lehnertz and C. E. Elger, "Can Epileptic Seizures be Predicted? Evidence from Nonlinear Time Series Analysis of Brain Electrical Activity," *Phys. Rev. Lett.* **80**, 5019–22 (1998).

LES88. J. Lee and H. E. Stanley, "Phase Transition in the Multifractal Spectrum of Diffusion-Limited Aggregation," *Phys. Rev. Lett.* **61**, 2945–48 (1988).

LEV49. N. Levinson, "A Second-Order Differential Equation with Singular Solutions," *Annals of Mathematics* **50**, 127–53 (1949).

LEV93. M. T. Levinson, "The Chaotic Oscilloscope," *Am. J. Phys.* **61**, 155–165 (1993).

LFO91. Y.T. Lau, J. M. Finn, and E. Ott, "Fractal Dimension in Nonhyperbolic Chaotic Scattering," *Phys. Rev. Lett.* **66**, 978–81 (1991).

LIB90. W. A. Lin and L. E. Ballentine, "Quantum Tunneling and Chaos in a Driven Anharmonic Oscillator," *Phys. Rev. Lett.* **65**, 2927–30 (1990).

LIC89. P. S. Linsay and A. W. Cumming, "Three-Frequency Quasiperiodicity, Phase Locking, and the Onset of Chaos," *Physica D* **40**, 196–217 (1989).

LIG86. J. Lighthill, "The Recently Recognized Failure of Predictability in Newtonian Dynamics," *Proc. Roy. Soc. Lond. A* **407**, 35–50 (1986).

LIY75. T.-Y. Li and J. A. Yorke, "Period Three Implies Chaos," *Amer. Math. Monthly* **82**, 985–992 (1975).

LKF90. P. Leboeuf, J. Kurchan, M. Feingold, and D. P. Arovas, "Phase-Space Localization: Topological Aspects of Quantum Chaos," *Phys. Rev. Lett.* **65**, 3076–79 (1990).

LMS97. Elisa Lanzara, Rosario N. Mangegna, Bernardo Spagnolo, and Rosalia Zangara, "Experimental study of a nonlinear system in the presence of noise: The stochastic resonance," *Am. J. Phys.* **65**, 341–49 (1997).

LNG97. Y.-C. Lai, Y. Nagai, and C. Grebogi, "Characterization of the Natural Measure by Unstable Periodic Orbits in Chaotic Attractors," *Phys. Rev. Lett.* **79**, 649–52 (1997).

LOR63. E. N. Lorenz, "Deterministic Nonperiodic Flow," *J. Atmos. Sci.* **20**, 130–41 (1963). (Reprinted in [Cvitanovic, 1984]).

LOR84. E. N. Lorenz, "The Local Structure of a Chaotic Attractor in Four Dimensions," *Physica D* **13**, 90–104 (1984).

LOR91. E. N. Lorenz, "Dimension of Weather and Climate Attractors," *Nature* **352**, 241–44 (1991).

LSW99. R. Larter, B. Speelman, and R. W. Worth, "A coupled ordinary differential equation lattice model for the simulation of epileptic seizures," *Chaos* **9**, 795–804 (1999).

LUB81. G. B. Lubkin, "Period-Doubling Route to Chaos Shows Universality," *Physics Today* **34**, 17–19 (1981).

LUB95. G. B. Lubkin, "Oscillating granular layers produce stripes, squares, hexagons, …," *Physics Today* **48** (10), 17–19 (October, 1995).

LYT98. M. L. Lyra and C. Tsallis, "Nonextensivity and Multifracticality in Low-Dimensional Dissipative Systems," *Phys. Rev. Lett.* **80**, 53–56 (1998).

MAF87. B. F. Madure and W. L. Freedman, "Self-Organizing Structures," *American Scientist* **75**, 252–59 (1987).

MAH97. J. Maron and S. Harrison, "Spatial Pattern Formation in an Insect Host-Parasitoid System," *Science* **278**, 1619–21 (1997).

MAM86a. S. Martin and W. Martienssen, "Circle Maps and Mode Locking in the Driven Electrical Conductivity of Barium Sodium Niobate Crystals," *Phys. Rev. Lett.* **56**, 1522–25 (1986).

MAM86b. S. Martin and W. Martienssen, "Small-signal Amplification in the Electrical Conductivity of Barium Sodium Niobate Crystals," *Phys. Rev. A* **34**, 4523–24 (1986).

MAN80. P. Manneville, "Intermittency, Self-Similarity and $1/f$-Spectrum in Dissipative Dynamical Systems," *J. Phys.* (Paris) **41**, 1235 (1980).

MAN85. B. B. Mandlebrot, "Self-Affine Fractals and Fractal Dimensions," *Phys. Scr.* **32**, 257–60 (1985).

MAN99. B. B. Mandelbrot, "A Multifractal Walk down Wall Street," *Scientific American* **280** (2), 70–73 (February, 1999).

MAP79. P. Manneville and Y. Pomeau, "Intermittency and the Lorenz Model," *Phys. Lett. A* **75**, 1–2 (1979).

MAS97. A. Mandell and K. Selz, "Entropy conservation as $h_{T\mu} \approx \overline{\lambda}_{\mu}^{+} d_{\mu}$ in neurobiological dynamical systems," *Chaos* **7**, 67–81 (1997).

MAY76. R. May, "Simple Mathematical Models with very Complicated Dynamics," *Nature* **261**, 459–67 (1976).

MBA91. J. F. Muzy, E. Bacry, and A. Arneodo, "Wavelets and Multifractal Formalism for Singular Signals: Application to Turbulence Data," *Phys. Rev. Lett.* **67**, 3515–18 (1991).

MBC99. K. Myneni , T. A. Barr, N. J. Corron, S. d. Pethel, "New Method for the Control of Fast Chaotic Oscillations," *Phys. Rev. Lett.* **83**, 2175–78 (1999).

MBE88. G. E. Mitchell, E. G. Bipluch, P. M. Endt, and J. F. Shringer, Jr., "Broken Symmetries and Chaotic Behavior in ^{26}Al," *Phys. Rev. Lett.* **61**, 1473–76 (1988).

MCT84. T. Matsumoto, L. O. Chua, and S. Tanaka, "Simplest Chaotic Non-autonomous Circuit," *Phys. Rev. A* **30**, 1155–57 (1984).

MCW94. S. C. Müller, P. Coullet, and D. Walgraef, "From oscillations to excitability: A case study in spatially extended systems," *Chaos* **4**, 439–42 (1994).

MEP87. E. Meron and I. Proccacia, "Gluing Bifurcations in Critical Flows: The Route to Chaos in Parametrically Excited Surface Waves," *Phys. Rev. A* **35**, 4008–11 (1987).

MES86. H. Meissner and G. Schmidt, "A Simple Experiment for Studying the Transition from Order to Chaos," *Am. J. Phys.* **54**, 800–4 (1986).

MES87. C. Meneveau and K. R. Sreenivasan, "Simple Multifractal Cascade Model for Fully Developed Turbulence," *Phys. Rev. Lett.* **59**, 1424–27 (1987).

MET87. T. M. Mello and N. B. Tufillaro, "Strange Attractors of a Bouncing Ball," *Am. J. Phys.* **55**, 316–20 (1987).

MGO85. S. W. McDonald, C. Grebogi, E. Ott, and J. A. Yorke, "Fractal Basin Boundaries," *Physica D* **17**, 125–153 (1985).

MHS90. G. B. Mindlin, X.-J. Hou, H. G. Solari, R. Gilmore, and N. B. Tufillaro, "Classification of Strange Attractors by Integers," *Phys. Rev. Lett.* **64**, 2350–3 (1990).

MIS81. M. Misiurewicz, *Publ. Math. I.H.E.S.* **53**, 17 (1981).

MLM89. M. Möller, W. Lange, F. Mitschke, N. B. Abraham, and U. Hübner, "Errors from Digitizing and Noise in Estimating Attractor Dimensions," *Phys. Lett. A* **138**, 176–82 (1989).

MMK91. S. A. Meyer, M. Morgenstern, S. Knudson, and D. Noid, "Novel Method for WKB Analysis of Multidimensional Systems," *Am. J. Phys.* **59**, 145–51 (1991).

MML88.   F. Mitschke, M. Möller, and W. Lange, "Measuring Filtered Chaotic Signals," *Phys. Rev. A* **37**, 4518–21 (1988).

MMS91.   P. Matthews, R. Mirollo, and S. Strogatz, "Dynamics of a Large System of Coupled Nonlinear Oscillators," *Physica D* **52**, 293–331 (1991).

MNC95.   A. Mekis, J. U. Nöckel, G. Chen, A. D. Stone, and R. K. Chang, "Ray Chaos and *Q* Spoiling in Lasing Droplets," *Phys. Rev. Lett.* **75**, 2682–85 (1995).

MOH85.   F. C. Moon and W. T. Holmes "Double Poincaré Sections of a Quasi-Periodically Forced, Chaotic Attractor," *Phys. Lett. A* **111**, 157–60 (1985).

MOL85.   F. C. Moon and G.-X. Li, "Fractal Basin Boundaries and Homoclinic Orbits for Periodic Motion in a Two-Well Potential," *Phys. Rev. Lett.* **55**, 1439–42 (1985).

MOO93.   H.-T. Moon, "Approach to chaos through instabilities," *Rev. Mod. Phys.* **65**, 1535–43 (1993).

MOW95.   F. Moss and K. Wiesenfeld, "The Benefits of Background Noise," *Scientific American* **273** (3) 66–69 (1995).

MPS85.   P. Martien, S. C. Pope, P. L. Scott, and R. S. Shaw, "The Chaotic Behavior of the Leaky Faucet," *Phys. Lett. A* **110**, 399–404 (1985).

MRB94.   F. L. Moore, J. C. Robinson, C. Bharucha, P. E. Williams, and M. G. Raizen, "Observation of Dynamical Localization in Atomic Momentum Transfer: A New Testing Ground for Quantum Chaos," *Phys. Rev. Lett.* **73**, 2974–77 (1994).

MSM95.   G. Metcalf, T. Shinbrot, J. J. McCarthy and J. M. Ottino, "Avalanche mixing of granular solids," *Nature* **374**, 39–41 (1995).

MSS73.   N. Metropolis, M. L. Stein, and P. R. Stein, "On Finite Limit Sets for Transformations of the Unit Interval," *J. Combinatorial Theory (A)* **15**, 25–44 (1973).

MTL83.   E. V. Mielczarek, J. S. Turner, D. Leiter, and L. Davis, "Chemical Clocks: Experimental and Theoretical Models of Nonlinear Behavior," *Am. J. Phys.* **51**, 32–42 (1983).

MTV86.   N. Margolis, T. Toffoli, and G. Vichniac, "Cellular-Automata Supercomputers for Fluid Dynamics Modeling," *Phys. Rev. Lett.* **56**, 1694–96 (1986).

MUL86.   G. Müller, "Nature of Quantum Chaos in Spin Systems," *Phys. Rev. A* **34**, 3345–55 (1986).

MUL97.   P. Muruganandam and M. Lakshmanan, "Bifurcation analysis of the travelling waveform of FitzHugh–Nagumo nerve conduction model equation," *Chaos* **7**, 476–87 (1997).

MUS95.   F. Melo, P. Umbanhowar, and H. L. Swinney, "Hexagons, Kinks, and Disorder in Oscillated Granular Layers," *Phys. Rev. Lett.* **75**, 3838–41 (1995).

MYR62.   P. J. Myrberg, "Sur l'Itération des Polynomes Réels Quadratiques," *J. Math. Pure Appl.* **41**, 339–51 (1962).

NAG94.   T. C. Newell, P. M. Alsing, A. Gavrielides, and V. Kovanis, "Synchronization of Chaotic Diode Resonators by Occasional Proportional Feedback," *Phys. Rev. Lett.* **72**, 1647–50 (1994).

NKG96. T. C. Newell, V. Kovanis, and A. Gavrielides, "Experimental Demonstration of Antimonotonicity: The Concurrent Creation and Destruction of Periodic Orbits in a Driven Nonlinear Electronic Resonator," *Phys. Rev. Lett.* **77**, 1747–50 (1996).

NPW84. L. Niemeyer, L. Pietronero, and H. J. Wiesmann, "Fractal Dimension of Dielectric Breakdown," *Phys. Rev. Lett.* **52**, 1033–36 (1984).

NRT78. S. E. Newhouse, D. Ruelle, and R. Takens, "Occurrence of Strange Axiom A Attractors near Quasi-Periodic Flows on T_m ($m = 3$ or more)," *Commun. Math. Phys.* **64**, 35 (1978).

NSC99. A. Neiman, L. Schimanksy-Geier, A. Cornell-Bell, and F. Moss, "Noise-Enhanced Phase Synchronization in Excitable Media," *Phys. Rev. Lett.* **83**, 4896–99 (1999).

NSM99. A. Neiman, L. Schimanksy-Geier, F. Moss, B. Shulgin, and J. J. Collins, "Synchronization of noisy systems by stochastic signals," *Phys. Rev. E* **60**, 284–92 (1999).

OCO83. R. F. O'Connell, "The Wigner Distribution Function—50th Birthday," *Found. Phys.* **13**, 83–93 (1983).

OGY90. E. Ott, C. Grebogi, and J. A. Yorke, "Controlling Chaos," *Phys. Rev. Lett.* **64**, 1196–99 (1990).

OLR88. J. M. Ottino, C. Leong, H. Rising, and P. Swanson, "Morphological Structure Produced by Mixing in Chaotic Flows," *Nature* **333**, 419–25 (1988).

OLS91. C. L. Olson and M. G. Olsson, "Dynamical symmetry breaking and chaos in Duffing's equation," *Am. J. Phys.* **59**, 907–11 (1991).

OLU98. J. S. Olafsena and J. S. Urbach, "Clustering, Order, and Collapse in a Driven Granular Monolayer," *Phys. Rev. Lett.* **81**, 4369–72 (1998).

OMH91. A. Ojha, S. Moon, B. Hoeling, and P. B. Siegel, "Measurements of the Transient Motion of a Simple Nonlinear System," *Am. J. Phys.* **59**, 614–19 (1991).

OMT92. J. M. Ottino, F. J. Muzzio, M. Tjahjadi, J. G. Frangione, S. C. Jano, and H. A. Kusch, "Chaos, Symmetry, and Self-Similarity: Exploiting Order and Disorder in Mixing Processes," *Science* **257**, 754–60 (1992).

OTS95. E. Ott and M. Spano, "Controlling Chaos," *Physics Today* **48** (5), 34–40 (1995).

OTT81. E. Ott, "Strange Attractors and Chaotic Motions of Dynamical Systems," *Rev. Mod. Phys.* **53**, 655–72 (1981).

OTT89. J. M. Ottino, "The Mixing of Fluids," *Scientific American* **260** (1), 56–67 (January, 1989).

OTT93. E. Ott and T. Tél, "Chaotic scattering: an introduction," *Chaos* **3**, 417–426 (1993).

OUS91a. Q. Ouyang and H. L. Swinney, "Transition from a Uniform State to Hexagonal and Striped Turing Patterns," *Nature* **352**, 610–12 (1991).

OUS91b. Q. Ouyang and H. L. Swinney, "Transition to Chemical Turbulence," *Chaos* **1**, 411–19 (1991).

PAH94. R. Palmer, W. B. Arthur, J. Holland, B. LeBaron, and P. Tayler, "Artificial economic life: a simple model of a stockmarket," *Physica D* **75**, 264–74 (1994).

PAN99. M. Paluš and D. Novotná, "Sunspot Cycle: A Driven Nonlinear Oscillator?" *Phys. Rev. Lett.* **83**, 3406–9 (1999).

PAR92. U. Parlitz, "Identification of True and Spurious Lyapunov Exponents from Time Series," *Int. J. Bifur. Chaos*, **2**, 155–65 (1992).

PAS87. K. Pawelzik and H. G. Schuster, "Generalized Dimensions and Entropies from a Measured Time Series," *Phys. Rev. A* **35**, 481–84 (1987).

PAV86. G. Paladin and A. Vulpiani, "Intermittency in chaotic systems and Renyi entropies," *J. Phys. A* **19**, L997–1001 (1986).

PCF80. N. H. Packard, J. P. Crutchfield, J. D. Farmer, and R. S. Shaw, "Geometry from a Time Series," *Phys. Rev. Lett.* **45**, 712–15 (1980).

PCF87. J. P. Pique, Y. Chen, R. W. Field, and J. L. Kinsey, "Chaos and Dynamics on 0.5–300 ps Time Scales in Vibrationally Exited Acetylene: Fourier Transform of Stimulated-Emission Pumping Spectrum," *Phys. Rev. Lett.* **58**, 475–78 (1987).

PEC91. L. Pecora and T. Carroll, "Driving Systems with Chaotic Signals," *Phys. Rev. A* **44**, 2374–83 (1991).

PEC91. L. Pecora and T. Carroll, "Pseudo-periodic Driving: Eliminating Multiple Domains of Attraction Using Chaos," *Phys. Rev. Lett.* **67**, 945–48 (1991).

PES77. Ya. B. Pesin, *Russ. Math. Surveys* **32**, 55 (1977).

PET89. M. Peterson, "Non-uniqueness in Singular Viscous Fingering," *Phys. Rev. Lett.* **62**, 284–87 (1989).

POM80. Y. Pomeau and P. Manneville, "Intermittent Transition to Turbulence in Dissipative Dynamical Systems," *Commun. Math. Phys.* **74**, 189–97 (1980).

POO89. R. Pool, "Ecologists Flirt with Chaos," *Science* **243**, 310–13 (1989).

PPG85. G. P. Puccioni, A. Poggi, W. Gadomski, J. R. Tredicce, and F. T. Arecchi, "Measurement of the Formation and Evolution of a Strange Attractor in a Laser," *Phys. Rev. Lett.* **55**, 339–41 (1985).

PPV86. G. Paladin, L. Peliti, and A. Vulpiani, "Intermittency as multifractality in history space," *J. Phys. A* **19**, L991–6 (1986).

PRF89. R. E. Prange and S. Fishman, "Experimental Realizations of Kicked Quantum Chaotic Systems," *Phys. Rev. Lett.* **63**, 704–7 (1989).

PRS83. I. Procaccia and H. G. Schuster, "Functional Renormalization Group Theory of Universal $1/f$-Noise in Dynamical Systems," *Phys. Rev. A* **28**, 1210–12 (1983).

PRT94. D. Prichard and J. Theiler, "Generating Surrogate Data for Time Series with Several Simultaneously Measured Variables," *Phys. Rev. Lett.* **73**, 951–54 (1994).

PST93. N. Platt, E. A. Spiegel, and C. Tresser, "On-Off Intermittency: A Mechanism for Bursting," *Phys. Rev. Lett.* **70**, 279–82 (1993).

PST97. A. Provenzale, E. A. Spiegel, and R. Thieberger, "Cosmic lacunarity," *Chaos* **7**, 82–88 (1997).

PTT87. I. Proccacia, S. Thomae, and C. Tresser, "First-Return Maps as a Unified Renormalization Scheme for Dynamical Systems," *Phys. Rev. A* **35**, 1884–1900 (1987).

RAI99. M. Raizen, "Quantum Chaos and Cold Atoms," in *Advances in Atomic, Molecular, and Optical Physics* **41**, 43–81. B. Bederson and H. Walther, eds. (Academic Press, San Diego, 1999).

RAO98. C. Robert, K. T. Alligood, E. Ott, and J. A. Yorke, "Outer Tangency Bifurcations of Chaotic Sets," *Phys. Rev. Lett.* **80**, 4867–70 (1998).

RAS99. K. Ramasubramanian and M. S. Sriram, "Alternative algorithm for the computation of Lyapunov spectra of dynamical systems," *Phys. Rev. E* **60**, R1126–29 (1999).

RBM95. J. C. Robinson, C. Bharucha, F. L. Moore, R. Jahnke, G. A. Georgakis, Q. Niu, and M. G. Raizen, and B. Sundaram, "Study of Quantum Dynamics in the Transition from Classical Stability to Chaos," *Phys. Rev. Lett.* **74**, 3963–66 (1995).

RBW96. N. Regez, W. Breymann, S. Weigert C. Kaufman, and G. Müller, "Hamiltonian Chaos IV," *Computers in Physics* **10**, 39–45 (1996). See, also SKM90, SKM91.

RCB95. F. Rödelsperger, A. Cenys, and H. Benner, "On-Off Intermittency in Spin-Wave Instabilities," *Phys. Rev. Lett.* **75**, 2594–97 (1995).

RCD94. M. T. Rosenstein, J. J. Collins, and C. J. De Luca, "Reconstruction expansion as a geometry-based framework for choosing proper delay times," *Physica D* **73**, 82–98 (1994).

RDC93. M. T. Rosenstein, J. J. Collins, and C. J. DeLuca, "A practical method for calculating largest Lyapunov exponents from small data sets," *Physica D* **65**, 117–34 (1993).

RED89. M. A. Rubio, C. A. Edwards, A. Dougherty, and J. P. Gollub, "Self-Affine Fractal Interfaces from Immiscible Displacements in Porous Media," *Phys. Rev. Lett.* **63**, 1685–87 (1989).

REN99. C. Reichardt and F. Nori, "Phase Locking, Devil's Staircases, Farey Trees, and Arnold Tongues in Driven Vortex Lattices with Periodic Pinning," *Phys. Rev. Lett.* **82**, 414–17 (1999).

RHO80. D. A. Russell, J. D. Hansen, and E. Ott, "Dimensions of Strange Attractors," *Phys. Rev. Lett.* **45**, 1175–78 (1980).

RIG98. K. Richardson, T. Imhoff, P. Grigg, and J. J. Collins, "Using electrical noise to enhance the ability of humans to detect subthreshold mechanical cutaneous stimuli," *Chaos* **8**, 599–603 (1998).

RIS87. G. H. Ristow, "A Quantum Mechanical Investigation of the Arnold Cat Map," Master's Thesis, School of Physics, Georgia Institute of Technology (1987).

RIZ99. R. De Los Rios and Y.-C. Zhang, "Universal $1/f$ Noise from Dissipative Self-Organized Criticality Models," *Phys. Rev. Lett.* **82**, 472–75 (1999).

RKZ00. V. Rom-Kedar and G. M. Zaslavsky, "Chaotic kinetics and transport (Overview)," *Chaos* **10** (1), 1–2 (2000).

RLG97.   P. Rohani, T. Lewis, D. Grünbaum, and G. Ruxton, "Spatial self-organization in ecology: pretty patterns or robust reality," *TREE* **12** (2), 70–74 (1997).

RMM92.   R. Roy, T. Murphy, T. Maier, Z. Gills, and E. R. Hunt, "Dynamical Control of a Chaotic Laser: Experimental Stabilization of a Globally Coupled System," *Phys. Rev. Lett.* **68**, 1259–62 (1992).

ROC94.   M. T. Rosenstein and J. J. Collins, "Visualizing the Effects of Filtering Chaotic Signals," *Computers and Graphics* **18**, 587–92 (1994).

ROH82.   R. W. Rollins and E. R. Hunt, "Exactly Solvable Model of a Physical System Exhibiting Universal Chaotic Behavior," *Phys. Rev. Lett.* **49**, 1295–98 (1982).

ROH84.   R. W. Rollins and E. R. Hunt, "Intermittent Transient Chaos at Interior Crises in the Diode Resonator," *Phys. Rev. A* **29**, 3327–34 (1984).

ROS76.   O. E. Rössler, "An equation for continuous chaos," *Phys. Lett. A* **57**, 397 (1976).

ROS82.   D. Rand, S. Ostlund, J. Sethna, and E. Siggia, "Universal Transition from Quasi-Periodicity to Chaos in Dissipative Systems," *Phys. Rev. Lett.* **49**, 132–35 (1982). (Reprinted in [Hao, 1984].)

RTV99.   M. Rabinovich, J. Torres, P. Varona, R. Huerta, and P. Weidman, "Origin of coherent structures in a discrete choatic medium," *Phys. Rev. E* **60**, R1130–33 (1999).

RUE90.   D. Ruelle, "Deterministic Chaos: The Science and the Fiction," *Proc. Roy. Soc. Lond. A* **427**, 241–48 (1990).

RUE94.   D. Ruelle, "Where can one hope to profitably apply the ideas of chaos," *Physics Today* **47** (7), 24–30 (1994).

RUT71.   D. Ruelle and F. Takens, "On the Nature of Turbulence," *Commun. Math. Phys.* **20**, 167–92 (1971).

SAL62.   B. Saltzman, "Finite Amplitude Free Convection as an Initial Value Problem-I," *J. Atmos. Sci.* **19**, 329–41 (1962).

SAN87.   L. M. Sander, "Fractal Growth," *Scientific American* **256**, 94–100 (January, 1987).

SAW86.   J. Salem and S. Wolfram, "Thermodynamics and Hydrodynamics of Cellular Automata," in [S. Wolfram, ed., 1986].

SAY91.   C. L. Sayers, "Statistical Inference Based Upon Nonlinear Science," *European Economic Review* **35**, 306–12 (1991).

SCC85.   C. L. Scofield and L. N. Cooper, "Development and Properties of Neural Networks," *Contemporary Physics* **26**, 125–45 (1985).

SCD97.   P. Schmelcher and F. K. Diakonos, "Detecting Unstable Periodic Orbits of Chaotic Dynamical Systems," *Phys. Rev. Lett.* **78**, 4733–36 (1997).

SCG91.   M. Schreiber and H. Grussbach, "Multifractal Wave Functions at the Anderson Transition," *Phys. Rev. Lett.* **67**, 607–10 (1991).

SCH83.   J. N. Schulman, "Chaos in Piecewise-Linear Systems," *Phys. Rev. A* **28**, 477–79 (1983).

SCH88. G. Schmidt, "Universality of Dissipative Systems," in [Hao, 1988], pp. 1–15.

SCK99. T. Shibata, T. Chawanya, and K. Kaneko, "Noiseless Collective Motion out of Noisy Chaos," *Phys. Rev. Lett.* **82**, 4424–27 (1999).

SCL99. W. L. Shew, H. A. Coy, and J. F. Lindner, "Taming chaos with disorder in a pendulum array," *Am. J. Phys.* **67**, 703–708 (1999).

SCS88. H. Sompolinksy, A. Crisanti, and H. J. Sommers, "Chaos in Random Neural Networks," *Phys. Rev. Lett.* **61**, 259–62 (1988).

SCS96. T. Schreiber and A. Schmitz, "Improved Surrogate Data for Nonlinearity Tests," *Phys. Rev. Lett.* **77**, 635–38 (1996).

SCT96. I. B. Schwartz and I. Triandaf, "Chaos and intermittent bursting in a reaction-diffusion process," *Chaos* **6**, 229–37 (1996).

SCW85. G. Schmidt and B. H. Wang, "Dissipative Standard Map," *Phys. Rev. A* **32**, 2994–99 (1985).

SDG86. Y. Sawado, A. Dougherty, and J. P. Gollub, "Dendritic and Fractal Patterns in Electrolytic Metal Deposits," *Phys. Rev. Lett.* **56**, 1260–63 (1986).

SDG91. J. C. Sommerer, W. L. Ditto, C. Grebogi, E. Ott, and M. L. Spano, "Experimental Confirmation of the Scaling Theory of Noise-Induced Crises," *Phys. Rev. Lett.* **66**, 1947–50 (1991).

SDG92. T. Shinbrot, W. Ditto, C. Grebogi, E. Ott, M. Spano, and J. A. Yorke, "Using the Sensitive Dependence of Chaos (the 'Butterfly Effect') to Direct Trajectories in an Experimental Chaotic System," *Phys. Rev. Lett.* **68**, 2863–66 (1992).

SEG89. J. Sacher, W. Elsässer, and E. Göbel, "Intermittency in the Coherence Collapse of a Semiconductor Laser with External Feedback," *Phys. Rev. Lett.* **63**, 2224–27 (1989).

SFM88. E. Sorensen, H. Fogedby, and O. Mouritsen, "Crossover from Nonequilibrium Fractal Growth to Equilibrium Compact Growth," *Phys. Rev. Lett.* **61**, 2770–73 (1988).

SFP97. A. Siahmakoun, V. A. French, and J. Patterson, "Nonlinear dynamics of a sinusoidally driven pendulum in a repulsive magnetic field," *Am. J. Phys.* **65** (5), 393–400 (1997).

SGO93. T. Shinbrot, C. Grebogi, E. Ott, and J. A. Yorke, "Using small perturbations to control chaos," *Nature* **363**, 411–17 (1993).

SGY97. T. Sauer, C. Grebogi, J. A. Yorke, "How Long Do Numerical Chaotic Solutions Remain Valid?" *Phys. Rev. Lett.* **79**, 59–62 (1997).

SHA81. R. Shaw, "Strange Attractors, Chaotic Behavior, and Information Flow," *Z. Naturf.* **36a**, 80–112 (1981).

SHA95. A. Sharma, "Assessing the magnetosphere's nonlinear behavior: Its dimension is low, its predictability, high," *Rev. Geophysics Suppl.* 645–50 (July, 1995).

SHE82. S. Shenker, "Scaling Behavior in a Map of a Circle onto Itself: Empirical Results," *Physica D* **5**, 405–11 (1982).

SHG84. M. Shapiro and G. Goelman, "Onset of Chaos in an Isolated Energy Eigenstate," *Phys. Rev. Lett.* **53**, 1714–17 (1984).

SHL85. J. Stavans, F. Heslot, and A. Libchaber, "Fixed Winding Number and the Quasiperiodic Route to Chaos in a Convective Fluid," *Phys. Rev. Lett.* **55**, 596–99 (1985).

SHM98. T. Shinbrot and F. J. Muzzio, "Reverse Buoyancy in Shaken Granular Beds," *Phys. Rev. Lett.* **81**, 4365–68 (1998).

SHO93. T. Shinbrot and J. M. Ottino, "Geometric Method to Create Coherent Structures in Chaotic Flows," *Phys. Rev. Lett.* **71**, 843–46 (1993).

SHU89. A. Shudo, "Algorithmic Complexity of the Eigenvalue Sequence of a Nonintegrable Hamiltonian System," *Phys. Rev. Lett.* **63**, 1897–901 (1989).

SID98. S. Sinha and W. L. Ditto, "Dynamics Based Computation," *Phys. Rev. Lett.* **81**, 2156–59 (1998).

SIL70. L. P. Sil'nikov, "A Contribution to the Problem of the Structure of an Extended Neighborhood of a Rough Equilibrium State of Saddle-Focus Type," *Math. USSR Sbornik* **10**, 91–102 (1970).

SIM99. N. S. Simonovic, "Calculations of periodic orbits: The monodromy method and application to regularized systems," *Chaos* **9**, 854–64 (1999).

SIN78. D. Singer, "Stable Orbits and Bifurcations of Maps of the Interval," *SIAM J. Appl. Math.* **35**, 260–7 (1978).

SKM90. SKM91. N. Srivastava, C. Kaufman, and G. Müller, "Hamiltonian Chaos," *Computers in Physics* **4**, 549–53 (1990) and "Hamiltonian Chaos II," *Computers in Physics* **5**, 239–43 (1991). "Hamiltonian Chaos III," *Computers in Physics* **6**, 84–8 (1991). See, also RBW96.

SMA63. S. Smale, "Diffeomorphisms with Many Periodic Points." In *Differential and Combinatorial Topology*, S. S. Cairns, ed. (Princeton University Press, Princeton, NJ, 1963).

SMA67. S. Smale, "Differentiable Dynamical Systems," *Bull. Amer. Math. Soc.* **73**, 747–817 (1967).

SOG88. H. G. Solari and R. Gilmore, "Relative Rotation Rates for Driven Dynamical Systems," *Phys. Rev. A* **37**, 3096–109 (1988).

SOG90. T. Shinbrot, E. Ott, C. Grebogi, and J. A. Yorke, "Using Chaos to Direct Trajectories to Targets," *Phys. Rev. Lett.* **65**, 3215–18 (1990).

SOO93a. J. C. Sommerer and E. Ott, "A physical system with qualitatively uncertain dynamics," *Nature* **365**, 136–140 (1993).

SOO93b. J. C. Sommerer and E. Ott, "Particles Floating on a Moving Fluid: A Dynamical Comprehensible Physical Fractal," *Science* **259**, 335–39 (1993).

SOS96. P. So, E. Ott, S. J. Schiff, D. T. Kaplan, T. Sauer, and C. Grebogi, "Detecting Unstable Periodic Orbits in Chaotic Experimental Data," *Phys. Rev. Lett.* **76**, 4705–8 (1996).

SRH87. Z. Su, R. W. Rollins, and E. R. Hunt, "Measurements of $f(\alpha)$ Spectrum of Attractors at Transitions to Chaos in Driven Diode Resonator Systems," *Phys. Rev. A* **36**, 3515–17 (1987).

SRS97. E. Simonott, M. Riani, C. Seife, M. Roberts, J. Twitty, and F. Moss, "Visual Perception of Stochastic Resonance," *Phys. Rev. Lett.* **78**, 1186–89 (1997).

SSD95. D. A. Sadovskif, J. A. Shaw, and J. B. Delos, "Organization of Sequences of Bifurcations of Periodic Orbits," *Phys. Rev. Lett.* **75**, 2120–23 (1995).

SSO98. C. G. Schroer, T. Sauer, E. Ott, and J. A. Yorke, "Predicting Chaos Most of the Time from Embeddings with Self-Intersections," *Phys. Rev. Lett.* **80**, 1410–13 (1998).

STF82. C. W. Smith, M. J. Tejwani, and D. A. Farris, "Bifurcation Universality for First-Sound Subharmonic Generation in Superfluid Helium-4," *Phys. Rev. Lett.* **48**, 492–94 (1982).

STM88. H. E. Stanley and P. Meakin, "Multifractal Phenomena in Physics and Chemistry," *Nature* **335**, 405–9 (1988).

STS93. S. H. Strogatz and I. Stewart, "Coupled Oscillators and Biological Synchronization," *Scientific American* **269** (6), 102–109 (1993).

STY98. T. D. Sauer, J. A. Tempkin, and J. A. Yorke, "Spurious Lyapunov Exponents in Attractor Reconstruction," *Phys. Rev. Lett.* **81**, 4341–44 (1998).

SUG95. H. B. Stewart, Y. Ueda, C. Grebogi, and J. A. Yorke, "Double Crises in Two-Parameter Dynamical Systems," *Phys. Rev. Lett.* **75**, 2478–2481 (1995).

SUW92. G. J. Sussman and J. Wisdom, "Chaotic Evolution of the Solar System," *Science* **257**, 56–62 (1992).

SWM81. B. Shraiman, C. E. Wayne, and P. C. Martin, "Scaling Theory for Noisy Period-Doubling Transitions to Chaos," *Phys. Rev. Lett.* **46**, 935–9 (1981).

SWS82. R. H. Simoyi, A. Wolf, and H. L. Swinney, "One-Dimensional Dynamics in a Multicomponent Chemical Reaction," *Phys. Rev. Lett.* **49**, 245–48 (1982).

SYC91. T. Sauer, J. A. Yorke, and M. Casdagli, "Embedology," *J. Stat. Phys.* **65**, 579–616 (1991).

TAH87. D. Tank and J. Hopfield, "Collective Computation in Neuronlike Circuits," *Scientific American* **257** (6), 104–14 (1987).

TAK81. F. Takens in *Dynamical Systems and Turbulence.* Vol. 898 of *Lecture Notes in Mathematics*, D. A. Rand and L. S. Young, eds. (Springer-Verlag, Berlin, 1981).

TEL87. T. Tél, "Dynamical Spectrum and Thermodynamic Functions of Strange Sets from an Eigenvalue Problem," *Phys. Rev. A* **36**, 2507–10 (1987).

TEL90. T. Tél, "Transient Chaos," in [Hao, 1990].

TEL92. J. Theiler, S. Eubank, A. Longtin, B. Galdrikian, and J. D. Farmer, "Testing for nonlinearity in time series: the method of surrogate data," *Physica D* **58**, 77–94 (1992).

TES85. J. Testa, "Fractal Dimension at Chaos of a Quasiperiodic Driven Tunnel Diode," *Phys. Lett. A* **111**, 243–45 (1985).

TGL91. J. Theiler, B. Galdrikian, A. Longtin, S. Eubank, and J. D. Farmer, in *Nonlinear Modeling and Forecasting*, Santa Fe Institute Studies in the Sciences of Complexity, Proc. Vol. XII, pp. 163–88. M. Casdagli and S. Eubank, eds. (Addison–Wesley, Reading, MA, 1991).

THE86. J. Theiler, "Spurious Dimension from Correlation Algorithms Applied to Limited Time-Series Data," *Phys. Rev. A* **34**, 2427–32 (1986).

THE90. J. Theiler, "Estimating Fractal Dimension," *J. Opt. Soc. Am. A* **7**, 1055–73 (1990).

TLA95. L. Tsimring, H. Levine, I. Aranson, E. Ben-Jacob, I. Cohen, O. Shochet, and W. Reynolds, "Aggregation Patterns in Stressed Bacteria," *Phys. Rev. Lett.* **75**, 1859–62 (1995).

TPJ82. J. Testa, J. Perez, and C. Jeffries, "Evidence for Universal Chaotic Behavior of a Driven Nonlinear Oscillator," *Phys. Rev. Lett.* **48**, 714–17 (1982).

TPW91. D. Y. Tang, J. Pujol, and C. O. Weiss, "Type III Intermittency of a Laser," *Phys. Rev. A* **44**, 35–38 (1991).

TRS97. I. Triandaf and I. B. Schwartz, "Karhunen–Loeve mode control of chaos in a reaction-diffusion process," *Phys. Rev. E* **56**, 204–212 (1997).

TSA88. C. Tsallis, "Possible Generalization of Boltzmann–Gibbs Statistics," *J. Stat. Phys.* **52**, 479–87 (1988).

TSB91. G. Tanner, P. Scherer, E. B. Bogomolny, B. Eckhardt, and D. Wintgen, "Quantum Eigenvalues from Classical Periodic Orbits," *Phys. Rev. Lett.* **67**, 2410–13 (1991).

TUA86. N. B. Tufillaro and A. M. Albano, "Chaotic Dynamics of a Bouncing Ball," *Am. J. Phys.* **54**, 939–44 (1986).

TUF89. N. B. Tufillaro, "Nonlinear and Chaotic String Vibrations," *Am J. Phys.* **57**, 408–14 (1989).

TUR52. A. M. Turing, "The Chemical Basis of Morphogenesis," *Phil. Trans. Roy. Soc. London B* **237**, 37–72 (1952).

TVS88. W. Y. Tam, J. A. Vastano, H. L. Swinney, and W. Horsthemke, "Regular and Chaotic Chemical Spatiotemporal Patterns," *Phys. Rev. Lett.* **61**, 2163–66 (1988).

TWB87. C. Tang, K. Wiesenfeld, P. Bak, S. Coppersmith, and P. Littlewood, "Phase Organization," *Phys. Rev. Lett.* **58**, 1161–64 (1987).

UMF85. D. K. Umberger and J. D. Farmer, "Fat Fractals on the Energy Surface," *Phys. Rev. Lett.* **55**, 661–64 (1985).

VAR98a. G. VanWiggeren and R. Roy, "Optical Communication with Chaotic Waveforms," *Phys. Rev. Lett.* **81**, 3547–50 (1998).

VAR98b. G. VanWiggeren and R. Roy, "Communication with Chaotic Lasers," *Science* **279**, 1198–1200 (1998).

VAS97. Stephen J. Van Hook and Michael F. Schatz, "Simple Demonstrations of Pattern Formation," *The Physics Teacher* **35**, 391–95 (1997).

VDP26. B. Van der Pol, "On Relaxation Oscillations," *Phil. Mag.* (7) **2**, 978–92 (1926).

VDP28. B. Van der Pol and J. Van der Mark, "The Heartbeat Considered as a Relaxation Oscillation and an Electrical Model of the Heart," *Phil. Mag.* **6**, 763–75 (1928).

VIL96. W. Vieira and P. Letelier, "Chaos around a Hénon–Heiles-Inspired Exact Perturbation of a Black Hole," *Phys. Rev. Lett.* **76**, 1409–12 (1996).

VOS89. R. F. Voss, "Random Fractals, Self-Affinity in Noise, Music, Mountains, and Clouds," *Physica D* **38**, 362–71 (1989).

WAH88. C. O. Weiss, N. B. Abraham, and U. Hübner. "Homoclinic and Heteroclinic Chaos in a Single-Mode Laser," *Phys. Rev. Lett.* **61**, 1587–90 (1988).

WAL87. J. Walker, "Fluid Interfaces, Including Fractal Flows can be Studied in a Hele–Shaw Cell," *Scientific American* **257** (5), 134–38 (November, 1987).

WEB86. C. O. Weiss and J. Brock, "Evidence for Lorenz-Type Chaos in a Laser," *Phys. Rev. Lett.* **57**, 2804–6 (1986).

WEI89. S. Weinberg, "Testing Quantum Mechanics," *Annals of Physics* **194**, 336–86 (1989).

WGC90. N. D. Whelan, D. A. Goodings, and J. K. Cannizzo, "Two Balls in One Dimension with Gravity," *Phys. Rev. A* **42**, 742–54 (1990).

WIB91. J. Wilkie and P. Brumer, "Time-Dependent Manifestations of Quantum Chaos," *Phys. Rev. Lett.* **67**, 1185–88 (1991).

WIG32. E. Wigner, "On the Quantum Correction for Thermodynamic Equilibrium," *Phys. Rev.* **40**, 749–59 (1932).

WIN74. A. Winfree, *SIAM–AMS Proceedings* **8**, 13 (1974).

WIS81. T. A. Witten, Jr. and L. M. Sander, "Diffusion-Limited Aggregation, a Kinetic Critical Phenomenon," *Phys. Rev. Lett.* **47**, 1400–03 (1981).

WIS87. J. Wisdom, "Chaotic Behavior in the Solar System," in *Dynamical Chaos*, M. Berry, I. Percival, and N. Weiss, eds. (Princeton University Press, Princeton, NJ, 1987). First published in *Proc. Roy. Soc. Lond. A* **413**, 1–199 (1987).

WKP84. R. W. Walden, P. Kolodner, A. Passner, and C. Surko, "Nonchaotic Rayleigh–Bénard Convection with Four and Five Incommensurate Frequencies," *Phys. Rev. Lett.* **53**, 242–45 (1984).

WKP95. F. Witkowski, K. Kavanagh, P. Penkoske, R. Plonsey, M. Spano, W. Ditto, and D. Kaplan, "Evidence for Determinism in Ventricular Fibrillation," *Phys. Rev. Lett.* **75**, 1230–33 (1995).

WOL84a. S. Wolfram, "Cellular Automata as Models of Complexity," *Nature* **341**, 419–24 (1984).

WOL84b. S. Wolfram, "Universality and Complexity in Cellular Automata," *Physica D* **10**, 1–35 (1984).

WOL86a. A. Wolf, "Quantifying Chaos with Lyapunov Exponents." In [Holden, 1986].

WOL86b. S. Wolfram, "Cellular Automaton Fluids 1: Basic Theory," *J. Stat. Phys.* **45**, 471–526 (1986).

WON88. Po-Zen Wong, "The Statistical Physics of Sedimentary Rock," *Physics Today* **41** (12), 24–32 (December, 1988).

WRI84. J. Wright, "Method for Calculating a Lyapunov Exponent," *Phys. Rev. A* **29**, 2924–27 (1984).

WSS85. A. Wolf, J. B. Swift, H. L. Swinney, and J. A. Vasano, "Determining Lyapunov Exponents from a Time Series," *Physica D* **7**, 285–317 (1985).

YEK82. W. J. Yeh and Y. H. Kao, "Universal Scaling and Chaotic Behavior of a Josephson-Junction Analog," *Phys. Rev. Lett.* **49**, 1888–91 (1982).

YGO85. J. A. Yorke, C. Grebogi, E. Ott, and L. Tedeschini-Lalli, "Scaling Behavior of Windows in Dissipative Dynamical Systems," *Phys. Rev. Lett.* **54**, 1095–98 (1985).

YLH99. D. Yu, W. Lu, and R. G. Harrison, "Detecting dynamical nonstationarity in time series data," *Chaos* **9**, 865–70 (1999).

ZAM97. D. Zanette and S. Manrubia, "Role of Intermittency in Urban Development: A Model of Large-Scale City Formation," *Phys. Rev. Lett.* **79**, 523–26 (1997).

ZAS99. G. M. Zaslavsky, "Chaotic Dynamics and the Origin of Statistical Laws," *Physics Today* **52** (8), 39–45 (1999).

ZDD95. J. Zakrzewski, K. Dupret, and D. Delande, "Statistical Properties of Energy Levels of Chaotic Systems: Wigner or Non-Wigner," *Phys. Rev. Lett.* **74**, 522–25 (1995).

ZEP91. X. Zeng, R. Eykholt, and R. Pielke, "Estimating the Lyapunov-Exponent Spectrum from Short Time Series of Low Precision," *Phys. Rev. Lett.* **66**, 3229–32 (1991).

ZOB98. R. Zorzenon dos Santos and A. Bernardes, "Immunization and Aging: A Learning Process in the Immune Network," *Phys. Rev. Lett.* **81**, 3034–37 (1998).

Index